Springer Series in Statistics

More information about this series at http://www.springer.com/series/692

Karl G. Jöreskog · Ulf H. Olsson
Fan Y. Wallentin

Multivariate Analysis with LISREL

 Springer

Karl G. Jöreskog
Department of Statistics
Uppsala University
Uppsala
Sweden

and

BI Norwegian Business School in Oslo
Oslo
Norway

Ulf H. Olsson
BI Norwegian Business School in Oslo
Oslo
Norway

Fan Y. Wallentin
Department of Statistics
Uppsala University
Uppsala
Sweden

ISSN 0172-7397   ISSN 2197-568X   (electronic)
Springer Series in Statistics
ISBN 978-3-319-33152-2   ISBN 978-3-319-33153-9   (eBook)
DOI 10.1007/978-3-319-33153-9

Library of Congress Control Number: 2016941611

Printed on acid-free paper

This Springer imprint is published by Springer Nature
The registered company is Springer International Publishing AG Switzerland

Preface

This book can be used by Master and PhD students and researchers in the economic, social, behavioral and many other sciences who need to have a basic understanding of multivariate statistical theory and methods for their analysis of multivariate data. It can also be used as a text book for courses on multivariate statistical analysis.

There are many books available on multivariate statistical analysis and many books have been written about structural equation modeling (SEM) and on LISREL[1]. But this book is unique in the sense of being the only one that covers both the statistical theory and methodology and how to do the analysis with LISREL. It does not only cover the typical uses of LISREL such as confirmatory factor analysis (CFA) and structural equation models (SEM) but also several other topics of multivariate analysis such as regression (univariate, multivariate, censored, logistic, and probit), generalized linear models, multilevel analysis, and principal components analysis. There is no other book with such a full and detailed coverage of all the models, methods and procedures one can use with LISREL.

Our experience is that although students and researchers in these disciplines can learn the basic statistical theory, they often do not know how to do the actual data analysis in practice. Therefore, this book focuses on both the basic statistical method and how to apply it using the LISREL program. Although there are many books on multivariate statistical analysis at various levels, most of them do not explain how to do the analysis in practice and they do not cover the interpretation of the results.

The book is topic-oriented. It describes the method and its basic statistical theory first and then it gives examples mostly with real data describing how to do the analysis with LISREL. This is followed by a discussion of the interpretation of the results in the context of the example. The focus is on learning by means of examples. There are many examples given in the book. We believe that much can be learned from each example. We also believe that researchers with their own data and problems can find a similar example in the book and use this as a guide for how to do the analysis.

This book is based on the idea of learning by examples. The examples, although sometimes small, are mostly real examples using real data. Each chapter and section begins with some general statistical theory presented at an intermediate level. Readers should be able to see how the examples that follow are special cases of this theory. The results are illustrated with sections of output from the LISREL program which are commented, discussed and interpreted in the context of the example. Readers are encouraged to do the examples themselves and verify the results given in the text.

The book is not a LISREL manual. More detailed information about LISREL is available in the various pdf documents that come with the LISREL program. Unlike these documents which either give a complete list of all LISREL commands in alphabetical order or illustrate how to do various analysis using the graphical users interface, this book focus on how to do it with syntax. Once the data is at hand, most analysis can be done with a few simple syntax lines. For example, multivariate regression can be done by the single command line

```
Regress Y1-Y3 on X1-X4
```

and factor analysis can be done by the single command line

---

[1]LISREL is a registered trademark of Scientific Software International.

Factor Analysis

Most material is new or based on courses that have been taught by the authors at the Norwegian School of Business (BI) in Oslo, Norway and in the Statistics Department at Uppsala University in Uppsala, Sweden. Various parts of the book have also been used in shorter courses and workshops given by the authors at universities and conferences around the world. Some parts of the book present revised material previously published by Scientific Software International (SSI). These publications are no longer available. All the material in the book is up-to-date for the latest available version of **LISREL**.

Although the aim is to give a thorough understanding of the basic statistical theory required for the application of each method and procedure, proofs of statements or theorems are not always included. These may be found in other books or articles referred to in the book.

For most examples we give an input file in both **SIMPLIS** syntax and **LISREL** syntax. For some examples, where the use of **SIMPLIS** syntax is inconvenient or impossible, we give only the **LISREL** syntax.

Readers who are familiar with mathematical notation and especially matrix notation should be able to read this book completely. Readers who are not familiar with mathematical and matrix notation should still be able to benefit from this book by skipping the mathematical theory and studying the examples and focusing on the **SIMPLIS** syntax.

All examples are listed in the Table of Contents. To do the examples you need **LISREL** 9.2 or a later version. Visit **www.ssicentral.com** for information about **LISREL** and information about ordering and prices.

If **LISREL** is installed on your computer, input and data files for the examples can be found in the following folders.

```
LISREL Examples\MVABOOK\Chapter 1
LISREL Examples\MVABOOK\Chapter 2
LISREL Examples\MVABOOK\Chapter 3
LISREL Examples\MVABOOK\Chapter 4
LISREL Examples\MVABOOK\Chapter 5
LISREL Examples\MVABOOK\Chapter 6
LISREL Examples\MVABOOK\Chapter 7
LISREL Examples\MVABOOK\Chapter 8
LISREL Examples\MVABOOK\Chapter 9
LISREL Examples\MVABOOK\Chapter 10
```

Ulf Henning Olsson has received financial support from the *Norwegian Non-fiction Literature Fund*. Some research for this book was funded by the Swedish Research Council for the project *Structural Equation Modeling with Ordinal Data* (Project number 421-2011-1727). Project Director Fan Y. Wallentin.

The authors thank BI Norwegian Business School for giving us the opportunity to work on this book project. The authors also wish to thank Stephen DuToit, Mathilda DuToit, and Dag Sörbom for allowing us to use some of their examples and for fixing many bugs in **LISREL** discovered in the process of running the examples given in the book.

Uppsala and Oslo, January 2016

Karl G Jöreskog, Ulf H Olsson, Fan Y. Wallentin

Contents

About the Authors

Karl G. Jöreskog is Professor Emeritus at Uppsala University, Sweden, and Senior Professor at the BI Norwegian Business School in Oslo. He has received three honorary doctorates: from the Faculty of Economics and Statistics at the University of Padua, Italy, 1993, from the Norwegian School of Economics, Bergen, Norway, 1996, and from the Faculty of Psychology at the Friedrich-Schiller-Universität, Jena, Germany, 2004. Professor Jöreskog is a member of the Swedish Royal Academy of Sciences, a Fellow of the American Statistical Association, and an Honorary Fellow of the Royal Statistical Society. He has received many awards including the American Psychological Association Distinguished Award for the Applications of Psychology and the Psychometric Society Award for Career Achievement to Educational Measurement. Together with Dag Sörbom he developed the LISREL computer program.

Ulf H. Olsson is Professor at Department of Economics and Provost at BI Norwegian Business School in Oslo with responsibility for research and academic resources. He has worked on structural equation modeling, statistical modeling and psychometrics and published several research articles in leading statistics and psychometric journals. Dr. Olsson has also authored textbooks on statistics and mathematics. In 2003 Olsson was awarded the BI Norwegian Business School's research prize.

Fan Y. Wallentin is Professor of Statistics at Uppsala University, Sweden. She received her Ph.D. in Statistics in 1997. She is a recipient of the Arnberg Prize from the Swedish Royal Academy of Sciences. Dr. Wallentin's program of research is on the theory and applications of latent variable modeling and other types of multivariate statistical analysis, particularly their applications in the social and behavioral sciences. She has published research articles in several leading statistics and psychometrics journals. She has taught courses on Structural Equation Models in Sweden, USA, China and several European countries. She has broad experience in statistical consultation for researchers in social and behavioral sciences.

Chapter 1

Getting Started

All data analyses begin with raw data in one form or another. In LISREL one can work with data in plain text (ASCII) form. But for most analysis with LISREL it is convenient to work with a **LISREL data system file** of the type **.lsf**. LISREL can import data from many formats such as SAS, SPSS, STATA, and EXCEL. LISREL can also import data in text format with spaces (*.dat or *.raw), commas (*.csv) or tab characters (*.txt) as delimiters between entries. The data is then stored as a LISREL data system file **.lsf**.

Section 1.1 describes how to create a LISREL data system file **.lsf** by data importation and how to set the attributes of the variables in the data, for example, whether a variable is categorical or continuous.

Sometimes the data can be split into several groups. Section 1.3 shows how to create different **.lsf** files for each group.

Section 1.6 shows how to define missing values in the **.lsf** file and some of the ways these can be handled.

It is often necessary to transform variables and/or construct new variables. Section 1.7 describes how this is done.

It should be emphasized that this chapter is *not* about modeling. It is about how to prepare the data for modeling. John Tukey once wrote "It is important to understand what you *can do* before you learn to measure how *well* you seem to have *done* it." (Tukey, 1977). It is with this spirit in mind we suggest that a fair amount of data screening should be done before the data is submitted to modeling. This includes the understanding of the characteristics of the distribution of the variables and the distribution of missing values over variables and cases. Graphs play an important rule in this and Section 1.2 describes some of the graphs that LISREL can produce.

1.1 Importing Data

This section illustrates how to import data and create a LISREL data system file **.lsf**. Most examples in later chapters of this book are based on LISREL data system files that have already been created.

To illustrate data importation we use a classic data originally published by Holzinger and Swineford (1939). They collected data on twenty-six psychological tests administered to seventh- and eighth-grade children in two schools in Chicago: the Pasteur School and the Grant-White School. The Pasteur

© Springer International Publishing Switzerland 2016
K.G. Jöreskog et al., *Multivariate Analysis with LISREL*, Springer Series in Statistics,
DOI 10.1007/978-3-319-33153-9_1

School had students whose parents had immigrated from Europe, mostly France and Germany. The students of the Grant-White School came from middle income American white families.

Nine of these tests are selected for this example. The nine tests are (with the original variable number in parenthesis):

VIS PERC Visual Perception (V1)

CUBES Cubes (V2)

LOZENGES Lozenges (V4)

PAR COMP Paragraph Comprehension (V6)

SEN COMP Sentence Completion (V7)

WORDMEAN Word meaning (V9)

ADDITION Addition (V10)

COUNTDOT Counting dots (V12)

S-C CAPS Straight-curved capitals (V13)

The nine test scores are preceded by four variables:

SCHOOL 0 for Pasteur School; 1 for Grant-White School

GENDER 0 for Boy; 1 for Girl

AGEYEAR Age in years

BIRTHMON Birthmonth: 1 = January, 2 = February ... 12 = December

The data file is an SPSS data file **hsschools.sav**. This file contains only the names of the variables and the data values. There are no missing values in this data set. Missing values are discussed in Section 1.6.

Using this data set we can learn

1. how to import data into a LISREL system data file **.lsf**

2. how to graph the data

3. how to divide the data into two groups (schools)

In later chapters we will use these data to illustrate

1. how to do exploratory factor analysis in one school

2. how to do confirmatory factor analysis in the other school

3. how to deal with non-normality

4. how to modify the model by means of path diagrams

5. how to interpret chi-square differences

6. how to estimate and test latent variable differences

When LISREL starts it opens with a screen like this:

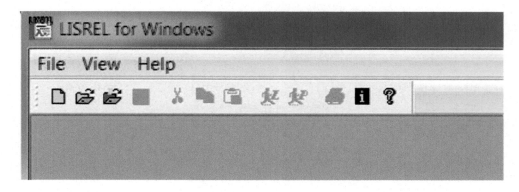

Select **File** and then **Import Data**. In the **Files of Type** select **SPSS Data File (.sav)** and browse your computer until you find the file **hsschools.sav**. Double click on this file name and accept the suggested file name **hsschools.lsf**. LISREL will then open this which looks like this, showing the first 15 rows of data, *i.e.* the first 15 cases.

	SCHOOL	GENDER	AGEYEAR	BIRTHMON	VISPERC	CUBES	LOZENGES	PARCOMP	SENCOMP	WORDMEAN	ADDITION	COUNTDOT	SCCAPS
1	0.000	0.000	13.000	2.000	20.000	31.000	3.000	7.000	23.000	9.000	78.000	115.000	229.000
2	0.000	1.000	13.000	8.000	32.000	21.000	17.000	5.000	12.000	9.000	87.000	125.000	285.000
3	0.000	1.000	13.000	2.000	27.000	21.000	15.000	3.000	7.000	3.000	75.000	78.000	159.000
4	0.000	0.000	13.000	3.000	32.000	31.000	24.000	8.000	18.000	17.000	69.000	106.000	175.000
5	0.000	1.000	12.000	3.000	29.000	19.000	7.000	8.000	16.000	18.000	85.000	126.000	213.000
6	0.000	1.000	14.000	2.000	32.000	20.000	18.000	3.000	12.000	6.000	100.000	133.000	270.000
7	0.000	0.000	12.000	2.000	17.000	24.000	8.000	10.000	24.000	20.000	108.000	124.000	175.000
8	0.000	1.000	12.000	3.000	34.000	25.000	15.000	11.000	17.000	9.000	78.000	103.000	132.000
9	0.000	1.000	13.000	1.000	27.000	23.000	12.000	8.000	23.000	19.000	104.000	93.000	265.000
10	0.000	1.000	12.000	6.000	21.000	21.000	6.000	8.000	20.000	18.000	95.000	91.000	157.000
11	0.000	0.000	12.000	3.000	22.000	23.000	16.000	6.000	14.000	11.000	86.000	114.000	155.000
12	0.000	0.000	12.000	12.000	35.000	24.000	23.000	8.000	18.000	19.000	85.000	103.000	149.000
13	0.000	1.000	12.000	8.000	34.000	18.000	33.000	8.000	16.000	16.000	135.000	104.000	211.000
14	0.000	1.000	12.000	9.000	36.000	22.000	14.000	14.000	16.000	11.000	118.000	94.000	160.000
15	0.000	0.000	12.000	7.000	35.000	23.000	29.000	15.000	22.000	21.000	92.000	87.000	211.000

Variable Types

LISREL has essentially two types of variables called ordinal and continuous. The term ordinal variable is used here in the sense of a categorical variable, *i.e.*, a variable that has a only few distinct values (maximum 15), whereas continuous variables typically have a wide range of values[1]. LISREL assumes that all variables are ordinal (categorical) by default. In this example the first four variables SCHOOL, GENDER, AGEYEAR and BIRTHMONTH are considered as ordinal (categorical) and the nine test scores VISPERC – SCCAPS as continuous. The latter variables must therefore be declared continuous in the .lsf file[2]

To do this, select **Data** in **LISREL System Data File Toolbar** when the .lsf file is displayed and select the **Define Variables** option. Then select the nine test score variables. Then click on **Variable Type**, select **Continuous** and click on the **OK** button twice and save the .lsf file as shown here.

---

[1]In Sections 7.4.1 and 7.4.2 in Chapter 7 we will learn about the more common use of ordinal variables where the observed variables are Likert scales with ordered categories

[2]There are also many situations where a binary variable should be treated as continuous. For example, in regression we may want to include a binary (dummy) variable as an independent variable. This variable must therefore be declared continuous in the .lsf file although it is categorical, see Chapter 2.

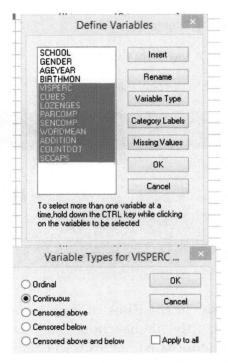

Category Labels

For the categorical variables it is convenient to define labels for the categories. These labels are limited to four characters. To do this, select **Data** in **LISREL System Data File Toolbar** when the **.lsf** file is displayed and select the **Define Variables** option. Then click on **Category Labels**, select the variable SCHOOL. Type 0 in the Value field and type PA in the Labels field and click on **Add**. Then type 1 in the Value field and type GW in the Labels field and click on **Add**. Then click **OK** button twice and save the **.lsf** file as shown here.

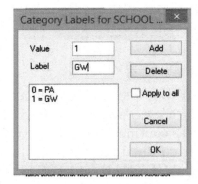

In the same way one can define category labels for GENDER.

1.2   Graphs

There are three types of graphs available in LISRELand within each category there are several types of graphs depending on whether the variables in the graphs are ordinal or continuous.

- Univariate Graphs

  – Bar Chart

  – Histogram

- Bivariate Graphs

  – 3D Bar Chart

  – Scatter Plot

  – Box and Whisker Plot

- Multivariate Graphs

We describe the univariate and bivariate graphs here.

To obtain a univariate graph of an ordinal variable, **leave all variables in the .lsf file unselected** and select Graphs in the **LISREL System Data File Toolbar**, then select the variable, for example SCHOOL, in the .lsf file, and select **Bar Chart** in the following screen

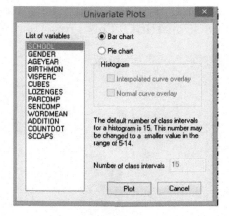

Click on **Plot**. This gives the following bar chart

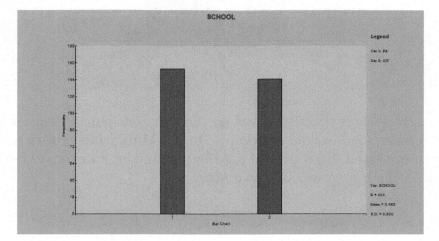

It is seen that the Pasteur school has approximately 156 students and the Grant-White school has approximately 145 students.

Alternatively, one can choose a pie chart in the **Univariate Plots** screen above. A pie chart of AGEYEAR, for example, gives the following pie chart

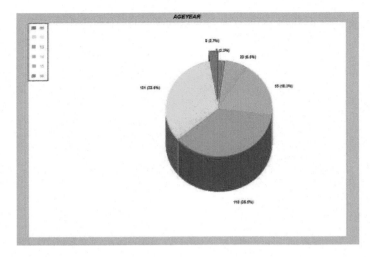

where we can see that most of the students are 12 or 13 years old and very few are 11 and 16 years.

To obtain a univariate graph of a continuous variable, **do not** select the variable in the **.lsf** file first, but select Graphs in the **LISREL System Data File Toolbar**, then select the variable, for example VISPERC, and and click on **Plot**. This gives the following **Histogram**.

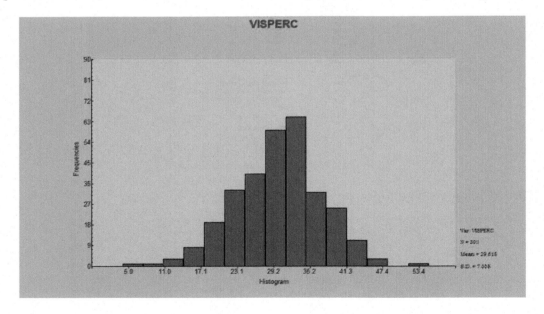

This shows that the variable is slightly skewed on the right side but not remarkable so. Later we will learn how to measure and evaluate skewness. In the **Univariate Plots** screen one can also choose to obtain an overlayed fitted normal distribution and/or an overlayed curve fitted to the middle of each bar in the histogram, as shown here

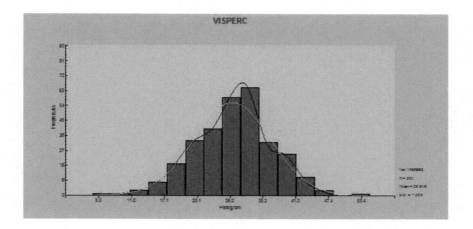

To obtain a bivariate plot with two categorical variables, select **Graphs** in the **LISREL System Data File Toolbar** and **Bivariate Graph** in the **Graphs** menu. Then select AGEYEAR as the independent variable (X variable) and SCHOOL as the dependent variable (Y variable). This gives a **3D Bar Chart** as follows

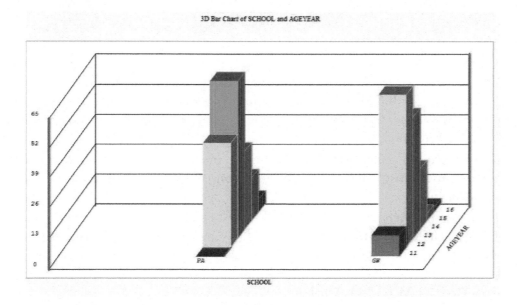

It is seen that most students in the Pasteur schools are 14 years old, whereas in the Grant-White school most students are 13 years old. In general, students in the Pasteur school tend to be older than the students in the Grant-White school.

To obtain a bivariate plot with two continuous variables, select **Graphs** in the **LISREL System Data File Toolbar** and **Bivariate Graph** in the **Graphs** menu. Then select VISPERC as the dependent variable, *i.e.*, the variable on the vertical axis, and SCCAPS as the independent variable, *i.e.*, the variable on the horizontal axis. This gives a **Scatter Plot** as follows

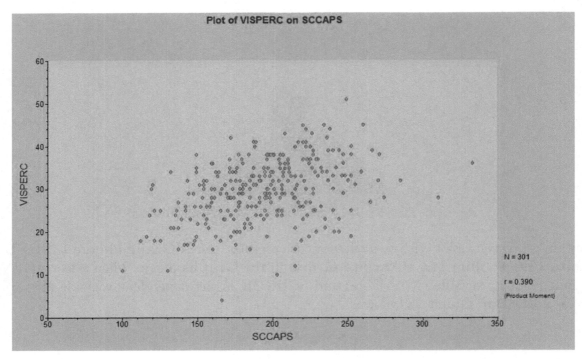

From this scatter plot it seems that there is no clear relationship between VISPERC and SCCAPS. It will be difficult to predict one variable from the other. The correlation is given in the lower right side of the plot as r = 0.390 based on a sample size of 301. This is Pearson correlation or product moment correlations. Later we will learn to evaluate the statistical significance of this and other types of correlations.

In many examples the so called Box and Whisker plot is very useful because it gives much detailed information about the conditional distribution of a continuous variable for given values of a categorical variable.

To obtain a bivariate plot with one continuous variable, here SENCOMP, and a categorical variable, here SCHOOL, select **Graphs** in the **LISREL System Data File Toolbar** and **Bivariate Graph** in the **Graphs** menu. Then select SENCOMP as the dependent variable, *i.e.*, the variable on the vertical axis, and SCHOOL as the independent variable, *i.e.*, the variable on the horizontal axis. This gives a **Box & Whisker Plot** as follows

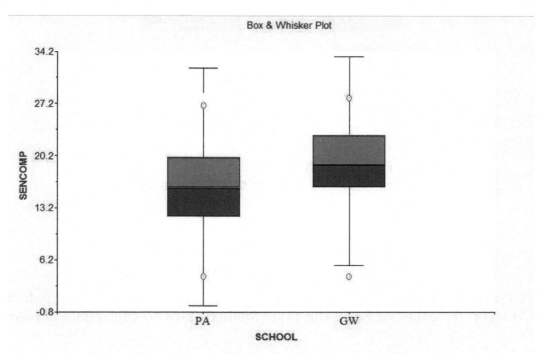

The box indicates the area between lower and upper quartile so that 25% of the observations are below the bottom border of the box and 25% of the observations are above the top border of the box. The border line between the two colored areas in the box is the mean. If the areas above and below the mean are of different sizes, this indicates that the distribution of the observations is skewed. The two unfilled circles are the largest and the smallest value in the data. The whiskers are located $\pm 1.5 \times IQR$, where IQR is the interquartile range, *i.e.* the distance between the upper and lower quartile which is the length of the box. Observations that are outside of the whiskers are sometimes called outliers.

The most striking feature of the Box and Whisker plot for SENCOMP is that the students in the Grant-White school have higher scores than the students in the Pasteur school. In particular, the mean score is larger for the Grant-White school than the Pasteur school. In Chapter 2 we will learn how to test the statistical significance of this difference.

1.3   Splitting the Data into Two Groups

In view of what we already know about the two schools it may be a good idea to split the data into two groups, one for the Pasteur school and one for the Grant-White school, rather than treating the two schools as one homogeneous group. This section illustrates how this can be done.

With the **hsschools.lsf** displayed select **Data** in the **LISREL System Data File Toolbar** and **Select Variables/Cases**. Then select **Select Cases**. Tick **Select Only Those Cases with** and tick **Equal To** and type 0 in the numeric field. Since all values in the selected school will have 0 value on the variable SCHOOL, this variable may well be deleted after selection. Therefore tick the box to do this. Then click on **Add**.The screen then looks like this:

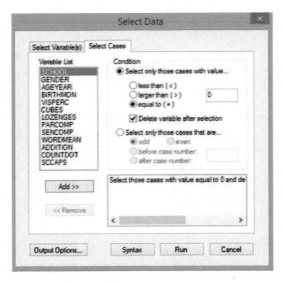

Then select **Output Options**. In the **Output** box, make sure it says **None Selected** under
Moment Matrix. Tick the box **Save the transformed data to file** and fill in **pasteur.lsf** in
the text box:

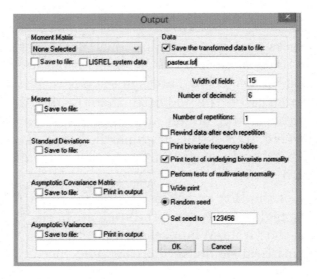

Click **OK**. Then click on **Run**. This produces a data file **pasteur.lsf**, where the first 15 rows of
data looks like this:

	GENDER	AGEYEAR	BIRTHMON	VISPERC	CUBES	LOZENGES	PARCOMP	SENCOMP	WORDMEAN	ADDITION	COUNTDOT	SCCAPS
1	0	13.	2.	20.	31.	3.	7.	23.	9.	78.	115.	229.
2	1	13.	8.	32.	21.	17.	5.	12.	9.	87.	125.	285.
3	1	13.	2.	27.	21.	15.	3.	7.	3.	75.	78.	159.
4	0	13.	3.	32.	31.	24.	8.	18.	17.	69.	106.	175.
5	1	12.	3.	29.	19.	7.	8.	16.	18.	85.	126.	213.
6	1	14.	2.	32.	20.	18.	3.	12.	6.	100.	133.	270.
7	0	12.	2.	17.	24.	8.	10.	24.	20.	108.	124.	175.
8	1	12.	3.	34.	25.	15.	11.	17.	9.	78.	103.	132.
9	1	13.	1.	27.	23.	12.	8.	23.	19.	104.	93.	265.
10	1	12.	6.	21.	21.	6.	8.	20.	18.	95.	91.	157.
11	0	12.	3.	22.	23.	16.	6.	14.	11.	86.	114.	155.
12	0	12.	12.	35.	24.	23.	8.	18.	19.	85.	103.	149.
13	1	12.	8.	34.	18.	33.	8.	16.	16.	135.	104.	211.
14	1	12.	9.	36.	22.	14.	14.	16.	11.	118.	94.	160.
15	0	12.	7.	35.	23.	29.	15.	22.	21.	92.	87.	211.

Follow the same procedure to obtain a data file **grantwhite.lsf**, where the first 15 rows of data
look like this:

	GENDER	AGEYEAR	BIRTHMON	VISPERC	CUBES	LOZENGES	PARCOMP	SENCOMP	WORDMEAN	ADDITION	COUNTDOT	SCCAPS
1	0	13.	1	23	19.	4	10.	17	10	69	82	156
2	1	11.	11.	33.	22.	17.	8.	17.	10.	65.	98.	195.
3	0	12.	7.	34.	24.	22.	11.	19.	19.	50.	86.	228.
4	0	11.	12.	29.	23.	9.	9.	19.	11.	114.	103.	144.
5	0	12.	6.	16.	25.	10.	8.	25.	24.	112.	122.	160.
6	1	12.	7.	30.	25.	20.	10.	23.	18.	94.	113.	201.
7	1	12.	9.	36.	33.	36.	17.	25.	41.	129.	139.	333.
8	1	11.	12.	28.	25.	9.	10.	18.	11.	96.	95.	174.
9	1	12.	6.	30.	25.	11.	11.	21.	8.	103.	114.	197.
10	1	12.	6.	20.	25.	6.	9.	21.	16.	89.	101.	178.
11	0	12.	1.	27.	26.	6.	10.	16.	13.	88.	107.	137.
12	0	12.	11.	32.	21.	8.	1.	7.	11.	103.	136.	154.
13	0	12.	10.	38.	31.	12.	10.	11.	14.	83.	108.	201.
14	1	12.	9.	17.	21.	6.	5.	10.	10.	99.	87.	147.
15	0	12.	10.	34.	28.	24.	14.	22.	26.	49.	84.	171.

These data files will be used in several later chapters in this book.

1.4 Introduction to LISREL Syntaxes

LISREL comes with several program modules, the most important ones being PRELIS and LISREL. The LISREL module is used for structural equation modeling (SEM), including confirmatory factor analysis for continuous and ordinal variables, models for relationships between latent variables, multiple group analysis, and general covariance structures. For most other purposes, one can use the PRELIS module.

PRELIS reads raw data and can be used for

- Data screening

- Data exploration

- Data summarization

- Computation of

    - Covariance matrix

    - Correlation matrix

    - Moment matrix

    - Augmented moment matrix

    - Asymptotic covariance matrix

With LISREL 9 it is not necessary to use PRELIS to compute these matrices as PRELIS is called from within LISREL for these purposes. PRELIS can also be used to create a LISREL data system file .lsf.

PRELIS is also used for various statistical analysis such as

- Multiple univariate and multivariate regression

- Logistic and probit regression

- Censored regression

described in Chapter 2, for principle components analysis described in Chapter 5 and for exploratory factor analysis described in Chapter 6.

This book presents multivariate statistical analysis as it can be done using simple syntax files. For this purpose, there are three syntaxes: PRELIS syntax, SIMPLIS syntax and LISREL syntax. These different syntaxes are introduced as we move along. For now, we give a brief introduction to the PRELIS syntax and demonstrate how one can do all the data descriptions in the previous section by a few simple PRELIS syntax commands.

The starting point is the data file **hsschools.lsf** created in Section 1.1. Here we show how the files **pasteur.lsf** and **grantwhite.lsf** can be obtained for each school and at the same time get information about the important characteristics of the data in simple tabular form.

To generate the file **pasteur.lsf**, use the following PRELIS syntax file **hsschools1.prl**:

```
!Creating a Data File for Pasteur School
SY=hsschools.lsf
SD SCHOOL = 0
OU RA=pasteur.lsf
```

The first line is a title line. One can include one or more title lines before the first line, *i.e.*, before the SY line. Such title lines are optional but they can be used to describe the data and the problem. The title can have any text but to avoid conflicts with real command lines it is recommended to begin with an exclamation mark (!).

The first real PRELIS syntax line is the SY line which is short for **System File**. This tells LISREL to read the data file **hsschools.lsf**. This line can also be written as

```
System File hsschools.lsf
```

The SD line is the **Select Delete Cases** command which tells LISREL to select the cases with SCHOOL = 0, *i.e.*, the Pasteur School sample, and delete the variable SCHOOL after selection. The last line is the **Output** command which must always be the last line in a PRELIS syntax file. The RA=pasteur.lsf on the OU lne tells LISREL to save the selected data in the file **pasteur.lsf**.

To run the file **hsschools1.prl** click the **P** (**Run PRELIS** button. This generates an output file **hsschools1.out** with some sections of output described as follows.

```
Total Sample Size(N) =      156


Univariate Distributions for Ordinal Variables

   GENDER Frequency Percentage
    Boy       74        47.4
    Girl      82        52.6


  AGEYEAR Frequency Percentage
    12        41        26.3
    13        62        39.7
    14        31        19.9
    15        17        10.9
    16         5         3.2
```

```
BIRTHMON Frequency Percentage
    1        9        5.8
    2       17       10.9
    3       15        9.6
    4       13        8.3
    5       16       10.3
    6       13        8.3
    7        8        5.1
    8       15        9.6
    9       16       10.3
   10        8        5.1
   11       11        7.1
   12       15        9.6
```

This reveals that

- There are 156 students in the Pasteur School.

- There are 74 boys (47.4%) and 82 girls (52.6%).

- Most of the students are 13 years old (39.7%) but some are 12 and a few are 16 years old.

- There are 17 students born in February, for example.

The next section of the output gives information about the continuous variables, *i.e.*, the nine test scores VISPERC - SCCAPS:

```
Univariate Summary Statistics for Continuous Variables
```

Variable	Mean	St. Dev.	Skewness	Kurtosis	Minimum	Freq.	Maximum	Freq.
VISPERC	29.647	7.110	-0.378	0.736	4.000	1	45.000	2
CUBES	23.936	4.921	0.695	0.255	14.000	1	37.000	3
LOZENGES	19.897	9.311	0.156	-1.074	2.000	1	36.000	5
PARCOMP	8.468	3.457	0.221	-0.052	0.000	1	18.000	1
SENCOMP	15.981	5.244	-0.132	-0.839	4.000	1	27.000	1
WORDMEAN	13.455	6.932	1.009	2.034	1.000	2	43.000	1
ADDITION	101.942	24.958	0.303	-0.395	47.000	1	171.000	1
COUNTDOT	111.263	19.577	0.362	0.091	70.000	1	166.000	1
SCCAPS	195.038	35.708	0.226	0.194	100.000	1	310.000	1

As shown this gives the mean, standard deviation, skewness, kurtosis, minimum, and maximum and the frequency of the minimum and maximum. It is seen that the mean, standard deviation, minimum and maximum varies considerably across the tests. This is a reflection of the fact that the tests have varying number of items.

The skewness and kurtosis are used to evaluate the degree of non-normality in the variables which may be important in the evaluation of the fit of models as explained in later chapters. In the output, there is also a section of tests of skewness of kurtosis which looks like this:

Test of Univariate Normality for Continuous Variables

	Skewness		Kurtosis		Skewness and Kurtosis	
Variable	Z-Score	P-Value	Z-Score	P-Value	Chi-Square	P-Value
VISPERC	-1.934	0.053	1.682	0.093	6.568	0.037
CUBES	3.360	0.001	0.782	0.434	11.900	0.003
LOZENGES	0.818	0.414	-5.966	0.000	36.257	0.000
PARCOMP	1.153	0.249	0.018	0.985	1.329	0.514
SENCOMP	-0.691	0.489	-3.585	0.000	13.332	0.001
WORDMEAN	4.555	0.000	3.207	0.001	31.028	0.000
ADDITION	1.567	0.117	-1.131	0.258	3.736	0.154
COUNTDOT	1.856	0.064	0.398	0.691	3.601	0.165
SCCAPS	1.176	0.240	0.645	0.519	1.798	0.407

The measures of skewness and kurtosis and their tests are explained in the Appendix (see Chapter 12). In practice, one only needs to examine the P-values given in this table. P-values which are very small, less than 0.05, say, indicate that there is some degree of non-normality. It is seen that CUBES and WORDMEAN are skewed variables, LOZENGES and SENCOMP have too small kurtosis and WORDMEAN has too high kurtosis. Probably WORDMEAN is the most non-normal variable since it is both skewed and kurtotic. In Chapter 7 we will learn about the consequences of non-normality and about ways to deal with this problem.

To obtain the data file for the Grant White School, use the following **PRELIS** input file **hsschools2.prl**:

```
!Creating a Data File for Grant White School
SY=hsschools.lsf
SD SCHOOL = 1
OU RA=grantwhite.lsf
```

Syntax for Reading the lsf file

In general, suppose the data file is called **data.lsf**. This name is not case sensitive, so one can use upper case **DATA.LSF** as well.

As we have already seen the **PRELIS** syntax for reading an **.lsf** file is

```
System File data.lsf
```

This can be shortened to

```
SY data.lsf
```

In **SIMPLIS** syntax, the command for reading an **.lsf** file is

```
Raw Data from File data.lsf
```

All the words `Raw Data from File` are needed.

In LISREL syntax the command for reading the .lsf file is

`RA=data.lsf`

More details about PRELIS, SIMPLIS, and LISREL syntax are given later in the book and in documents that come with the LISREL program.

1.5   Estimating Covariance or Correlation Matrices

To estimate a covariance matrix put `MA=CM` on the `OU` line in **hsschools1.prl** or **hsschools2.prl**. This is not useful in this case since there is no meaningful interpretation of the covariances between the categorical variables GENDER, AGEYEAR, and BIRTHMON and the test scores VISPERC - SCCAPS. However, one can select any subset of variables and estimate the covariance matrix for these variables. For example, to estimate the covariance matrix for the nine test scores for the Grant White School, add the line

`SE VISPERC-SCCAPS`

in **hsschools2.prl** and add `MA=CM` on the `OU` line.

The output file gives the covariance matrix in the following form

```
Covariance Matrix
```

	VISPERC	CUBES	LOZENGES	PARCOMP	SENCOMP	WORDMEAN
VISPERC	47.801					
CUBES	10.013	19.758				
LOZENGES	25.798	15.417	69.172			
PARCOMP	7.973	3.421	9.207	11.393		
SENCOMP	9.936	3.296	11.092	11.277	21.616	
WORDMEAN	17.425	6.876	22.954	19.167	25.321	63.163
ADDITION	17.132	7.015	14.763	16.766	28.069	33.768
COUNTDOT	44.651	15.675	41.659	7.357	19.311	20.213
SCCAPS	124.657	40.803	114.763	39.309	61.230	79.993

```
Covariance Matrix
```

	ADDITION	COUNTDOT	SCCAPS
ADDITION	565.593		
COUNTDOT	293.126	440.792	
SCCAPS	368.436	410.823	1371.618

The output file also gives certain characteristics of covariance matrix:

```
Total Variance = 2610.906 Generalized Variance = 0.106203D+17

Largest Eigenvalue = 1734.725 Smallest Eigenvalue = 3.665

Condition Number = 21.756
```

In this book a sample covariance matrix will always be denoted $\mathbf{S}$. The total variance is the sum of the diagonal elements of $\mathbf{S}$ and the generalized variance is the determinant of $\mathbf{S}$ which equals the product of all the eigenvalues of $\mathbf{S}$. The largest and smallest eigenvalues of $\mathbf{S}$ are also given. These quantities are useful in principal components analysis, see Chapter 5. The condition number is the square root of the ratio of the largest and smallest eigenvalue. A large condition number indicates multicollinearity in the data. If the condition number is larger than 30, **LISREL** gives a warning. This might indicate that one or more variables are linear or nearly linear combinations of other variables. If this is intentional, the warning may be ignored.

The means and standard deviations of the variables are also given in the output file as follows

Means

	VISPERC	CUBES	LOZENGES	PARCOMP	SENCOMP	WORDMEAN
	--------	--------	--------	--------	--------	--------
	29.579	24.800	15.966	9.952	18.848	17.283

Means

	ADDITION	COUNTDOT	SCCAPS
	--------	--------	--------
	90.179	109.766	191.779

Standard Deviations

	VISPERC	CUBES	LOZENGES	PARCOMP	SENCOMP	WORDMEAN
	--------	--------	--------	--------	--------	--------
	6.914	4.445	8.317	3.375	4.649	7.947

Standard Deviations

	ADDITION	COUNTDOT	SCCAPS
	--------	--------	--------
	23.782	20.995	37.035

To estimate the correlation matrix instead of the covariance matrix put **MA=KM** on the **OU** line instead. Since the nine test scores are declared as continuous variables in the **.lsf** file, the correlations will the be product-moment correlations, also called Pearson correlations. There are also several other types of correlations that can be estimated, such as polychoric, polyserial, Spearman-rank, and Kendall-tau correlations but these are only relevant when some or all variables are ordinal.

The correlation matrix for the Grant White School, as obtained in the output file is:

Correlation Matrix

	VISPERC	CUBES	LOZENGES	PARCOMP	SENCOMP	WORDMEAN
VISPERC	1.000					
CUBES	0.326	1.000				
LOZENGES	0.449	0.417	1.000			
PARCOMP	0.342	0.228	0.328	1.000		
SENCOMP	0.309	0.159	0.287	0.719	1.000	
WORDMEAN	0.317	0.195	0.347	0.714	0.685	1.000
ADDITION	0.104	0.066	0.075	0.209	0.254	0.179
COUNTDOT	0.308	0.168	0.239	0.104	0.198	0.121
SCCAPS	0.487	0.248	0.373	0.314	0.356	0.272

Correlation Matrix

	ADDITION	COUNTDOT	SCCAPS
ADDITION	1.000		
COUNTDOT	0.587	1.000	
SCCAPS	0.418	0.528	1.000

One can also save the covariance or correlation matrix in a file. For example, to estimate the covariance matrix for the nine test score for the Pasteur School, add both MA=CM and cm=pasteur_npv.cm on the OU line in **hsschools1.prl**. The saved covariance matrix **pasteur_npv.cm** is

```
0.50552D+02   0.97127D+01   0.24215D+02   0.29957D+02   0.15355D+02   0.86686D+02
0.10289D+02   0.10689D+01   0.39773D+01   0.11954D+02   0.11400D+02   0.22955D+01
0.21400D+01   0.13041D+02   0.27503D+02   0.21032D+02   0.54681D+01   0.12518D+02
0.15960D+02   0.26318D+02   0.48056D+02   0.64118D+01  -0.18875D+02  -0.45737D+01
0.22330D+02   0.15825D+02   0.35504D+02   0.62288D+03   0.19951D+02   0.31008D+01
0.24414D+02   0.88762D+01   0.12199D+02   0.28976D+02   0.19727D+03   0.38325D+03
0.76336D+02   0.31693D+02   0.96256D+02   0.16866D+02   0.30207D+02   0.47221D+02
0.23936D+03   0.25538D+03   0.12751D+04
```

This gives the elements below and in the diagonal of $\underline{S}$ as one long string of numbers, where each number is given in scientific notation to get as much accuracy as possible. Users rarely need to look at these types of files.

The "classical" way of doing structural equation modeling with LISREL was to use a two-step procedure as follows.

Step 1 Use PRELIS to estimate and save the covariance matrix to a file.

Step 2 Read the covariance matrix into LISREL and estimate the model.

Although this two-step procedure can still be used, it is no longer necessary in LISREL 9 since LISREL can read the data file and estimate the model in one step.

All the procedures described in this section can be done with one single PRELIS syntax file using so called stacked input, see file **hsschools3.prl**:

```
!Creating a Data File for Pasteur School
SY=hsschools.lsf
SD SCHOOL = 0
SE VISPERC - SCCAPS
OU MA=CM RA=pasteur_npv.lsf cm=pasteur_npv.cm
!Creating a Data File for Grant White School
SY=hsschools.lsf
SD SCHOOL = 1
SE VISPERC - SCCAPS
OU MA=CM RA=grantwhite_npv.lsf cm=grantwhite_npv.cm
```

This file will split the data into two schools and save the **.lsf** for the nine test scores for each school as well as the covariance matrices for each school.

1.6   Missing Values

Missing values and incomplete data are almost unavoidable in the social, behavioral, medical and most other areas of investigation. In particular, missing values are very common in longitudinal data and repeated measurements data, where individuals are followed and measured repeatedly over several points in time.

One can distinguish between three types of incomplete data:

- Unit nonresponse, for example, a person does not respond at all to an item in a questionnaire.

- Subject attrition, for example, when a person falls out of a sample after some time in a longitudinal follow-up study.

- Item nonresponse, for example, a person respond to some but not all items in a question-naire.

The literature, *e.g.*, Schafer (1997) distinguishes between three mechanisms of nonresponse.

MCAR Missing completely at random

MAR Missing at random

MNAR Missing not at random

Let z_{ij} be any element on the data matrix. Informally, one can define these concepts as

MCAR $Pr(z_{ij} = missing)$ does not depend on any variable in the data.

MAR $Pr(z_{ij} = missing)$ may depend on other variables in the data but not on z_{ij}. Example: A missing value of a person's income may depend on his/her age and education but not on his/her actual income.

MNAR $Pr(z_{ij} = missing)$ depends on z_{ij}. Example: In a questionnaire people with higher income tend not to report their income.

LISREL has several ways of dealing with missing values:

1. Listwise deletion

2. Pairwise deletion

3. Imputation by matching

4. Multiple imputation

   - EM
   - MCMC

5. Full Information Maximum Likelihood (FIML)

Of these methods the first three are *ad hoc* procedures whereas the last two are based on probability models for missingness. As a consequence, the *ad hoc* methods may lead to biased estimates under MAR and can only be recommended under MCAR.

Listwise deletion means that all cases with missing values are deleted. This leads to a complete data matrix with no missing values which is used to estimate the model. This procedure can lead to a large loss of information in that the resulting sample size is much smaller than the original. Listwise deletion can give biased, inconsistent, and inefficient estimates under MAR. It should only be used under MCAR.

Pairwise deletion means that means and variances are estimated using all available data for each variable and covariances are estimated using all available data for each pair of variables. The means, variances and covariances are then combined to form a mean vector and a covariance matrix which are used to estimate the model. While some efficiency is obtained compared to listwise deletion, it is difficult to specify a sample size N to be used in the estimation of the model, since the variances and covariances are all based on different sample sizes and there is no guarantee that the covariance matrix will be positive definite which is required by the maximum likelihood method. Although pairwise deletion is available in **LISREL**, it is not recommended. Its best use is for data screening for then it gives the most complete information about the missing values in the data.

Imputation means that real values are substituted for the missing values. Various *ad.hoc.* procedures for imputation have been suggested in the literature. One such is imputation by matching which is available in **LISREL**. It is based on the idea that individuals who have similar values on a set of matching variables may also be similar on a variable with missing values. This will work well if the matching variables are good predictors of the variable with missing values.

Methods 4 and 5 are both based on the assumption of multivariate normality and missingness under MAR. Method 4 uses multiple imputation methods to generate a complete data matrix. The multiple imputation procedure implemented in **LISREL** is described in details in Schafer (1997) and uses the EM algorithm and the method of generating random draws from probability distributions via Markov chains (MCMC). Formulas are given in Section 16.1.11. The EM algorithm generates one single complete data matrix whereas the MCMC method generates several complete data matrices and uses the average of these. As a consequence, the MCMC method is more reliable than the EM algorithm. In both cases, the complete data matrix can be used to estimate the mean vector and the covariance matrix of the observed variables which can be used to estimate the model. However, in **LISREL** 9 it is not necessary to do these steps separately as they are done automatically as will be described in what follows.

Method 5 is based on the following idea. If the variables have a multivariate normal distribution all subsets of the variables also have that distribution. So the likelihood function for the observed values can be evaluated for each observation without using any missing values. Formulas are given in Section 16.1.10.

Method 5 is the recommended method for dealing with the problem of missing data but it can only be used together with a specified model. So we have to postpone the discussion of this method until later chapters.

To illustrate the concepts of missing values and some ways to deal with this problem, we consider another example and data set.

A medical doctor offered all his patients diagnosed with prostate cancer a treatment aimed at reducing the cancer activity in the prostate. The severity of prostate cancer is often assessed by a plasma component known as prostate specific antigen (PSA), an enzyme that is elevated in the presence of prostate cancer. The PSA level was measured regularly every three months. The data contains five repeated measurements of PSA. The age of the patient is also included in the data. Not every patient accepted the offer at the first visit to the doctor and several patients chose to enter the program after the first occasion. Some patients, who accepted the initial offer, are absent at some later occasions for various reasons. Thus, there are missing values in the data.

The data file is **psavar.lsf** where the first 16 rows of data are shown here

psavar.LSF	PSA0	PSA3	PSA6	PSA9	PSA12	Age
1	30.400	28.000	26.900	25.200	13.600	69.000
2	27.800	26.700	20.500	18.700	18.800	58.000
3	26.600	21.800	17.800	17.900	14.500	53.000
4	24.800	24.500	20.200	19.800	18.800	61.000
5	33.700	30.300	25.400	27.300	20.100	63.000
6	26.500	24.600	20.900	-9.000	18.900	49.000
7	26.200	24.400	21.800	22.200	18.400	63.000
8	24.800	19.500	18.000	16.100	12.500	49.000
9	28.400	-9.000	22.500	19.400	22.900	63.000
10	26.100	-9.000	23.300	22.000	14.600	56.000
11	28.800	31.300	-9.000	23.100	22.800	68.000
12	29.800	-9.000	25.600	24.500	21.000	67.000
13	22.900	23.900	-9.000	19.400	15.600	47.000
14	30.100	27.700	25.700	20.400	20.800	56.000
15	26.500	-9.000	-9.000	20.000	17.400	57.000

The value -9 is used here to indicate a missing value. This is called a missing value code. One can have one or more missing value code(s). The missing code(s) must be specified in the **.lsf** file. If this is not done the missing value code(s) are treated as real values in the data which is likely to give very wrong results.

To specify the missing value code in the .lsf file, proceed as follows. Select **Data** in the **LISREL System Data File Toolbar**, select **Define Variable** and click on **Missing Values**. In the section **Global Missing Value** type -9 in the numerical field:

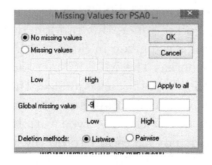

Now click **OK** twice and save the **.lsf** file.

In this kind of data it is almost inevitable that there are missing values. For example, a patient may be on vacation or ill or unable to come to the doctor for any reason at some occasion or a patient may die and therefore will not come to the doctor after a certain occasion. For example, a quick perusal of the data shows that

- Patients 9 and 10 are missing at 3 months

- Patient 15 is missing at 3 and 6 months

- Patient 16 is missing at 0, 3, and 12 months

For this data we illustrate listwise deletion, imputation by matching, imputation by EM, and imputation by MCMC and in each case we show how one can get a new **.lsf** file. While these procedures can all be done by using the **Graphical User's Interface**, we will use simple PRELIS syntax files here

Listwise Deletion

To generate an **.lsf** file for the listwise sample, use the following PRELIS syntax file **psavar1.prl**:

```
!Creating psavar_listwise.lsf
SY=psavar.LSF
OU ra=psavar_listwise.lsf
```

This just reads the data and produces the **.lsf** file for the listwise sample. Since listwise deletion is the default in PRELIS when there are missing values in the data, nothing needs to be specified in the PRELIS syntax file.

Some sections of the output file are shown and discussed here

```
Number of Missing Values per Variable

    PSA0       PSA3       PSA6       PSA9      PSA12       Age
 --------   --------   --------   --------   --------   --------
    17         14         13         12         11          0
```

This shows that there are most (17) missing values at the first visit to the doctor and fewest (11) missing at 12 months. Age is recorded for all patients.

```
Distribution of Missing Values

Total Sample Size(N) =    100

Number of Missing Values    0    1    2    3
          Number of Cases   46   43    9    2
```

This shows that there are only 46 patients with complete data on all occasions. So the listwise sample has only 46 cases, i.e., Since the sample size is 100, 54 cases have been deleted. It is also seen that 43 patients have 1 missing value, 9 have 2 missing values, and 2 have 3 missing

values. Thus, no patient have 4 or 5 missing values. The **.lsf file** for the listwise sample is
psavar_listwise.lsf. This has complete data on 46 patients.

Imputation by Matching

It is not likely that data is MCAR. More likely, the mechanism is MAR since the probability
of missingness may depend on Age. So one idea is to use imputation by matching and use Age
as a matching variable, see **psavar2.prl**:

```
!Creating psavar_machimputed.lsf
SY=psavar.LSF
IM (PSA0 - PSA12) (Age) XN
OU ra=psavar_machimputed.lsf
```

Here the line

```
IM (PSA0 - PSA12) (Age) XN
```

means impute the missing values on the variables PSA0 - PSA12 by matching on the variable Age.
The **XN** option tells **LISREL** to list all successful imputations. The general form of **IM** command is

```
IM Ivarlist Mvarlist VR=n XN XL
```

where **Ivarlist** is a set of variables whose missing values should be imputed and **Mvarlist** is a
set of matching variables. **VR**, **XN**, and **XL** are explained below.

The imputation procedure is as follows. Let $y_1, y_2, \ldots, y_p$ denote the variables in **Ivarlist** and
let $x_1, x_2, \ldots, x_q$ denote the variables in **Mvarlist**. To begin, let us assume that there is only a
single variable y in **Ivarlist** whose missing values are to be imputed and that y is not included
in **Mvarlist**. Let $z_1, z_2, \ldots, z_q$ be the standardized $x_1, x_2, \ldots, x_q$, *i.e.*, for each case c

$$z_{cj} = (x_{cj} - \bar{x}_j)/s_j \qquad j = 1, 2, \ldots, q ,$$

where $\bar{x}_j$ and s_j are the estimated mean and standard deviation of x_j. These are estimated from
all complete data on x_j.

The imputation procedure is as follows.

1. Find the first case a with a missing value on y and no missing values on $x_1, x_2, \ldots, x_q$. If
   no such case exists, imputation of y is impossible. Otherwise, proceed to impute the value
   y_a as follows.

2. Find *all* cases b which have no missing value on y and no missing values on $x_1, x_2, \ldots, x_q$,
   and which minimizes

$$\sum_{j=1}^{q} (z_{bj} - z_{aj})^2 . \qquad (1.1)$$

3. Two cases will occur

   - If there is a single case b satisfying 2, y_a is replaced by y_b.

- Otherwise, if there are $n > 1$ matching cases b *with the same minimum value* of (1.1), denote their y-values by $y_1^{(m)}, y_2^{(m)}, \ldots, y_n^{(m)}$. Let

$$\bar{y}_m = (1/n) \sum_{i=1}^{n} y_i^{(m)}, \quad s_m^2 = [1/(n-1)]) \sum_{i=1}^{n} (y_i^{(m)} - \bar{y}_m)^2,$$

be the mean and variance of the y-values of the matching cases. Then, imputation will be done only if

$$\frac{s_m^2}{s_y^2} < v, \tag{1.2}$$

where s_y^2 is the total variance of y estimated from all complete data on y, and v is the value VR specified on the MI command. This may be interpreted to mean that the matching cases predict the missing value with a reliability of at least $1 - v$. The default value of VR is VR=.5, *i.e.*, $v = .5$. Larger values than this are not recommended. Smaller values may be used if one requires high precision in the imputation. For each value imputed, LISREL gives the value of the variance ratio and the number of cases on which s_m^2 is based.

If condition (1.2) is satisfied, then y_a is replaced with the mean $\bar{y}_m$ if y is continuous or censored, or with the value on the scale of y closest to $\bar{y}_m$ if y is ordinal. Otherwise, no imputation is done and y_a is left as a missing value.

4. This procedure is repeated for the next case a for which y_a is missing, and so on, until all possible missing values on y have been imputed.

If Ivarlist contains several variables, they will be imputed in the order they are listed. This is of no consequence if no variables in Ivarlist is included in Mvarlist. Ideally, Ivarlist contains the variables with missing values and Mvarlist contains variables without missing values.

This procedure has the advantage that it does not make any distributional assumption unlike Methods 3 - 5 which depend on the assumption of multivariate normality. Furthermore, it has the advantage that it gives the same results under linear transformation of the matching variables. Thus, if age is a matching variable, age can be in years or months, or represented by the year of birth, and the resulting imputed data will be the same in each case. Another advantage is that the results of the imputation will be the same regardless of the order of cases in the data. A disadvantage is that it does not always impute all the missing values, so one would typically use listwise deletion after imputation.

Imputation of missing values should be done with utmost care and control, since missing values will be replaced by other values that will be treated as real observed values. If possible, use matching variables which are *not* to be used in the LISREL modeling. Otherwise, if the matching variables are included in the LISREL model, it is likely that the imputation will affect the result of analysis. This should be checked by comparing with the result obtained without imputation.

For each variable to be imputed, the output lists all the cases with missing values. If imputation is successful, it gives the value imputed, the number of matching cases NM and the variance ratio VR. If the imputation is not successful, it gives the reason for the failure. This can be that no matching case was found or that the variance ratio was too large. The XN option on the IM command will make LISREL list only successful imputations, and the XL option makes LISREL skip the entire listing of cases.

The output always gives the number of missing values per variable, both before and after imputation. Here we show some sections of the output file **psavar2.out**:

```
Number of Missing Values per Variable

     PSA0        PSA3        PSA6        PSA9       PSA12        Age
  --------    --------    --------    --------    --------    --------
      17          14          13          12          11           0
```

This is the number of missing values per variable before imputation. After this comes the listing of the successful imputations:

```
Imputations for       PSA0

Case    16 imputed  with value    33.200 (Variance Ratio = 0.000), NM=     1
Case    33 imputed  with value    29.800 (Variance Ratio = 0.000), NM=     1
Case    67 imputed  with value    38     (Variance Ratio = 0.016), NM=     2
Case    68 imputed  with value    33.400 (Variance Ratio = 0.365), NM=     4
Case    73 imputed  with value    26.800 (Variance Ratio = 0.000), NM=     1
Case    83 imputed  with value    29.150 (Variance Ratio = 0.327), NM=     4
Case    99 imputed  with value    38     (Variance Ratio = 0.008), NM=     3


Imputations for       PSA3

Case    12 imputed  with value    31.500 (Variance Ratio = 0.000), NM=     1
Case    15 imputed  with value    37.350 (Variance Ratio = 0.212), NM=     2
Case    16 imputed  with value    31.200 (Variance Ratio = 0.000), NM=     1
Case    32 imputed  with value    35.900 (Variance Ratio = 0.000), NM=     1
Case    34 imputed  with value    22.800 (Variance Ratio = 0.000), NM=     3
Case    99 imputed  with value    35.867 (Variance Ratio = 0.055), NM=     3


Imputations for       PSA6

Case    11 imputed  with value    35.900 (Variance Ratio = 0.000), NM=     1
Case    15 imputed  with value    34.600 (Variance Ratio = 0.000), NM=     1
Case    20 imputed  with value    27.600 (Variance Ratio = 0.000), NM=     1
Case    32 imputed  with value    33.900 (Variance Ratio = 0.000), NM=     1
Case    44 imputed  with value    22.167 (Variance Ratio = 0.085), NM=     3
Case    68 imputed  with value    30.800 (Variance Ratio = 0.261), NM=     4
Case    74 imputed  with value    32.133 (Variance Ratio = 0.185), NM=     3
Case    89 imputed  with value    34.600 (Variance Ratio = 0.000), NM=     2
Case    95 imputed  with value    29     (Variance Ratio = 0.000), NM=     1


Imputations for       PSA9

Case     6 imputed  with value    17.500 (Variance Ratio = 0.096), NM=     2
Case    22 imputed  with value    28.400 (Variance Ratio = 0.000), NM=     1
Case    28 imputed  with value    19.833 (Variance Ratio = 0.250), NM=     3
Case    30 imputed  with value    33.575 (Variance Ratio = 0.006), NM=     4
Case    60 imputed  with value    15.400 (Variance Ratio = 0.000), NM=     1
Case    70 imputed  with value    29.400 (Variance Ratio = 0.000), NM=     1
Case   100 imputed  with value    25.100 (Variance Ratio = 0.000), NM=     1
```

Imputations for   PSA12

```
Case   16 imputed  with value    24.750 (Variance Ratio = 0.026), NM=    4
Case   19 imputed  with value    24.325 (Variance Ratio = 0.420), NM=    4
Case   23 imputed  with value    24.750 (Variance Ratio = 0.000), NM=    1
Case   35 imputed  with value    26.400 (Variance Ratio = 0.086), NM=    3
Case   51 imputed  with value    24.750 (Variance Ratio = 0.027), NM=    4
Case   69 imputed  with value     9.600 (Variance Ratio = 0.000), NM=    1
Case   84 imputed  with value    24.750 (Variance Ratio = 0.000), NM=    1
```

After these imputations the number of missing variables per variable is

Number of Missing Values per Variable After Imputation

```
    PSA0      PSA3      PSA6      PSA9      PSA12      Age
 --------  --------  --------  --------  --------  --------
    10         8         4         5         4         0
```

It is seen that the number of missing values has been considerably reduced for each variable. The distribution of missing values after imputation is now

Distribution of Missing Values

Total Sample Size(N) =   100

```
Number of Missing Values      0     1     2     3
           Number of Cases   76    18     5     1
```

It is seen that after the imputation there are 76 cases without missing values. So the listwise sample size after imputation will be 76. More details about the patterns of missing values are obtained in the missing data map:

Missing Data Map

```
Frequency PerCent    Pattern
       76     76.0    0 0 0 0 0 0
        5      5.0    1 0 0 0 0 0
        6      6.0    0 1 0 0 0 0
        1      1.0    1 1 0 0 0 0
        2      2.0    0 0 1 0 0 0
        1      1.0    1 0 1 0 0 0
        2      2.0    0 0 0 1 0 0
        1      1.0    1 0 0 1 0 0
        1      1.0    1 1 0 1 0 0
        1      1.0    0 0 1 1 0 0
        3      3.0    0 0 0 0 1 0
        1      1.0    1 0 0 0 1 0
```

The missing data map gives all possible patterns of missing data and their sample frequencies in absolute and percentage form. Each column under `Pattern` corresponds to a variable. A 0 means a complete data and a 1 means a missing data.

In this example, although we know from the distribution of missing values that 18 cases have only 1 missing value, we don't know on which variable they are missing. But the missing data map tells us that 5 are missing on PSA0, 6 are missing on PSA3, 2 are missing on PSA6, 2 are missing on PSA9, and 3 are missing on PSA12, thus $5 + 6 + 2 + 2 + 3 = 18$. The last column is the variable Age which has no missing values.

Multiple Imputation by EM

Another way to impute missing values is multiple imputation using the EM algorithm. This will impute all missing values and create a complete data matrix. To do this kind of imputation, use the **PRELIS** syntax file **psavar3.prl**:

```
!Creating psavar_eminputed.lsf
SY=psavar.LSF
EM CC = 0.00001 IT = 25 TC = 0
OU ra=psavar_emimputed.lsf
```

The third line is the command for doing multiple imputation by EM. Here CC is a convergence criterion with default value 0.00001, IT is the maximum number allowed (default = 200), and TC is a constant which specifies how cases with missing values on *all* variables should be treated. TC = 0 means that their values will be replaced by means, TC = 1 means that these cases will be deleted, and TC = 2 means that their values will be left as missing values. Although each of the parameters CC, IT, and TC have default values, they must nevertheless be specified, *i.e.*, an empty EM line will not work.

The resulting imputed data file is **psavar_emimputed.lsf** which has complete data on all 100 cases.

To perform multiple imputation with the MCMC method instead, just replace the line, see file **psavar4.prl**

```
EM CC = 0.00001 IT = 25 TC = 0
```

with

```
MC CC = 0.00001 IT = 25 TC = 0
```

1.7   Data Management

Often the initial data is not such that it can be analyzed directly. One can add or delete cases in the **.lsf** file directly and one can also delete variables. This section describes how one can add new variables which are functions of other variables.

We illustrate this with some data that will be analyzed in Chapter 3.

In 1951, all British doctors were sent a brief questionnaire about whether they smoked tobacco. Ten years later the number of deaths from coronary heart disease for smokers and non-smokers was recorded. The original data come from a study by Sir Richard Doll. A table of the number of deaths and the number of person-years of observation in different age groups was published by Breslow and Day (1987, p. 112). A similar table was given by Dobson and Barnett (2008, p. 168). This is given here in Table 1.1.

Table 1.1: Deaths from Coronary Heart Disease among British Doctors. Originally published in Dobson and Barnett (2008, Table 9.1). Published with kind permission of © Taylor and Francis Group, LLC 2008. All Rights Reserved

Age Group	Smokers		Non-smokers	
	Deaths	Person-years	Deaths	Person-years
35–44	32	52407	2	18790
45–54	104	43248	12	10673
55–64	206	28612	28	5710
65–74	186	12663	28	2585
75–84	102	5317	31	1462

For the analysis with **LISREL** in Section 3.3.1 in Chapter 3, we need some additional variables. Begin by typing a text file like this, **corheart1.dat**, say

```
SMOKE AGECAT DEATHS PYEARS
   1      1     32    52407
   1      2    104    43248
   1      3    206    28612
   1      4    186    12633
   1      5    102     5317
   0      1      2    18790
   0      2     12    10673
   0      3     28     5710
   0      4     28     2585
   0      5     31     1462
```

where **SMOKE** is a dummy variable equal to 1 for smokers and equal to 0 for non-smokers, **AGECAT** is coded 1, 2, 3, 4, 5 and treated as a continuous variable, **DEATHS** is the number of deaths, and **PYEARS** is the number of person-years. All numbers are taken from Table 1.1.

For the analysis in this example we need to do a number of simple calculations with the data in this **dat** file. There are at least three different ways to do these calculations:

1. Since this is a very small data set, one can do these calculations by using a hand calculator or by using an external spread sheet, such as Microsoft Excel, and then import the results to a **LISREL** system file **.lsf** as described in Section 1.1.

2. One can use **PRELIS** syntax.

3. One can use **LISREL**'s **Compute Dialog Box** in the graphical users interface, see the file **Graphical Users Interface.pdf** that comes with the program.

We illustrate each of these alternatives in turn.

1: Calculation by Hand

The file **corheart1.dat** is appended by four new variables and called **corheart2.dat**:

```
SMOKE AGECAT DEATHS PYEARS    DTHRATE   AGESQ SMKAGE LNPYEARS
  1     1      32    52407   0.0006116    1     1    10.866796
  1     2     104    43248   0.0024047    4     2    10.674706
  1     3     206    28612   0.0071998    9     3    10.261581
  1     4     186    12633   0.0147233   16     4     9.444068
  1     5     102     5317   0.0191837   25     5     8.578665
  0     1       2    18790   0.0001064    1     0     9.841080
  0     2      12    10673   0.0011243    4     0     9.275473
  0     3      28     5710   0.0049037    9     0     8.649974
  0     4      28     2585   0.0108317   16     0     7.857481
  0     5      31     1462   0.0212038   25     0     7.287560
```

where `DTHRATE` is the death rate equal to the number of deaths divided by the number of person-years, `AGESQ` is `AGECAT` squared, `SMKAGE` is the product of `SMOKE` and `AGECAT`, and `LNPYEARS` is the natural logarithm of `PYEARS`.

To import **corheart2.dat** to a LISREL system file select **File $\Rightarrow$ Import Data**, enter 8 as the **Number of Variables**, tick the box **Variable names at top of file**, and click **OK**. This gives a LISREL system file **corheart2.lsf** which looks like this (to see six decimals, select **Edit $\Rightarrow$ Format**, enter 12 as Width and 6 as Number of Decimals):

corheart2.LSF	SMOKE	AGECAT	DEATHS	PYEARS	DTHRATE	AGESQ	SMKAGE	LNPYEARS
1	1.000000	1.000000	32.000000	52407.000000	0.000612	1.000000	1.000000	10.866796
2	1.000000	2.000000	104.000000	43248.000000	0.002405	4.000000	2.000000	10.674706
3	1.000000	3.000000	206.000000	28612.000000	0.007200	9.000000	3.000000	10.261581
4	1.000000	4.000000	186.000000	12633.000000	0.014723	16.000000	4.000000	9.444068
5	1.000000	5.000000	102.000000	5317.000000	0.019184	25.000000	5.000000	8.578665
6	0.000000	1.000000	2.000000	18790.000000	0.000106	1.000000	0.000000	9.841080
7	0.000000	2.000000	12.000000	10673.000000	0.001124	4.000000	0.000000	9.275473
8	0.000000	3.000000	28.000000	5710.000000	0.004904	9.000000	0.000000	8.649974
9	0.000000	4.000000	28.000000	2585.000000	0.010832	16.000000	0.000000	7.857481
10	0.000000	5.000000	31.000000	1462.000000	0.021204	25.000000	0.000000	7.287560

2: Using PRELIS Syntax

If the data is large, it is better and easier to do these calculations using PRELIS syntax instead. The following PRELIS syntax file, see file **corheart2.prl**, will produce a file **corheart2.lsf** which is identically the same as the one generated previously

```
!Creating the .lsf file by PRELIS syntax
da ni=4
ra=corheart1.dat lf
new DTHRATE=DEATHS*PYEARS**-1
new AGESQ=AGECAT**2
```

```
new SMKAGE=SMOKE*AGECAT
new LNPYEARS=PYEARS
log LNPYEARS
co all
ou ra=corheart2.lsf
```

This is a slightly different **PRELIS** syntax file, compared with those previously used. This is because it reads a text file **.dat** instead of a **.lsf** file. So we must tell **LISREL** first how many variables there are. This is done by the line

```
da ni=4
```

The line

```
ra=corheart1.dat lf
```

tells **LISREL** to read the data file **corheart1.dat**. Note that this is an **ra** line and not an **sy** line. The keyword **lf** tells **LISREL** that the names of the of the variables are in the first line of the **.dat** file. **new** is the command for computing a new variable. So

```
new DTHRATE=DEATHS*PYEARS**-1
new AGESQ=AGECAT**2
new SMKAGE=SMOKE*AGECAT
```

will compute the new variables DTHRATE, AGESQ, and SMKAGE. The logarithm transformation needs a special explanation. First we make a copy of PYEARS and call it LNPYEARS. Then we compute the log of LNPYEARS.

3: Using the Compute Dialog Box

No matter how large the data is one can use **LISREL**'s **Compute Dialog Box** to do the calculations.

First import **corheart1.dat** to obtain the following **.lsf** file **corheart.lsf**:

corheart.LSF	SMOKE	AGECAT	DEATHS	PYEARS
1	1.000	1.000	32.000	52407.000
2	1.000	2.000	104.000	43248.000
3	1.000	3.000	206.000	28612.000
4	1.000	4.000	186.000	12633.000
5	1.000	5.000	102.000	5317.000
6	0.000	1.000	2.000	18790.000
7	0.000	2.000	12.000	10673.000
8	0.000	3.000	28.000	5710.000
9	0.000	4.000	28.000	2585.000
10	0.000	5.000	31.000	1462.000

Select **Compute** in the **Transformation** menu:

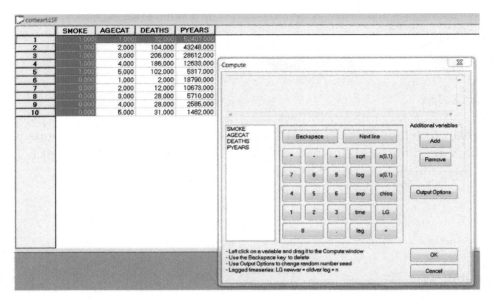

Click on **Add** and type `DTHRATE`:

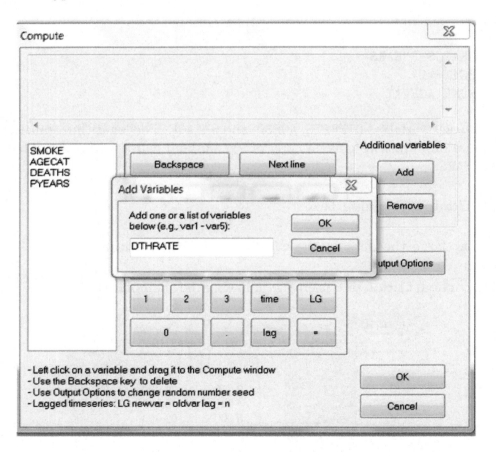

Click **OK**. This will add the variable `DTHRATE` in the list of variables on the left side of the compute screen. To compute the values of this variable drag the variable `DTHRATE` to the compute screen and use the symbols in the calculator as shown (the death rate is defined the number of deaths divided by the number of person-years, but since no division sign is available in the calculator, we multiply by the inverse of `PYEARS` instead as indicated by `PYEARS**-1`):

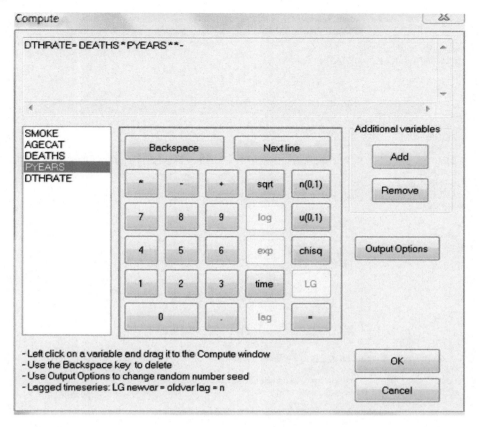

Then click **OK** and the **.lsf** file now looks like

	SMOKE	AGECAT	DEATHS	PYEARS	DTHRATE
1	1.000000	1.000000	32.000000	52407.000000	0.000611
2	1.000000	2.000000	104.000000	43248.000000	0.002405
3	1.000000	3.000000	206.000000	28612.000000	0.007200
4	1.000000	4.000000	186.000000	12633.000000	0.014723
5	1.000000	5.000000	102.000000	5317.000000	0.019184
6	0.000000	1.000000	2.000000	18790.000000	0.000106
7	0.000000	2.000000	12.000000	10673.000000	0.001124
8	0.000000	3.000000	28.000000	5710.000000	0.004904
9	0.000000	4.000000	28.000000	2585.000000	0.010832
	0.000000	5.000000	31.000000	1462.000000	0.021204

Note that the death rate increases with age for both smokers and non-smokers and the death rate is higher for smokers than non-smokers for all age groups except the age category 5 (the 75-84 years group), where the reverse is true.

In the GLM model we therefore include two more variables, `AGESQ` as the square of `AGACAT` and `SMKAGE` as the product of `SMOKE` and `AGECAT`. To compute the values of these variables, we use the compute screen again:

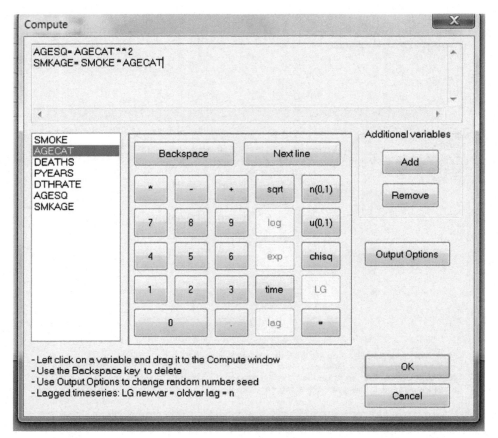

After clicking **OK** in the compute screen, the new **.lsf** file looks like this:

	SMOKE	AGECAT	DEATHS	PYEARS	DTHRATE	AGESQ	SMKAGE
1	1.000000	1.000000	32.000000	52407.000000	0.000611	1.000000	1.000000
2	1.000000	2.000000	104.000000	43248.000000	0.002405	4.000000	2.000000
3	1.000000	3.000000	206.000000	28612.000000	0.007200	9.000000	3.000000
4	1.000000	4.000000	186.000000	12633.000000	0.014723	16.000000	4.000000
5	1.000000	5.000000	102.000000	5317.000000	0.019184	25.000000	5.000000
6	0.000000	1.000000	2.000000	18790.000000	0.000106	1.000000	0.000000
7	0.000000	2.000000	12.000000	10673.000000	0.001124	4.000000	0.000000
8	0.000000	3.000000	28.000000	5710.000000	0.004904	9.000000	0.000000
9	0.000000	4.000000	28.000000	2585.000000	0.010832	16.000000	0.000000
10	0.000000	5.000000	31.000000	1462.000000	0.021204	25.000000	0.000000

To perform the estimation of the GLM model, we must also compute the **offset** variable $\ln(n_i)$, *i.e.*, the logarithm of PYEARS. To do this is somewhat counterintuitive. To keep the original variable PYEARS, click **Add** in the compute screen and type LNPYEARS, generate a line LNPYEARS=PYEARS, select the variable LNPYEARS in the list of variables on the left and click on **LOG**. The compute screen now looks like this:

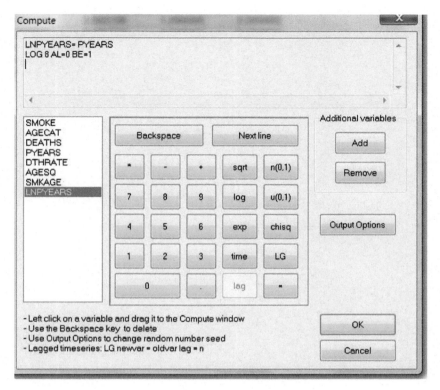

After clicking **OK** in the compute screen, the new **.lsf** file looks like this:

	SMOKE	AGECAT	DEATHS	PYEARS	DTHRATE	AGESQ	SMKAGE	LNPYEARS
1	1.000000	1.000000	32.000000	52407.000000	0.000611	1.000000	1.000000	10.866796
2	1.000000	2.000000	104.000000	43248.000000	0.002405	4.000000	2.000000	10.674706
3	1.000000	3.000000	206.000000	28612.000000	0.007200	9.000000	3.000000	10.261581
4	1.000000	4.000000	186.000000	12633.000000	0.014723	16.000000	4.000000	9.444068
5	1.000000	5.000000	102.000000	5317.000000	0.019184	25.000000	5.000000	8.578665
6	0.000000	1.000000	2.000000	18790.000000	0.000106	1.000000	0.000000	9.841080
7	0.000000	2.000000	12.000000	10673.000000	0.001124	4.000000	0.000000	9.275473
8	0.000000	3.000000	28.000000	5710.000000	0.004904	9.000000	0.000000	8.649974
9	0.000000	4.000000	28.000000	2585.000000	0.010832	16.000000	0.000000	7.857481
10	0.000000	5.000000	31.000000	1462.000000	0.021204	25.000000	0.000000	7.287560

Chapter 2

Regression Models

There are several different types of regression models, for example,

- Linear regression

- Logistic regression

- Probit regression

- Censored regression

These types of regression models are covered in this chapter. Other types of regression models are obtained as *Generalized Linear Models* which are covered in Chapter 3. Still other types of regression models are covered in *Multilevel Analysis* in Chapter 4.

All of these types of regression have in common that there is a single dependent variable y and a set of explanatory variables $\mathbf{x}' = (x_1, x_2, \ldots, x_q)$, called covariates, and a regression equation of the form

$$E(y|\mathbf{x}) = F(\mathbf{x}, \boldsymbol{\theta}) , \tag{2.1}$$

where E is the expectation operator, see Sections 11.3 and 11.4 in the appendix (Chapter 11). F is a function of both $\mathbf{x}$ and $\boldsymbol{\theta}$, where $\boldsymbol{\theta}' = (\theta_1, \theta_2, \ldots, \theta_t)$ is a set of unknown parameters.

The objective of regression analysis is to estimate the parameters $\boldsymbol{\theta}$ from a random sample of N independent observations $(y_i, \mathbf{x}_i)$, $i = 1, 2, \ldots, N$, thereby establishing the relationship between y and the x-variables. This relationship can be used to predict y from future observations of $\mathbf{x}$.

2.1 Linear Regression

The most common type of regression analysis is linear regression:

$$y = \alpha + \gamma_1 x_1 + \gamma_2 x_2 + \cdots + \gamma_q x_q + z , \tag{2.2}$$

where $\alpha, \gamma_1, \gamma_2, \ldots \gamma_q$ are parameters to be estimated, and z, is an error term or residual with mean 0 and variance σ^2. The error term z is assumed to be independent of $x_1, x_2, \ldots, x_q$. To see the correspondence with (2.1), the model may be written

$$E(y \mid \mathbf{x}) = \alpha + \gamma_1 x_1 + \gamma_2 x_2 + \cdots + \gamma_q x_q = \alpha + \boldsymbol{\gamma}'\mathbf{x} , \tag{2.3}$$

© Springer International Publishing Switzerland 2016
K.G. Jöreskog et al., *Multivariate Analysis with LISREL*, Springer Series in Statistics,
DOI 10.1007/978-3-319-33153-9_2

where $\mathbf{x}' = (x_1, x_2, \ldots, x_q)$ is a vector of observed random variables and $\boldsymbol{\gamma}' = (\gamma_1, \gamma_2, \ldots, \gamma_q)$ is a parameter vector called the regression vector.

The parameter α is called an intercept, the γs are called regression coefficients or slopes, and σ^2 is called the error or residual variance. The usual interpretation of α and γ_j is that α is the mean of y when all x-variables are zero and γ_j is the expected change in y if x_j increases by one unit and all the other x-variables are held constant. However, these interpretations are only meaningful if the x-variables can take zero values and if the units of measurement of y and x_j have some definite meaning.

The parameter vector corresponding to $\boldsymbol{\theta}$ in (2.1) is $\boldsymbol{\theta}' = (\alpha, \gamma_1, \gamma_2, \ldots, \gamma_q, \sigma^2)$. The characteristic feature of a linear regression model is that equation (2.2) is linear in the parameters α, γ_1, γ_2, ..., γ_q. A linear regression model may, however, be non-linear in variables. For example,

$$y = \alpha + \gamma_1 x_1 + \gamma_2 x_1^2 + z , \tag{2.4}$$

is also a linear regression model because it is linear in the parameters α, γ_1, and γ_2, although it is non-linear in x_1. To see the linearity, rewrite (2.4) as

$$y = \alpha + \gamma_1 x_1 + \gamma_2 x_2 + z , \tag{2.5}$$

where $x_2 = x_1^2$.

Consider a random sample of N observations, for example, individuals, regions, or time points, of the variables y, x_1, x_2, ..., x_q. These observations are usually represented as a data matrix of the form

$$\begin{array}{c|cccc}
y_1 & x_{11} & x_{12} & \cdots & x_{1q} \\
y_2 & x_{21} & x_{22} & \cdots & x_{2q} \\
\cdot & \cdot & \cdot & \cdots & \cdot \\
\cdot & \cdot & \cdot & \cdots & \cdot \\
\cdot & \cdot & \cdot & \cdots & \cdot \\
y_N & x_{N1} & x_{N2} & \cdots & x_{Nq}
\end{array}$$

The columns correspond to variables and the general term for the rows are *cases*.

For all N observations the regression model may be written

$$y_i = \alpha + \gamma_1 x_{i1} + \gamma_2 x_{i2} + \ldots + \gamma_q x_{iq} + z_i , \quad i = 1, 2, \ldots, N . \tag{2.6}$$

which may be written in matrix form as

$$\mathbf{y} = \mathbf{1}\alpha + \mathbf{X}\boldsymbol{\gamma} + \mathbf{z} , \tag{2.7}$$

where $\mathbf{y} = (y_1, y_2, \ldots, y_N)'$, $\mathbf{z} = (z_1, z_2, \ldots, z_N)'$, $\mathbf{1}$ is an $N \times 1$ vector of ones, and $\mathbf{X} = (x_{ij})$ is a matrix of order $N \times q$. The matrix $\mathbf{X}$ contains the values of the covariates.

The usual formulation of the regression model (2.7) is as

$$\mathbf{y} = \mathbf{X}_\star \boldsymbol{\gamma}^\star + \mathbf{z} . \tag{2.8}$$

The matrix $\mathbf{X}_\star = \begin{pmatrix} \mathbf{1} & \mathbf{X} \end{pmatrix}$ contains an extra column of ones. This variable is often called *intcept*. It is the invisible variable multiplying α in (2.6). Let

$$\boldsymbol{\gamma}^\star = \begin{pmatrix} \alpha \\ \boldsymbol{\gamma} \end{pmatrix} , \tag{2.9}$$

so that

$$\mathbf{y} = \begin{pmatrix} \mathbf{1} & \mathbf{X} \end{pmatrix} \begin{pmatrix} \alpha \\ \boldsymbol{\gamma} \end{pmatrix} + \mathbf{z} . \tag{2.10}$$

2.1.1 Estimation and Testing

Ordinary least squares (OLS) estimates $\boldsymbol{\gamma}^\star$ by minimizing the residual sum of squares $\mathbf{z}'\mathbf{z}$ with respect to $\boldsymbol{\gamma}^\star$. Thus, the estimate $\hat{\boldsymbol{\gamma}}^\star$ must satisfy

$$\partial \mathbf{z}'\mathbf{z}/\partial \boldsymbol{\gamma}^\star = -2\mathbf{X}'_\star\mathbf{z} = -2\mathbf{X}'_\star(\mathbf{y} - \mathbf{X}_\star\hat{\boldsymbol{\gamma}}^\star) = \mathbf{0} , \tag{2.11}$$

which shows that $\hat{\boldsymbol{\gamma}}^\star$ is the solution of the *normal equations*

$$\mathbf{X}'_\star\mathbf{X}_\star\hat{\boldsymbol{\gamma}}^\star = \mathbf{X}'_\star\mathbf{y} . \tag{2.12}$$

If $\mathbf{X}_\star$ has full column rank, the solution is unique and given by

$$\hat{\boldsymbol{\gamma}}^\star = (\mathbf{X}'_\star\mathbf{X}_\star)^{-1}\mathbf{X}'_\star\mathbf{y} \tag{2.13}$$

If $\mathbf{X}_\star$ has rank $r < q + 1$, see the appendix in Chapter 11, there are r independent linear combinations of $\alpha, \gamma_1, \gamma_2, \ldots, \gamma_q$ which can be estimated and these can be chosen in various ways. This problem typically occurs when dummy coded variables are used to represent the categories of categorical variables or groups of observations. For example, if there are three categorical variables, then

$$\mathbf{X}_\star = \begin{pmatrix} 1 & 1 & 0 & 0 \\ 1 & 0 & 1 & 0 \\ 1 & 0 & 0 & 1 \end{pmatrix} , \tag{2.14}$$

which has rank 3.

For further development it is convenient to give the solution in a different notation. The matrix

$$\mathbf{A} = (1/N)\mathbf{X}'_\star\mathbf{X}_\star = \begin{pmatrix} 1 & \bar{\mathbf{x}}' \\ \bar{\mathbf{x}} & \mathbf{S} + \bar{\mathbf{x}}\bar{\mathbf{x}}' \end{pmatrix} , \tag{2.15}$$

is called *the augmented moment matrix*. Here $\bar{\mathbf{x}}$ and $\mathbf{S}$ are the sample mean vector and covariance matrix of the covariates $\mathbf{x}$ defined as

$$\bar{\mathbf{x}} = (1/N)\mathbf{X}'\mathbf{1} \tag{2.16}$$

$$\mathbf{S} = (1/N)\mathbf{X}'\mathbf{X} - \bar{\mathbf{x}}\bar{\mathbf{x}}' . \tag{2.17}$$

Using these, the normal equations (2.12) take the form

$$\hat{\alpha} + \bar{\mathbf{x}}'\hat{\boldsymbol{\gamma}} = \bar{y} , \tag{2.18}$$

$$\mathbf{S}\hat{\boldsymbol{\gamma}} = \mathbf{s}_{\mathbf{x}y} , \tag{2.19}$$

where $\bar{y}$ is the mean of y and $\mathbf{s}_{\mathbf{x}y}$ is the vector of covariances between y and $\mathbf{x}$. Equations (2.18) and (2.19) show that $\hat{\alpha}$ and $\hat{\boldsymbol{\gamma}}$ can be obtained directly from the means, variances and covariances of y and $\mathbf{x}$ without the use of the full data. If $\mathbf{S}$ is positive definite, see the appendix in Chapter 11, the solutions can be obtained explicitly as

$$\hat{\boldsymbol{\gamma}} = \mathbf{S}^{-1}\mathbf{s}_{\mathbf{x}y} , \tag{2.20}$$

$$\hat{\alpha} = \bar{y} - \bar{\mathbf{x}}'\hat{\boldsymbol{\gamma}} = \bar{y} - \bar{\mathbf{x}}'\mathbf{S}^{-1}\mathbf{s}_{\mathbf{x}y} . \tag{2.21}$$

So far, we have not made any distributional assumptions. For testing of hypotheses about $\boldsymbol{\gamma}$ one usually assumes that z_i is normally distributed. Let $\mathbf{x}_i$ be the ith row of $\mathbf{X}$. Under normality the model (2.6) states that y_i is distributed as $N(\alpha + \mathbf{x}_i\boldsymbol{\gamma}, \sigma^2)$, $i = 1, 2, \ldots, N$. The logarithm of the likelihood for the sample is

$$\ln L = -\frac{N}{2}\ln 2\pi\sigma^2 - \frac{1}{2\sigma^2}\sum(y_i - \alpha - \mathbf{x}_i\boldsymbol{\gamma})^2 \,. \tag{2.22}$$

This shows that maximum likelihood estimates of α and $\boldsymbol{\gamma}$ are obtained by maximizing the second term in (2.22) which is the same as minimizing the residual sum of squares. Hence the maximum likelihood estimates are the same as the ordinary least squares (OLS) estimates. The maximum likelihood estimate of σ^2 is

$$\tilde{\sigma}^2 = \frac{1}{N}\sum(y_i - \hat{\alpha} - \mathbf{x}_i\hat{\boldsymbol{\gamma}})^2 \,, \tag{2.23}$$

but LISREL gives the unbiased estimate of σ^2:

$$\hat{\sigma}^2 = \frac{1}{N - 1 - q}\sum(y_i - \hat{\alpha} - \mathbf{x}_i\hat{\boldsymbol{\gamma}})^2 \,. \tag{2.24}$$

To test $H_0 : \boldsymbol{\gamma} = \mathbf{0}$ against $H_1 : \boldsymbol{\gamma} \neq \mathbf{0}$ with α free in both models, one evaluates $-2\ln L$ at the minimum under both H_0 and H_1:

$$-2\ln L_0 = N\ln 2\pi\tilde{\sigma}_0^2 + \frac{1}{\tilde{\sigma}_0^2}\sum(y_i - \bar{y})^2 \tag{2.25}$$

$$-2\ln L_1 = N\ln 2\pi\tilde{\sigma}_1^2 + \frac{1}{\tilde{\sigma}_1^2}\sum(y_i - \hat{\alpha} - \mathbf{x}_i\hat{\boldsymbol{\gamma}})^2 \tag{2.26}$$

where

$$\tilde{\sigma}_0^2 = \frac{1}{N}\sum(y_i - \bar{y})^2 \tag{2.27}$$

and

$$\tilde{\sigma}_1^2 = \frac{1}{N}\sum(y_i - \hat{\alpha} - \mathbf{x}_i\hat{\boldsymbol{\gamma}})^2 \,. \tag{2.28}$$

Hence,

$$-2\ln L_0 = N(1 + \ln 2\pi\tilde{\sigma}_0^2) \,, \tag{2.29}$$

and

$$-2\ln L_1 = N(1 + \ln 2\pi\tilde{\sigma}_1^2) \,. \tag{2.30}$$

Then one can use

$$\chi^2 = 2\ln L_1 - 2\ln L_0 = N\ln\frac{\tilde{\sigma}_0^2}{\tilde{\sigma}_1^2} \tag{2.31}$$

with q degrees of freedom to test H_0.

The *coefficient of determination* or *squared multiple correlation* R^2 is

$$R^2 = 1 - \frac{\tilde{\sigma}_1^2}{\tilde{\sigma}_0^2} \,. \tag{2.32}$$

This R^2 satisfies $0 \leq R^2 \leq 1$ and $R^2 = 1$ only if there is a perfect linear relationship between y and $\mathbf{x}$. R^2 is interpreted as the proportion of variance of y explained or accounted for by $\mathbf{x}$. R^2 increases (strictly speaking it cannot decrease) when one or more x-variables are added in the regression equation.

Let RSS be the residual sum of squares

$$\text{RSS} = \sum \hat{z}_i^2 = \sum (y_i - \hat{\alpha} - \mathbf{x}_i \hat{\boldsymbol{\gamma}})^2 \tag{2.33}$$

Using RSS one can test the hypothesis that a subset q_0 of the q regression coefficients in $\boldsymbol{\gamma}$ are zero. Let RSS_0 be the residual sum of squares for the submodel that includes does not include the q_0 regression coefficients. If the hypothesis holds, then

$$F = \frac{(\text{RSS}_0 - \text{RSS})/q_0}{\text{RSS}/(N - q - 1)} \;, \tag{2.34}$$

has an $F(q_0, N - q - 1)$-distribution. The quantities involved in this test are often summarized in an Analysis of Variance (AOV) table. Let R_0^2 be the unadjusted squared multiple correlation for the model with only q_0 regressors and R^2 be the unadjusted squared multiple correlation for the model with the $q - q_0$ regressors. Then F in (2.34) can be expressed as

$$F = \frac{(R^2 - R_0^2)/q_0}{(1 - R^2)/(N - 1 - q)} \;. \tag{2.35}$$

Assuming that $\mathbf{X}_\star$ has rank $q + 1$, the estimated covariance matrix of $\hat{\alpha}, \hat{\gamma}_1, \hat{\gamma}_2, \ldots, \hat{\gamma}_q$ is

$$\hat{\sigma}^2 (\mathbf{X}_\star' \mathbf{X}_\star)^{-1} \;, \tag{2.36}$$

where $\hat{\sigma}^2$ is given by (2.24). The square roots of the diagonal elements of (2.36) are the standard errors of $\hat{\alpha}, \hat{\gamma}_1, \hat{\gamma}_2, \ldots, \hat{\gamma}_q$. Denote the standard error of $\hat{\gamma}_i$ by $se(\hat{\gamma}_i)$. Under the normality assumption it can be shown that $\hat{\gamma}_i / se(\hat{\gamma}_i)$ has a t-distribution with $d = N - 1 - q$ degrees of freedom if $\gamma_i = 0$. This can be used to test the hypothesis that $\gamma_i = 0$.

2.1.2   Example: Cholesterol

Table 2.1 shows data on serum cholesterol (millimoles per liter), age (years), and body mass index, BMI (weight in kilograms divided by the square of height in meters) for 30 women. Does the level of cholesterol increase with age, and if so, by how much? Is there an effect of BMI after controlling for the effect of age?

2.1.3   Importing Data

For most analysis with **LISREL** it is convenient to work with a **LISREL data system file** of the type .lsf. LISREL can import data from many formats such as SAS, SPSS, STATA, and EXCEL. LISREL can also import data in text format with spaces (*.dat or *.raw), commas (*.csv) or tab characters (*.txt) as delimiters between entries. The data is then stored as a **LISREL** data system file .lsf.

This section illustrates how to import data and create a **LISREL** data system file. Most other examples in this book are based on **LISREL** data system files that have already been created.

The data matrix in this example has 30 rows corresponding to the 30 women and 3 columns labeled Chol, Age, and BMI. For the present illustration suppose that the data is available in a text file with tabs as delimiters and with the three labels in the first line. Furthermore, suppose

Table 2.1: Cholesterol(Chol), Age, and Body Mass Index(BMI) for 30 Women. Originally published in Dobson and Barnett (2008, Table 6.18). Published with kind permission of © Taylor and Francis Group, LLC 2008. All Rights Reserved

Chol	Age	BMI	Chol	Age	BMI
5.94	52	20.7	6.48	65	26.3
4.71	46	21.3	8.83	76	22.7
5.86	51	25.4	5.10	47	21.5
6.52	44	22.7	5.81	43	20.7
6.80	70	23.9	4.65	30	18.9
5.23	33	24.3	6.82	58	23.9
4.97	21	22.2	6.28	78	24.3
8.78	63	26.2	5.15	49	23.8
5.13	56	23.3	2.92	36	19.6
6.74	54	29.2	9.27	67	24.3
5.95	44	22.7	5.57	42	22.7
5.83	71	21.9	4.92	29	22.5
5.74	39	22.4	6.72	33	24.1
4.92	58	20.2	5.57	42	22.7
6.69	58	24.4	6.25	66	27.3

the data file is **cholesterol.txt**. Here we illustrate how this data file can be converted to a LISREL data system file.

LISREL for Windows starts up by opening a main window with three menus. Use the **Import Data** option from the **File** menu of the main window of LISRELSelect **Tab Delimited Data(*.txt)** from the **Files of type** drop-down list box at the right bottom of the window. Select the file **cholesterol.txt**. Click on the **Open** button:

In the next dialog box, enter the 3 as the **Number of Variables**, tick the box **Variable names at top of file**, and click **OK**:

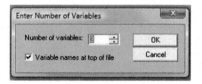

Accept the name **cholesterol.lsf** and click **Save**. The resulting file is a spread sheet that looks as follows, showing cases 1–15:

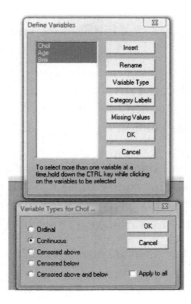

LISREL for Windows - [CHOLESTEROL.LSF]

	Chol	Age	BMI
1	5.940	52.000	20.700
2	4.710	46.000	21.300
3	5.860	51.000	25.400
4	6.520	44.000	22.700
5	6.800	70.000	23.900
6	5.230	33.000	24.300
7	4.970	21.000	22.200
8	8.780	63.000	26.200
9	5.130	56.000	23.300
10	6.740	54.000	29.200
11	5.950	44.000	22.700
12	5.830	71.000	21.900
13	5.740	39.000	22.400
14	4.920	58.000	20.200
15	6.690	58.000	24.400

LISREL has essentially two types of variables called ordinal and continuous. The term ordinal variable is used here in the sense of a discrete variable, *i.e.*, a variable that has a only few distinct values (maximum 15), whereas continuous variables typically have a wide range of values. **LISREL** assumes that all variables are ordinal (discrete) by default. For linear regression models all variables should be continuous even though some may be dummy-coded binary variables. The three variables in this example must therefore be declared continuous in the **lsf** file. To do this, select the **Data** button when the **lsf** file is displayed and select the **Define Variables** option. Then select the variables to be included in the analysis, *i.e.*, all three variables in this example. Then click on **Variable Type**, select **Continuous** and click on the **OK** button twice and save the **.lsf** file.

Consider estimating the regression of `Chol` on `Age` and BMI. The regression equation is

$$\text{Chol}_i = \alpha + \gamma_1 \text{Age}_i + \gamma_2 \text{BMI}_i + z_i, \ i = 1, 2, \ldots, 30 \tag{2.37}$$

There are several ways to perform linear regression with LISREL.One way is as follows. When the **.lsf** file is visible, select **Regressions...** in the **Statistics** Menu, select `Chol` as a Y variable and `Age` and `BMI` as X variables.

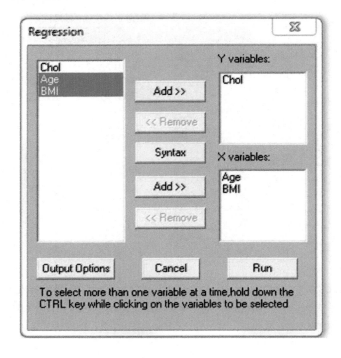

Then click **Run**.

This way of doing regression analysis is most useful if one wants to estimate one single regression equation. For a deeper analysis, it is more useful to write a **PRELIS** syntax file like this (see file **Cholesterol1a.prl**):

```
System file Cholesterol.lsf
Regress Chol on Age BMI
```

The first line is the command for reading a **LISREL** data system file. This can be shortened to

```
SY=Cholesterol.lsf
```

The second line is the command for performing a linear regression. The word **Regress** can be abbreviated to `RG`.These commands are case insensitive so **regress**, `RG`, `Rg`, and `rg` are all synonymous. The word **on** on the **Regress** line is important. It separates the dependent variable from the list of covariates.

This **PRELIS** syntax file has only two lines but one can include one or more title lines before the first line. Such title lines are optional but they can be used to describe the data and the problem.

The first section of the output **Cholesterol1a.out** gives the following information about all the variables in the data set, *i.e.*, all the variables in the file **Cholesterol.lsf**.

Univariate Summary Statistics for Continuous Variables

Variable	Mean	St. Dev.	Skewness	Kurtosis	Minimum	Freq.	Maximum	Freq.
Chol	6.005	1.309	0.671	1.591	2.920	1	9.270	1
Age	50.700	14.770	0.028	-0.718	21.000	1	78.000	1
BMI	23.180	2.268	0.526	0.659	18.900	1	29.200	1

In this sample the mean cholesterol level is 6.01 and the minimum and maximum cholesterol values are 2.92 and 9.27, respectively. The standard deviation of Chol is 1.31. The resulting estimated regression equation is

```
     Chol =   - 0.740 + 0.0410*Age + 0.201*BMI + Error, R² = 0.465
Standerr      (1.896) (0.0136)        (0.0888)
t-values       -0.390   3.006          2.269
P-values        0.699   0.006          0.031
```

```
Error Variance = 0.984
```

The first line is the regression equation with estimated regression coefficients. The estimated intercept is -0.740 and the regression coefficient of Age is 0.0410. The standard errors are given in the second line in parenthesis. The third line gives the t-values, *i.e.*, the ratio between the estimate and its standard error. Using the t-distribution with $d = N - 1 - q = 27$ degrees of freedom, one can compute the probability of obtaining a t-value larger in magnitude than the obtained t-value. This probability is the P-value given in the fourth line. In practice, one only needs to look at the P-value to determine whether a regression coefficient is statistically significant.

The estimated regression coefficient of Age is $\hat{\gamma}_1 = 0.0410$. Since its P-value is small this means that there is strong evidence that the level of cholesterol increases with age. Holding BMI constant, the cholesterol level increases by 0.041 millimoles per liter per year on average.

Obviously, the usual interpretation of the intercept -0.740 as the mean cholesterol level at Age and BMI zero does not make sense here since nobody has zero values on these variables. The youngest person in the sample is 21 years old and the smallest BMI value in the sample is 19.8. The negative value of $\hat{\alpha}$ is a reflection of the fact that if one were to extrapolate the regression line to zero values of Age and BMI, the predicted cholesterol level would be negative. This shows that the estimated regression line does not hold for small values of Age and BMI.

The estimated regression coefficient of BMI is $\hat{\gamma}_2 = 0.201$ with P-value 0.031. Hence, BMI may have an additional effect on cholesterol but the evidence is not as strong as that of age. One can examine this evidence further by testing the hypothesis $H_0 : \gamma_2 = 0$ against $H_1 : \gamma_2 \neq 0$. To do so, one can estimate two regression equations in the same PRELIS syntax file, see file **Cholesterol1b.prl**:

```
System file Cholesterol.lsf
Regress Chol on Age
Regress Chol on Age BMI
Output MA=CM
```

This includes an additional line

```
Output MA=CM
```

One can put various keywords on the Output line to request various kinds of output. In this case the MA=CM is a request to compute the covariance matrix CM. MA is short for Matrix and CM is short for Covariance Matrix. Other matrices that can be requested are, for example, KM for Correlation Matrix and AM for Augmented Moment Matrix.

The output file gives the following results:

```
    Chol = 3.296 + 0.0534*Age + Error, R² = 0.363
Standerr  (0.705) (0.0134)
t-values   4.676   3.999
P-values   0.000   0.000


    Chol =  - 0.740 + 0.0410*Age + 0.201*BMI + Error, R² = 0.465
Standerr      (1.896) (0.0136)        (0.0888)
t-values      -0.390   3.006           2.269
P-values       0.699   0.006           0.031
```

When only **Age** is used as an explanatory variable, its regression coefficient is estimated at 0.0534. This is different from the previous value 0.0410 because **Age** and **BMI** are correlated. The second estimated regression equation is the same as before.

First consider the chi-square test of the hypothesis $\gamma_2 = 0$. This is given in the output as

```
The following chi-squares test the hypothesis that all
regression coefficients are zero except the intercept.
Variable        -2lnL  Chi-square   df  Covariates
--------     ----------  ----------   --  ----------
    Chol      86.729      13.553     1   Age
    Chol      81.495      18.788     2   Age BMI
```

For the first equation, chi-square is 13.553 with 1 degree of freedom for testing that $\gamma_1 = 0$, *i.e.*, that the effect of **Age** is zero. This is highly significant and in agreement with the P-value for $\hat{\gamma}_1$ which is 0.000. For the second equation, chi-square is 18.788 with 2 degrees of freedom for testing $\gamma_1 = \gamma_2 = 0$, *i.e.*, that both regression coefficients are be zero. This chi-square is also highly significant. The difference in chi-square, $18.788 - 13.553 = 5.235$ is another chi-square with 1 degree of freedom for testing that the additional effect of **BMI** is zero. This chi-square is statistically significant at the 5% level but not at the 1% level, which is also in agreement with the P-value of **BMI** in the second equation.

Another kind of test can be obtained from the Analysis of Variance Table, where the lines correspond to the first and the second equation.

```
Analysis of Variance Table
    Regression d.f.      Residual d.f.           F  Covariates
    ---------------      -------------           -  ----------
        18.067    1         31.636   28     15.990  Age
        23.132    2         26.571   27     11.753  Age BMI
```

The first two columns give the sum of squares due to regression and its degrees of freedom. The next two columns give the residual sum of squares (RSS) and its degrees of freedom. The F-value is computed by (2.34). To test the additional effect of BMI compute

$$F = \frac{(31.636 - 26.571)/1}{26.571/27} = 5.15$$

and use this as an $F(1, 27)$ statistic. This is statistically significant at 5% level but not at the 1% level, which is also in agreement with previous results. For samples less than 30 the F-statistic is more accurate than the chi-square test.

Interestingly, the value $F = 5.15$ can be obtained from the two R^2's shown in the output. These are 0.363 for the first equation and 0.465 for the second equation. Then F can be computed as

$$F = \frac{(0.465 - 0.363)/1}{(1 - 0.465)/27} = 5.15$$

The conclusion from these tests is that there is probably an additional effect of BMI after controlling for age but the evidence from this sample is not very strong. Further studies are needed. We return to this issue in Section 2.1.5.

At the end of the output file, the following covariance matrix, means and standard deviations are given

```
Covariance Matrix
            Chol        Age         BMI
          --------    --------    --------
Chol         1.714
Age         11.658     218.148
BMI          1.589      13.511       5.144


Means
            Chol        Age         BMI
          --------    --------    --------
             6.005      50.700      23.180


Standard Deviations
            Chol        Age         BMI
          --------    --------    --------
             1.309      14.770       2.268
```

2.1.4   Checking the Assumptions

Every statistical method and analysis is based on some assumptions and regression analysis is no exception. These assumptions should be checked when ever possible. Violations of these assumptions may lead to biased regression estimates or biased standard errors or other problems in the analysis.

Primarily, the assumptions of linear regression analysis are

1. the error term z is independent of the covariates $\mathbf{x}$.

2. the mean of y is linear in $\mathbf{x}$.

3. the residuals $z_1, z_2, \cdots, z_N$ are identically and independently distributed as $N(0, \sigma^2)$.

This section discusses various possibilities for checking these assumptions. Here we focus on tests that can easily be used with **LISREL**. For other tests, see *e.g.*, Green (2000).

Omitted Variables Bias

Assumption 1 is that the error term z is independent of $\mathbf{x}$. This is a fundamental assumption. Violations of this is sometimes called *omitted variables bias*. Consider the two estimated regression equations in the **Cholesterol** example:

```
Chol =    3.296 + 0.0534*Age + Error
Chol =  - 0.740 + 0.0410*Age + 0.201*BMI + Error
```

The difference between the two regression coefficients of Age 0.0534–0.0410 is an indication of omitted variables bias because age may be correlated with other variables like height and weight which are omitted from the first regression equation. If this is the case, then the error term in the regression may be correlated with `Age`, thus violating this fundamental assumption.

The error term z is an aggregate of all variables influencing y which are not included in the regression equation.

Consider this in a slightly more general form. Suppose, for example, that both x_1 and x_2 influence y so that the true regression relationship is

$$y = \alpha + \gamma_1 x_1 + \gamma_2 x_2 + z ,$$

with z independent of x_1 and x_2. But suppose we do not realize that x_2 is an important determinant of y. So we set out to estimate the regression of y on x_1 alone. Let $\phi_{11} = Var(x_1)$. The covariance between y and x_1 is $\gamma_1 \phi_{11} + \gamma_2 \phi_{21}$, where ϕ_{21} is the covariance of x_1 and x_2. The regression coefficient of y on x_1 is

$$\frac{\gamma_1 \phi_{11} + \gamma_2 \phi_{21}}{\phi_{11}} = \gamma_1 + \frac{\gamma_2 \phi_{21}}{\phi_{11}} ,$$

which is not equal to γ_1 if $\gamma_2 \neq 0$ and x_1 and x_2 are correlated. The difference $\gamma_2 \phi_{21}/\phi_{11}$ is the omitted variables bias. This can be made arbitrarily large by suitable choice of γ_2 or ϕ_{21}. Studies should be planned and designed so that this bias can be avoided. Unfortunately, it is not possible to check this assumption using the same data that is used to estimate the regression equation, for the estimated residuals $\hat{z}_i = y_i - \hat{\alpha} - \hat{\boldsymbol{\gamma}}' \mathbf{x}_i$ are automatically uncorrelated with each x_j by construction. This follows from equation (2.11).

Checking Linearity

Non-linearity can take different functional forms but as an approximation one can use a quadratic form. Assumption 2 can be examined by adding quadratic terms in the regression equation. If the linear model holds, one expects the estimated regression coefficients of these quadratic terms to be non-significant. If the linear regression equation has two covariates x_1 and x_2, say, one can add the terms x_1^2, x_2^2, and $x_1 x_2$.

To illustrate this for the **Cholesterol Example** consider the following **PRELIS** syntax file (see file **Cholesterol2a.prl**)

```
Checking Linearity
System file Cholesterol.lsf
New Age2 = Age*Age
New BMI2 = BMI*BMI
```

```
New AgeBMI = Age*BMI
Continuous Age2 BMI2 AgeBMI
Regress Chol on Age BMI Age2 BMI2 AgeBMI
```

New is the command for defining a new variable. The line

```
New Age2 = Age*Age
```

defines the new variable Age2 as the square of Age. This can also be written

```
New Age2 = Age**2
```

Similarly, the other two lines define BMI2 as the square of BMI and AgeBMI as the product of Age and BMI. The line

```
Continuous Age2 BMI2 AgeBMI
```

is used to define the three new variables as continuous variables; otherwise they are assumed to be ordinal variables[1].

The new variables Age2, BMI2, and AgeBMI are automatically added in the **.lsf** file. To see this, close the **.lsf** file (if it is open) and open it again.

The resulting estimated regression equation is:

```
     Chol =  - 18.908 + 0.0129*Age + 1.795*BMI + 0.000501*Age2 - 0.0322*BMI2
Standerr     (14.865)  (0.202)       (1.283)     (0.000852)      (0.0313)
t-values     -1.272    0.0638        1.400       0.587           -1.028
P-values      0.215    0.950         0.174       0.562            0.314


           - 0.00110*AgeBMI + Error, R² = 0.513
           (0.00925)
           -0.119
            0.906
```

```
Error Variance = 1.008
```

As judged by the P-values for the quadratic terms, none of the three regression coefficients are statistically significant. One can also test this by using an F statistic

$$F = \frac{(0.513 - 0.465)/3}{(1 - 0.513)/24} = 0.8$$

Here 0.513 is the R^2 for the estimated quadratic regression and 0.465 is the R^2 for the linear regression. The value $F = 0.8$ is not significant compared to the $F(3, 24)$-distribution. This gives support for the linearity of the regression.

Assumption 3 consists of three different assumptions:

3a the residuals are normally distributed.

3b the residuals have the same variance for all values of **x**.

3c the residuals are independent of one another.

Each of these should be checked.

---

[1]this will not do any harm in this example since LISREL automatically treats all variables as continuous if they have more than 15 distinct values.

Checking Normality of Residuals

The normality of residuals can be checked by various graphs or tests. The estimated sample residuals $\hat{z}_i = y_i - \hat{\alpha} - \mathbf{x}_i \hat{\gamma}$ can be obtained by putting `Res=` at the end of the `Regress` line and give a name of a variable for the residual after the equal sign, for example, see file **Cholesterol2b.prl**

```
System file Cholesterol.lsf
Regress Chol on Age BMI Res=CholRes
Output MA=CM Ra=CholesterolwithRes.lsf
```

The variable `CholRes` will then be added to the variables in the file **CholesterolwithRes.lsf**. The first 10 lines of the appended LISREL data system file looks like this

CholesterolwithRes.LSF	Chol	Age	BMI	CholRes
1	5.940	52.000	20.700	0.381
2	4.710	46.000	21.300	-0.724
3	5.860	51.000	25.400	-0.604
4	6.520	44.000	22.700	0.886
5	6.800	70.000	23.900	-0.141
6	5.230	33.000	24.300	-0.275
7	4.970	21.000	22.200	0.379
8	8.780	63.000	26.200	1.663
9	5.130	56.000	23.300	-1.116
10	6.740	54.000	29.200	-0.612

To examine the normality of the residuals one can plot them against normal quantiles (normal scores), see Section 14. These are obtained by adding the line

```
CholNsc = Nsc(CholRes)
```

and a new data system file `CholesterolExtended.lsf`, say, including the the normal scores, can be created by adding `RA=CholesterolExtended.lsf` on the output line. The PRELIS syntax file now looks like, see file **Cholesterol2c.prl**:

```
System file CholesterolwithRes.lsf
New CholNsc = CholRes
NS CholNsc
Output RA=CholesterolExtended.lsf
```

and the extended data file looks like this

CholesterolExtended.LSF	Chol	Age	BMI	CholRes	CholNsc
1	5.940	52.000	20.700	0.381	0.741
2	4.710	46.000	21.300	-0.724	-0.542
3	5.860	51.000	25.400	-0.604	-0.365
4	6.520	44.000	22.700	0.886	0.981
5	6.800	70.000	23.900	-0.141	-0.119
6	5.230	33.000	24.300	-0.275	-0.199
7	4.970	21.000	22.200	0.379	0.638
8	8.780	63.000	26.200	1.663	1.311
9	5.130	56.000	23.300	-1.116	-1.311
10	6.740	54.000	29.200	-0.612	-0.451
11	5.050	44.000	22.700	0.316	0.281

Select the **Graphs** menu and the **Bivariate Plot** option. Tick the **Scatter Plot** box and Select
`CholRes` as Y variable and `CholNsc` as X variable

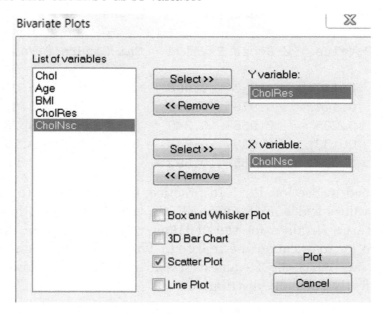

This gives the following scatter plot

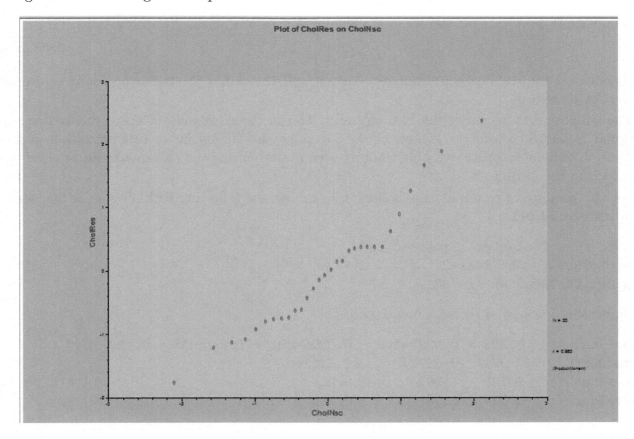

If the residuals are perfectly normal the points in this scatter plot would fall on a straight line
(the Y = X line)

Another possibility to evaluate the normality of the residuals is available in the output file
which provides a test of zero skewness and kurtosis:

Test of Univariate Normality for Continuous Variables

	Skewness		Kurtosis		Skewness and Kurtosis	
Variable	Z-Score	P-Value	Z-Score	P-Value	Chi-Square	P-Value
Chol	1.589	0.112	1.670	0.095	5.311	0.070
Age	0.070	0.944	-0.976	0.329	0.957	0.620
BMI	1.268	0.205	0.955	0.340	2.520	0.284
CholRes	1.504	0.133	0.603	0.546	2.626	0.269
CholNsc	0.000	1.000	0.135	0.892	0.018	0.991

These tests are explained in Section 12.1. In practice, one only needs to examine the P-values given in this table. P-values which are very small, less than 0.05, say, indicate lack of normality. In this example, the three P-values for the **CholRes** variable, suggests that the hypothesis of zero skewness and zero kurtosis cannot be rejected. This is an indication that normality of the residuals holds. Note also that the P-values for **CholNsc** are very close to 1 indicating that **CholNsc** is almost perfectly normally distributed.

Checking Homoscedasticity

Assumption 3b implies that the residual variance is constant across all observations, *i.e.*,

$$E(z_i^2 \mid \mathbf{x}) = \sigma^2 \qquad (2.38)$$

This is called the *assumption of homoscedasticity*. If this is not the case, the residuals are said to be heteroscedastic.

The simplest way to check this assumption is to run the regression of the estimated squared residual $\hat{z}_i^2$ on all covariates. If none of the covariates are significant in this regression, it is an indication of homoscedasticity. This can be tested directly using the chi-square test or the F test provided in the output.

For the cholesterol example, this is done by running the following **PRELIS** syntax file, see file **Cholesterol2d.prl**

```
System file CholesterolExtended.lsf
New CholRes2 = CholRes**2
Regress CholRes2 on Age BMI
```

The resulting estimated regression equation is

```
CholRes2 =   - 0.305 + 0.0339*Age - 0.0228*BMI + Error, R² = 0.139
Standerr      (2.381) (0.0171)       (0.111)
t-values     -0.128   1.981         -0.205
P-values      0.899   0.057          0.839
```

which shows that both regression coefficients are not statistically significant (at the 5% level). This gives support for homoscedasticity.

To get more power of the test, one can also include the quadratic terms **Age2**, **BMI2**, and the product **AgeBMI** in the regression equation, see file **Cholesterol2d1.prl**. This gives the following estimated regression equation

```
CholRes2 = 6.564   - 0.227*Age - 0.0744*BMI + 0.000774*Age2 - 0.00734*BMI2
Standerr  (18.997) (0.258)      (1.639)      (0.00109)       (0.0400)
t-values   0.346   -0.878       -0.0454       0.711          -0.184
P-values   0.733    0.388        0.964        0.484           0.856

          + 0.00788*AgeBMI + Error, R² = 0.188
            (0.0118)
             0.666
             0.511
```

Since none of the five P-values are small, this gives stronger support for homoscedasticity than the previous regression. The chi-square given in the output for testing that all five regression coefficients are zero is 6.232 with 5 degrees of freedom which clearly indicates that the hypothesis cannot be rejected. An alternative chi-square test was developed by White (1980). He suggests regressing `CholRes2` only on the quadratic terms to obtain $R^2 = 0.159$. The test is easily obtained since the chi-square value is $nR^2 = 30 \times 0.159 = 4.77$. According to White (1980) this chi-square also has an asymptotic χ^2 distribution with 5 degrees of freedom if homoscedasticity holds.

Checking Autocorrelation

Assumption 3c is usually checked by testing for autocorrelation in the error term. Autocorrelation is usually not a problem in cross-sectional data. Here it is shown how one can lag a variable and estimate the regression on the lagged variable.

The estimated residuals $\hat{z}_i$ and the lagged residuals $\hat{z}_{i-1}$ are shown in the following table

$$
\begin{vmatrix}
z_1 & - \\
z_2 & z_1 \\
z_3 & z_2 \\
\cdot & \cdot \\
\cdot & \cdot \\
z_N & z_{N-1}
\end{vmatrix}
$$

Note that the first value in the second column is missing.

To obtain the lagged residual, include a line, see file **Cholesterol2e.prl**:

```
LG ChRes_1 = CholRes LAG=1
```

The autoregression result is

```
CholRes = 0.0521 - 0.243*ChRes1 + Error, R² = 0.0577
Standerr (0.176)  (0.189)
t-values  0.296   -1.286
P-values  0.769    0.209
```

Since the autoregression coefficient -0.243 is not statistically significant, this suggest that there is no autocorrelation present in the residual `CholRes`. Since the variances of the residual and the lagged residual are approximately equal, -0.243 is an approximate estimate of the residual autocorrelation.

2.1.5   The Effect of Increasing the Sample Size

Suppose another researcher has data on 117 other individuals where the variables Chol, Age, and BMI have been measured in the same way. The data is given in the file **Cholesterol2.lsf**[2]. Estimating the regression of Chol on Age and BMI using these data gives the following results:

```
      Chol =  - 0.753 + 0.0410*Age + 0.202*BMI + Error,  R² = 0.464
Standerr       (0.924) (0.00667)     (0.0432)
t-values       -0.815   6.144         4.671
P-values        0.417   0.000         0.000

Error Variance = 0.934
```

It is seen that while the regression coefficients are almost the same as before, their standard errors are smaller and as a consequence the P-values are smaller. The regression coefficients are estimated more precisely. In this sample the effect of BMI is highly significant and it is clear that BMI has an additional effect on Chol even after controlling for Age. This is often expressed by saying that we have more power for rejecting the hypothesis that $\gamma_2 = 0$. It is typically what happens when the sample size increases.

2.1.6   Regression using Means, Variances, and Covariances

As shown in (2.18) and (2.19), the regression estimates can be obtained from the means, variances, and covariances without the use of the individual data. The statistical term for this is that the means, variances, and covariances are sufficient statistics for estimating the regression coefficients, which means that the individual data provides no further information than what is already available in the means, variances, and covariances. This holds under the assumption of normality. The case of non-normal data can be can be handled by Robust Estimation, see Section 7.3.3 and the appendix in Chapter 16.

To estimate a regression equation from means, variances, and covariances, one must use a different syntax file, see file **cholesterol3.spl**:

```
Observed Variables: Chol Age BMI
Means:                  6.005    50.700     23.180
Covariance Matrix:      1.714
                       11.658   218.148
                        1.589    13.511     5.144
Sample Size: 30
Regress Chol on Age BMI
```

Only one blank space is needed between entries and a line feed character is equivalent to a blank space. So the same **SIMPLIS** syntax file can also be written as:

```
Observed Variables: Chol Age BMI
Means: 6.005 50.700 23.180
Covariance Matrix: 1.714 11.658 218.148 1.589 13.511 5.144
```

---

[2]This data set has been generated from random normal variables scaled to have the same means, variances and covariances as the data in **Cholesterol.lsf**.

```
Sample Size: 30
Regress Chol on Age BMI
```

With many variables it is inconvenient to put the means and the covariance matrix in the SIMPLIS syntax file. It is better to save these in files first and then read them from these files. To estimate the mean vector and covariance matrix and save them in files, one can use the following syntax file:

```
System File Cholesterol.lsf
Output MA=CM ME=cholesterol.me CM=cholesterol.cm
```

This saves the mean vector in file **cholesterol.me** and the covariance matrix in file **cholesterol.cm** using maximal accuracy. One can then use the following SIMPLIS syntax file to estimate the regression equation, see file **cholesterol4.spl**:

```
Observed Variables: Chol Age BMI
Means from File cholesterol.me
Covariance Matrix from File cholesterol.cm
Sample Size: 30
Regress Chol on Age BMI
```

The output file gives the following results

```
Estimated Equations

    Chol =  - 0.740 + 0.0410*Age + 0.201*BMI + Error, R² = 0.465
Standerr     (1.896) (0.0136)      (0.0888)
Z-values     -0.390   3.006         2.269
P-values      0.697   0.003         0.023


Error Variance = 0.916
```

This gives exactly the same results as given in the beginning of section 1.1.

2.1.7 Standardized Solution

In many cases, particularly in the social and behavioral sciences, the units of measurement in the observed variables have no important meaning. In this situation it is common to analyze the variables in standardized form. This answers the question: What will the regression coefficients be if all variables have variance 1? In this particular example, one can answer this question by analyzing the correlation matrix instead of the covariance matrix, but to analyze correlation matrices to obtain standardized solutions is not generally recommended.

To obtain the correlation matrix of Chol, Age, and BMI and save this in the file **cholesterol.km**, use the following PRELIS syntax file

```
System File cholesterol.lsf
Output MA=KM km=cholesteril.km
```

Then the following **SIMPLIS** syntax file can be used to obtain the standardized regression coefficients, see file **cholesterol5.spl**:

```
Observed Variables: Chol Age BMI
Correlation Matrix from File cholesterol.km
Sample Size: 30
Regress Chol on Age BMI
```

This gives the following results:

```
      Chol = 0.462*Age + 0.349*BMI + Error, R² = 0.465
 Standerr  (0.154)       (0.154)
 t-values   3.006         2.269
 P-values   0.003         0.023


 Error Variance = 0.535
```

No intercept is estimated since only the correlation matrix is provided. Note that the regression coefficients and the standard error are different from the unstandardized solution but the t-values and P-values are the same. Also, the error variance is different but R^2 is the same. In the standardized solution the error variance is in fact equal to $1 - R^2$.

A common misunderstanding is that standardized regression coefficients must be smaller than 1 in magnitude. However, the standardized regression coefficients are not correlations; so they can be larger than 1 in magnitude. In fact, the standardized regression coefficients can be made arbitrarily large by suitable choice of the correlations in the correlation matrix, subject only to the condition that this is positive definite. Here is an example, see file **ficticious.spl**:

```
Observed Variables: Y X1 X2
Correlation Matrix
 1.000
 0.600  1.000
 0.200  0.900  1.000
Sample Size: 30
Regress Y on X1 X2
Number of Decimals: 3
```

which gives

```
        Y = 2.211*X1 - 1.789*X2 + Error, R² = 0.968
 Standerr  (0.0785)    (0.0785)
 Z-values   28.174      -22.808
 P-values   0.000        0.000
```

Note that both regression coefficients are larger than 1 in magnitude, yet all variables are standardized.

2.1.8 Predicting y When ln(y) is Used as the Dependent Variable

To improve the normality and heteroskedasticity of the residual, a transformation of the dependent variable can be used. The family of Box-Cox transformations includes several possibilities by selecting suitable values of α, γ, and λ:

$$f_\lambda(y) = \alpha + \gamma \frac{y^\lambda - 1}{\lambda}, \quad y > 0, \quad \lambda \neq 0 \tag{2.39}$$

$$f_\lambda(y) = \ln(\alpha + \gamma y), \quad y > 0, \quad \lambda = 0 \tag{2.40}$$

For example, $\lambda = 1$ gives a linear transformation, $\lambda = \frac{1}{2}$ gives a square root transformation, and $\lambda = -1$ gives a reciprocal transformation. Perhaps the most useful transformation is with $\lambda = 0$ which gives a logarithmic transformation.

Suppose we want to use $\ln(y)$ instead of y as the dependent variable and we set out to estimate the equation

$$\ln(y) = \alpha + \gamma_1 x_1 + \gamma_2 x_2 + \cdots + \gamma_q x_q + z, \tag{2.41}$$

assuming that z is normally distributed with mean 0 and variance σ^2. This implies that the conditional distribution y is lognormal with mean

$$E(y|\mathbf{x}) = e^{E(\ln(y)|\mathbf{x}) + \frac{1}{2}\sigma^2} = e^{\alpha + \gamma_1 x_1 + \gamma_2 x_2 + \cdots + \gamma_q x_q + \frac{1}{2}\sigma^2}. \tag{2.42}$$

Suppose the estimated regression equation is

$$\ln(y) = \hat{\alpha} + \hat{\gamma}_1 x_1 + \hat{\gamma}_2 x_2 + \cdots + \hat{\gamma}_q x_q + z, \tag{2.43}$$

with estimated residual variance $\hat{\sigma}^2$. This can be used to predict $\ln(y)$ from future values of x_1, x_2, ..., x_q. The predicted value of $\ln(y)$ is

$$\widehat{\ln(y)} = \hat{\alpha} + \hat{\gamma}_1 x_1 + \hat{\gamma}_2 x_2 + \cdots + \hat{\gamma}_q x_q. \tag{2.44}$$

But what is the predicted value of y? It is not $e^{\widehat{\ln(y)}}$ as one might think, but

$$\hat{y} = e^{\hat{\alpha} + \hat{\gamma}_1 x_1 + \hat{\gamma}_2 x_2 + \cdots + \hat{\gamma}_q x_q} e^{\frac{\hat{\sigma}^2}{2}} = e^{\widehat{\ln(y)}} e^{\frac{\hat{\sigma}^2}{2}} \tag{2.45}$$

This is illustrated in the next example.

2.1.9 Example: Income

The data file **income.lsf** contains a subset of data from the USA March 1995 Population Survey. The respondents are a sample of persons who are between the ages of 21 and 65 and who held full time positions in 1994 with an annual income of US\$1 or more.

The variables are

Age age in years

Gender = 0 for female; = 1 for male

Marital = 1 for married; = 0 otherwise

Hours number of hours worked last week at all jobs

Citizen = 1 for native Americans; = 0 for foreign born

Degree = 1 for master's degree, professional school degree, or doctoral degree; = 0 otherwise

Group = 1 for respondents with professional specialty in the educational sector; = 0 for operators, fabricators, and laborers in the construction sector

Income = the personal income in thousands of US$ during 1994

The following PRELIS command file (file **income1.prl**)

```
sy=income.lsf
new LnInc=Income
log LnInc
co all
ou ra=income.lsf
```

can be used to append the variable `LnInc` equal to the natural logarithm of `Income` in the file **income.lsf**. After this, the first few lines of this file looks like this

	Age	Gender	Marital	Hours	Citizen	Degree	Group	Income	LnInc
1	59.000	0.000	0.000	40.000	1.000	0.000	1.000	24691.000	10.114
2	56.000	1.000	1.000	40.000	1.000	0.000	1.000	31023.000	10.342
3	64.000	0.000	1.000	12.000	1.000	0.000	1.000	33830.000	10.429
4	30.000	1.000	1.000	40.000	1.000	0.000	0.000	15201.000	9.629
5	27.000	0.000	1.000	40.000	1.000	0.000	1.000	21500.000	9.976
6	49.000	0.000	1.000	40.000	1.000	1.000	1.000	43678.000	10.685
7	41.000	1.000	1.000	40.000	1.000	0.000	0.000	40300.000	10.604
8	36.000	0.000	1.000	40.000	1.000	0.000	1.000	22299.000	10.012

The regression of `LnInc` can be estimated using the following PRELIS file (file **income2.prl**):

```
!Regression of LnInc
System File income.lsf
Regress LnInc on Age - Group
```

The estimated regression equation is

```
Estimated Equations


   LnInc = 8.240     + 0.0169*Age + 0.233*Gender + 0.0713*Marital + 0.0133*Hours
Standerr  (0.0603)   (0.00102)     (0.0288)        (0.0225)         (0.000652)
t-values  136.694    16.567         8.104           3.165            20.350
P-values  0.000      0.000          0.000           0.002            0.000




         + 0.245*Citizen + 0.426*Degree + 0.196*Group + Error, R^2 = 0.214
           (0.0338)        (0.0285)        (0.0316)
            7.247           14.923          6.225
            0.000           0.000           0.000


Error Variance = 0.620
```

Table 2.2: Six Profiles

Person	Alice	Bertil	Cindy	David	Henry	Linda
Age	30	30	40	40	40	40
Gender	0	1	0	1	1	0
Marital	0	1	0	0	1	1
Hours	40	40	35	35	35	35
Citizen	0	1	0	1	1	1
Degree	0	1	0	0	1	1
Group	0	1	0	1	1	1
$\widehat{\ln(y)}$	9.279	10.450	9.381	10.055	10.553	10.320
$exp(\widehat{\ln(y)})$	10711	34555	11867	23283	38284	30327
$\hat{y}$	14603	47112	16179	31745	52197	41348

Consider the following six persons and their profiles: The values in the last three rows can be calculated as follows. First, write a data file **Profiles.dat**:

```
30  0  0  40  0  0  0
30  1  1  40  1  1  1
40  0  0  35  0  0  0
40  1  0  35  1  0  1
40  1  1  35  1  1  1
40  0  1  35  1  1  1
```

Then run the following PRELIS syntax file **Profiles1.prl**:

```
!Profiles Calculation
da ni=7
la
Age Gender Marital Hours Citizen Degree Group
ra=profiles.dat
new A=8.240+0.0169*Age+0.233*Gender+0.0713*Marital
new B=0.0133*Hours+0.245*Citizen+0.426*Degree+0.196*Group
new LNY=A+B
new EST=LNY
exp EST
new INCOME=1.3634*EST
co all
sd A B
ou ra=profiles_extended.dat
```

The calculated values appear in the last three columns in **profiles_extended.dat**.

Table 2.2 shows that Henry has a predicted income of \$ 52197 whereas Linda has a predicted income of \$ 41348, yet both of them have the same values on all predictors. However, the difference in income \$ 10849 may not reflect a gender bias because there may be other variables that can explain the difference. For example, they may live in different regions or states where the level of incomes are different. In Chapter 4 we consider a multilevel analysis of the same data that takes such differences into account.

2.1.10 ANOVA and ANCOVA

Suppose there are G groups or categories of a categorical variable or levels of factors and we want to compare their means on a response variable y. We may also wish to know if the sensitivity of the analysis can be increased by the use of the covariates, $x_1, x_2, \ldots, x_q$. These analyses are usually done by one-way analysis of variance ($ANOVA$) or analysis of covariance ($ANCOVA$), respectively.

Analysis of variance (ANOVA) can be done by forming dummy variables $d_1, d_2, \ldots, d_{G-1}$ and regressing y on these dummy variables. Analysis of covariance is done by regressing y on the covariates and the dummy variables.

The dummy variables represent group memberships or categories of a categorical variable or levels of factors such that $d_{ig} = 1$ if case i belongs to group g and $d_{ig} = 0$ otherwise. The raw data is of the form:

Case	y	x	d_1	d_2	$\cdots$	d_G
1	y_1	x_1	d_{11}	d_{12}	$\cdots$	d_{1G}
2	y_2	x_2	d_{21}	d_{22}	$\cdots$	d_{2G}
$\vdots$	$\vdots$	$\vdots$	$\vdots$	$\vdots$	$\ddots$	$\vdots$
N	y_N	x_N	d_{N1}	d_{N2}	$\cdots$	d_{NG}

Anyone of the groups or categories is used as a reference category. Here we assume that this is the last one. So the dummy variable d_G is not used in the analysis.

The hypothesis that all group means are equal is the same as the hypothesis that $\gamma_1 = \gamma_2 = \cdots = \gamma_{G-1} = 0$. For the analysis of variance a formal F statistic can be computed as

$$F = \frac{R^2/(G-1)}{(1-R^2)/(N-G)} , \qquad (2.46)$$

where R^2 is the squared multiple correlation in the regression of y on $d_1, d_2, \ldots, d_{G-1}$. Each γ_i measures the mean difference $\mu_i - \mu_G$, the significance of which can be tested with the corresponding t-value used as a t-statistic.

The analysis of covariance ($ANCOVA$) can be done in a similar way. First, regress the response variable y on one or more covariates, $x_1, x_2, \ldots, x_q$, to obtain the squared multiple correlation R^2_{yx}. Second, regress y on $x_1, x_2, \ldots, x_q$ and $d_1, d_2, \ldots, d_{G-1}$ to obtain the squared multiple correlation R^2_{yxd}. Then

$$F = \frac{(R^2_{yxd} - R^2_{yx})/(G-1)}{(1-R^2_{yxd})/(N-G-1)} , \qquad (2.47)$$

can be used as an F statistic with $(G-1)$ and $(N-G-1)$ degrees of freedom for testing the hypothesis that the group means, adjusted for mean differences in the covariates, are zero.

Both the $ANOVA$ and the $ANCOVA$ considered above assume that the within-group variances of y are equal. In addition, $ANCOVA$ assumes that the regression coefficients of y on $x_1, x_2, \ldots, x_q$ are the same for each group. Test of such assumptions are developed in Section 2.1.12.

2.1.11 Example: Biology

Two investigators are interested in academic achievement. Investigator A is primarily concerned with the effects of three different types of study objectives on student achievement in freshman biology. The three types of objectives are:

- General - Students are told to know and understand everything in the text.

- Specific - Students are provided with a clear specification of the terms and concepts they are expected to master and of the testing format.

- Specific with study time allocations - The amount of time that should be spent on each topic is provided in addition to specific objectives that describe the type of behavior expected on examinations.

The dependent variable is the biology achievement test.

Investigator B is interested in predicting student achievement using a recently developed aptitude test. This investigator is not particularly concerned with differences in the mean level of achievement resulting from the use of different types of study objective, but rather with making a general statement of the predictive efficiency of the test across a population of freshmen biology students of whom one-third have been exposed to general study objectives, one-third to specific behavioral objectives, and one-third to specific behavioral objectives with study time allocations. More specifically, he wants to know (1) whether the regression of achievement scores on aptitude scores is statistically significant and (2) the regression equation.

A population of freshmen students scheduled to enroll in biology is defined, and 30 students are randomly selected. Investigator B obtains aptitude test scores (X) for all students before investigator A randomly assigns 10 students to each of the three treatments. Treatments are administered, and scores on the dependent variable are obtained for all students.

The data is given in the file **biology.lsf**, the first few lines of which looks like this

biology.LSF	Biology	Aptitude	D1	D2	D3
1	15.000	29.000	1.000	0.000	0.000
2	19.000	49.000	1.000	0.000	0.000
3	21.000	48.000	1.000	0.000	0.000
4	27.000	35.000	1.000	0.000	0.000
5	35.000	53.000	1.000	0.000	0.000
6	39.000	47.000	1.000	0.000	0.000
7	23.000	46.000	1.000	0.000	0.000
8	38.000	74.000	1.000	0.000	0.000
9	33.000	72.000	1.000	0.000	0.000
10	50.000	67.000	1.000	0.000	0.000
11	20.000	22.000	0.000	1.000	0.000
12	34.000	24.000	0.000	1.000	0.000

Investigator B examines the results of his colleague's ANOVA and suggests that a more powerful technique, one that controls for aptitude differences among subjects, might reveal significant treatment effects; ANCOVA is one such technique.

For the ANOVA part of the problem, we run the regression of y on d_1 and d_2. For the ANCOVA, we run the regression of y on x and the regression of y on x, d_1 and d_2. All three regressions can be estimated in one single run using the following PRELIS syntax file, see file **biology1.prl**.

```
System file biology.lsf
Regress Biology on D1 D2
Regress Biology on Aptitude
Regress Biology on Aptitude D1 D2
```

The three estimated regression equations are given in the output as

```
Estimated Equations
```

```
 Biology = 36.000 - 6.000*D1 + 3.000*D2 + Error, R² = 0.106
Standerr  (3.619)  (5.118)      (5.118)
t-values   9.948   -1.172        0.586
P-values   0.000    0.251        0.562
```

```
Error Variance = 130.963
```

```
 Biology = 9.413 + 0.519*Aptitude + Error, R² = 0.396
Standerr  (6.203) (0.121)
t-values   1.518   4.286
P-values   0.140   0.000
```

```
Error Variance = 85.309
```

```
 Biology = 8.044 + 0.571*Aptitude - 7.712*D1 + 4.141*D2 + Error, R² = 0.575
Standerr  (5.805) (0.106)          (3.610)    (3.602)
t-values   1.386   5.357           -2.136      1.150
P-values   0.177   0.000            0.042      0.260
```

```
Error Variance = 64.640
```

The squared multiple correlation is, $R^2 = 0.106$, for the first equation. The hypothesis that the three group means are equal (equivalent to the hypothesis that γ_1 and γ_2 are both zero) is tested via the F statistic (2.46) with 2 and 27 degrees of freedom. F becomes

$$F = \frac{0.106/2}{(1-0.106)/27} = 1.60$$

This is not significant so no differences in group means on the response variable y are detected.

For the ANCOVA part, the regression of y on x alone gives $R^2 = 0.396$ and the regression of y on x, d_1, and d_2 gives $R^2 = 0.575$.

The F statistic (2.47) with 2 and 26 degrees of freedom for testing the equality of adjusted means is

$$\frac{(0.575 - 0.396)/2}{(1 - 0.575)/26} = 5.47$$

which is significant at the 5% level. Thus there is some evidence that the group means of the Biology Achievement Score are different when they are adjusted for differences in the Aptitude Test Score. By controlling for x we get a more powerful test for testing group differences on y than we get by using y alone.

2.1.12   Conditional Regression

A common situation in regression is when there are observed categorical variables in addition to the y and $\mathbf{x}$ variables used in the regression. These categorical variables often represent group memberships. In this situation one can consider the regression of y on $\mathbf{x}$ for each category of the categorical variable and investigate the extent to which this regression is the same across all levels of the categorical variable. We give three examples of this.

2.1.13   Example: Birthweight

Table 2.3 gives the birthweight in grams and gestational age in weeks for 12 boys and 12 girls.

Table 2.3: Birthweight (grams) and gestational age (weeks) for 12 boys and girls. Originally published in Dobson and Barnett (2008, Table 2.3). Published with kind permission of © Taylor and Francis Group, LLC 2008. All Rights Reserved

\ Boys		Girls	
Age	Birthweight	Age	Birthweight
40	2968	40	3317
38	2795	36	2729
40	3163	40	2935
35	2925	38	2754
36	2625	42	3210
37	2847	39	2817
41	3292	40	3126
40	3473	37	2539
37	2628	36	2412
38	3176	38	2991
40	3421	39	2875
38	2975	40	3231

In the file **birthweight1.lsf** shown in Figure 2.1 the variables are: SEX coded as 1 for boys and 2 for girls, BWEIGHT = birth weight recorded in kilograms instead of grams and AGE35 = gestational age in weeks - 35. Although we are interested in the intercept in the regression of birth weight on age, the intercept at zero is of no interest since nobody has age zero; by measuring age from 35 weeks we can study differences in birth weight at 35 weeks between boys and girls[3].

---

[3]The number of boys and girls need not be the same; the order of the rows in the lsf file is irrelevant as long as there is a unique code identifying the boys and girls and these codes need not be 1 and 2 but can be any two distinct numbers.

birthweight1.lsf

	SEX	AGE35	BWEIGHT
1	1,000	5,000	2,968
2	1,000	3,000	2,795
3	1,000	5,000	3,163
4	1,000	0,000	2,925
5	1,000	1,000	2,625
6	1,000	2,000	2,847
7	1,000	6,000	3,292
8	1,000	5,000	3,473
9	1,000	2,000	2,628
10	1,000	3,000	3,176
11	1,000	5,000	3,421
12	1,000	3,000	2,975
13	2,000	5,000	3,317
14	2,000	1,000	2,729
15	2,000	5,000	2,935
16	2,000	3,000	2,754
17	2,000	7,000	3,210
18	2,000	4,000	2,817
19	2,000	5,000	3,126
20	2,000	2,000	2,539
21	2,000	1,000	2,412
22	2,000	3,000	2,991
23	2,000	4,000	2,875
24	2,000	5,000	3,231

Figure 2.1: LSF File for Birthweight Example

Conditional regression can be obtained easily using the following simple PRELIS syntax file **birthweight1.prl**):

```
System File birthweight1.lsf
Regress BWEIGHT on AGE35 by SEX
```

Note the keyword `by`. This instructs LISREL to estimate a regression equation for each value of SEX. The output gives the following two estimated regression equations:

```
For SEX = 1, Sample Size = 12:

 BWEIGHT = 2.651   + 0.112*AGE35 + Error, R² = 0.546
Standerr  (0.122)  (0.0323)
t-values   21.669   3.466
P-values   0.000    0.005

Error Variance = 0.0404

For SEX = 2, Sample Size = 12:
```

```
 BWEIGHT = 2.422  + 0.130*AGE35 + Error, R² = 0.712
Standerr  (0.108)  (0.0262)
t-values   22.374    4.978
P-values    0.000    0.000


Error Variance = 0.0249
```

The following chi-squares test the hypothesis that all regression coefficients are zero except the intercept.

Variable	-2lnL	Chi-square	df	Covariates
BWEIGHT	-6.649	9.468	1	AGE35
BWEIGHT	-12.461	14.958	1	AGE35

2.1.14   Testing Equal Regressions

A regression equation with a single explanatory variable has three parameters: one intercept, one slope, and one error variance. Each of these can differ between groups. This section develop tests of equal intercepts, equal slopes, and equal error variances.

Let y_{ij} and x_{ij} be the birth weight and age, respectively, of child j in group i, where $i = 1$ for boys and $i = 2$ for girls. The two regression equations estimated previously can be written as

$$y_{ij} = \alpha_i + \gamma_i x_{ij} + z_{ij} \, , i = 1, 2 \, , \tag{2.48}$$

with $E(z_{ij}^2) = \sigma_i^2$. This model has six parameters, namely α_1, α_2, γ_1, γ_2, σ_1^2, and σ_2^2. Consider testing the hypotheses $\alpha_1 = \alpha_2$, $\gamma_1 = \gamma_2$, $\sigma_1^2 = \sigma_2^2$. For this purpose we rewrite (2.48) as

$$y_{ij} = \mu + \gamma_1 d_{ij} + \gamma_2 x_{ij} + \gamma_3 d_{ij} x_{ij} + z_{ij} \, , \tag{2.49}$$

where the dummy variable $d_{ij} = 1$ if $i = 1$, i.e., for boys, and $d_{ij} = 0$ if $i = 2$, i.e., for girls. We now have the following correspondence between the parameters in (2.48) and (2.49): $\alpha_1 = \mu + \gamma_1$, $\alpha_2 = \mu$, $\gamma_1 = \gamma_2 + \gamma_3$, $\gamma_2 = \gamma_2$ so that $\alpha_1 = \alpha_2 \iff \gamma_1 = 0$ and $\gamma_1 = \gamma_2 \iff \gamma_3 = 0$. One can therefore test these hypotheses by estimating (2.49).

For this purpose we create the lsf file **birthweight2.lsf** shown in Figure 2.2

Here the variables BOYS and BOYSAGE correspond to the variables d_{ij} and $d_{ij} x_{ij}$ in (2.49).

The **PRELIS** syntax file **birthweight2.prl**:

```
System File birthweight2.lsf
Regress BWEIGHT on BOYS AGE35 BOYSAGE
```

gives the following estimated regression equation

```
 BWEIGHT = 2.422  + 0.228*BOYS + 0.130*AGE35 - 0.0184*BOYSAGE
Standerr  (0.124)  (0.166)      (0.0300)       (0.0418)
t-values   19.537   1.378        4.347         -0.441
P-values    0.000    0.183        0.000          0.664
```

The following chi-squares test the hypothesis that all

	AGE35	BWEIGHT	BOYS	BOYSAGE
1	5,000	2,968	1,000	5,000
2	3,000	2,795	1,000	3,000
3	5,000	3,163	1,000	5,000
4	0,000	2,925	1,000	0,000
5	1,000	2,625	1,000	1,000
6	2,000	2,847	1,000	2,000
7	6,000	3,292	1,000	6,000
8	5,000	3,473	1,000	5,000
9	2,000	2,628	1,000	2,000
10	3,000	3,176	1,000	3,000
11	5,000	3,421	1,000	5,000
12	3,000	2,975	1,000	3,000
13	5,000	3,317	0,000	0,000
14	1,000	2,729	0,000	0,000
15	5,000	2,935	0,000	0,000
16	3,000	2,754	0,000	0,000
17	7,000	3,210	0,000	0,000
18	4,000	2,817	0,000	0,000
19	5,000	3,126	0,000	0,000
20	2,000	2,539	0,000	0,000
21	1,000	2,412	0,000	0,000
22	3,000	2,991	0,000	0,000
23	4,000	2,875	0,000	0,000
24	5,000	3,231	0,000	0,000

Figure 2.2: Alternative LSF File for Birthweight Example

```
regression coefficients are zero except the intercept.
Variable        -2lnL  Chi-square   df  Covariates
--------     ----------  ----------   --  ----------

 BWEIGHT      -18.414      24.751     3   BOYS AGE35 BOYSAGE
```

Neither the estimated coefficient of BOYS nor that of BOYSAGE are significant. This suggests that both the intercept and the slope in the regression may be the same for boys and girls.

Although there is a one-to-one correspondence between the parameters α_1, α_2, γ_1, and γ_2, in model (2.48) and the parameters μ, γ_1, γ_2, and γ_3 in model (2.49), the model (2.49) assumes that $\sigma_1^2 = \sigma_2^2$. To test the hypothesis $\sigma_1^2 = \sigma_2^2$ one can use the three previously obtained $-2\ln L$ and compute -18.038-(-6.462-12.273) = 18.735 - 18.038 = 0.657 and use this as a χ^2 with one degree of freedom. This suggests that the hypothesis of equal error variances might hold as well. The estimate of the common error variance is 0.0326. So the final conclusion will be that the regression equation is the same for boys and girls.

2.1.15   Example: Math on Reading by Career

The next example illustrates the case of several categories with unequal number of observations. Yet it is small enough to be able to consider some details.

In the study of cognitive ability, verbal skill is considered by many to be the core feature of intelligence. Individual differences in verbal skills are strongly associated with performance in many different mental tasks. Although this association has been observed in many different settings, the strength of the correlation varies considerably across characteristics of students.

For scientific and practical reasons, it is of interest to know whether the prediction of Math from Reading is essentially the same for different types of students.

- Is the association stronger for students with professional goals than for students interested in technical careers?

- Does it differ according to characteristics of the families?

- How does previous educational experience moderate the relationship?

If the regression relationship varies systematically across levels of a covariate, then it is important to take the covariate into consideration when making predictions.

In this example we consider data on Math and Reading scores for students choosing different career goals. Can Mathematics Achievement be predicted by Reading Ability, and, if so, is this prediction different for different career groups? The career groups are

1. Trades (Tr)

2. Police or Security (P)

3. Business Management (B)

4. Sales (S)

5. Military Service (M)

6. Teacher Training (T)

7. Industrial Operations (I)

8. Undecided (U)

9. Real Estate Management (R)

The model is
$$y_{ij} = \alpha_i + \gamma_i x_{ij} + z_{ij} \; , i = 1, 2, \ldots, 9 \; , j = 1, 2, \ldots, n_i \; , \tag{2.50}$$
where y_{ij} and x_{ij} are the Math and Reading scores for student j in career group i and z_{ij} is the error in the regression.

The data is given in the file **Math on Reading by Career.lsf**.

The overall regression of Math on Reading is obtained by running the PRELIS syntax file **Math on Reading by Career.prl**

```
System file 'Math on Reading by Career.lsf'
Regress Math on Reading
```

If the name of the **.lsf** file contain spaces or other special characters enclose the name within single quotes as shown in the first line.

The estimated overall regression is

```
    Math = 37.969 + 0.640*Reading + Error, R² = 0.693
Standerr  (0.754)   (0.0354)
t-values   50.356    18.084
P-values    0.000     0.000
```

indicating a strong effect of Reading on Math. By adding `Res=Math_res` at the end of the Regress line one can obtain the residuals in this regression. This will add the variable `Math_res` containing the residuals as the last variable in the lsf file. The first few lines of the resulting lsf file looks like this

Math on Reading by Career.psf	Career	Math	Reading	Math_res
1	1,000	48,710	15,240	0,985
2	1,000	43,490	6,330	1,468
3	1,000	44,080	15,000	-3,492
4	1,000	47,500	23,000	-5,193
5	1,000	63,880	34,670	3,716
6	1,000	45,620	15,430	-2,227
7	1,000	43,770	12,690	-2,323
8	1,000	49,490	13,200	3,070
9	1,000	42,890	13,940	-4,003
10	1,000	49,690	8,910	6,017
11	1,000	42,230	9,190	-1,623
12	1,000	56,150	17,190	7,176
13	2,000	57,870	29,270	1,163
14	2,000	47,020	14,000	0,088
15	2,000	38,260	7,450	-4,479
16	2,000	47,260	10,280	2,710
17	2,000	48,750	15,630	0,775
18	2,000	46,050	25,820	-8,448
19	2,000	37,500	3,460	-2,684
20	2,000	38,080	4,410	-2,713
21	2,000	63,220	41,140	-1,086
22	2,000	60,570	33,370	1,238
23	2,000	48,800	19,030	-1,352

To plot `Math_res` on Career using a Box-Whisker plot, select **Graphs** and **Bivariate Graphs** and select `Math_res` as the Y variable and Career as the X variable. Then select **Box Whisker Plot** and click on **Plot**:

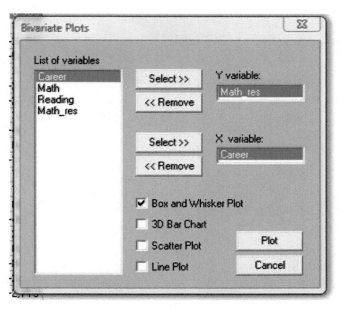

giving the following Box-Whisker plot

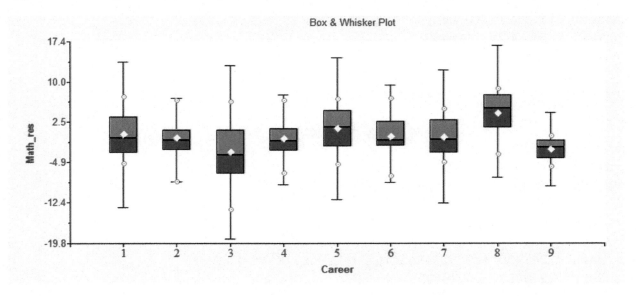

This indicates that there may be some differences in the regression of Math on Reading for different careers. Note particularly Career 3 (Business Management) and Career 8 (Undecided).

The PRELIS syntax file **Math on Reading by Career1.prl**

```
System file 'Math on Reading by Career.lsf'
Regress Math on Reading by Career
```

gives one estimated regression equation for each career:

```
For Career = 1, Sample Size = 12:

    Math = 38.033 + 0.655*Reading + Error, R² = 0.591
Standerr  (2.924)  (0.172)
t-values  13.008   3.803
P-values  0.000    0.003
```

```
For Career = 2, Sample Size = 23:

      Math = 36.149 + 0.707*Reading + Error, R² = 0.864
Standerr  (1.456)  (0.0613)
t-values   24.820   11.541
P-values    0.000    0.000


For Career = 3, Sample Size = 19:

      Math = 40.512 + 0.383*Reading + Error, R² = 0.376
Standerr  (2.811)  (0.120)
t-values   14.411   3.200
P-values    0.000   0.005


For Career = 4, Sample Size = 24:

      Math = 36.085 + 0.736*Reading + Error, R² = 0.855
Standerr  (1.162)  (0.0646)
t-values   31.062   11.402
P-values    0.000    0.000


For Career = 5, Sample Size = 22:

      Math = 39.527 + 0.630*Reading + Error, R² = 0.581
Standerr  (2.201)  (0.120)
t-values   17.961   5.266
P-values    0.000   0.000


For Career = 6, Sample Size = 9:

      Math = 34.430 + 0.827*Reading + Error, R² = 0.821
Standerr  (2.948)  (0.146)
t-values   11.678   5.661
P-values    0.000   0.000


For Career = 7, Sample Size = 12:

      Math = 37.439 + 0.653*Reading + Error, R² = 0.752
Standerr  (2.910)  (0.119)
t-values   12.864   5.507
P-values    0.000   0.000
```

```
For Career = 8, Sample Size = 17:

    Math = 41.436 + 0.676*Reading + Error, R² = 0.755
Standerr  (2.169)  (0.0994)
t-values  19.107   6.799
P-values  0.000    0.000

For Career = 9, Sample Size = 9:

    Math = 35.292 + 0.647*Reading + Error, R² = 0.935
Standerr  (1.602)  (0.0644)
t-values  22.027   10.053
P-values  0.000    0.000
```

It is seen that both intercepts and slopes vary considerably across careers. For a better interpretation of the intercepts one can subtract the overall mean of Reading, 18.967, from the Reading scores. Then the intercept can be interpreted as the average Math score for each career at the overall mean of Reading. Still better, would be to subtract the career mean of Reading from each Reading score, so that the mean Reading score is zero for each career. Then the intercept can be interpreted as the mean Math score when Reading is at the mean for each career. To consider these ideas we use the data file **Math on Reading with groupcentered Reading.lsf** where Read_gc is the group-centered Reading score. Use the PRELIS syntax file **Math on Reading by Career2.prl**:

```
System file 'Math on Reading with groupcentered Reading.lsf'
Regress Math on Read_gc by Career
Output YE=RegressionEstimates.dat
```

The output file gives 9 estimated regression equations where the slope and error variances are the same as in the previous run but where the intercepts represent the group means of Math after correcting for group differences in Reading. These regression estimates are summarized in the file **RegressionEstimates.dat** which looks as like this:

Career	Intcept	Read_gc	Ssize	ErrorVar	R2
1.00000	48.12502	0.65535	12.00000	18.10355	0.59126
2.00000	50.91783	0.70698	23.00000	11.12696	0.86381
3.00000	48.80947	0.38348	19.00000	22.36225	0.37586
4.00000	47.05791	0.73624	24.00000	10.16140	0.85527
5.00000	50.38227	0.63006	22.00000	13.06497	0.58099
6.00000	49.46778	0.82712	9.00000	14.71577	0.82070
7.00000	52.43333	0.65280	12.00000	12.67347	0.75204
8.00000	54.70530	0.67551	17.00000	15.20631	0.75502
9.00000	49.79998	0.64692	9.00000	4.35736	0.93522

It is seen that the intercept varies from 47.058 for career Sales to 54.705 for career Undecided and the slope varies from 0.383 for Business Management to 0.827 for career Teacher Training. One can test whether the intercepts and/or slopes are the same across careers by introducing a dummy variable for each career as shown in Section 2.1.13.

If there are many groups, this procedure is not practical. Instead one can deal with this problem by regarding the intercepts and slopes as random regression coefficients and estimate their variances as is done in *Multilevel Analysis* in Chapter 4.

2.1.16   Instrumental Variables and Two-Stage Least Squares

Consider the estimation of the the linear relationship between a dependent variable y and a set of explanatory variables $\mathbf{x}' = (x_1, x_2, \ldots, x_q)$:

$$y = \alpha + \gamma_1 x_1 + \gamma_2 x_2 + \cdots + \gamma_q x_q + u \,, \tag{2.51}$$

or in matrix form

$$y = \alpha + \boldsymbol{\gamma}'\mathbf{x} + u \,, \tag{2.52}$$

where u is a random error term, α is an intercept term, and $\boldsymbol{\gamma}' = (\gamma_1, \gamma_2, \ldots, \gamma_p)$ is a vector of coefficients to be estimated.

As shown in equation (2.20), if u is uncorrelated with $x_1, \ldots, x_q$, and the sample covariance matrix of $\mathbf{x}$ is positive definite, ordinary least squares (OLS) can be used to obtain a consistent estimate of $\boldsymbol{\gamma}$, yielding the wellknown solution

$$\hat{\boldsymbol{\gamma}} = \mathbf{S}_{xx}^{-1}\mathbf{s}_{xy} \,, \tag{2.53}$$

where $\hat{\boldsymbol{\gamma}}$ is a $q \times 1$ vector of estimated γ's, $\mathbf{S}_{xx}$ is the $q \times q$ sample covariance matrix of the x-variables and $\mathbf{s}_{xy}$ is the $q \times 1$ vector of sample covariances between the x-variables and y.

If u is *correlated* with one or more of the x_i, however, the OLS estimate in (2.53) is not consistent, *i.e.*, it is biased even in large samples. The bias can be positive or negative, large or small depending on the correlations between u and $\mathbf{x}$. But suppose some *instrumental variables* $\mathbf{z}' = (z_1, \ldots, z_r)$ are available, where $r \geq q$. An instrumental variable is a variable which is uncorrelated with u but correlated with y. Then the following two-stage least squares (TSLS) estimator:

$$\hat{\boldsymbol{\gamma}} = (\mathbf{S}'_{zx}\mathbf{S}_{zz}^{-1}\mathbf{S}_{zx})^{-1}\mathbf{S}'_{zx}\mathbf{S}_{zz}^{-1}\mathbf{s}_{zy} \,, \tag{2.54}$$

can be used to estimate $\boldsymbol{\gamma}$ consistently, where $\mathbf{S}_{zx}$ is the $r \times q$ matrix of sample covariances between the z-variables and the x-variables, $\mathbf{S}_{zz}$ is the $r \times r$ sample covariance matrix of the z-variables, and $\mathbf{s}_{zy}$ is the $r \times 1$ vector of sample covariances between the z-variables and y.

The usual way of deriving (2.54) is a two-step procedure, see, *e.g.*, Goldberger (1964) or Theil (1971):

Step 1 Estimate the *OLS* regression of each x_i on $\mathbf{z}$, yielding $\mathbf{x} = \hat{\mathbf{B}}\mathbf{z} + \mathbf{v}$, where $\hat{\mathbf{B}} = \mathbf{S}'_{zx}\mathbf{S}_{zz}^{-1}$.

Step 2 Replace $\mathbf{x}$ in (2.52) with $\hat{\mathbf{x}} = \hat{\mathbf{B}}\mathbf{z}$ and estimate the *OLS* regression of y on $\hat{\mathbf{x}}$. Combining the results of Steps 1 and 2 gives (2.54).

However, this procedure is unnecessarily complicated. Equation (2.54) shows that this estimator can be computed in one step once the variances and covariances of the variables involved has been computed and, as will be shown, LISREL has a very simple command for estimating an equation by TSLS. This has the advantage that it does not require any distributional assumptions. It is valid if the estimated relationship is linear, the required variances and covariances exist, and the inverses in (2.54) exist.

The covariance matrix of $\hat{\boldsymbol{\gamma}}$ can be estimated as

$$(N - 1 - q)^{-1}\hat{\sigma}_{uu}(\mathbf{S}'_{zx}\mathbf{S}^{-1}_{zz}\mathbf{S}_{zx})^{-1} \,, \tag{2.55}$$

where

$$\hat{\sigma}_{uu} = s_{yy} - 2\hat{\boldsymbol{\gamma}}'\mathbf{s}_{xy} + \hat{\boldsymbol{\gamma}}'\mathbf{S}_{xx}\hat{\boldsymbol{\gamma}} \tag{2.56}$$

is a consistent estimate of the variance of u. The standard errors of the estimated γ's are the square roots of the diagonal elements of (2.55).

Every x-variable which is uncorrelated with u may serve as an instrumental variable z. If all x-variables are uncorrelated with u, the x-variables themselves serve as instrumental variables. It may be easily verified that if $\mathbf{z} = \mathbf{x}$, then (2.54) reduces to (2.53) and (2.55) reduces to the well-known OLS formula

$$(N - 1 - q)^{-1}\hat{\sigma}_{uu}\mathbf{S}^{-1}_{xx} \,.$$

For every x-variable which is correlated with u there must be at least one instrumental variable outside the set of x-variables. Usually there is exactly one instrumental variable for each x-variable which is correlated with u, but it is possible to use more than one. In practice, use all available instrumental variables.

The most common application of TSLS is to estimate systems of non-recursive equations where there are several jointly dependent y-variables. A non-recursive system, most often referred to as an interdependent system, is a system of equations in which the dependent variables cannot be ordered in a sequence. Such a system may have reciprocal causation or loops. The simplest example is given in the following figure, where y_1 depends on y_2 and y_2 depends on y_1.

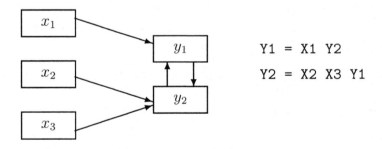

This kind of model were used in time series econometrics in the 1950's and 60's (see $e.g.$, Klein & Goldberger, 1955) to analyze large systems of interdependent equations. In econometric terminology, the y-variables are endogenous, the x-variables are exogenous variables or lagged endogenous variables and the errors are random disturbance terms.

In a general setting, if there are p jointly dependent endogenous variables and q exogenous variables, the equations to be estimated may be written

$$\begin{aligned}
y_i = {} & \alpha_i + \beta_{i1}y_1 + \beta_{i2}y_2 + \cdots + \beta_{i,i-1}y_{i-1} \\
& + \beta_{i,i+1}y_{i+1} + \beta_{i,i+2}y_{i+2} + \cdots + \beta_{i,p}y_p \\
& + \gamma_{i1}x_1 + \gamma_{i2}x_2 + \cdots + \gamma_{iq}x_q + u_i \\
& i = 1, 2, \ldots, p \,.
\end{aligned}$$

Any coefficient β_{ij} and γ_{ij} may be specified to be $a\ priori$ zero. To estimate these equations one can use all x-variables as instrumental variables but additional instrumental variables may be required for identification. A necessary condition for identification of an equation is that the number of x-variables excluded from the equation is at least as great as the number of y-variables included in the right side of the equation. This is the so-called *order condition*. In other words, for every y-variable on the right side of an equation there must be at least one x-variable excluded from that equation. There is also a sufficient condition for identification, the so-called *rank condition*, but this is often difficult to apply in practice. For further information on identification of interdependent systems, see, *e.g.*, Goldberger (1964, pp. 313–318).

To estimate a linear equation by TSLS, write

`Regress Y-variable ON Y-variables X-variables WITH Z-variables`

where `Y-variable` is the left-hand variable y_i, `Y-variables` are the y-variables on the right-hand side, `X-variables` are the x variables on the right-hand side and `Z-variables` are the instrumental variables. In the formulation of TSLS in connection with equation (2.54) the right-hand y-variables and the x-variables are combined into the vector of x-variables. The word `ON` must follow the name of the left-hand y-variable and the word `WITH` separates the right-hand variables from the instrumental variables. The words `ON` and `WITH` are not case sensitive so they could also be written `on` and `with`.

The estimated residual $\hat{u} = y - \hat{\gamma}'\mathbf{x}$ may be computed for each case in the sample by including `RES=varname` at the end of the *Regress* command in **PRELIS**, where `varname` is the name the user assigns for the residual. After the regression equation has been estimated, this variable may be augmented to the **.lsf** file and used in another **PRELIS** run. Once the residual is available in the raw data, one can examine its behavior in the sample by importing the raw data into **PRELIS** and viewing various univariate and bivariate graphs. This was illustrated in Section 2.1.4.

If $q = p$ and $\mathbf{S}_{zx}$ is nonsingular, the residual $\hat{u}$ is uncorrelated with the instrumental variables $\mathbf{z}$, just as the residual in ordinary least squares (*OLS*) is uncorrelated with the explanatory variables $\mathbf{x}$. If $q > p$ and the rank of $\mathbf{S}_{zx}$ is p, there will be p linear combinations of $\mathbf{z}$ that are uncorrelated with the residual $\hat{u}$.

2.1.17 Example: Income and Money Supply

The following example provided by Gujarati (1995, p. 686) gives the simultaneous relationship between income and money supply. The model is

$$y_1 = \alpha_1 + \beta_1 y_2 + \gamma_1 x_1 + \gamma_2 x_2 + u_1 \qquad (2.57)$$

$$y_2 = \alpha_2 + \beta_2 y_1 + u_2 \qquad (2.58)$$

where the endogenous variables are

$y_1 =$ income

$y_2 =$ money supply

and the exogenous variables are

$x_1 =$ investment expenditure

$x_2 =$ government spending on goods and services

Table 2.4: Selected Macro Economic Data for United States 1970–1991. y_1–x_2 in billions of dollars. x_3 in percent. Originally published in Gujarati (1995, Table 20.2). Published with kind permission of © McGraw-Hill Education 1995. All Rights Reserved

Year	y_1	y_2	x_1	x_2	x_3
1970	1010.7	628.1	150.3	208.5	6.562
1971	1097.2	717.2	175.5	224.3	4.511
1972	1207.0	805.2	205.6	249.3	4.466
1973	1349.6	861.0	243.1	270.3	7.178
1974	1458.6	908.6	245.8	305.6	7.926
1975	1585.9	1023.3	226.0	364.2	6.122
1976	1768.4	1163.7	286.4	392.7	5.266
1977	1974.1	1286.6	358.3	426.4	5.510
1978	2232.7	1388.7	434.0	469.3	7.572
1979	2488.6	1496.7	480.2	520.3	10.017
1980	2708.0	1629.5	467.6	613.1	11.374
1981	3030.6	1792.9	558.0	697.8	13.776
1982	3149.6	1951.9	503.4	770.9	11.084
1983	3405.0	2186.1	546.7	840.0	8.750
1984	3777.2	2374.3	718.9	892.7	9.800
1985	4038.7	2569.4	714.5	969.9	7.660
1986	4268.6	2811.1	717.6	1028.2	6.030
1987	4539.9	2910.8	749.3	1065.6	6.050
1988	4900.0	3071.1	793.6	1109.0	6.920
1989	5250.8	3227.3	832.3	1181.6	8.040
1990	5522.2	3339.0	799.5	1273.6	7.470
1991	5677.5	3439.8	721.1	1332.7	5.490

Yearly data for 1970–1991 is given in Table 2.4:

Equation (2.57) states that income is determined by money supply, investment expenditure, and government expenditure. Equation (2.58) states that money supply is determined by the Federal Reserve System on the basis of income. Obviously, there is a simultaneous relationship between income and money supply. Note that u_1 and u_2 are only assumed to be uncorrelated with x_1 and x_2, so none of the equations (2.57) and (2.58) can be estimated by OLS. For equation (2.58), the necessary condition for identification suggests that it is over-identified since neither x_1 nor x_2 is included in that equation. However, (2.57) is under-identified since both x_1 and x_2 are included in that equation. To estimate this equation by TSLS, one must find an instrumental variable in addition to x_1 and x_2. Such a variable is

x_3 = interest rate on 6-month Treasury bills in %

which is available in the data in Table 2.4. Although not included in the model, x_3 can be used as an instrumental variable in addition to x_1 and x_2 to estimate (2.57).

For the analysis with **LISREL**, we omit the variable Year and divide the four variables y_1–x_2 by 100. After importing this data to a **LISREL** data system file **incomemoney.lsf**, it looks like this

	Y1	Y2	X1	X2	X3
1	10.107	6.281	1.503	2.085	6.562
2	10.972	7.172	1.755	2.243	4.511
3	12.070	8.052	2.056	2.493	4.466
4	13.496	8.610	2.431	2.703	7.178
5	14.586	9.086	2.458	3.056	7.926
6	15.859	10.233	2.260	3.642	6.122
7	17.684	11.637	2.864	3.927	5.266
8	19.741	12.866	3.583	4.264	5.510
9	22.327	13.887	4.340	4.693	7.572
10	24.886	14.967	4.802	5.203	10.017
11	27.080	16.295	4.676	6.131	11.374
12	30.306	17.929	5.580	6.978	13.776
13	31.496	19.519	5.034	7.709	11.084
14	34.050	21.861	5.467	8.400	8.750
15	37.772	23.743	7.189	8.927	9.800
16	40.387	25.694	7.145	9.699	7.660
17	42.686	28.111	7.176	10.282	6.030
18	45.399	29.108	7.493	10.656	6.050
19	49.000	30.711	7.936	11.090	6.920
20	52.503	32.273	8.323	11.816	8.040
21	55.222	33.390	7.995	12.736	7.470
22	56.775	34.398	7.211	13.327	5.490

Both equations can be estimated by TSLS using the following simple **PRELIS** syntax file **incomemoney1.prl**:

```
!Income and Money Supply Model Estimated by TSLS
Systemfile incomemoney.lsf
Regress Y1 on Y2 X1 X2 with X1-X3
Regress Y2 on Y1 with X1-X2
```

The resulting estimated equations are

```
Estimated Equations

        Y1 = 0.773 + 0.493*Y2 + 0.292*X1 + 2.698*X2 + Error, R² = 0.996
Standerr  (0.700) (0.411)      (0.416)      (0.957)
t-values   1.104   1.200        0.701        2.818
P-values   0.283   0.245        0.492        0.011


Error Variance = 1.076


Instrumental Variables: X1 X2 X3

        Y2 = 0.346 + 0.614*Y1 + Error, R² = 0.996
Standerr  (0.305) (0.00906)
t-values   1.132   67.793
P-values   0.270   0.000


Error Variance = 0.404


Instrumental Variables: X1 X2
```

The results for y_2 agrees with Gujarati (1995, Equation 20.5.5) except for differences in standard errors. **LISREL** uses (2.55) to obtain standard errors.

As expected, the estimate of β_2 is highly significant. In the income equation (2.57) only $\hat{\gamma}_2$ is statistically significant at the 5% level.

2.1.18   Example: Tintner's Meat Market Model

Tintner (1952, pp. 176–179) formulated a model for the American meat market:

$$y_1 = \beta_1 y_2 + \gamma_1 x_1 + u_1 \,, \tag{2.59}$$

$$y_1 = \beta_2 y_2 + \gamma_2 x_2 + \gamma_3 x_3 + u_2 \,, \tag{2.60}$$

where the endogenous variables are

$y_1 =$ meat consumption per capita (pounds)

$y_2 =$ meat price (1935–39 = 100)

and the exogenous variables are

$x_1 =$ disposable income per capita (dollars)

$x_2 =$ unit cost of meat processing (1935–39 = 100)

$x_3 =$ cost of agricultural production (1935–39 = 100)

Note that both equations (2.59) and (2.60) have y_1 on the left side. This is so because (2.59) represents the demand for meat and (2.60) represents the supply of meat and in a free market these are supposed to be equal. Note also that in (2.59), u_1 and u_2 are correlated with y_1 and y_2. Therefore the equations cannot be estimated consistently by OLS. The only assumption is that both u_1 and u_2 are uncorrelated with x_1, x_2, and x_3. So x_1, x_2, and x_3 can be used as instrumental variables to estimate both equations.

The sample covariance matrix based on annual data for United States 1919–1941 ($N = 23$) is given in Table 2.5.[4]

Table 2.5: Covariance Matrix for Variables in Tintner's Meat Market Model

x_1	x_2	x_3	y_1	y_2
3792.439				
164.169	115.218			
554.762	33.217	119.409		
166.905	-24.385	44.721	62.252	
379.754	38.651	56.171	-16.020	71.886

To estimate the equations (2.59) and (2.60) directly by TSLS one can use the following SIMPLIS syntax file, see file **TINTNER1a.SPL**

---

[4]This was obtained by dividing the numbers in Goldberger (1964, p. 321) by 22

```
Estimating Tintner's Demand and Supply Functions
Observed Variables: X1-X3 Y1 Y2
Covariance Matrix from File TINTNER.COV
Sample Size = 23
Regress Y1 on Y2 and X1 with X1-X3
Regress Y1 on Y2 X2 and X3 with X1-X3
End of Problem
```

The output gives the two estimated equations as

```
Estimating Tintner's Demand and Supply Functions

Estimated Equations

        Y1 =   - 1.579*Y2 + 0.202*X1 + Error, R² = 0.422
Standerr        (0.625)     (0.0662)
Z-values        -2.528       3.052
P-values         0.011       0.002


Error Variance = 35.957

        Y1 =   - 0.321*Y2 - 0.278*X2 + 0.603*X3 + Error, R² = 0.707
Standerr        (0.293)     (0.114)     (0.152)
Z-values        -1.097      -2.428       3.965
P-values         0.273       0.015       0.000


Error Variance = 18.222
```

2.1.19   Example: Klein's Model I of US Economy

Klein's (1950) Model I is a classical econometric model which has been used extensively as a benchmark problem for studying econometric methods. It is an eight-equation system based on annual data for the United States in the period between the two World Wars. It is dynamic in the sense that elements of time play important roles in the model. The three behavioral equations of Klein's Model are

$$
\begin{aligned}
C_t &= a_1 P_t + a_2 P_{t-1} + a_3 W_t + u_1 \\
I_t &= b_1 P_t + b_2 P_{t-1} + b_3 K_{t-1} + u_2 \\
W_t^* &= c_1 E_t + c_2 E_{t-1} + c_3 A_t + u_3
\end{aligned}
$$

In addition to these stochastic equations, the model includes five identities (definitional equations):

$$
\begin{aligned}
P_t &= Y_t - W_t \\
Y_t &= C_t + I_t + G_t - T_t \\
K_t &= K_{t-1} + I_t
\end{aligned}
$$

$$W_t = W_t^* + W_t^{**}$$
$$E_t = Y_t + T_t - W_t^{**}$$

The endogenous variables are

$$
\begin{aligned}
C_t &= \text{Aggregate Consumption } (y_1)\\
I_t &= \text{Net Investment } (y_2)\\
W_t^* &= \text{Private Wage Bill } (y_3)\\
P_t &= \text{Total Profits } (y_4)\\
Y_t &= \text{Total Income } (y_5)\\
K_t &= \text{End-of-Year Capital Stock } (y_6)\\
W_t &= \text{Total Wage Bill } (y_7)\\
E_t &= \text{Total Production of Private Industry } (y_8)
\end{aligned}
$$

The predetermined variables are the exogenous variables

$$
\begin{aligned}
W_t^{**} &= \text{Government Wage Bill } (x_1)\\
T_t &= \text{Taxes } (x_2)\\
G_t &= \text{Government Non-Wage Expenditures } (x_3)\\
A_t &= \text{Time in Years From 1931 } (x_4)
\end{aligned}
$$

and the lagged endogenous variables P_{t-1} (x_5), K_{t-1} (x_6), and E_{t-1} (x_7). All variables except A_t are in billions of 1934 dollars. Annual time series data for 1921–1941 are given in Table 2.6, which has been computed from Theil's (1971) Table 9.1. These data are given in file **KLEIN.LSF**.

In the consumption function there are 2 y-variables included on the right side, namely P_t and W_t and 6 x-variables excluded namely W_t^{**}, T_t, G_t, A_t, K_{t-1}, and E_{t-1} so that the order condition is fulfilled. Similarly, it can be verified that the order condition is met also for the other two equations.

To estimate the consumption function, we use C_t as the y-variable, P_t, P_{t-1}, and W_t as x-variables, and all the predetermined variables as z-variables, in the sense of equation (2.51). A PRELIS command file to do this is **KLEIN1.PRL**:

```
Estimating Klein's Consumption Function
Systemfile KLEIN.LSF
Regress C on P P_1 W with W** T G A P_1 K_1 E_1
```

The estimated consumption equation is given in the output as:[5]

```
        C = 16.555 + 0.0173*P + 0.216*P_1 + 0.810*W + Error, R² = 0.977
Standerr  (1.468)   (0.131)    (0.119)      (0.0447)
t-values  11.277    0.132      1.814        18.111
P-values  0.000     0.897      0.086        0.000
```

---

[5]These results agree with those given in Goldberger (1964, p. 336) and Theil (1971, p. 458), except that these authors do not give the estimate of the intercept and they divide by N in (2.55) instead of $N \_ 1 \_ q$

Table 2.6: Time Series Data for Klein's Model I. This table was constructed from Theil (1971, Table 9.1) using the identities of the model

t	C_t	P_{t-1}	W_t^*	I_t	K_{t-1}	E_{t-1}	W_t^{**}
1921	41.9	12.7	25.5	-0.2	182.8	44.9	2.7
1922	45.0	12.4	29.3	1.9	182.6	45.6	2.9
1923	49.2	16.9	34.1	5.2	184.5	50.1	2.9
1924	50.6	18.4	33.9	3.0	189.7	57.2	3.1
1925	52.6	19.4	35.4	5.1	192.7	57.1	3.2
1926	55.1	20.1	37.4	5.6	197.8	61.0	3.3
1927	56.2	19.6	37.9	4.2	203.4	64.0	3.6
1928	57.3	19.8	39.2	3.0	207.6	64.4	3.7
1929	57.8	21.1	41.3	5.1	210.6	64.5	4.0
1930	55.0	21.7	37.9	1.0	215.7	67.0	4.2
1931	50.9	15.6	34.5	-3.4	216.7	61.2	4.8
1932	45.6	11.4	29.0	-6.2	213.3	53.4	5.3
1933	46.5	7.0	28.5	-5.1	207.1	44.3	5.6
1934	48.7	11.2	30.6	-3.0	202.0	45.1	6.0
1935	51.3	12.3	33.2	-1.3	199.0	49.7	6.1
1936	57.7	14.0	36.8	2.1	197.7	54.4	7.4
1937	58.7	17.6	41.0	2.0	199.8	62.7	6.7
1938	57.5	17.3	38.2	-1.9	201.8	65.0	7.7
1939	61.6	15.3	41.6	1.3	199.9	60.9	7.8
1940	65.0	19.0	45.0	3.3	201.2	69.5	8.0
1941	69.7	21.1	53.3	4.9	204.5	75.7	8.5

t	T_t	A_t	P_t	K_t	E_t	W_t	Y_t	G_t
1921	7.7	-10.0	12.4	182.6	45.6	28.2	40.6	6.6
1922	3.9	-9.0	16.9	184.5	50.1	32.2	49.1	6.1
1923	4.7	-8.0	18.4	189.7	57.2	37.0	55.4	5.7
1924	3.8	-7.0	19.4	192.7	57.1	37.0	56.4	6.6
1925	5.5	-6.0	20.1	197.8	61.0	38.6	58.7	6.5
1926	7.0	-5.0	19.6	203.4	64.0	40.7	60.3	6.6
1927	6.7	-4.0	19.8	207.6	64.4	41.5	61.3	7.6
1928	4.2	-3.0	21.1	210.6	64.5	42.9	64.0	7.9
1929	4.0	-2.0	21.7	215.7	67.0	45.3	67.0	8.1
1930	7.7	-1.0	15.6	216.7	61.2	42.1	57.7	9.4
1931	7.5	0.0	11.4	213.3	53.4	39.3	50.7	10.7
1932	8.3	1.0	7.0	207.1	44.3	34.3	41.3	10.2
1933	5.4	2.0	11.2	202.0	45.1	34.1	45.3	9.3
1934	6.8	3.0	12.3	199.0	49.7	36.6	48.9	10.0
1935	7.2	4.0	14.0	197.7	54.4	39.3	53.3	10.5
1936	8.3	5.0	17.6	199.8	62.7	44.2	61.8	10.3
1937	6.7	6.0	17.3	201.8	65.0	47.7	65.0	11.0
1938	7.4	7.0	15.3	199.9	60.9	45.9	61.2	13.0
1939	8.9	8.0	19.0	201.2	69.5	49.4	68.4	14.4
1940	9.6	9.0	21.1	204.5	75.7	53.0	74.1	15.4
1941	11.6	10.0	23.5	209.4	88.4	61.8	85.3	22.3

Note that one cannot estimate this equation by regressing C on P_t, P_{t-1}, and W_t, as u_1 is not uncorrelated with P_t and W_t.

Instead of

```
Regress C on P P_1 W with W** T G A P_1 K_1 E_1
```

one can write

```
Equation C = P P_1 W with W** T G A P_1 K_1 E_1
```

and the word `Equation` can be abbreviated as `Eq`. Upper case or lower case can be used. The intercept term in the equation will always be estimated by **PRELIS**.

One can estimate all three behavioral equations in Klein's Model in one run, see file **KLEIN2.PRL**):

```
Estimating Klein's Model I with TSLS
Systemfile KLEIN.LSF
Regress C on P P_1 W with W** T G A P_1 K_1 E_1
Regress I on P P_1 K_1 with W** T G A P_1 K_1 E_1
Regress W* on E E_1 A with W** T G A P_1 K_1 E_1
```

This gives the following result.

```
Estimated Equations


        C = 16.555 + 0.0173*P + 0.216*P_1 + 0.810*W + Error, R² = 0.977
Standerr  (1.468)   (0.131)    (0.119)     (0.0447)
t-values  11.277    0.132      1.814       18.111
P-values  0.000     0.897      0.086       0.000


        I = 20.278 + 0.150*P + 0.616*P_1 - 0.158*K_1 + Error, R² = 0.885
Standerr  (8.383)   (0.193)   (0.181)     (0.0402)
t-values  2.419     0.780     3.404       -3.930
P-values  0.026     0.445     0.003       0.001


       W* = 1.500 + 0.439*E + 0.147*E_1 + 0.130*A + Error, R² = 0.987
Standerr  (1.276)  (0.0396)  (0.0432)    (0.0324)
t-values  1.176    11.082    3.398       4.026
P-values  0.255    0.000     0.003       0.001
```

2.2   General Principles of SIMPLIS Syntax

Except for the side examples in Sections 2.1.6 and 2.1.7, we have been able to do all analysis using the **PRELIS** command

```
Regress Y-variables on X-variables
```

with the extension

```
Regress Y-variables on X-variables by C
```

for conditional regression, *i.e.*, regressions for each value of a categorical variable C, as explained in Section 2.1.12 and the extension

```
Regress Y-variables on Y-variables X-variables with Z-variables
```

for two-stage least squares (TSLS) using `Z-variables` as instrumental variables as explained in Section 2.1.16.

In previous sections we considered regression equations with a single dependent variable y. Although one can estimate several regression equations in one run using **PRELIS**, these equations are estimated for each equation separately. For example, it is not possible to test if the residuals in different equations are uncorrelated. To be able to test such hypotheses and other hypotheses about the coefficients in the linear equations, this approach is generalized to the multivariate case by considering equations for several jointly dependent variables $\mathbf{y}' = (y_1, y_2, \ldots, y_p)$, a more general command language, namely the **SIMPLIS** syntax which is described briefly in this Section or the more advanced **LISREL** syntax described in Section 2.3.1 should be used. The latter is based on matrix notation.

Here we consider only models for directly observed variables. Models for latent variables are considered in Chapters 7–10.

SIMPLIS is a free form English command language based on the following principles

- Name all observed variables

- Formulate the model to be estimated by paths or relationships (equations)

- It is not necessary to be familiar with the general **LISREL** model or any of its submodels.

- No Greek or matrix notations are required.

- There are no complicated options to learn.

- Anyone who can formulate the model as a path diagram can use the **SIMPLIS** command language immediately.

A typical **SIMPLIS** command file consists of four parts:

1. Title (optional)

2. Data description

3. Model specification

4. Output Specification

Title

Title lines are optional but may be used to describe the kind of analysis intended. Any number of title lines may be used. **LISREL** assumes that all lines are title lines until it finds a line begining with words, or `Observed Variables`. To avoid conflicts with title lines beginning with these words or any other words used in the **SIMPLIS** syntax, put an exclamation sign ! in front of it.

Data Description

LISREL reads a data matrix by writing

`Raw Data from File filename`

For most analysis with **LISREL** it is convenient to use a **LISREL** data system file, **.lsf** file for short. **LISREL** can import data from many sources such as SPSS, SAS, STATA, and EXCEL. **LISREL** can also import data in text format with spaces (*.dat or *.raw), commas (*.csv) or tab characters (*.txt) as delimiters. How to import data into a **.lsf** is described in Section 1.1. The **.lsf** file can also be generated in other ways as illustrated in the following sections.

LISREL can also read data directly from a text file with spaces as delimeters. In this case put a line

`Observed Variables`

first and put the names of the variables on the same line or on a new line. With many variables several lines may be used. Alternatively, the names of the variables may instead be given in the first line of the text file with the data.

Model Specification

In this section generic names of variables like **Y**, **X1**, **X2**, are used for the observed variables. In the real examples that are used in the following sections, these variables have real names.

The model specification begins with the following line

`Relationships`

or a line

`Equations`

followed by a number of relationships defining the equations of the model to be estimated.

To specify an equation like
$$y = \gamma_1 x_1 + \gamma_2 x_2 + \gamma_3 x_3 + z$$
write

`Y = X1 X2 X3`

To include an intercept in the equation write

`Y = CONST X1 X2 X3`

The variable `CONST` is a variable which is equal to 1 for every individual case. This variable is always available in **SIMPLIS**; it need not be in the data. It is used to to specify that an intercept term or a mean should be estimated. For example,

`Y = CONST X`

is used to specify the regression of **Y** on **X**:

$$Y = \alpha + \gamma X .$$

α is the coefficient of `CONST` just like γ is the coefficient of **X**. One can also use `CONST` to estimate a mean. For example,

```
Y = CONST
```

will estimate the mean of `Y` as the coefficient of `CONST`.

The data for **SIMPLIS** may also be a covariance or correlation matrix available in the literature or which has been computed by **PRELIS** or external sources. One can also include the means of the variables.

To read a covariance or correlation matrix, write

```
Covariance Matrix from File filename
```

or

```
Correlation Matrix from File filename
```

To read means or standard deviations, write

```
Means from File filename
```

or

```
Standard Deviations from File filename
```

If the number of variables are reasonably small one can put the covariance matrix directly in the **SIMPLIS** syntax file itself as illustrated in the example of Tintner's Meat Market in Section 2.1.18.

Output Specification

The output consists of the default output containing parameter estimates with standard errors, z-values and P-values. It also includes various measures of fit including chi-squares. One can also request additional output, for example,

```
Print Residuals
Standardized Solution
```

One can also specify the method of estimation, for example,

```
Method of Estimation: Unweighted Least Squares
```

or

```
Method of Estimation: Diagonally Weighted Least Squares
```

The default method of estimation is Maximum Likelihood.

Instead of this way of specifying additional output requests, one can specify them on an `Options` line, for example,

```
Options: RS SC ULS
```

With **SIMPLIS** one can handle many different types of models. Some typical ones are illustrated here in path diagrams together with the corresponding relationships lines. In presenting and discussing a **LISREL** model it is often useful to draw a path diagram. The path diagram effectively communicates the basic conceptual ideas of the model.

The general rules for path diagrams are:

- Observed variables are enclosed in squares or rectangles. Error variables need not be included in the path diagram but may be indicated by a one-way arrow.

- A one-way arrow between two variables indicate a postulated direct influence of one variable on another. A two-way arrow between two variables indicates that these variables may be correlated without any assumed direct relationship.

- There is a fundamental distinction between independent variables and dependent variables. In the path diagram, dependent variables have one or more one-way errors pointing to them whereas independent variables have no one-way arrow pointing to them. Two-way arrows between dependent variables are not allowed.

- For each dependent variable in the path diagram, there is an equation specified by `Relationships`.

- The error term on each dependent variable is automatically added. All error terms are assumed to be mutually uncorrelated by default. If necessary, correlations between error terms can be specified by `Set` or `Let` lines, For example, if the errors on `Y1` and `Y2` should be correlated, write a line

  `Let the errors of Y1 and Y2 correlate`

- Independent variables are automatically correlated and their covariance matrix is automatically estimated.

Example 1: Regression Model

The regression model is the simplest type of **LISREL** model.
A regression equation of the form

$$y = \alpha + \gamma_1 x_1 + \gamma_2 x_2 + \gamma_3 x_3 + z$$

is written as a path diagram (left) and in **SIMPLIS** syntax (right) as

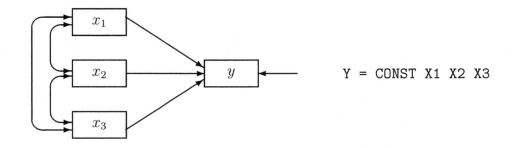

The two-way arrows on the left indicates that the x-variables may be correlated. But this is default in **SIMPLIS** so these two-way arrows may be omitted in the path diagram. Similarly, the one-way arrow on the right indicates the error term z. But **LISREL** assumes that there is always an error term on every dependent variable[6]. So this one-way arrow may also be omitted in the path diagram. Intercept terms are not included in the path diagram but will be estimated if CONST is included in the relationship.

So the path diagram can be broken down to its rudimental parts and looks like this

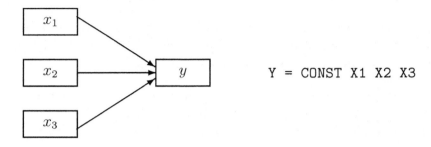

For the following examples we give the path diagram in this rudimental form only.

Example 2: Bivariate Regression

Several regression equations can be estimated simultaneously. Suppose two variables y_1 and y_2 depend on three variables x_1, x_2, and x_3 in such a way that y_1 depend x_1 and x_2 and y_2 depend on x_2 and x_3. The equations are

$$y_1 = \gamma_{11}x_1 + \gamma_{12}x_2 + \gamma_{13}x_3 + z_1$$

$$y_2 = \gamma_{21}x_1 + \gamma_{22}x_2 + \gamma_{23}x_3 + z_2 \, ,$$

with $\gamma_{13} = \gamma_{21} = 0$. A path diagram for this is

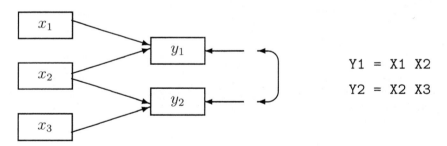

---

[6]To specify an equation without an error term, write `Set the error on Y to zero`. This defines y as a linear combination of x_1, x_2 and x_3 with coefficients which may be specified as fixed numbers or as parameters to be estimated.

Note that the lack of an arrow from x_3 to y_1 is equivalent to $\gamma_{13} = 0$ and the lack of an arrow from x_1 to y_2 is equivalent to $\gamma_{21} = 0$. All error terms are assumed to be uncorrelated by default. To specify correlated error terms, include a line

```
Let the errors of Y1 and Y2 correlate
```

The relationships in the example are written as

```
Y1 = X1 X2
Y2 = X2 X3
```

```
Y1 =    X1 X2 0*X3
Y2 = 0*X1 X2    X3
```

The latter form illustrates how one can specify regression coefficients to be 0.

Example 3: Recursive System

A recursive system is a system of equations such that the dependent variables are ordered in a sequence. In the following figure y_1, y_2, and y_3 are dependent variables. y_1 comes first since it does not depend on any dependent variables, y_2 comes second since it depends on y_1 but not on y_3, y_3 comes third since it depends on both y_1 and y_2.

Models of this kind were used in sociology under the name of path analysis, see *e.g.*, Duncan (1975).

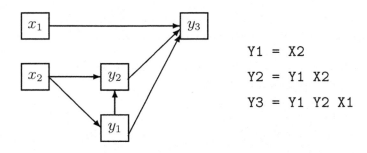

```
Y1 = X2

Y2 = Y1 X2

Y3 = Y1 Y2 X1
```

The relationships are written

```
Y1 = X2
Y2 = Y1 X2
Y3 = Y1 Y2 X1
```

Alternatively, they can be written as

```
Y1 =       0*X1    X2
Y2 = Y1  0*X1    X2
Y3 = Y1 Y2 X1 0*X2
```

Example 4: Non-Recursive System

Non-recursive systems, as considered in Section 2.1.16 can also be estimated using the **SIMPLIS** command language. In this way, the models can not only be estimated by TSLS but also with the maximum likelihood method or other methods available in **LISREL**. First we consider how the equations should be written in **SIMPLIS** syntax. Consider the hypothetical model given previously in 2.1.16.

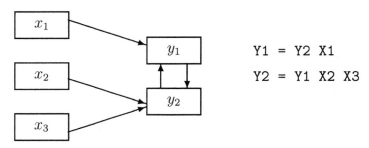

```
Y1 = Y2 X1
Y2 = Y1 X2 X3
```

The relationships are

```
Y1 = Y2 X1
Y2 = Y1 X2 X3
```

Alternatively these can be written as

```
Y1 = Y2   X1 0*X2 0*X3
Y2 = Y1 0*X1   X2   X3
```

2.2.1   Example: Income and Money Supply Using SIMPLIS Syntax

Recall the example Income and Money Supply in Section 2.1.17. The model can be estimated by the following **SIMPLIS** syntax file **incomemoney1a.spl**

```
!Income and Money Supply Estimated by ML using SIMPLIS syntax
Raw Data from File incomemoney.lsf
Equations
Y1 = CONST Y2 X1 X2
Y2 = CONST Y1
End of Problem
```

```
        Structural Equations
```

$$Y1 = 3.505 - 2.165*Y2 + 1.504*X1 + 8.703*X2 , R^2 = 0.996$$

```
Standerr  (2.164) (1.720)   (1.015)    (3.975)
Z-values   1.619  -1.259     1.482      2.189
P-values   0.105   0.208     0.138      0.029
```

$$Y2 = 0.345 + 0.614*Y1, Errorvar.= 0.385 , R^2 = 0.999$$

```
Standerr  (0.286) (0.00844)          (0.114)
Z-values   1.207  72.776             3.390
P-values   0.228   0.000             0.001
```

NOTE: R^2 for Structural Equations are Hayduk's (2006) Blocked-Error R^2

In non-recursive systems of equations it is unclear how R^2 should be calculated and interpreted since the total variance of a y-variable may depend not only on the explanatory (exogenous) x-variables and the error variance but can depend on other y-variables in the system. In this example, the error variance in the first equation depends on the variance of y_2 which in turn depends on the variance of y_1. There is no unique way of dividing the total variance of y_1 between the error variance and the variance contributed by the x-variables. However, Hayduk (2006) suggests a way to resolve this problem so that an R^2 can be calculated and interpreted for each structural equation. We use this in **LISREL**.

Another way to calculate and interpret R^2's is to use the R^2's for the reduced form equations. This example shows the differences between these two alternatives.

```
        Reduced Form Equations


       Y1 = 0.645*X1 + 3.734*X2, Errorvar.= 1.219, R² = 0.995
Standerr   (0.330)      (0.209)
Z-values    1.956       17.873
P-values    0.050        0.000




       Y2 = 0.396*X1 + 2.295*X2, Errorvar.= 0.406, R² = 0.995
Standerr   (0.203)      (0.128)
Z-values    1.956       17.953
P-values    0.050        0.000
```

The following log-likelihood values and goodness-of-fit chi-squares show that the model fits very well.

```
                      Log-likelihood Values


                    Estimated Model           Saturated Model
                    ---------------           ---------------

Number of free parameters(t)        13                     14
-2ln(L)                        100.162                 99.754
AIC (Akaike, 1974)*            126.162                127.754
BIC (Schwarz, 1978)*           140.346                143.029


*LISREL uses AIC= 2t - 2ln(L) and BIC = tln(N)- 2ln(L)



                    Goodness-of-Fit Statistics


Degrees of Freedom for (C1)-(C2)                    1
Maximum Likelihood Ratio Chi-Square (C1)            0.408 (P = 0.5232)
Browne's (1984) ADF Chi-Square (C2_NT)             0.404 (P = 0.5251)
```

2.2.2   Example: Prediction of Grade Averages

The following example illustrates bivariate regression. The model has two dependent variables y_1 and y_2 and three explanatory variables x_1, x_2, and x_3.

Finn (1974) presents the data. These data represent the scores of fifteen freshmen at a large midwestern university on five educational measures. The five measures are

$y_1 =$ grade average for required courses taken (GRAVEREQ)

$y_2 =$ grade average for elective courses taken (GRAVELEC)

$x_1 =$ high-school general knowledge test, taken previous year (KNOWLEDGE)

$x_2 =$ IQ score from previous year (IQPREVYR)

$x_3 =$ educational motivation score from previous year (ED MOTIV)

We examine the predictive value of x_1, x_2 and x_3 in predicting the grade averages y_1 and y_2.

The data is in the file **GradeAverages.lsf**. To estimate the regression of y_1 and y_2 on x_1, x_2, and x_3 use the following PRELIS command file (see file **gradeaverages1.prl**).

```
Prediction of Grade Averages
Systemfile GradeAverages.lsf
Regress GRAVEREC - GRAVELEC on KNOWLEDGE - 'ED MOTIV'
```

The two regression equations are estimated as (see **gradeaverages1.out**)

```
GRAVEREC =   - 5.619 + 0.0854*KNOWLEDG + 0.00822*IQPREVYR - 0.0149*ED MOTIV
Standerr     (5.614) (0.0270)            (0.0485)           (0.112)
t-values     -1.001   3.168               0.169             -0.134
P-values      0.337   0.008               0.868              0.896

             + Error, R² = 0.568

Error Variance = 0.327
```

```
GRAVELEC =   - 20.405 + 0.0472*KNOWLEDG + 0.145*IQPREVYR + 0.126*ED MOTIV
Standerr     (5.398)   (0.0259)           (0.0467)          (0.107)
t-values     -3.780    1.823              3.117             1.170
P-values      0.003    0.093              0.009             0.265
```

$$+ \text{Error, } R^2 = 0.685$$

```
Error Variance = 0.302
```

This analysis assumes that the two residuals z_1 and z_2 are uncorrelated. To investigate whether this is the case, one can estimate the two residuals and add them in **GradeAverages.lsf**. This is done by the following PRELIS syntax file **gradeaverages2.prl**:

```
!Bivariate Regression of Grade Averages: Estimating Residuals
Systemfile GradeAverages.lsf
Regress GRAVEREC on KNOWLEDGE - 'ED MOTIV' Res = Z_1
Regress GRAVELEC on KNOWLEDGE - 'ED MOTIV' Res = Z_2
Continuous All
Output MA=CM WP
```

The WP on the Output line tells PRELIS to use up to 132 columns in the printed output file. Otherwise, 80 columns will be used and the covariance matrix will be printed in sections.

The output gives the same estimates as before but also the following covariance matrix

```
Covariance Matrix
              GRAVEREC   GRAVELEC   KNOWLEDG   IQPREVYR   ED MOTIV   Z_1        Z_2
              --------   --------   --------   --------   --------   --------   --------
GRAVEREC       0.594
GRAVELEC       0.483      0.754
KNOWLEDG       3.993      3.626     47.457
IQPREVYR       0.426      1.757      4.100     10.267
ED MOTIV       0.499      0.716      6.261      0.557      2.694
Z_1            0.257      0.169      0.000      0.000      0.000      0.257
Z_2            0.169      0.237      0.000      0.000      0.000      0.169      0.237
```

This shows that the estimated covariance between z_1 and z_2 is 0.169 corresponding to a correlation of $0.169/\sqrt{0.257 \times 0.237} = 0.68$. To see whether this is significantly different from 0, one should use SIMPLIS or LISREL syntax. A SIMPLIS syntax file to estimate the residual covariance with its standard error is **gradeavarages3a.spl**:

```
SIMPLIS Syntax for Bivariate Regression of Grade Averages
Raw Data from file GradeAverages.lsf
Relationships
GRAVEREC - GRAVELEC = KNOWLEDGE - 'ED MOTIV'
Set the Error Covariance between GRAVEREC and GRAVELEC Free
End of Problem
```

This gives the following estimated residual covariance

```
Error Covariance for GRAVELEC and GRAVEREC = 0.169
                                            (0.0903)
                                             1.875
```

2.2.3   Example: Prediction of Test Scores

This is an example of multivariate regression with two groups. The model has three y-variables and five x-variables.

Dr William D. Rohwer of the University of California at Berkeley collected data on children in kindergarten to predict test scores from paired-associated (PA) learning-proficiency tests developed by him, see Timm (1975, p. 281).

The variables are

SES $= 0$ for children with low SES and $= 1$ for children with high SES

SAT Score on a Student Achievement Test

PPVT Score on the Peabody Picture Vocabulary Test

Raven Score on the Raven Progressive Matrices Test

n performance on a 'named' PA task

s performance on a 'still' PA task

ns performance on a 'named still' PA task

na performance on a 'named action' PA task

ss performance on a 'sentence still' PA task

The data file **rohwer.lsf** has 37 children with low SES and 32 children with high SES. Timm (1975) published the data on PPVT and n - ss for the high SES group in Table 4.3.1 and the data on PPVT - ss for the low SES group in Table 4.7.1.

Consider predicting the three test scores SAT, PPVT, and Raven from the five PA tasks n, s, ns, na, ss. Do the same PA tasks predict the same test scores, or are different predictors needed for each test score? Are the effective predictors the same for low and high SES students?

Model 1 assumes that the intercepts and the regression coefficients are the same for the high and low SES group. This can be estimated without the use of SES using the **PRELIS** syntax file **rohwer1.prl**.

Model 2 allows the intercepts and the regression coefficient to be different between the two groups. This can be estimated by using conditional multivariate regression, see file **rohwer2.prl**:

```
!Conditional Regression by SES
Systemfile rohwer.lsf
Regress SAT PPVT Raven on n - ss by SES
```

Inspecting the regression results for each value of SES and considering the coefficients which are significant at the 5% level, we get the pattern of significant regression coefficients shown in Table 2.7, where 0 indicates a non-significant value and 1 indicates a significant value.

To see if the differences noted in Table 2.7 are statistically significant, one should test the hypotheses of equal regression coefficients between the two groups. This can be done by multiple group analysis as described in Chapter 10. We shall therefore revisit this example in that chapter.

Table 2.7: Pattern of Significant Regression Coefficients

	High SES					Low SES				
	n	s	ns	na	ss	n	s	ns	na	ss
SAT	0	0	0	0	1	1	0	1	1	0
PPVT	0	0	0	1	0	0	0	0	1	0
Raven	0	0	0	0	0	0	1	0	0	0

The previous analysis assumes that the residuals in the three regressions are uncorrelated. To see whether this is the case, one can use **SIMPLIS** syntax. One can test this on each group separately but here we consider the case which assumes that the regression coefficients are the same for High and Low SES whereas the intercepts are allowed to differ between the groups.

This model can be estimated using the following **SIMPLIS** syntax file **rohwer1a.spl**:

```
!Rohwer Data: Regression on Pooled Data
Raw Data from File rohwer.lsf
SAT PPVT Raven = CONST SES n - ss
Set the error covariance matrix of SAT - Raven free
```

This gives the following estimated equations

```
     SAT =  - 1.767    + 8.798*SES + 1.605*n + 0.0257*s  - 2.627*ns + 2.106*na + 0.930*ss
Standerr     (13.316)   (6.754)     (1.037)   (0.873)     (0.872)    (0.881)    (0.828)
Z-values     -0.133     1.303       1.548     0.0295      -3.012     2.390      1.123
P-values      0.894     0.193       0.122     0.976        0.003     0.017      0.262
R² = 0.295
```

```
    PPVT = 31.349 + 16.877*SES + 0.00233*n - 0.351*s - 0.299*ns + 1.294*na + 0.479*ss
Standerr   (5.381)  (2.729)      (0.419)     (0.353)   (0.353)    (0.356)    (0.335)
Z-values   5.825    6.183        0.00555    -0.995    -0.848      3.633      1.431
P-values   0.000    0.000        0.996       0.320     0.397      0.000      0.153
R² = 0.628
```

```
   Raven = 10.747 + 1.586*SES + 0.0149*n + 0.181*s + 0.112*ns - 0.00970*na - 0.00446*ss
Standerr   (1.460)  (0.741)     (0.114)    (0.0958)  (0.0957)   (0.0966)     (0.0908)
Z-values   7.359    2.141       0.131      1.892     1.166     -0.100       -0.0491
P-values   0.000    0.032       0.896      0.059     0.244      0.920        0.961
R² = 0.211
```

These results indicates that the High SES group has higher scores on SAT – Raven. The effect of SES is significant for PPVT and Raven. Also note that the R^2 is much higher for PPVT than for SAT and Raven. This suggests that PPVT can be predicted better from n – ss than SAT and Raven.

The estimated residual covariances are estimated as

```
Error Covariance for PPVT and SAT = 58.660
                                   (33.229)
                                     1.765
```

```
Error Covariance for Raven and SAT = 23.533
                                    (9.282)
                                     2.535


Error Covariance for Raven and PPVT = 8.381
                                     (3.707)
                                      2.261
```

which shows that the residual covariance is significant for Raven and SAT and for Raven and PPVT but not for PPVT and SAT.

2.2.4   Example: Union Sentiment of Textile Workers

This is an example of a recursive model estimated by maximum likelihood using SIMPLIS syntax. The meaning of a recursive system was explained in Section 2.2. The model used here corresponds to the figure used in that example.

McDonald and Clelland (1984) analyzed data on union sentiment of southern non-union textile workers. After transformation of one variable and treatment of outliers, Bollen (1989a, pp 82-83) reanalyzed a subset of the variables according to the model shown in Figure 2.3, which is the same as the figure used in Section 2.2.

The variables are:

$y_1 = $ deference (submissiveness) to managers

$y_2 = $ support for labor activism

$y_3 = $ sentiment towards unions

$x_1 = $ logarithm of years in textile mill

$x_2 = $ age

The covariance matrix is

y_1	y_2	y_3	x_1	x_2
14.610				
−5.250	11.017			
−8.057	11.087	31.971		
−0.482	0.677	1.559	1.021	
−18.857	17.861	28.250	7.139	215.662

In the previous example we have used names of variables like GRAVEREQ and IQPREVYR. While this is a recommended practice, some users may prefer to use generic names such as Y1, Y2, Y3, X1, and X2 as in Figure 2.3.

The model in Figure 2.3 can be easily specified as relationships (equations):

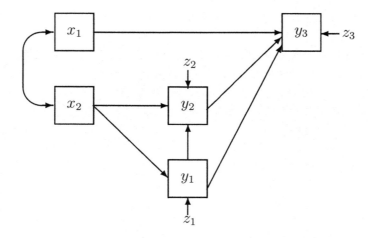

Figure 2.3: Path Diagram for Union Sentiment Model

```
Relationships
Y1 = X2
Y2 = Y1 X2
Y3 = Y1 Y2 X1
```

These three lines express the following three statements, respectively:

- Y1 depends on X2

- Y2 depends on Y1 and X2

- Y3 depends on Y1, Y2, and X1

It is also easy to formulate the model by specifying its paths:

```
Paths
X1 -> Y3
X2 -> Y1 Y2
Y1 -> Y2 Y3
Y2 -> Y3
```

These lines express the following statements:

- There is one path from X1 to Y3

- There are two paths from X2, one to Y1 and one to Y2

- There are two paths from Y1, one to Y2 and one to Y3

- There is one path from Y2 to Y3

Altogether there are six paths in the model corresponding to the six one-way (unidirected) arrows in Figure 2.3.

The input file for this problem is straightforward (file **union1a.spl**):

```
Title: Union Sentiment of Textile Workers

       Variables: Y1 = deference (submissiveness) to managers
                  Y2 = support for labor activism
                  Y3 = sentiment towards unions
                  X1 = Years in textile mill
                  X2 = age

Observed Variables: Y1 - Y3 X1 X2
Covariance Matrix:
   14.610
   -5.250   11.017
   -8.057   11.087   31.971
   -0.482    0.677    1.559    1.021
  -18.857   17.861   28.250    7.139 215.662
Sample Size 173
Relationships
   Y1 = X2
   Y2 = Y1 X2
   Y3 = Y1 Y2 X1
End of Problem
```

The first nine lines are title lines that define the problem and the variables. To indicate the beginning of the title lines, one may use the word **Title**, although this is not necessary. The first real command line begins with the words **Observed Variables**. Note that Y1, Y2, and Y3 may be labeled collectively as **Y1-Y3**. The observed variables, the covariance matrix, and the sample size are specified in the syntax file. Normally these are read from files. The model is specified as relationships.

The model makes a fundamental distinction between two kinds of variables: dependent and independent. The *dependent variables* are supposed to be explained by the model, i.e., variation and covariation in the dependent variables are supposed to be accounted for by the *independent variables*. The dependent variables are on the left side of the equal sign. They correspond to those variables in the path diagram which have one-way arrows pointing towards them. Variables which are not dependent are called independent. The distinction between dependent and independent variables is already inherent in the labels: x-variables are independent and y-variables are dependent. Note that y-variables can appear on the right side of the equations.

The output gives the following estimated relationships:

```
        Structural Equations


      Y1 =  - 0.0874*X2, Errorvar.= 12.961, R² = 0.113
Standerr     (0.0187)              (1.402)
Z-values     -4.664                 9.247
P-values      0.000                 0.000


      Y2 =  - 0.285*Y1 + 0.0579*X2, Errorvar.= 8.488 , R² = 0.230
Standerr     (0.0619)   (0.0161)              (0.918)
Z-values     -4.598      3.597                 9.247
P-values      0.000      0.000                 0.000
```

```
      Y3 =   - 0.218*Y1 + 0.850*Y2 + 0.861*X1, Errorvar.= 19.454, R² = 0.390
Standerr     (0.0974)    (0.112)     (0.341)                  (2.104)
Z-values     -2.235       7.555       2.526                    9.247
P-values      0.025       0.000       0.012                    0.000
```

The reduced form equations are also given in the output:

```
        Reduced Form Equations


      Y1 = 0.0*X1 - 0.0874*X2, Errorvar.= 12.961, R² = 0.113
Standerr            (0.0188)
Z-values            -4.650
P-values             0.000


      Y2 = 0.0*X1 + 0.0828*X2, Errorvar.= 9.538, R² = 0.134
Standerr            (0.0161)
Z-values             5.135
P-values             0.000


      Y3 = 0.861*X1 + 0.0894*X2, Errorvar.= 28.320, R² = 0.112
Standerr   (0.342)     (0.0184)
Z-values    2.518       4.858
P-values    0.012       0.000
```

From the P-values we conclude that all estimated coefficients are statistically significant.

The model was fitted by the maximum likelihood method and the output gives the following chi-square measures of fit as explained in Sections 16.1.3–16.1.5.

```
Degrees of Freedom for (C1)-(C2)                  3
Maximum Likelihood Ratio Chi-Square (C1)          1.259 (P = 0.7390)
Browne's (1984) ADF Chi-Square (C2_NT)            1.255 (P = 0.7398)
```

These chi-squares indicate that the model fits very well, which together with the fact that all coefficients in the three equations are statistically significant, give some validity to the model

2.3   The General Multivariate Linear Model

All previous models can be formulated in matrix form as one single equation:

$$\mathbf{y} = \boldsymbol{\alpha} + \mathbf{B}\mathbf{y} + \boldsymbol{\Gamma}\mathbf{x} + \mathbf{z} \,, \tag{2.61}$$

where $\mathbf{y}' = (y_1, y_2, \ldots, y_p)$ is a set of observed random jointly dependent variables, $\mathbf{x}' = (x_1, x_2, \ldots, x_q)$ is a set of observed explanatory variables with mean vector $\boldsymbol{\kappa}$ and covariance matrix $\boldsymbol{\Phi}$, and $\mathbf{z}' = (z_1, z_2, \ldots, z_p)$ is a vector of error terms. The vector $\boldsymbol{\alpha}$ is a vector intercepts, and the matrices $\mathbf{B}$ and $\boldsymbol{\Gamma}$ are parameter matrices called structural parameters. The square matrix $\mathbf{B}$ has zeros in the diagonal, since a y-variable on the left side cannot appear on the right side in the same equation. The parameter matrices involved in this model are

$$\boldsymbol{\alpha}, \ \boldsymbol{\kappa} = \mathsf{E}(\mathbf{x}), \ \mathbf{B}, \ \boldsymbol{\Gamma}, \ \boldsymbol{\Phi} = \mathsf{Cov}(\mathbf{x}), \ \boldsymbol{\Psi} = \mathsf{Cov}(\mathbf{z}) \ .$$

Each of these may contain fixed zeros (or other fixed values) in specified positions. The assumptions are

(i) $(\mathbf{I} - \mathbf{B})$ is nonsingular,

(ii) $\mathsf{E}(\mathbf{z}) = \mathbf{0}$,

(iii) $\mathbf{z}$ is uncorrelated with $\mathbf{x}$.

In most models the mean vector $\boldsymbol{\kappa}$ and covariance matrix $\boldsymbol{\Phi}$ are unconstrained and will be estimated by $\bar{\mathbf{x}}$ and $\mathbf{S}_{xx}$, the observed sample mean vector and sample covariance matrix of $\mathbf{x}$.

The error covariance matrix $\boldsymbol{\Psi}$ is not necessarily diagonal but is sometimes assumed to be diagonal, i.e., the residuals z_i and z_j in two different equations may or may not be correlated.

The structural equation (2.61) should be distinguished from the *reduced form equation*

$$\mathbf{y} = \mathbf{A}\boldsymbol{\alpha} + \mathbf{A}\boldsymbol{\Gamma}\mathbf{x} + \mathbf{A}\mathbf{z} \ , \tag{2.62}$$

where $\mathbf{A} = (\mathbf{I} - \mathbf{B})^{-1}$. Equation (2.62) is obtained from (2.61) by solving for $\mathbf{y}$.

The reduced form equation (2.62) is the regression of $\mathbf{y}$ on $\mathbf{x}$. In contrast, the structural equations in (2.61) are not necessarily regression equations because the error term z_i in the i-th equation is not necessarily uncorrelated with all the y-variables appearing on the right in that equation. The coefficients in the reduced form are functions of the structural parameters $\mathbf{B}$ and $\boldsymbol{\Gamma}$, and in general these are not in a one-to-one correspondence.

On these assumptions, the mean vector and covariance matrix of the observed variables $\mathbf{y}$ and $\mathbf{x}$ are

$$\boldsymbol{\mu} = \mathsf{E}\left(\begin{array}{c} \mathbf{y} \\ \mathbf{x} \end{array} \right) = \left(\begin{array}{c} \mathbf{A}\boldsymbol{\alpha} + \mathbf{A}\boldsymbol{\Gamma}\boldsymbol{\kappa} \\ \boldsymbol{\kappa} \end{array} \right) \ . \tag{2.63}$$

$$\boldsymbol{\Sigma} = \mathsf{Cov}\left(\begin{array}{c} \mathbf{y} \\ \mathbf{x} \end{array} \right) = \left(\begin{array}{cc} \mathbf{A}\boldsymbol{\Gamma}\boldsymbol{\Phi}\boldsymbol{\Gamma}'\mathbf{A}' + \mathbf{A}\boldsymbol{\Psi}\mathbf{A}' & \\ \boldsymbol{\Phi}\boldsymbol{\Gamma}'\mathbf{A}' & \boldsymbol{\Phi} \end{array} \right) \ . \tag{2.64}$$

There are three important special cases:

1. When $\mathbf{B} = \mathbf{0}$ the structural and reduced form equations are identical and each equation represents an actual regression equation. This includes the case of multivariate linear regression or multivariate analysis of variance or covariance (MANOVA or MANCOVA) with $\boldsymbol{\Psi}$ diagonal or fully symmetric.

2. When $\mathbf{B}$ is *sub-diagonal* (or when the y-variables can be ordered so that $\mathbf{B}$ becomes sub-diagonal) and $\boldsymbol{\Psi}$ is diagonal, the structural equation model is a *recursive system*. Models of this kind were used in sociology, under the name of path analysis, see *e.g.*, Duncan (1975).

3. When the structural equations cannot be ordered such that $\mathbf{B}$ becomes subdiagonal, the system is non-recursive. These models are common in econometrics, where they are referred to as *interdependent systems* or *simultaneous equations*.

2.3.1 Introductory LISREL Syntax

The number of y- and x-variables in the model are denoted **NY** and **NX**, respectively. The parameter matrices

$$\boldsymbol{\alpha}, \; \boldsymbol{\kappa} = \mathsf{E}(\mathbf{x}), \; \mathbf{B}, \; \boldsymbol{\Gamma}, \; \boldsymbol{\Phi} = \mathsf{Cov}(\mathbf{x}), \; \boldsymbol{\Psi} = \mathsf{Cov}(\mathbf{z}) \; .$$

in the model (2.61) are specified as **AL, KA, BE, GA, PH**, and **PS**, respectively, and any element of these matrices can be specified by giving its row and column index. For example, **GA(2,3)** refers to the element γ_{23} in the matrix $\boldsymbol{\Gamma}$. Each element of these matrices can be specified to be a free parameter to be estimated or a fixed parameter with value 0 or any other fixed value. An entire parameter matrix can be specified to be fixed or free. To simplify this process, each of the parameter matrices has a default form and a default fixed or free status as follows

Name	Math Symbol	LISREL Name	Order	Possible Forms	Default Form	Default Mode
ALPHA	α	AL	NY $\times$ 1	FU	FU	FI
KAPPA	κ	KA	NX $\times$ 1	FU	FU	FI
BETA	$\mathbf{B}$	BE	NY $\times$ NY	ZE,SD,FU	ZE	FI
GAMMA	$\boldsymbol{\Gamma}$	GA	NY $\times$ NX	ID,DI,FU	FU	FR
PHI	$\boldsymbol{\Phi}$	PH	NX $\times$ NX	ID,DI,SY,ST	SY	FR
PSI	$\boldsymbol{\Psi}$	PS	NY $\times$ NY	ZE,DI,SY	SY	FR

The meaning of the possible form values is as follows:

ZE $\mathbf{0}$ (zero matrix)

ID $\mathbf{I}$ (identity matrix)

DI a diagonal matrix

SD a full square matrix with fixed zeros in and above the diagonal and all elements under the diagonal free (refers to $\mathbf{B}$ only)

SY a symmetric matrix that is not diagonal

ST a symmetric matrix with fixed ones in the diagonal (a correlation matrix; refers to $\boldsymbol{\Phi}$ only)

FU a rectangular or square non-symmetric matrix

Note the following

- For the symmetric matrices $\boldsymbol{\Phi}$ and $\boldsymbol{\Psi}$, only the lower triangular part are stored in memory. Symmetrically situated elements are equal: **PH(1,2) = PH(2,1)**.

- The zero ($\mathbf{0}$) and identity ($\mathbf{I}$) matrices are not physically stored in memory and their elements cannot be referenced.

- All elements in a matrix may be specified fixed or free by the notation: **LISREL NAME = MATRIX FORM, MATRIX MODE**. For example: **PH=DI,FI** specifies that $\boldsymbol{\Phi}$ is diagonal with fixed diagonal elements.

- Individual elements may be fixed with the word **FI** or freed with the the word **FR**, followed by a list of the elements separated by spaces. For example: **FI BE(3,1) BE(4,1) BE(4,2)**.

- Later specifications override earlier specifications. For example:
BE=SD

FIX BE(3,1) BE(4,1) BE(4,2)

The first statement sets all subdiagonal elements of **B** free; the second leaves only the elements in the first subdiagonal free.

- For the symmetric $\mathbf{\Phi}$ matrix, the specification PH=ST means that the diagonal elements of $\mathbf{\Phi}$ are fixed at 1 and the off-diagonal elements are free. The specifications PH=ST,FI and PH=ST,FR are contradictory and not permitted.

- The value of each parameter, whether free or fixed, is zero by default. Fixed parameters may be set at a specific value by the word VA followed by the value (with decimal point) and the list of elements to be set to that value. If the value is an integer, the decimal point may be omitted. For example: VA 0.065 GA(4,2) BE(6,1) PS(5,4). Any statement of this type overrides any previous VA statement with the same referents.

To see how the LISREL syntax works, we will reconsider one typical example of each type of model from the previous sections.

2.3.2   Univariate Regression Model

Recall the example **Income** from section 2.1.9 and the LISREL data file **income.lsf** after the variable LnInc (the logarithm of Income) has been added as the last variable. To run the regression of LnInc on Age-Group one can use the following LISREL syntax file **income2b1.lis**:

```
Regression of LnInc
DA NI=9
RA=income.lsf
SE
LnInc Age Gender Marital Hours Citizen Degree Group /
MO NY=1 NX=7 AL=FR KA=FR
OU
```

The first line is a title line. The DA line says that the number of variables in the data file is 9. The RA line is the command for reading the data file **income.lsf**. Since the variable Income is not included in the regression equation (LnInc is used instead Income), only 8 of the variables in the data file are used. An awkward requirement in LISREL syntax is that the y-variables must come first in the data. Here the y-variable is LnInc which is the last variable in **income.lsf**. To select the variables to be used in the regression equation and put them with the y-variable first, one uses an SE line to put the variables in the order one wants. Put SE on a single line and the list of the variables on the next line. If the number of variables in this list is less than the number of variables in the data file, put a forward slash / at the end.

The MO line specifies that there is 1 y-variable and 7 x-variable. Furthermore, BE, GA, PH, and PS are default on the MO line, so $\mathbf{B}(1 \times 1) = 0$, $\mathbf{\Gamma}(1 \times 7)$ (the regression coefficients) is a vector of free parameters to be estimated, $\mathbf{\Phi}(7 \times 7)$ (the covariance matrix of the x-variables) is a free symmetric matrix to be estimated, $\mathbf{\Psi}(1 \times 1)$ (the residual variance) is a free parameter to be estimated. To estimate the intercept α and the mean vector $\boldsymbol{\kappa}$ of the x-variables one must specify AL=FR KA=FR on the MO line. Otherwise, these would be zero.

The OU line is always the last line in a LISREL syntax file. It is used to request various output in addition to the standard output. In this example the OU line is left empty.

When **SIMPLIS** or **LISREL** syntax is used, **LISREL** will regard the model as a mean and covariance structure as defined in Sections 16.1.3 – 16.1.5. **LISREL** will fit all free parameters of the model to the sample mean vector $(\bar{\mathbf{y}}, \bar{\mathbf{x}})$ and sample covariance matrix

$$\mathbf{S} = \begin{pmatrix} \mathbf{S}_{yy} & \\ \mathbf{S}_{xy} & \mathbf{S}_{xx} \end{pmatrix} .$$

of $(\mathbf{y}, \mathbf{x})$ by minimizing some fit function, see Section 16.1.5. The default fit function is ML which gives maximum likelihood estimates under the assumption that $(\mathbf{y}, \mathbf{x})$ has a multivariate normal distribution, see Section 16.1.2. As a mean and covariance structure, the regression model is a saturated[7] model because the number of free parameters to be estimated is the same as the number of independent elements in $(\bar{\mathbf{y}}, \bar{\mathbf{x}})$ and

$$\mathbf{S} = \begin{pmatrix} \mathbf{S}_{yy} & \\ \mathbf{S}_{xy} & \mathbf{S}_{xx} \end{pmatrix} .$$

In this case there is an explicit solution for the parameter estimates. This will be demonstrated in the general case of multivariate regression in the next section. The model here is a special case of this.

The estimated regression coefficients with their standard errors and z-values are given in the 1×7 matrix GAMMA:

```
LISREL Estimates (Maximum Likelihood)
```

```
        GAMMA
```

	Age	Gender	Marital	Hours	Citizen	Degree
LnInc	0.017	0.233	0.071	0.013	0.245	0.426
	(0.001)	(0.029)	(0.023)	(0.001)	(0.034)	(0.028)
	16.579	8.110	3.167	20.365	7.252	14.934

```
        GAMMA
```

	Group
LnInc	0.196
	(0.032)
	6.230

---

[7]In general, a saturated model fits the data perfectly and has as many free parameters as the number of observable means, variances and covariances. Saturated models, sometime called just-identified models, can provide useful information in the search for a more parsimonious model in which the parameters of the saturated model are constrained, for example, constrained to be zero.

The intercept is given in the 1×1 matrix `ALPHA`:

```
ALPHA

      LnInc

     --------

      8.240
     (0.060)
     136.795
```

The estimated standard errors differs slightly between the **LISREL** output and the **PRELIS** output, because **PRELIS** uses the more exact formulas (2.24) and (2.36), whereas **LISREL** uses the asymptotic[8] formula (16.57). Consequently, **PRELIS** uses t-values whereas **LISREL** uses Z-values. The standard errors in **PRELIS** and **LISREL** differ by a scale factor of $\sqrt{N/(N-1-q)}$, where N is the sample size and q is the number of x-variables. Here $N = 6062$ and $q = 7$. Since N is very large in this example, these differences are very small.

Why would anyone use the **LISREL** syntax in **income2b1.lis** when one can use the much simpler two-line (apart from the title line) **PRELIS** syntax

```
Systemfile income.lsf
Regress lnInc on Age - Group
```

Well, suppose we want to test the hypotheses $\gamma_2 = \gamma_5$. This can be done by adding the line

```
EQ GA(1,2) GA(1,5)
```

in **income2b1.lis**, see file **income2b2.lis** The output file **income2b2.out** gives the following log-likelihood values and chi-square for testing the hypotheses:

Log-likelihood Values

	Estimated Model	Saturated Model
Number of free parameters(t)	43	44
-2ln(L)	49580.723	49580.651
AIC (Akaike, 1974)*	49666.723	49668.651
BIC (Schwarz, 1978)*	49955.244	49963.882

*LISREL uses AIC= 2t - 2ln(L) and BIC = tln(N)- 2ln(L)

Goodness-of-Fit Statistics

Degrees of Freedom for (C1)-(C2)	1
Maximum Likelihood Ratio Chi-Square (C1)	0.0722 (P = 0.7882)
Browne's (1984) ADF Chi-Square (C2_NT)	0.0722 (P = 0.7882)

The log-likelihood values are explained in Section 16.1.3 and the chi-squares `C1` and `C2_NT` are explained in Section 16.1.5. To test the hypothesis $\gamma_2 = \gamma_5$ one can use $\chi^2 = 0.0722$ with 1 degree of freedom. This has a $P = 0.7882$ so the hypothesis cannot be rejected.

Suppose we want to test the hypothesis $\gamma_1 = \gamma_6$ instead. This can be done by replacing

```
EQ GA(1,2) GA(1,5)
```

---

[8]asymptotic means something valid in large samples

by

EQ GA(1,1) GA(1,6)

see file **income2b3.lis**. This gives a C1=199.420 with a P-value of 0.0000, so that this hypothesis is strongly rejected.

These two small exercises suggest that not all regression coefficients are equal. Would they be equal if all variables were standardized? This can be tested by adding MA=KM on the DA line and replacing the EQ line by

EQ GA(1,1) - GA(1,6)

see file **income2b4.lis**. The resulting χ^2 (C1) is 201.188 with 6 degrees of freedom. Its P-value is 0.0000. So it is clear that also this hypothesis is rejected.

The MA on the DA line is short for *Type of matrix to be analyzed*, and KM means correlation matrix. The default is MA=CM, where CM means covariance matrix.

2.3.3   Multivariate Linear Regression

Following the setup in Section 2.3, the multivariate linear regression model is

$$\mathbf{y} = \boldsymbol{\alpha} + \boldsymbol{\Gamma}\mathbf{x} + \mathbf{z} , \tag{2.65}$$

The mean vector $\boldsymbol{\mu}$ and covariance matrix $\boldsymbol{\Sigma}$ of the observed variables $(\mathbf{y}, \mathbf{x})'$ are

$$\boldsymbol{\mu} = \begin{pmatrix} \boldsymbol{\alpha} + \boldsymbol{\Gamma}\boldsymbol{\kappa} \\ \boldsymbol{\kappa} \end{pmatrix} , \tag{2.66}$$

$$\boldsymbol{\Sigma} = \begin{pmatrix} \boldsymbol{\Gamma}\boldsymbol{\Phi}\boldsymbol{\Gamma}' + \boldsymbol{\Psi} & \\ \boldsymbol{\Phi}\boldsymbol{\Gamma}' & \boldsymbol{\Phi} \end{pmatrix} . \tag{2.67}$$

This is a special case of (2.63) and (2.64) with $\mathbf{B} = \mathbf{0}$, so that $\mathbf{A} = \mathbf{I}$.

Suppose that $\boldsymbol{\alpha}$, $\boldsymbol{\kappa}$, and $\boldsymbol{\Gamma}$ are unconstrained and that the covariance matrices $\boldsymbol{\Phi}$ and $\boldsymbol{\Psi}$ of $\mathbf{x}$ and $\mathbf{z}$ are full symmetric matrices. Let $(\bar{\mathbf{y}}, \bar{\mathbf{x}})$ be the sample mean vector and let

$$\mathbf{S} = \begin{pmatrix} \mathbf{S}_{yy} & \\ \mathbf{S}_{xy} & \mathbf{S}_{xx} \end{pmatrix} . \tag{2.68}$$

be the sample covariance matrix. Assuming that $(\mathbf{y}, \mathbf{x})$ have a multivariate normal distribution and that $\mathbf{S}_{xx}$ is non-singular, the maximum likelihood solution is given explicitly as

$$\hat{\boldsymbol{\alpha}} = \bar{\mathbf{y}} - \mathbf{S}_{yx}\mathbf{S}_{xx}^{-1}\bar{\mathbf{x}} , \tag{2.69}$$

$$\hat{\boldsymbol{\kappa}} = \bar{\mathbf{x}} , \tag{2.70}$$

$$\hat{\boldsymbol{\Gamma}} = \mathbf{S}_{yx}\mathbf{S}_{xx}^{-1} , \tag{2.71}$$

$$\hat{\boldsymbol{\Phi}} = \mathbf{S}_{xx} , \tag{2.72}$$

$$\hat{\boldsymbol{\Psi}} = \mathbf{S}_{yy} - \mathbf{S}_{yx}\mathbf{S}_{xx}^{-1}\mathbf{S}_{xy} , \tag{2.73}$$

where $\mathbf{S}_{yx} = \mathbf{S}'_{xy}$.

Using these estimates one can verify that

$$\hat{\boldsymbol{\mu}} = \begin{pmatrix} \hat{\boldsymbol{\mu}}_y \\ \hat{\boldsymbol{\mu}}_x \end{pmatrix} = \begin{pmatrix} \hat{\boldsymbol{\alpha}} + \hat{\boldsymbol{\Gamma}}\hat{\boldsymbol{\kappa}} \\ \hat{\boldsymbol{\kappa}} \end{pmatrix} = \begin{pmatrix} \bar{\mathbf{y}} \\ \bar{\mathbf{x}} \end{pmatrix} . \tag{2.74}$$

and that

$$\hat{\boldsymbol{\Sigma}} = \begin{pmatrix} \hat{\boldsymbol{\Sigma}}_{yy} \\ \hat{\boldsymbol{\Sigma}}_{xy} & \hat{\boldsymbol{\Sigma}}_{xx} \end{pmatrix} = \begin{pmatrix} \hat{\boldsymbol{\Gamma}}\hat{\boldsymbol{\Phi}}\hat{\boldsymbol{\Gamma}}' + \hat{\boldsymbol{\Psi}} \\ \hat{\boldsymbol{\Phi}}\hat{\boldsymbol{\Gamma}}' & \hat{\boldsymbol{\Phi}} \end{pmatrix} = \begin{pmatrix} \mathbf{S}_{yy} \\ \mathbf{S}_{xy} & \mathbf{S}_{xx} \end{pmatrix} = \mathbf{S} . \tag{2.75}$$

This shows that the model is saturated. That equations (2.69)–(2.73) are indeed maximum likelihood estimates under normality may be shown by noting that the likelihood of an observation $(\mathbf{y}_i, \mathbf{x}_i)$ is

$$L(\mathbf{y}_i, \mathbf{x}_i) = L(\mathbf{y}_i \mid \mathbf{x}_i)L(\mathbf{x}_i) \tag{2.76}$$

and that

$$\mathbf{y}_i \mid \mathbf{x}_i \sim N(\boldsymbol{\mu}_y + \boldsymbol{\Sigma}_{yx}\boldsymbol{\Sigma}_{xx}^{-1}(\mathbf{x} - \boldsymbol{\mu}_x), \boldsymbol{\Sigma}_{yy} - \boldsymbol{\Sigma}_{yx}\boldsymbol{\Sigma}_{xx}^{-1}\boldsymbol{\Sigma}_{xy}) \tag{2.77}$$

and

$$\mathbf{x}_i \sim N(\boldsymbol{\mu}_x, \boldsymbol{\Sigma}_{xx}) \tag{2.78}$$

We leave the proof to the reader.

The above derivations hold for the model where all parameters are unconstrained. Using the **LISREL** syntax, one can test various hypotheses about the parameters. The most common types of hypotheses are

- That specified elements of $\boldsymbol{\Gamma}$ are equal to 0 (or equal to any other fixed value) or that some elements of $\boldsymbol{\Gamma}$ are equal (equality constraints).

- That $\boldsymbol{\Psi}$ is diagonal (or that some elements of $\boldsymbol{\Psi}$ are 0) indicating uncorrelated error terms.

Each such hypothesis leads to a model where the number of parameters to be estimated (here denoted t) is less than the number of means, variances, and covariances of the observed variables $\mathbf{y}$ and $\mathbf{x}$(here denoted s). In general, for such models there is no explicit solution. Instead the solution must be determined by an iterative procedure as described in the appendix (see Chapter 17). The hypothesis can be tested by a chi-square statistic χ^2 with degrees of freedom $d = s - t$. The degrees of freedom represents the number of independent constraints imposed on the parameters. If p is the number of y-variables and q is the number of x-variables, then $s = p + q + [(p + q)(p + q + 1)]/2$ so that

$$d = p + q + [(p + q)(p + q + 1)]/2 - t \tag{2.79}$$

2.3.4   Example: Prediction of Test Scores with LISREL Syntax

To illustrate multivariate linear regression and testing of hypotheses, we will reconsider the example **Prediction of Test Scores** from Section 2.2.3.

Suppose we want to estimate the regression of SAT, PPVT and Raven on SES, n, s, ns, na, and ss. Then $p = 3$ and $q = 6$. Assuming that the three error terms are correlated, we begin by testing the hypothesis that

$$\Gamma = \begin{pmatrix} 0 & 0 & 0 & x & x & 0 \\ x & 0 & 0 & 0 & x & 0 \\ x & 0 & x & 0 & 0 & 0 \end{pmatrix}, \tag{2.80}$$

where 0 is a fixed zero and x is free parameter.

This model can be estimated and tested using the following LISREL syntax file **rohwer1b1.lis**

```
!Rohwer Data: Regression on Pooled Data
!Testing hypothesis on GAMMA
DA NI=9
RA=rohwer.lsf
SE
SAT PPVT Raven SES n s ns na ss
MO NY=3 NX=6 AL=FR KA=FR PS=SY,FR
PA GA
0 0 0 1 1 0
1 0 0 0 1 0
1 0 1 0 0 0
OU
```

Here

- The DA line specifies that there are 9 variables in the data file **rohwer.lsf**.

- Since the variables in the data file are in the order

  SES SAT PPVT Raven n s ns na ss

  they must be reordered such that the three y-variables SAT PPVT Raven come first. This is done by SE line.

- The MO line specifies that there are 3 y-variables and 6 x-variables and that both α and κ should be estimated. To estimate a full symmetric matrix Ψ one must specify PS=SY,FR. Otherwise Ψ will be diagonal.

- To specify the hypothesis on Γ one can use a pattern matrix, where 0 represents a fixed parameter and 1 represents a free parameter. All fixed parameters are zero by default.

The output file **rohwer1b1.out** gives the following chi-square values

```
Degrees of Freedom for (C1)-(C2)                  12
Maximum Likelihood Ratio Chi-Square (C1)          14.950 (P = 0.2442)
Browne's (1984) ADF Chi-Square (C2_NT)            14.269 (P = 0.2838)
```

The degrees of freedom equals the number of fixed zero elements in Γ. Both chi-square values indicate that the hypothesis on Γ cannot be rejected suggesting that many of the variables

```
SES n s ns na ss
```

do not have any significant effect on

```
SAT PPVT Raven
```

Of the remaining effects all but one are statistically significant, which can be seen from the estimated **GAMMA** matrix

```
       GAMMA
```

	SES	n	s	ns	na	ss
SAT	- -	- -	- -	-2.239 (0.779) -2.875	2.980 (0.699) 4.266	- -
PPVT	16.751 (2.559) 6.545	- -	- -	- -	1.268 (0.207) 6.141	- -
Raven	1.216 (0.666) 1.826	- -	0.231 (0.076) 3.030	- -	- -	- -

Here - - represent a fixed zero value. Note that

- n and ss have no significant effect at all on any of the dependent variables.

- SES has an effect only on PPVT

- Only s has an effect on Raven.

Next assume that Γ is free and consider the hypothesis that Ψ is diagonal, *i.e.*, that the three error terms are uncorrelated. Since PSI is diagonal by default, this model can be estimated and tested by the following simpler LISREL syntax file, see file **rohwer1b2.lis**:

```
!Rohwer Data: Regression on Pooled Data
!Testing hypothesis on PSI
DA NI=9
RA=rohwer.lsf
SE
SAT PPVT Raven SES n s ns na ss
MO NY=3 NX=6 AL=FR KA=FR
OU
```

This gives the following chi-square values

```
Degrees of Freedom for (C1)-(C2)               3
Maximum Likelihood Ratio Chi-Square (C1)       16.399 (P = 0.0009)
Browne's (1984) ADF Chi-Square (C2_NT)         17.832 (P = 0.0005)
```

suggesting that the hypothesis is rejected. The degrees of freedom 3 equals the number of off-diagonal elements in Ψ.

2.3.5   Recursive Systems

Following the setup in Section 2.3, a recursive system is a model with $\mathbf{B}$ subdiagonal, *i.e.*, all elements in the diagonal and above the diagonal of $\mathbf{B}$ are zero, whereas all elements below the diagonal may be a free parameter to be estimated or specified as a fixed value. If all elements below the diagonal are free parameters to be estimated, the system is called a complete recursive system.

2.3.6   Example: Union Sentiment of Textile Workers with LISREL Syntax

An example of this is the example **Union Sentiment of Textile Workers** considered in Section 2.2.4, which will be used here again.

This example introduces another fundamental distinction in **LISREL** namely that only the sample covariance matrix $\mathbf{S}$ is available; no individual data is available; the means of the variables are also not available. As a consequence, it is not possible to estimate the parameter vectors $\boldsymbol{\alpha}$ and $\boldsymbol{\kappa}$. Only the parameter matrices $\mathbf{B}$, $\boldsymbol{\Gamma}$, $\boldsymbol{\Phi}$, and $\boldsymbol{\Psi}$ can be estimated.

To estimate this model using **LISREL** syntax, one can use the following input file, see file **union1b.lis**:

```
DA NI=5 NO=173
LA
Y1 Y2 Y3 X1 X2
CM=union.cov
MO NY=3 NX=2 BE=SD
FI GA(1,1) GA(2,1) GA(3,2)
OU
```

where the file **union.cov** contain the following values

```
 14.610
 -5.250  11.017
 -8.057  11.087  31.971
 -0.482   0.677   1.559    1.021
-18.857  17.861  28.250    7.139 215.662
```

being the elements of the sample covariance matrix $\mathbf{S}$. Note that

- $\mathbf{B}$ is specified as sub-diagonal on the MO line

- $\boldsymbol{\Gamma}$ is default on the MO line

- The default $\boldsymbol{\Gamma}$ means that $\boldsymbol{\Gamma}$ is a full free 3×2 matrix. However, the model in Figure 2.3 requires that $\gamma_{11} = \gamma_{21} = \gamma_{32} = 0$. This is specified by the line

```
FI GA(1,1) GA(2,1) GA(3,2)
```

All fixed elements are zero, unless some other value is specified on a VA line.

- PS is default on the MO line which means that $\boldsymbol{\Psi}$ is diagonal,*i.e.*, the three error terms are uncorrelated.

The output file **union1b.out** gives the estimates of $\mathbf{B}$ and $\boldsymbol{\Gamma}$ as

```
        BETA

                Y1          Y2          Y3
            --------    --------    --------
    Y1        - -         - -         - -

    Y2      -0.285        - -         - -
            (0.062)
            -4.598

    Y3      -0.218       0.850        - -
            (0.097)     (0.112)
            -2.235       7.555

        GAMMA

                X1          X2
            --------    --------
    Y1        - -       -0.087
                        (0.019)
                        -4.664

    Y2        - -        0.058
                        (0.016)
                         3.597

    Y3       0.861        - -
            (0.341)
             2.526
```

As judged by the z-values, all the estimated parameters in $\mathbf{B}$ and $\boldsymbol{\Gamma}$ are statistically significant. The output file also gives the following chi-squares for testing the hypothesis $\gamma_{11} = \gamma_{21} = \gamma_{32} = 0$.

```
Degrees of Freedom for (C1)-(C2)                        3
Maximum Likelihood Ratio Chi-Square (C1)                1.259 (P = 0.7390)
Browne's (1984) ADF Chi-Square (C2_NT)                  1.255 (P = 0.7398)
```

Hence, the hypothesis cannot be rejected.

2.3.7   Non-Recursive Systems

A non-recursive system is a system of equations where the dependent y-variables cannot be ordered in a sequence. We considered such systems in Section 2.1.16, where it was shown how one can estimate each equation separately by two-stage least squares (TSLS) provided there are a sufficient number of instrumental variables available in the data. However, provided the model is identified, such systems can also be estimated by the maximum likelihood method.

In the general setting of Section 2.3, a recursive model is characterized by a matrix $\mathbf{B}$ having non-zero elements both below and above the diagonal. Recall that $\mathbf{B}$ is a square non-symmetric matrix with zeros on the diagonal. The simplest example is a two-equation system where $\mathbf{B}$ is of order 2×2 with $\beta_{21} \neq 0$ and $\beta_{12} \neq 0$.

2.3.8   Example: Income and Money Supply with LISREL syntax

To illustrate how a non-recursive system can be estimated by the maximum likelihood method using LISREL syntax, we will revisit the example **Income and Money Supply** introduced in Section 2.1.17. In that section the model was estimated using TSLS. In Section 2.2.1 we used SIMPLIS syntax to obtain maximum likelihood estimates. Here we show how the same estimates can be obtained using LISREL syntax.

Using the notation in equations (2.57) and (2.58), the model parameter matrices $\mathbf{B}$ and $\mathbf{\Gamma}$ are

$$\mathbf{B} = \begin{pmatrix} 0 & \beta_1 \\ \beta_2 & 0 \end{pmatrix} \quad \mathbf{\Gamma} = \begin{pmatrix} \gamma_1 & \gamma_2 \\ 0 & 0 \end{pmatrix}$$

A LISREL syntax file for estimating this model by maximum likelihood is **incomemoney1b.lis**:

```
!Income and Money Supply Model Estimated by ML
!Using LISREL Syntax
da ni=5
ra=incomemoney.lsf
mo ny=2 nx=2 be=fi al=fr ka=fr
fr be(1,2) be(2,1)
fi ga(2,1) ga(2,2)
ou
```

The maximum likelihood estimates are given in the output as

```
        BETA

                Y1              Y2
            --------        --------
   Y1          - -           -2.165
                             (1.847)
```

```
Y2         0.614            - -
          (0.009)
          67.794



   GAMMA

              X1              X2
           --------        --------
   Y1       1.504           8.703
           (1.089)         (4.268)
            1.381           2.039


   Y2        - -             - -
```

These estimates are exactly the same as were obtained using **SIMPLIS** syntax in Section 2.2.1 but they are very different from the TSLS estimates obtained in Section 2.1.17. Some coefficients have opposite signs. This can be seen in Table 2.8.

Table 2.8: TSLS and ML Estimates for Income and Money Supply

	TSLS	ML
β_1	0.493	-2.165
β_2	0.614	0.614
γ_1	0.292	1.504
γ_2	2.698	8.709

Which estimates are best? TSLS usually works better than ML in small samples, whereas ML has better large sample properties. Also, TSLS does not depend on any strong distributional assumptions, whereas ML assumes multivariate normality, an assumption which does not hold for this data. Time series data has often strong autocorrelation which makes the assumption of independent observations questionable. Furthermore, time series data often has strong multicollinearity. For this data set, this is revealed in the output file **income and money1b.out** as

```
   Covariance Matrix

              Y1              Y2              X1              X2
           --------        --------        --------        --------
   Y1       235.456
   Y2       144.314          88.836
   X1        34.531          21.264           5.407
```

```
        X2       56.757      34.865        8.324       13.758
```

Total Variance = 343.457 Generalized Variance = 1.706

Largest Eigenvalue = 342.789 Smallest Eigenvalue = 0.055

Condition Number = 78.631

WARNING: The Condition Number indicates severe multicollinearity.

One or more variables may be redundant.

The total variance is the sum of the diagonal elements of $\mathbf{S}$ and the generalized variance is the determinant of $\mathbf{S}$ which equals the product of all the eigenvalues of $\mathbf{S}$. The largest and smallest eigenvalues of $\mathbf{S}$ are also given. These quantities are useful in principal components analysis. The condition number is the square root of the ratio of the largest and smallest eigenvalue. A large condition number indicates multicollinearity in the data. If the condition number is larger than 30, LISREL gives a warning. This might indicate that one or more variables are linear or nearly linear combinations of other variables. If this is intentional, the warning may be ignored.

2.3.9   Direct, Indirect, and Total Effects

Once again, consider the model shown in the following path diagram

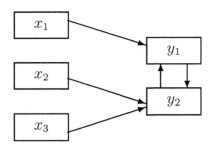

In this model x_1 does not have an effect on y_2. However, since x_1 has an effect on y_1 and y_1 has an effect on y_2, it can still have an indirect effect on y_2. Similarly, although x_2 and x_3 do not affect y_1 directly, they can have an indirect effect on y_1 mediated by y_2. To explain this, assume first that we have only y_1 and y_2 that have an effect on each other as shown in the following figure.

One cycle for y_1 consists of one path to y_2 and a return to y_1. The effect of one cycle on y_1 is $\beta_{21}\beta_{12}$. After two cycles the effect will be $\beta_{21}^2\beta_{12}^2$, after three cycles $\beta_{21}^3\beta_{12}^3$, etc. The total effect on y_1 will be the sum of the infinite geometric series

$$\beta_{21}\beta_{12} + \beta_{21}^2\beta_{12}^2 + \beta_{21}^3\beta_{12}^3 \cdots$$

which is $\beta_{21}\beta_{12}/(1 - \beta_{21}\beta_{12})$ if $\beta_{21}\beta_{12} < 1$.

In general, the total effect of $\mathbf{y}$ on itself is

$$\mathbf{B} + \mathbf{B}^2 + \mathbf{B}^3 + ... = (\mathbf{I} - \mathbf{B})^{-1} - \mathbf{I}\,, \tag{2.81}$$

provided the infinite series converges. Similarly, one finds that the total effect of $\mathbf{x}$ on $\mathbf{y}$ is

$$(\mathbf{I} + \mathbf{B} + \mathbf{B}^2 + \mathbf{B}^3 + \cdots)\mathbf{\Gamma} = (\mathbf{I} - \mathbf{B})^{-1}\mathbf{\Gamma}\,. \tag{2.82}$$

A necessary and sufficient condition for convergence of these series, i.e., for stability of the system, is that all the eigenvalues of $\mathbf{B}$ are within the unit circle. This is the same condition as is employed to test stationarity of a time series which is most often called the unit root test. In general the eigenvalues of $\mathbf{B}$ are complex numbers somewhat difficult to compute. However, a sufficient condition for convergence is that the largest eigenvalue of $\mathbf{BB}'$ is less than one. This is easy to verify. In a two-equation system it can be shown that the largest eigenvalue of $\mathbf{BB}'$ is the largest of β_{12}^2 and β_{21}^2. LISREL prints the largest eigenvalue of $\hat{\mathbf{B}}\hat{\mathbf{B}}'$ under the name of Stability Index.

The general formulas for direct, indirect and total effects are given as follows:

Decomposition of Effects

	$\mathbf{x} \longrightarrow \mathbf{y}$	$\mathbf{y} \longrightarrow \mathbf{y}$
Direct	$\mathbf{\Gamma}$	$\mathbf{B}$
Indirect	$(\mathbf{I} - \mathbf{B})^{-1}\,\mathbf{\Gamma} - \mathbf{\Gamma}$	$(\mathbf{I} - \mathbf{B})^{-1} - \mathbf{I} - \mathbf{B}$
Total	$(\mathbf{I} - \mathbf{B})^{-1}\,\mathbf{\Gamma}$	$(\mathbf{I} - \mathbf{B})^{-1} - \mathbf{I}$

It is seen that the total effect of $\mathbf{x}$ on $\mathbf{y}$ is the same as the regression matrix in the reduced form (2.62).

In LISREL syntax, one can get the estimates of the indirect and total effects and their standard errors by putting EF on the OU line. In SIMPLIS syntax one includes a line

```
Options: EF
```

For the example **Income and Money Supply** these estimates are

```
Total and Indirect Effects

       Total Effects of X on Y

              X1           X2
          --------     --------
   Y1       0.645        3.734
          (0.322)      (0.204)
            2.002       18.293


   Y2       0.396        2.295
          (0.198)      (0.125)
            2.002       18.376
```

Indirect Effects of X on Y

	X1	X2
Y1	-0.858	-4.969
	(0.806)	(3.958)
	-1.065	-1.255
Y2	0.396	2.295
	(0.198)	(0.125)
	2.002	18.376

Total Effects of Y on Y

	Y1	Y2
Y1	-0.571	-0.929
	(0.195)	(0.316)
	-2.930	-2.937
Y2	0.264	-0.571
	(0.119)	(0.195)
	2.206	-2.930

Largest Eigenvalue of B*B' (Stability Index) is   4.689

Indirect Effects of Y on Y

	Y1	Y2
Y1	-0.571	1.236
	(0.195)	(1.404)
	-2.930	0.880
Y2	-0.351	-0.571
	(0.120)	(0.195)
	-2.918	-2.930

Since the Stability Index is larger than 1, it seems that the system is diverging. However, this is just because of the poor maximum likelihood estimate of $\beta_{12} = -2.165$. If the stability index is calculated from the TSLS estimates instead, it becomes 0.377.

2.4   Logistic and Probit Regression

2.4.1   Continuous Predictors

Logistic regression is one type of non-linear regression model which can be used when the response variable is binary. A binary variable is a variable that takes only two values, for example, dead or alive, male or female, present or absent, and an event occurred or not occurred. A general term is success or failure usually coded as 1 and 0.

A binary random variable y has the density function

$$f(y) = \pi^y (1 - \pi)^{1-y}; y = 0, 1 , \tag{2.83}$$

where $\pi = Pr(y = 1)$, so that $Pr(y = 0) = 1 - \pi$. The expected value of y is $E(y) = \pi$.

The linear regression model in Section 2.1 is not suitable for a binary response variable because the linear predictor can take any values and not just the values 0 and 1. The logistic regression model is

$$E(y|\mathbf{x}) = \frac{1}{1 + e^{-(\alpha + \boldsymbol{\gamma}'\mathbf{x})}} . \tag{2.84}$$

Note that this can also be written

$$E(y|\mathbf{x}) = \pi(\mathbf{x}) = \frac{e^{\alpha + \boldsymbol{\gamma}'\mathbf{x}}}{1 + e^{\alpha + \boldsymbol{\gamma}'\mathbf{x}}} , \tag{2.85}$$

where $\pi(\mathbf{x}) = E(y = 1|\mathbf{x}) = Pr(y = 1|\mathbf{x})$ is the conditional probability of $y = 1$ for given $\mathbf{x}$. Equation (2.84) is equivalent to

$$\ln\left(\frac{\pi(\mathbf{x})}{1 - \pi(\mathbf{x})}\right) = \alpha + \boldsymbol{\gamma}'\mathbf{x} . \tag{2.86}$$

Equation (2.84) is a non-linear regression model in the sense of (2.1). The quantity

$$O(\mathbf{x}) = \frac{\pi(\mathbf{x})}{1 - \pi(\mathbf{x})} , \tag{2.87}$$

is called the odds ratio. It is the odds of 1 vs 0 of y. Equation (2.86) states that the logarithm of the odds ratio, $\ln[O(\mathbf{x})]$ is a linear function of the parameters α and $\boldsymbol{\gamma}$. Compare (2.84) with (2.3). Note that there is no parameter corresponding to σ^2 in the logistic regression model.

The interpretation of α is the value of $\ln[O(\mathbf{x})]$ when all x-variables are 0. This is only meaningful if all x-variables can take the value 0. Otherwise, subtract the mean or some other typical value from x. The interpretation of γ_j is as follows. If x_j increases by one unit, this increases the value of $\ln[O(\mathbf{x})]$ by γ_j if $\gamma_j > 0$. Otherwise, if $\gamma_j < 0$ and x_j increases by one unit, then $\ln[O(\mathbf{x})]$ decreases by γ_j.

If $\boldsymbol{\gamma} = 0$ in equations (2.84)-(2.86), the model is called an Intercept-Only model. Then the probability $Pr(y = 1) = \frac{e^{\alpha}}{1+e^{\alpha}}$ is the same regardless of $\mathbf{x}$-values.

Equation (2.84) is just one possible regression model for a binary response variable. In fact, let $u = \alpha + \boldsymbol{\gamma}'\mathbf{x}$, and let $f(u)$ be any continuous density function defined for $-\infty < u < +\infty$. The corresponding distribution function is

$$F(u) = \int_{-\infty}^{u} f(s)ds .$$

Then
$$E(y|\mathbf{x}) = F(\alpha + \boldsymbol{\gamma}'\mathbf{x}) . \tag{2.88}$$

is also a suitable regression model for a binary variable y.

The logistic regression model takes

$$F(u) = \Psi(u) = \frac{1}{1 + e^{-u}} , \tag{2.89}$$

where $\Psi(u)$ is the standard logistic distribution function. The most common alternative is the probit regression model which takes

$$F(u) = \Phi(u) = \int_{-\infty}^{u} \frac{1}{\sqrt{2\pi}} e^{-\frac{1}{2}t^2} dt , \tag{2.90}$$

where $\Phi(u)$ is the distribution function of the standard normal distribution.

The functions $\Phi(u)$ and $\Psi(u)$ are very close. In fact, Lord and Novick (1968,p. 299) noted that

$$|\Phi(u) - \Psi(1.7u)| < 0.01 ,$$

for all values of u. Because of this, it is expected that the results of probit and logistic regression are approximately related by a scale factor.

Let $(y_i, \mathbf{x}_i)$, $i = 1, 2, \ldots, N$ be a random sample of N independent observations of y and $\mathbf{x}$, where each y_i is 0 or 1. The logarithm of the likelihood function is

$$\ln(L) = \sum_{i=1}^{N} \{y_i \ln(\pi(\mathbf{x}_i) + (1 - y_i) \ln[1 - \pi(\mathbf{x}_i)]\} \tag{2.91}$$

This is to be maximized with respect to the parameters $\boldsymbol{\theta} = (\alpha, \boldsymbol{\gamma})$.

2.4.2 Example: Credit Risk

A bank wanted to investigate the credit risk of its loan takers and collected data on the following variables

AGE age in years

ED = Education, a four category scale

EMPLOY = Years of employment at current job

ADDRESS number of years at current address

INCOME monthly income from work

DEPTINC income from other sources than work

CREDDEPT credit dept

OTHRDEPT other dept

DEFAULT = 1 if the loan taker defaulted on a loan, = 0, otherwise

The data is in the file **Credit RISK.lsf**.

To estimate a logistic regression of DEFAULT, use the following **PRELIS** syntax file **Credit Risk1.prl**:

```
!Logistic Regression for Credit Risk
System File 'Credit Risk.LSF'
LR DEFAULT ON AGE - OTHRDEPT
OU AP
```

PRELIS gives the resulting estimated logistic regression equation in a similar way as for ordinal regression:

```
DEFAULT = -1.554 + 0.0344*AGE + 0.0906*ED - 0.258*EMPLOY - 0.105*ADDRESS
Standerr          (0.0174)     (0.123)     (0.0332)        (0.0232)
Z-values           1.981        0.736      -7.787          -4.521
P-values           0.048        0.462       0.000           0.000

         - 0.00857*INCOME + 0.0673*DEBTINC + 0.626*CREDDEBT
           (0.00796)         (0.0305)         (0.113)
           -1.077             2.205            5.545
            0.282             0.027            0.000

         + 0.0627*OTHRDEPT + Error, R² = 0.825
           (0.0775)
            0.809
            0.418
```

but it should be understood that the right-hand side of the equation is the logodds of the left hand variable. It is seen that the most useful variables for predicting the credit risk of a future loan applicant are AGE, EMPLOY, ADDRESS, DEPTINC, and CREDDEPT. One can calculate the probability that an individual will default on a loan from equation (2.88).

PRELIS gives the following measures of fit of the model

```
-2lnL for Full Model                                       551.669
-2lnL for Intercept-Only Model                             804.364
Chi-Square for Testing Intercept-Only Model                252.695
Degrees of Freedom                                               8
```

as well as

```
Pseudo-R²
---------

McFadden                                                     0.314
McFadden Adjusted                                            0.294
Cox and Snell                                               0.303
Nagelkerke                                                  0.444
```

These pseudo-R^2s are defined in the next section.

2.4.3 Pseudo-R^2s

In logit and probit regression one cannot compute an R^2 in the usual way, since the dependent variable is binary. Instead, it is common to report so called pseudo-R^2s. Various such pseudo-R^2s have been suggested in the literature. The **LISREL** output file gives four such R^2s which are described as follows. Let L_1 and L_0 be the estimated likelihoods for the full and the intercept-only models, respectively. As before, q is the number of x-variables and N is the sample size. Then

McFadden

$$R^2 = 1 - \frac{\ln L_1}{\ln L_0} \tag{2.92}$$

McFadden Adjusted

$$R^2 = 1 - \frac{\ln L_1 - q}{\ln L_0} \tag{2.93}$$

Cox and Snell

$$R^2 = 1 - \left(\frac{L_1}{L_0}\right)^{(2/N)} \tag{2.94}$$

Nagelkerke

$$R^2 = \frac{1 - \left(\frac{L_1}{L_0}\right)^{(2/N)}}{1 - L_1^{(2/N)}} \tag{2.95}$$

These pseudo-R^2s are often small even when other measures indicate that the model fits well. For a discussion of these and other pseudo-R^2s, see *e.g.*, Cragg and Uhler (1970), McFadden (1973), Maddala (1983) and Nagelkerke (1991).

2.4.4 Categorical Predictors

If the random variables $y_1, \cdots, y_n$ are independently distributed with the same π, then the sum $\sum_{j=1}^n y_j$, is the number of successes in n trials. This has a binomial distribution with probability function

$$f(y) = \binom{n}{y} \pi^y (1 - \pi)^{n-y}, \quad y = 0, 1, \cdots, n.$$

In a more general situation, if $y_1, \cdots, y_N$ are independent variables corresponding to the number of successes in N different subgroups, where

$$f(y_i) = \binom{n_i}{y_i} \pi_i^{y_i} (1 - \pi_i)^{n_i - y_i}, \quad y_i = 0, 1, \cdots, n_i,$$

in subgroup i. Then the log-likelihood function is

$$\ln L = \sum_{i=1}^N \left[y_i \ln(\frac{\pi_i}{1 - \pi_i}) + n_i \ln(1 - \pi_i) + \ln \binom{n_i}{y_i} \right]. \tag{2.96}$$

The data may be viewed as frequencies for N binomial distributions:

	Subgroup			
	1	2	⋯	N
Successes	y_1	y_2	⋯	y_N
Failures	$n_1 - y_1$	$n_2 - y_2$	⋯	$n_N - y_N$
Totals	n_1	n_2	⋯	n_N

Data of this form is very common in the literature. For example, if residents of a city are asked whether they are satisfied with the public transportation system in that city, the responses for subpopulations are (fictitious data) may be

Subgroup	Response Yes	Response No	Total
Male, low income	59	47	106
Male, high income	43	83	126
Female, low income	70	75	145
Female, high income	47	64	111

Here the subgroups are defined by the rows rather than by the columns.

The proportion of successes is $p_i = y_i/n_i$ in each subgroup. Since $E(y_i) = n_i\pi_i$, $E(p_i) = \pi_i$. The model for the response probabilities as functions of the explanatory variables is

$$\ln(\frac{\pi_i(\mathbf{x})}{1 - \pi_i(\mathbf{x})}) = \alpha + \boldsymbol{\gamma}'\mathbf{x}_i . \tag{2.97}$$

In the probit case this is

$$\Phi^{-1}[(\pi_i(\mathbf{x})] = \alpha + \boldsymbol{\gamma}'\mathbf{x}_i . \tag{2.98}$$

Alternatively, these equations can be written as

$$\pi_i(\mathbf{x}) = \frac{e^{\alpha+\boldsymbol{\gamma}'\mathbf{x}_i}}{1 + e^{\alpha+\boldsymbol{\gamma}'\mathbf{x}_i}} . \tag{2.99}$$

$$\pi_i(\mathbf{x}) = \Phi(\alpha + \boldsymbol{\gamma}'\mathbf{x}_i) . \tag{2.100}$$

The subgroups are often defined as combinations of categorical variables and the data is often presented as a contingency table. This is illustrated in the following example.

2.4.5 Example: Death Penalty Verdicts

Table 2.9 shows the number of death penalty verdicts for cases involving multiple murders in Florida between 1976 and 1987, classified by the race of the defendant and the race of the victim. Does the probability of a death penalty depend on the race of the defendant and/or the race of the victim?

Define the variables as follows

dp $= 1$ for a death penalty verdict; $= 0$ otherwise

vr $= 1$ if the victim is black; $= 0$ if the victim is white

Table 2.9: Observed Death Penalty Verdicts. Reproduced from Radelet and Pierce (1991, Table 4). Published with kind permission of © Michael L. Radelet and Glenn L. Pierce 1991. All Rights Reserved.

		Victim's Race			
		White		Black	
		Defendant's Race		Defendant's Race	
		White	Black	White	Black
Death	Yes	53	11	0	4
Penalty	No	414	37	16	139
Totals		467	48	16	143

$dr = 1$ if the defendant is black; $= 0$ if the defendant is white

To estimate the logistic regression of dp on vr and dr, first create the data file **dpv.lsf**:

dpv.LSF	COUNT	DP	VR	DR
1	53.	1.	0.	0.
2	11.	1.	0.	1.
3	0.	1.	1.	0.
4	4.	1.	1.	1.
5	414	0.	0.	0.
6	37.	0.	0.	1.
7	16.	0.	1.	0.
8	139	0.	1.	1.

Go to **Define Variables** and define COUNT as a weight variable (weight cases) and vr and dr as continuous variables.

Then the logistic regression can be estimated using the following **PRELIS** syntax file **dpv1.prl**:

```
!Logistic Regression of Death Penalty Verdicts
sy=dpv.lsf
lr DP on VR DR
ou
```

The output gives the estimated logistic regression equation as

```
Univariate Logit Regression for DP
      DP =  - 2.059  - 2.404*VR + 0.868*DR
Standerr     (0.146)   (0.601)     (0.367)
z-values    -14.121    -4.003      2.364
P-values      0.000     0.000      0.018
```

which suggests that the race of the victim might have an effect on the probability of a death penalty but not the race of the dependant. The output also gives the following information about the deviances $-2 \ln L$.

```
-2lnL for Full Model                              418.957
-2lnL for Intercept-Only Model                    440.843
Chi-Square for Testing Intercept-Only Model        21.886
Degrees of Freedom                                      2
```

The Intercept-Only model is a model that states that the probability of a cell in Table 2.9 is the same regardless of the race of the defendant and the victim. The chi-square 21.886 with 2 degrees of freedom tells us that the intercept-only model does not fit, which means that both coefficient of VR and DR cannot be zero. This is in agreement with the estimated regression result, which shows that VR is highly significant. The output also gives four pseudo-rsquares:

```
Pseudo-R²
---------
McFadden                                                  0.050
McFadden Adjusted                                         0.041
Cox and Snell                                            0.032
Nagelkerke                                               0.067
```

These small values suggest that the logistic regression model fits much better then the Intercept-Only Model, which again means that at least one of the variables VR or DR has an effect on DP.

To estimate the probit regression of DP on VR and DR just change **lr** to **pr** in the **PRELIS** syntax file, see file **dpv2.prl**. Although the estimated regression coefficients are roughly half of those for the logistic equation, the P-values, the deviances $-2\ln L$ and the pseudo-rsquares are approximately the same. So the conclusion is the same regardless of whether one uses probit or logit regression.

The chi-square and the pseudo-rsquares obtained in **dpv1.out** only tell that the logistic regression model fits better than the Intercept-Only model. They do not tell whether the model fits the data or not.

To take a closer look at the fit of the logistic model, we calculate the probabilities and expected frequencies (according to the model) of each cell in Table 2.9 using equation 3.23. These calculations can be done with **LISREL** as follows. First create a small data matrix **dpv0.dat**:

```
COUNT DP VR DR TOTALS
  53   1  0  0   467
  11   1  0  1    48
   0   1  1  0    16
   4   1  1  1   143
```

This consists of the first five lines of **dpv.dat** with one column added corresponding to the column totals in Table 2.9 The probabilities and expected frequencies can be calculated using the following **PRELIS** syntax file **dpv3.prl**:

```
da ni=5
ra=dpv0.dat lf
new a=-2.059-2.404*VR+0.868*DR
new ea=a
exp ea
new b=1+ea
new pi=ea*b**-1
new E1=pi*TOTALS
new E2=TOTALS-E1
sd DP VR DR
co all
ou ra=dpvext.dat wi=8 nd=3
```

The probabilities and expected frequencies appear in the last three columns of **dpvext.dat**:

```
53.000 467.000  -2.059    0.128   1.128   0.113   52.839 414.161
11.000  48.000  -1.191    0.304   1.304   0.233   11.188  36.812
 0.000  16.000  -4.463    0.012   1.012   0.011    0.182  15.818
 4.000 143.000  -3.595    0.027   1.027   0.027    3.822 139.178
```

Rounding the expected frequencies to the nearest integer and entering them in a table corresponding to Table 2.9 gives the following table of expected frequencies:

Table 2.10: Expected Death Penalty Verdicts. Reproduced from Radelet and Pierce (1991, Table 4). Published with kind permission of © Michael L. Radelet and Glenn L. Pierce 1991. All Rights Reserved.

		Victim's Race			
		White		Black	
		Defendant's Race		Defendant's Race	
		White	Black	White	Black
Death	Yes	53	11	0	4
Penalty	No	414	37	16	139
Totals		467	48	16	143

Comparing Tables 2.9 and 2.10, it is seen that there are no difference. Thus it is clear that the logistic regression model fits very well.

According to these results, the probability of a death penalty is 0.011 if a white kills a black, but 0.233 if a black kills a white. Thus, the latter probability is 21 times larger than the former.

2.4.6   Extensions of Logistic and Probit Regression

The type of logistic and probit regression models considered here can be extended to the polytomous case, where the response variable is ordinal, *i.e.*, has more than two categories. This is considered in Chapters 6 - 7.

Another extension is considered in Chapter 3 where other link functions than the probit and logit can be used.

2.5   Censored Regression

A censored variable has a large fraction of observations at the minimum or maximum. Because the censored variable is not observed over its entire range ordinary estimates of the mean and variance of a censored variable will be biased. Ordinary least squares (OLS) estimates of its regression on a set of explanatory variables will also be biased. These estimates are not consistent, *i.e.*, the bias does not become smaller when the sample size increases. This section explains how maximum likelihood estimates can be obtained using **PRELIS**. The maximum likelihood estimates are consistent, *i.e.*, the bias is small in large samples.

Examples of censored variables are

Econometrics The first example of censored regression appears to be that of Tobin (1958). This is a study of the demand for capital goods such as automobiles or major household appliances. Households are asked whether they purchased such a capital good in the last 12 months. Many households report zero expenditures. However, among those households that made such an expenditure, there will be a wide variation in the amount of money spent.

Greene (2000) p. 205 lists several other examples of censored variables:

1. The number of extramarital affairs (Fair, 1978)
2. The number of hours worked by a woman in the labor force (Quester & Greene, 1982)
3. The number of arrests after release from prison (Witte, 1980)
4. Vacation expenditures (Melenberg & van Soest, 1996)

Biomedicine or Epidemiology Censored variables are common in biomedical, epidemiological, survival and duration studies. For example, in a five year follow-up study, time to death or time to recovery after surgery, medical treatment or diagnosis, are censored variables if, after five years, many patients are still alive or not yet recovered.

Educational Testing If a test is too easy or too difficult there will be a large number of examinees with all items or no items correctly answered.

In econometrics dependent censored variables are often called limited dependent variables and censored regression is sometimes called the tobit model[9].

2.5.1   Censored Normal Variables

A censored variable can be defined as follows. Let $y^\star$ be normally distributed with mean μ and variance σ^2. An observed variable y is censored below if

$$\begin{aligned} y &= c \text{ if } y^\star \le c \\ &= y^\star \text{ otherwise} , \end{aligned}$$

where c is a given constant. This is illustrated in Figure 2.4.

Let ϕ be the density function and Φ the cumulative distribution functions of the standard normal distribution. Then the density function of y can be expressed as

$$f(y) = \left[\Phi \left(\frac{c - \mu}{\sigma} \right) \right]^j \left[\frac{1}{\sqrt{2\pi}\sigma} e^{-\frac{1}{2}(\frac{y-\mu}{\sigma})^2} \right]^{1-j} , \qquad (2.101)$$

where $j = 1$ if $y = c$ and $j = 0$, otherwise. This may be regarded as a mixture of a binary variable and a normally distributed variable.

The mean and variance of y are (see, e.g., Greene, 2000, p. 907)

$$E(y) = \pi c + (1 - \pi)(\mu + \lambda\sigma) , \qquad (2.102)$$

---

[9]This model was first discussed by Tobin (1958). Goldberger (1964, p. 253) gave it this nickname (Tobin's probit) in analogy with the probit model.

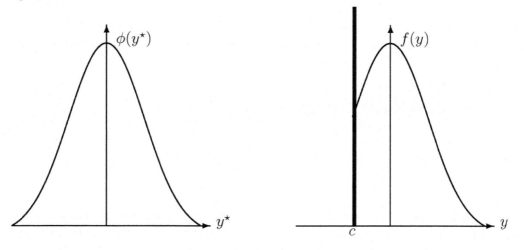

Figure 2.4: Normal Variable $y^\star$ and Censored Variable y

$$Var(y) = (1 - \pi)[(1 - \delta) + (\alpha - \lambda)^2 \pi]\sigma^2 , \tag{2.103}$$

where

$$\alpha = \frac{c - \mu}{\sigma} , \tag{2.104}$$

$$\pi = \Phi(\alpha) , \tag{2.105}$$

$$\lambda = \frac{\phi(\alpha)}{1 - \Phi(\alpha)} , \tag{2.106}$$

$$\delta = \lambda^2 - \lambda\alpha . \tag{2.107}$$

A consequence of (2.102) and (2.103) is that the sample mean and variance of y are not consistent estimates of μ and σ^2. The bias of the mean $E(y) - \mu$ as a function of c is shown in Figure 2.5 for $\mu = 0$ and $\sigma = 1$.

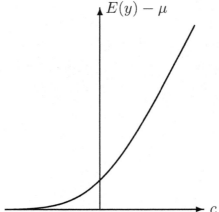

Figure 2.5: Bias $E(y) - \mu$ as a function of c

An observed variable y is censored above if

$$\begin{aligned} y &= c \text{ if } y^\star \geq c \\ &= y^\star \text{ otherwise} , \end{aligned}$$

A variable can be censored both below and above. In all three cases, the mean μ and variance σ^2 can be estimated by maximum likelihood as described in the appendix (see Chapter 13).

2.5.2 Censored Normal Regression

Consider estimating the regression equation

$$y^\star = \alpha + \boldsymbol{\gamma}'\mathbf{x} + z \,, \tag{2.108}$$

where α is an intercept term and $\boldsymbol{\gamma}$ is a vector of regression coefficients on the explanatory variables $\mathbf{x}$. The error term z is assumed to be normally distributed with mean 0 and variance ψ^2. If $y^\star$ is observed as y throughout its entire range, the estimation of (2.108) is straightforward. However, if the observed variable y is censored below or above, then ordinary least squares (OLS) estimates of y on $\mathbf{x}$ are biased. However, α and $\boldsymbol{\gamma}$ can be estimated by maximum likelihood as described in the appendix (see Chapter 13) and these maximum likelihood estimates are unbiased in large samples.

To begin with a hypothetical example with the case of 3 censored variables and 4 explanatory variables is illustrated. Let Y1 Y2 Y3 be the names of the censored variables and let X1 X2 X3 X4 be the names of the regressors.

Censored regression of y_1 on x_1, x_2, x_3, x_4 is obtained by the PRELIS command

```
CR Y1 on X1 X2 X3 X4
```

One can select any subset of y-variables and any subset of x-variables to be included in the equation. Thus, one can obtain the regression for all the censored variables simultaneously. For example, the command

```
CR Y1 Y2 Y3 on X1 X2 X3 X4
```

will estimate three regression equations, namely the regression equation of each y_i on all x_j. Note the word on (or ON) separating the censored variables from the regressors.

One can have several CR commands in the same input file. For example,

```
CR Y1 on X1
CR Y1 on X1 X2
CR Y1 on X1 X2 X3
CR Y1 on X1 X2 X3 X4
```

will introduce one regressor at a time in the order x_1, x_2, x_3, x_4.

General rules:

- All y and x-variables appearing on CR lines must be declared continuous before the first CR command, or else they must have at least 16 different values.

- A censored regression equation can only be estimated from raw data. If there are missing values in the data, the estimation will be based on all cases without missing values on the variables included in the regression equation. Thus, the number of cases used depends on the variables included in the equation. Alternatively, one can impute missing values by multiple imputation before estimating the regression equation.

- If several regression equations are estimated, the regression residuals in each equation are assumed to be uncorrelated.

2.5.3   Example: Affairs

Fair (1978) published an example of censored regression. His study concerns the number of extramarital affairs and its determinants. From a large data set, the results of which were published in the July 1970 issue of *Psychology Today*, Fair extracted 601 observations on men and women who were then currently married for the first time. His data set consisting of 15 variables is available on the Internet at *http://fairmodel.econ.yale.edu/rayfair/workss.htm*. For present purposes the following nine variables are used.

GENDER	0 = female, 1 = male
AGE	in years
YEARS	number of years married[10]
CHILDREN	0 = no, 1 = yes
RELIGIOUS	1 = anti, ..., 5 = very religious
EDUCATION	number of years of schooling, 9 = grade school, 12 = high school, 20 = PhD
OCCUPATION	Hollingshead scale of 1 to 7
HAPPINESS	self rating of quality of marriage, 1 = very unhappy, ..., 5 = very happy
AFFAIRS	number of affairs in the past year, 1, 2, 3, 4–10 coded as 7, monthly, weekly, and daily coded as 12

These variables have been selected from the data set on the Internet. A text (ASCII) file is given in **affairs.dat** and the corresponding LSF file is given in **affairs1.lsf**. Some of these variables are ordinal. For the purpose of data screening they are all declared ordinal in **affairs1.lsf**. A data screening is obtained by running the following simple **LISREL** command file (**affairs0.prl**).

```
!Data Screening of Affairs Data
sy=affairs1.lsf
ou
```

The data screening reveals interesting characteristics of the distribution of the variables.

AGE	Frequency	Percentage		YEARS	Frequency	Percentage
17.5	6	1.0		0.1	11	1.8
22	117	19.5		0.4	10	1.7
27	153	25.5		0.8	31	5.2
32	115	19.1		1.5	88	14.6
37	88	14.6		4	105	17.5
42	56	9.3		7	82	13.6
47	23	3.8		10	70	11.6
52	21	3.5		15	204	33.9
57	22	3.7				

---

[10]It is unclear how this was coded. In addition to integer values, there are values 0.12, 0.42, 0.75, and 1.50 on this variable.

GENDER	Frequency	Percentage
0	315	52.4
1	286	47.6

CHILDREN	Frequency	Percentage
0	171	28.5
1	430	71.5

RELIGIOUS	Frequency	Percentage
1	48	8.0
2	164	27.3
3	129	21.5
4	190	31.6
5	70	11.6

EDUCATION	Frequency	Percentage
9	7	1.2
12	44	7.3
14	154	25.6
16	115	19.1
17	89	14.8
18	112	18.6
20	80	13.3

OCCUPATION	Frequency	Percentage
1	113	18.8
2	13	2.2
3	47	7.8
4	68	11.3
5	204	33.9
6	143	23.8
7	13	2.2

HAPPINESS	Frequency	Percentage
1	16	2.7
2	66	11.0
3	93	15.5
4	194	32.3
5	232	38.6

AFFAIRS	Frequency	Percentage
0	451	75.0
1	34	5.7
2	17	2.8
3	19	3.2
7	42	7.0
12	38	6.3

It is seen that 75% of the respondents report having no extramarital affairs. Thus, the dependent variable is highly censored at 0. It is also seen that 38 persons (6.3%) report having extramarital affairs monthly, weekly, or daily (code 12). To estimate censored regression equations, all variables in the equation must be continuous. This can be specified by including the line

```
co all
```

in the **LISREL** command file. Furthermore, we have changed one value on AFFAIRS from 12.000 to 12.001, since we want to treat AFFAIRS as censored below[11]. Otherwise, **LISREL** will treat AFFAIRS as censored both below and above. This change of a single data value has no other effects whatsoever. The data file with this change is given in **affairs2.lsf** in which all variables are declared continuous.

One can use long names of variables but **LISREL** truncates all variable names to 8 characters. To estimate the censored regression of AFFAIRS using all the other variables as explanatory variables, use the following **LISREL** command file (**affairs1.prl**).

```
Censored Regression of Affairs
sy=affairs2.lsf
cr AFFAIRS on GENDER - HAPPINESS
ou
```

The output reveals the following.

---

[11]This is done primarily because in Fair (1978) and Greene (2000) the AFFAIRS variable has been treated as censored below. Table 2.11 gives the results for the case when AFFAIRS is treated as censored both below and above.

Total Sample Size =   601

Univariate Summary Statistics for Continuous Variables

Variable	Mean	St. Dev.	Skewness	Kurtosis	Minimum	Freq.	Maximum	Freq.
GENDER	0.476	0.500	0.097	-1.997	0.000	315	1.000	286
AGE	32.488	9.289	0.889	0.232	17.500	6	57.000	22
YEARS	8.178	5.571	0.078	-1.571	0.125	11	15.000	204
CHILDREN	0.715	0.452	-0.958	-1.087	0.000	171	1.000	430
RELIGIOUS	3.116	1.168	-0.089	-1.008	1.000	48	5.000	70
EDUCATION	16.166	2.403	-0.250	-0.302	9.000	7	20.000	80
OCCUPATION	4.195	1.819	-0.741	-0.776	1.000	113	7.000	13
HAPPINESS	3.932	1.103	-0.836	-0.204	1.000	16	5.000	232
AFFAIRS	1.456	3.299	2.347	4.257	0.000	451	12.001	1

Variable AFFAIRS is censored below
It has 451 (75.04%) values = 0.000
Estimated Mean and Standard Deviation based on 601 complete cases.
Mean = -6.269 (0.056)
Standard Deviation = 9.420 (0.007)

Estimated Censored Regression based on 601 complete cases.
 AFFAIRS = 7.609 + 0.946*GENDER - 0.193*AGE + 0.533*YEARS + 1.019*CHILDREN

Standerr	(3.936)	(1.071)	(0.0816)	(0.148)	(1.289)
Z-values	1.933	0.883	-2.362	3.610	0.790
P-values	0.053	0.377	0.018	0.000	0.429

        - 1.699*RELIGIOUS + 0.0254*EDUCATION + 0.213*OCCUPATION

	(0.409)	(0.229)	(0.324)
	-4.159	0.111	0.658
	0.000	0.912	0.510

        - 2.273*HAPPINESS + Error, Rŷ = 0.220

	(0.419)
	-5.431
	0.000

Error Variance = 69.239

As judged by the t-values, the effects of GENDER, CHILDREN, EDUCATION, and OCCU-PATION are not statistically significant. The number of extramarital affairs seems to increase with number of years of marriage, and decrease when age, religiousness, and happiness increase.

In this analysis, the AFFAIRS variable is treated as continuous. This means that the values 0, 1, 2, 3, 7, and 12 that this variable takes are assumed to be numbers on an interval scale. One could also treat AFFAIRS as censored both below and above. Another way is to treat AFFAIRS as an ordinal variable with six categories[12] or as a binary variable where 0 is used in one category

---

[12]It may be better to use the exact counts (number of extramarital affairs) and treat this as a Poisson variable, but such data is not available.

and all values larger than 0 are used in the other category. One can then use logistic or probit regression. The results obtained from these analyses are not directly comparable because the variable $y^\star$ is scaled differently for different methods. However, they can be made comparable by multiplying the regression coefficients and their standard error estimates by a suitable scale factor. The probit regressions (columns 4 and 6 in Table 2.11) are scaled such that the error variance is 1 (standard parameterization). Using this as a standard, we must scale the other solutions by $1/\hat{\psi}$. If AFFAIRS is treated as censored below (column 2 in Table 2.11), this scale factor is $1/\sqrt{12.421} = 0.284$, see file **affairs2.out**. If AFFAIRS is treated as censored below and above (column 3 in Table 2.11), this scale factor is $1/\sqrt{123.39} = 0.09002$, see file **affairs3.out**. For the logistic regressions (columns 5 and 7 in Table 2.11) the scale factor is $\sqrt{3}/\pi = 0.55133$, because the variance of the standard logistic distribution is $\pi^2/3$. The t-values are not affected by this scaling. After this scaling the results are shown in Table 2.11.

Table 2.11: Estimated Regression Coefficients with Different Methods

Variable	Censored		Ordinal		Binary	
	Below	&Above	Probit	Logit	Probit	Logit
HAPPINESS	-.273(.049)	-.281(.052)	-.284(.049)	-.278(.047)	-.270(.052)	-.254(.049)
RELIGIOUS	-.207(.049)	-.208(.051)	-.209(.049)	-.200(.048)	-.187(.052)	-.181(.049)
YEARS	.065(.016)	.067(.017)	.067(.016)	.067(.016)	.058(.017)	.056(.016)
AGE	-.019(.009)	-.020(.010)	-.021(.010)	-.023(.009)	-.020(.010)	-.019(.010)

It is seen that all methods give similar results. For most practical purposes these results are the same. The binary methods do not make use of all information in the AFFAIRS variable. Nevertheless, the results are very close to the other methods which make use of all available information.

Which of these methods should be used to estimate the model? The question arises because of the nature of the dependent variable. This is not fully continuous (as if it were an amount of money spent). Neither is it fully ordinal as if the responses were classified as never, sometimes, and often. It is somewhere in between. Censored regression is a method for continuous variables and probit and logit regressions are methods for ordinal variables.

2.5.4   Example: Reading and Spelling Tests

The file **readspel.lsf** contains scores on 11 reading and spelling tests for 90 school children used in a study of the meta-phonological character of the Swedish language. It is of particular interest to predict one of these tests, V23, using the other 10 variables as predictors and to determine which of these variables are the best predictors. However, a data screening of **readspel.lsf** reveals that V23 may be censored both below and above. Hence, we must use censored regression to estimate the prediction equation. The LISREL command file is (**readspel1.prl**)

```
Eleven Reading and Spelling Tests
sy=READSPEL.LSF
co all
cr V23 on V01 - V22 V24 V25
ou
```

The output shows

```
Variable V23 is censored below and above.
It has 3 ( 3.33%) values = 3.000 and 21 (23.33%) values = 16.000

Estimated Mean and Standard Deviation based on 90 complete cases.
Mean = 13.142 (0.286)
Standard Deviation = 4.397 (0.021)

Estimated Censored Regression based on 90 complete cases.
     V23 =   - 1.724 - 0.0731*V01 + 0.122*V02 + 0.384*V07 - 0.129*V08
Standerr      (2.414) (0.0815)      (0.0900)      (0.262)      (0.213)
z-values      -0.714  -0.897         1.359        1.463       -0.605
P-values       0.475   0.370         0.174        0.144        0.545

          + 0.0954*V09 + 0.117*V10 + 0.208*V21 + 0.0679*V22 + 0.0276*V24
            (0.0688)      (0.0838)     (0.177)      (0.149)      (0.163)
             1.388         1.403        1.178        0.456        0.170
             0.165         0.161        0.239        0.648        0.865

          + 0.208*V25 + Error, R² = 0.491
            (0.158)
             1.319
             0.187
```

None of the predictors are statistically significant. However, in terms of the t-values, the most important predictors seem to be V02, V07, V09, and V10. Using only these as predictors (see file **readspel2.prl**) gives the following prediction equation.

```
     V23 = 1.652 + 0.210*V02 + 0.283*V07 + 0.0775*V09 + 0.169*V10
Standerr  (1.843) (0.0663)      (0.151)      (0.0656)      (0.0801)
z-values   0.896   3.170         1.877        1.181         2.114
P-values   0.370   0.002         0.060        0.238         0.035

          + Error, R² = 0.462
```

Here it is seen that the effects V02 and V10 are statistically significant at the 5% level.

2.6   Multivariate Censored Regression

Consider two ordinal variables y_1 and y_2 with underlying continuous variables $y_1^\star$ and $y_2^\star$, respectively. The equations to be estimated are

$$y_1^\star = \alpha_1 + \boldsymbol{\gamma}_1' \mathbf{x} + z_1 , \tag{2.109}$$

$$y_2^\star = \alpha_2 + \boldsymbol{\gamma}_2' \mathbf{x} + z_2 , \tag{2.110}$$

where α_1 and α_2 are intercept terms, $\boldsymbol{\gamma}_1$ and $\boldsymbol{\gamma}_2$ are vectors of regression coefficients, and z_1 and z_2 are error terms. It is assumed that z_1 and z_2 are independent of $\mathbf{x}$ and have a bivariate normal distribution with means zero and covariance matrix

$$\begin{pmatrix} \psi_1^2 & \\ \psi_{21} & \psi_2^2 \end{pmatrix}.$$

y_1 and y_2 are assumed to be censored both above and below, such that

$$\begin{aligned} y_1 &= c_{1L} \text{ if } y_1^\star \leq c_{1L} \\ &= y_1^\star \text{ if } c_{1L} < y_1^\star < c_{1U} \\ &= c_{1U} \text{ if } y_1^\star \geq c_{1U}, \end{aligned}$$

$$\begin{aligned} y_2 &= c_{2L} \text{ if } y_2^\star \leq c_{2L} \\ &= y_2^\star \text{ if } c_{2L} < y_2^\star < c_{2U} \\ &= c_{2U} \text{ if } y_2^\star \geq c_{2U}, \end{aligned}$$

where c_{1L}, c_{1U}, c_{2L}, and c_{2U} are constants.

Let

$$z_1^\star = (y_1^\star - \alpha_1 - \boldsymbol{\gamma}_1'\mathbf{x})/\psi_1 \tag{2.111}$$

$$z_2^\star = (y_2^\star - \alpha_2 - \boldsymbol{\gamma}_2'\mathbf{x})/\psi_2 \tag{2.112}$$

Then $z_1^\star$ and $z_2^\star$ have a standard bivariate normal distribution with correlation $\rho = \psi_{21}/\psi_1\psi_2$. The density of $z_1^\star$ and $z_2^\star$ is the density $f(u, v)$ at the end in the appendix (see Chapter 13) with

$$a = (c_{1L} - \alpha_1 - \boldsymbol{\gamma}_1'\mathbf{x})/\psi_1 \tag{2.113}$$

$$b = (c_{1U} - \alpha_1 - \boldsymbol{\gamma}_1'\mathbf{x})/\psi_1 \tag{2.114}$$

$$c = (c_{2L} - \alpha_2 - \boldsymbol{\gamma}_2'\mathbf{x})/\psi_2 \tag{2.115}$$

$$d = (c_{2U} - \alpha_2 - \boldsymbol{\gamma}_2'\mathbf{x})/\psi_2 \tag{2.116}$$

This density depends on the parameter vector

$$\boldsymbol{\theta} = (\alpha_1, \boldsymbol{\gamma}_1', \psi_1, \alpha_2, \boldsymbol{\gamma}_2', \psi_2, \rho)' \tag{2.117}$$

as well as on $\mathbf{x}$.

Let $(y_{1i}, y_{2i}, \mathbf{x}_i)$, $i = 1, 2, \ldots, N$ be a random sample of size N. The likelihood L_i of this observation can be computed using the results in the appendix (see Chapter 13). The maximum likelihood estimate of $\boldsymbol{\theta}$ is the value of $\boldsymbol{\theta}$ that maximizes $\ln L = \sum_{i=1}^{N} \ln L_i$. For practical purposes the following two-step procedure is more convenient and computationally much faster.

Step 1: Estimate α_1, $\boldsymbol{\gamma}_1$, and ψ_1 by univariate censored regression of y_1 on $\mathbf{x}$ and α_2, $\boldsymbol{\gamma}_2$, and ψ_2 by univariate censored regression of y_2 on $\mathbf{x}$ as described in Section 2.5.4.

Step 2: For given estimates obtained in Step 1, estimate ρ by maximizing $\ln L$. This gives $\hat{\rho}$. Compute the estimate of ψ_{21} as $\hat{\psi}_{21} = \hat{\psi}_1\hat{\psi}_2\hat{\rho}$.

The multivariate case is handled in a similar way. In Step 1, we estimate each univariate censored regression, including the standard deviations of the error terms ψ_i. In Step 2 we estimate the correlation of the error terms for each pair of variables. The covariance matrix of the error terms can then be computed from these estimates.

The general syntax for multivariate censored regression is

```
CR Y-varlist ON X-varlist
```

where `Y-varlist` is a list of censored variables and `X-varlist` is a list of explanatory variables. The meaning of this is that *each* variable in `Y-varlist` is regressed on *all* variables in `X-varlist`. If `Y-varlist` and/or `X-varlist` contains a set of consecutive variables one can use - to denote a range of variables. For example, suppose that the data set consists of the six variables Y1 Y2 Y3 X1 X2 X3. Then

```
CR Y1-Y3 ON X1-X3
```

will perform multivariate censored regression of Y1, Y2, and Y3 on X1, X2, and X3. All variables in the `Y-varlist` and `X-varlist` must be continuous variables.

LISREL can distinguish between univariate and multivariate censored regression. For example,

```
CR Y1-Y3 ON X1-X3
```

will do multivariate censored regression, whereas

```
CR Y1 on X1-X3
CR Y2 on X1-X3
CR Y3 on X1-X3
```

will do three univariate censored regressions. The difference is that in the second case LISREL will estimate the error variances but not the error covariances, whereas in the first case LISREL will estimate the whole error covariance matrix called *Residual Covariance Matrix*.

Each of `Y-varlist` and `X-varlist` may contain a subset of variables from the data set. Further explanation is needed if one has `MA=CM` on the `OU` line in a LISREL syntax file, for this refers to the covariance matrix of all the variables in the data set. If `MA=CM` on the `OU` line and the LISREL syntax file contains `CA`, `CB`, `CE`, or `CR` lines, then LISREL will treat all variables in the data set as censored variables. The reason for this is that LISREL cannot estimate the covariance between a censored variable and an uncensored variable. However, this does no harm because LISREL can handle an uncensored variable, *i.e.*, a variable that has only one observation at the minimum and maximum, as a special case of a censored variable. For the same reason, if `Y` is an uncensored variable,

```
CR Y ON X-Varlist
```

will give the same result as

```
RG Y ON X-Varlist
```

namely OLS regression.

The `CR` command can also be used without the `ON X-varlist`, thus

```
CR Y-varlist-
```

This gives estimates of the mean vector and covariance matrix of the variables in `Y-varlist`.

2.6.1   Example: Testscores

For this example I use a subset of the variables in the file **readspel.lsf** described in Section 2.5.4. This file contains scores on 11 reading and spelling tests for 90 school children used in a study of the meta-phonological character of the Swedish language. The variables used here are V01, V02, V07, V21, and V23. They are available in the file **TESTSCORE.LSF**. The sample size is 90.

I estimate the bivariate censored regression of V21 and V23 on V01, V02, and V07 using the following LISREL syntax file (**TESTSCORE1.PRL**).

```
SY=TESTSCORE.LSF
CR V21 V23 ON V01 V02 V07
OU
```

The output gives the following information about the distribution of the variables.

```
Univariate Summary Statistics for Continuous Variables
```

Variable	Mean	St. Dev.	Skewness	Kurtosis	Minimum	Freq.	Maximum	Freq.
V01	21.789	7.856	-1.117	0.333	0.000	2	30.000	6
V02	14.622	7.048	-0.173	-0.568	0.000	2	28.000	5
V07	11.489	3.069	-0.175	2.673	0.000	1	20.000	2
V21	13.352	2.998	-1.956	3.424	2.330	1	15.330	41
V23	12.013	3.402	-1.141	0.735	2.436	3	15.436	21

If we regard a variable to be censored if it has more than one observation at the minimum or maximum, then we see that all five variables are censored. But V21 and V23 are more severely censored because they have many observations at the maximum.

The results of the censored regression is given in the output file as

```
Variable V21 is censored above
It has 41 (45.56%) values = 15.330
```

```
Estimated Mean and Standard Deviation based on 90 complete cases.
Mean = 15.022 (0.324)
Standard Deviation = 4.759 (0.023)
```

```
Estimated Censored Regression based on 90 complete cases.
      V21 = 6.387 + 0.125*V01 + 0.243*V02 + 0.198*V07 + Error, R² = 0.420
Standerr  (1.745) (0.0838)     (0.0933)      (0.198)
z-values   3.660   1.497        2.598         1.002
P-values   0.000   0.134        0.009         0.316
```

```
Variable V23 is censored below and above.
```

```
It has 3 ( 3.33%) values = 2.436 and 21 (23.33%) values = 15.436
```

Estimated Mean and Standard Deviation based on 90 complete cases.
Mean = 12.013 (0.285)
Standard Deviation = 3.402 (0.022)

Estimated Censored Regression based on 90 complete cases.
     V23 = 5.981 + 0.0371*V01 + 0.206*V02 + 0.192*V07 + Error, R^2 = 0.361
Standerr (1.142) (0.0597)   (0.0642)   (0.125)
z-values   5.238   0.621   3.211   1.539
P-values   0.000   0.534   0.001   0.124

Residual Correlation Matrix

                 V21   V23

                --------   --------
      V21   1.000
      V23   0.215   1.000

Residual Covariance Matrix

                 V21   V23

                --------   --------
      V21   13.124
      V23   2.664   11.692

Only V02 has a significant effect. V01 and V07 are not significant for either V21 or V23. The residual correlation is 0.215. As will be seen in the next example the correlation between V21 and V23 is 0.437 if we treat both as censored.

Now consider a different problem. Suppose we want to do a factor analysis of the five variables V01, V02, V07, V21, and V23 and test the hypothesis that there is one common factor. In the first step I treat all the variables as multivariate censored and estimate the mean vector and covariance matrix of all the variables. This is done using the following **LISREL** syntax file (**TESTSCORE2.PRL**).

```
SY=TESTSCORE.LSF
CE ALL
OU MA=CM CM=TESTSCORE.CM
```

If the sample size had been larger, I would also estimate the asymptotic covariance matrix. The line **CE ALL** declares all variables as censored. This is OK even if some variables are not censored. As explained in Section 2.5.4, an uncensored variable is simply treated as a special case of censoring when there is only one observation at the maximum or minimum. The results are:

Correlation Matrix

	V01	V02	V07	V21	V23
V01	1.000				
V02	0.761	1.000			
V07	0.641	0.595	1.000		
V21	0.485	0.653	0.607	1.000	
V23	0.492	0.576	0.478	0.553	1.000

Covariance Matrix

	V01	V02	V07	V21	V23
V01	61.719				
V02	41.829	49.676			
V07	15.284	12.962	9.421		
V21	13.038	11.997	3.618	8.988	
V23	13.854	14.288	5.051	5.751	11.573

Means

V01	V02	V07	V21	V23
21.789	14.622	11.489	13.352	12.013

Standard Deviations

V01	V02	V07	V21	V23
7.856	7.048	3.069	2.998	3.402

These results are summarized in the bottom half of Table 2.12, where the numbers in parentheses are the number of observations at the minimum and maximum. For comparison, the corresponding statistics for the case when all variables are treated as uncensored, are given in the top part of the table.

Table 2.12: Basic Statistics Estimated in Two Different Ways

Treating all Variables as Uncensored

Var	Min	Max	Mean	St Dev	Correlations				
V01	0(2)	30(6)	21.789	7.836	1.000				
V02	0(2)	28(5)	14.622	7.048	0.755	1.000			
V07	0(1)	20(2)	11.489	3.069	0.634	0.599	1.000		
V21	4(1)	17(41)	15.022	2.998	0.554	0.568	0.393	1.000	
V23	3(3)	16(21)	12.578	3.402	0.518	0.596	0.484	0.564	1.000

Treating all Variables as Censored

Var	Min	Max	Mean	St Dev	Correlations				
V01	0(2)	30(6)	21.789	7.856	1.000				
V02	0(2)	28(5)	14.622	7.048	0.761	1.000			
V07	0(1)	20(2)	11.489	3.069	0.641	0.595	1.000		
V21	4(1)	17(41)	13.352	2.998	0.485	0.653	0.607	1.000	
V23	3(3)	16(21)	12.013	3.402	0.492	0.576	0.478	0.553	1.000

Chapter 3

Generalized Linear Models

The normal linear model from Section 2.1 belongs to an extended family of linear models often called generalized linear models (GLM). Elements of this family can be traced back to Sir Ronald A. Fisher, but it was Nelder and Wedderburn (1972) who provided an unifying approach. Books covering this topic include McCullagh and Nelder (1983) and Dobson and Barnett (2008).

In general one could say that GLM is made up of three parts: the response variable, the systematic part corresponding to the linear part consisting of the explanatory variable, and the link function which transforms the mean of the response variable to linear form.

Let $\mathbf{y}' = (y_1, y_2, \ldots, y_N)$ be a vector of independent random variables. Recall from Section 2.1 that the normal linear model is

$$\mathbf{y} \sim N(\boldsymbol{\mu}, \sigma^2 \mathbf{I}) , \tag{3.1}$$

where

$$\boldsymbol{\mu} = E(\mathbf{y}) = \mathbf{X}_\star \boldsymbol{\gamma} \tag{3.2}$$

In the following we omit the subscript on $\mathbf{X}$ and simply write $\mathbf{X}$ instead of $\mathbf{X}_\star$, where $\mathbf{X}$ may contain a column of ones.

Generalized linear models generalizes the linear model to the situation where $\mathbf{y}$ may not be normally distributed but such that

$$\mathbf{g}(\boldsymbol{\mu}) = \mathbf{g}[E(\mathbf{y})] = \mathbf{X}\boldsymbol{\gamma} , \tag{3.3}$$

where $\mathbf{g}' = (g_1, g_2, \ldots, g_N)$ is a vector of monotonic functions $g_i(u)$ called **link functions**.

The normal linear model corresponds to the special case where $g_i(u) = u$.

3.1 Components of Generalized Linear Models

The components of a GLM model are:

Random Component The elements of $\mathbf{y}$ are independently distributed according to *some specific* distribution $f(\mathbf{y}, \boldsymbol{\theta})$ with $E(\mathbf{y}) = \boldsymbol{\mu}(\boldsymbol{\theta})$, where $\boldsymbol{\theta}$ is a parameter vector defining the distribution.

Systematic Component The predictor $\boldsymbol{\eta}$ is linear in the covariates $x_1, x_2, \ldots, x_q$:

$$\boldsymbol{\eta} = \mathbf{X}\boldsymbol{\gamma} .$$

© Springer International Publishing Switzerland 2016
K.G. Jöreskog et al., *Multivariate Analysis with LISREL*, Springer Series in Statistics,
DOI 10.1007/978-3-319-33153-9_3

Link Function The link function relates the elements of $\boldsymbol{\eta}$ to the elements of $\boldsymbol{\mu}$:

$$\eta_i = g_i(\mu_i) \, , i = 1, 2, \ldots, N$$

where $g_i(u_i)$ is a monotonic differentiable function.

3.2 Exponential Family Distributions

The individual distribution functions $f(y_i)$ are assumed to be of the form

$$f(y; \theta, \psi) = exp[yb(\theta, \psi)] + c(\theta, \psi) + d(y, \psi)] \, , \tag{3.4}$$

where b and c are functions of θ and ψ but not of y, and d is a function of y and ψ but not of θ. Equation (3.4) is often called the canonical form of the exponential family distribution. Special cases of this exponential family distribution include the Normal, Poisson, Bernoulli, Binomial, and Gamma distributions as shown in the Table 3.1.

Table 3.1: Some Special Cases of the Exponential Family Distributions

	Normal	Poisson	Bernoulli	Binomial	Gamma
Parameter	μ	θ	π	π	θ
Range	$(-\infty, \infty)$	$0(1)\infty$	$0, 1$	$0(1)n$	$(0, \infty)$
ψ	σ^2	-	-	n	α
$b(\theta, \psi)$	μ/σ^2	$\log \theta$	$\log[\pi/(1-\pi)]$	$\log[\pi/(1-\pi)]$	$-\theta$
$c(\theta, \psi)$	$c_1^\star(\theta, \psi)$	$-\theta$	$\log(1-\pi)$	$n\log(1-\pi)$	$c_2^\star(\theta, \psi)$
$d(y, \psi)$	$-y^2/(2\sigma^2)$	$logy!$	$-\infty$	$\log \binom{n}{y}$	$(\alpha-1)\log y$
$E(y)$	μ	θ	π	$n\pi$	α/θ
$Var(y)$	σ^2	θ	$\pi(1-\pi)$	$n\pi(1-\pi)$	α/θ^2

$$c_1^\star(\theta, \psi) = -(\mu/2\sigma^2) - (1/2)\log(2\pi\sigma^2)$$

$$c_2^\star(\theta, \psi) = \log \theta^\alpha - \log \Gamma(\alpha)$$

Some distributions in this family contain an extra parameter ψ, often called a nuisance parameter, which is not included in the linear form and which is usually not of any interest. For example, such nuisance parameters are σ^2 for the Normal distribution and α for the Gamma distribution.

3.2.1 Distributions and Link Functions

Let $\mu = E(y)$. The most common distributions and link functions used in GLM are given in Table 3.2.

Other link functions for the Bernoulli and Binomial distributions are the Probit and Complementary Log-Log (CLL) as shown in Table 3.3.

Table 3.2: Most Common Distributions and Link Functions in GLM

Distributions	Link Function	Inverse Link Function
Normal	$\eta = \mu$	$\mu = \eta$
Poisson	$\eta = \log\theta$	$\theta = e^\eta$
Bernoulli	$\eta = \log[\pi/(1-\pi)]$	$\pi = \frac{e^\eta}{1+e^\eta}$
Binomial	$\eta = \log[\pi/(1-\pi)]$	$\pi = \frac{e^\eta}{1+e^\eta}$
Gamma	$\eta = \alpha^{-1}$	$\alpha = \eta^{-1}$
Inverse Gaussian	$\eta = \alpha^{-2}$	$\alpha = 1/\sqrt{\eta}$

Table 3.3: Other Link Functions for the Bernoulli and Binomial Distributions

Probit	$\eta = \Phi^{-1}(\pi)$	$\pi = \Phi(\eta)$
CLL	$\eta = \log[-\log(1-\pi)]$	$\pi = 1 - e^{-e^\eta}$

The distributions and link functions available in **LISREL** are shown in Table 3.4. The last three columns are link functions used for analysis of contingency tables with ordinal or nominal categories. These will be explained in Section 3.6.

Since the Normal distribution with the Identity link function can be handled by linear regression as described in Section 2.1 and the Bernoulli and Binomial distributions with the Logit and Probit link functions can be handled by logistic and probit regression as described in Section 2.4, we will focus on other cases in this chapter.

3.3   The Poisson-Log Model

The Poisson distribution is the distribution of the number of occurrences y of a rare event. A classical example is Bortkiewics' (1898) study of the number of Prussian soldiers kicked by horses in 14 cavalry corps over a 20-year period. Another classical example is the number of cargo

Table 3.4: Distributions and Link Functions Available in **LISREL**

Distributions	CLL	Identity	Log	Logit	Probit	OCLL	Oligit	Oprobit
Bernoulli	x		x	x	x			
Binomial				x				
Gamma			x					
Inverse Gaussian			x					
Multinomial	x		x	x	x	x	x	x
Negative Binomial			x					
Normal		x						
Poisson			x					

ships damaged by waves which is analyzed by McCullagh and Nelder (1983). We will consider two examples, one with continuous covariates and one with both continuous and categorical covariates. The first example is the number of deaths of British doctors in different age groups. This example is considered in Section 3.3.1. The other example is the number of awards earned by students in high school which is considered in Section 3.3.2.

The density of the Poisson distribution with parameter $\theta > 0$ is

$$f(y; \theta) = \frac{\theta^y e^{-\theta}}{y!} \quad y = 0, 1, 2, \ldots \tag{3.5}$$

The mean $E(y)$ and variance $E(y - \theta)^2$ are both equal to θ.

Let $y_1, y_2, \ldots, y_N$ be independent random Poisson variables with the same parameter θ. The log-likelihood of y_i is

$$\ln(L_i) = y_i \ln(\theta) - \theta - \ln(y_i!) . \tag{3.6}$$

The last term is a function of observations only, so it is omitted in what follows. The total log-likelihood is

$$\ln(L) = \sum \ln(L_i) = \left(\sum y_i\right) \ln(\theta) - N\theta , \tag{3.7}$$

which is maximum when $\theta = \bar{y}$ so that the maximum likelihood estimate of θ is $\hat{\theta} = \bar{y}$.

Let $y_1, y_2, \ldots, y_N$ be independent random Poisson variables with parameters $\theta_1, \theta_2, \ldots, \theta_N$, where y_i denotes the number of events observed from n_i exposures for the ith covariate pattern $\mathbf{x}_i$. Then the Poisson regression model is

$$E(y_i) = \mu_i = n_i \theta_i = n_i e^{\mathbf{x}_i' \boldsymbol{\gamma}} , \tag{3.8}$$

where $\mathbf{x}_i'$ is the ith row of $\mathbf{X}$. The link function is the function of μ_i which makes it linear in $\mathbf{x}_i$. Thus,

$$g(\mu_i) = \ln(\mu_i) = \ln(n_i) + \mathbf{x}_i' \boldsymbol{\gamma} . \tag{3.9}$$

The term $\ln(n_i)$ is called an **offset** variable. This is a known constant which can be viewed as a covariate whose coefficient is fixed equal to 1.

The log-likelihood of y_i is

$$\ln(L_i) = y_i \ln(\mu_i) - \mu_i - \ln(y_i!) . \tag{3.10}$$

The whole log-likelihood of all the observations is

$$\ln(L) = \sum [y_i \ln(\mu_i) - \mu_i] , \tag{3.11}$$

This is to be maximized with respect to $\boldsymbol{\gamma}$, considering that μ_i is a non-linear function of $\boldsymbol{\gamma}$ in (3.8).

Let $\hat{\boldsymbol{\gamma}}$ be the value of $\boldsymbol{\gamma}$ which maximizes $\ln(L)$ and let $\hat{\mu}_i = n_i \hat{\theta}_i = n_i e^{\mathbf{x}_i' \hat{\boldsymbol{\gamma}}}$. Then the log-likelihood of the fitted model is

$$\ln(L_0) = \sum [y_i \ln(\hat{\mu}_i) - \hat{\mu}_i] , \tag{3.12}$$

If the μ_i are unconstrained then $\ln(L)$ is maximized when $\theta_i = y_i / n_i$ or $\mu_i = n_i y_i$, so that

$$\ln(L_1) = \sum [y_i \ln(n_i y_i) - n_i y_i] . \tag{3.13}$$

To test the model one can use the likelihood ratio chi-square χ^2_{LR}:

$$\chi^2_{\mathrm{LR}} = 2[\ln(L_1) - \ln(L_0)] = 2 \sum [y_i \ln(y_i / \hat{\mu}_i) - (y_i - \hat{\mu}_i)] . \tag{3.14}$$

This is used with a χ^2 distribution with $N - q - 1$ degrees of freedom, where q is the number of genuine covariates, *i.e.*, excluding the intercept[1].

The χ^2_{LR} in (3.14) is also called a Deviance chi-square and denoted D. Let d_i be the deviance residual[2]

$$d_i = sign(y_i - \hat{\mu}_i)\sqrt{2[y_i \ln(y_i/\hat{\mu}_i) - (y_i - \hat{\mu}_i)]} \, . \tag{3.15}$$

Then

$$D = \sum d_i^2 \, . \tag{3.16}$$

If $y_i \approx \hat{\mu}_i$, then a Taylor series gives

$$y_i \ln(y_i/\hat{\mu}_i) = (y_i - \hat{\mu}_i) + \frac{1}{2}\frac{(y_i - \hat{\mu}_i)^2}{\mu_i} + \cdots$$

If $y_i \approx \hat{\mu}_i$ for all i, then D is approximately equal to

$$2\sum[(y_i - \hat{\mu}_i) + \frac{1}{2}\frac{(y_i - \hat{\mu}_i)^2}{\mu_i} - (y_i - \hat{\mu}_i)] = \sum \frac{(y_i - \hat{\mu}_i)^2}{\mu_i}$$

which is the Pearson chi-square χ^2_{P}:

$$\chi^2_{\text{P}} = \sum r_i^2 \, , \tag{3.17}$$

where

$$r_i = (y_i - \hat{\mu}_i)/\sqrt{\hat{\mu}_i} \, . \tag{3.18}$$

So if the model fits the data well χ^2_{LR} and χ^2_{P} are close.

The residuals d_i and r_i can be used to detect cases that contribute maximally to χ^2_{LR} or χ^2_{P}. However, neither d_i nor r_i have variance 1, nor are they normally distributed, so their sum of squares are not distributed as χ^2 for small N. Various sorts of standardizations of d_i and r_i have been proposed in the literature and some programs, including LISREL, report standardized as well as unstandardized residuals.

3.3.1   Example: Smoking and Coronary Heart Disease

In 1951, all British doctors were sent a brief questionnaire about whether they smoked tobacco. Ten years later the number of deaths from coronary heart disease for smokers and non-smokers was recorded. The original data come from a study by Sir Richard Doll. A table of the number of deaths and the number of person-years of observation in different age groups was published by Breslow and Day (1987, p. 112). A similar table was given by Dobson and Barnett (2008, p. 168). This is reproduced here in Table 3.5.

- Is the death rate higher for smokers than non-smokers?

- If so, by how much?

- Is the differential effect related to age?

A plot of the death rate against age is given in Figure 3.1.

---

[1]Strictly speaking $q + 1$ is the rank of the matrix $\mathbf{X}$ which includes a column of ones

[2]If $y_i \approx \hat{\mu}_i$, it can be shown that $2[y_i \ln(y_i/\hat{\mu}_i) - (y_i - \hat{\mu}_i)]$ is non-negative

Table 3.5: Deaths from Coronary Heart Disease among British Doctors. Originally published in Dobson and Barnett (2008, Table 9.1). Published with kind permission of © Taylor and Francis Group, LLC 2008. All Rights Reserved

	Smokers		Non-smokers	
Age Group	Deaths	Person-years	Deaths	Person-years
35–44	32	52407	2	18790
45–54	104	43248	12	10673
55–64	206	28612	28	5710
65–74	186	12663	28	2585
75–84	102	5317	31	1462

The plot indicates that there is some evidence of non-linearity and of interaction between age and smoking. This would suggest a model like this

$$\ln(\text{DEATHS}) = \ln(\text{PYEARS}) + \alpha + \gamma_1 \text{SMOKE} + \gamma_2 \text{AGECAT} + \gamma_3 \text{AGECAT}^2 + \gamma_4 \text{SMOKE} \times \text{AGECAT}$$

In Section 1.7 in Chapter 1 we created the data file **corheart.lsf** necessary to do the analysis. This file looks like this:

corheart.LSF

	SMOKE	AGECAT	DEATHS	PYEARS	DTHRATE	AGESQ	SMKAGE	LNPYEARS
1	1.000000	1.000000	32.000000	52407.000000	0.000611	1.000000	1.000000	10.866796
2	1.000000	2.000000	104.000000	43248.000000	0.002405	4.000000	2.000000	10.674706
3	1.000000	3.000000	206.000000	28612.000000	0.007200	9.000000	3.000000	10.261581
4	1.000000	4.000000	186.000000	12633.000000	0.014723	16.000000	4.000000	9.444068
5	1.000000	5.000000	102.000000	5317.000000	0.019184	25.000000	5.000000	8.578665
6	0.000000	1.000000	2.000000	18790.000000	0.000106	1.000000	0.000000	9.841080
7	0.000000	2.000000	12.000000	10673.000000	0.001124	4.000000	0.000000	9.275473
8	0.000000	3.000000	28.000000	5710.000000	0.004904	9.000000	0.000000	8.649974
9	0.000000	4.000000	28.000000	2585.000000	0.010832	16.000000	0.000000	7.857481
10	0.000000	5.000000	31.000000	1462.000000	0.021204	25.000000	0.000000	7.287560

We are now ready to perform the estimation of the GLM model. This can be done by using the **Graphical User Interface**. For details, see the file **General Linear Modeling.pdf**.

Select **Generalized Linear Model** in the **Statistics** menu. This will open the **Title and Options** dialog box. Fill in the title and tick the **Residual file** box as shown here:

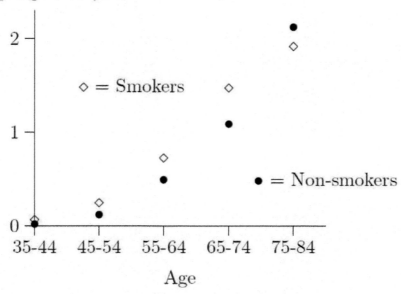

Figure 3.1: Deaths from Coronary Heart Disease per Person-Years

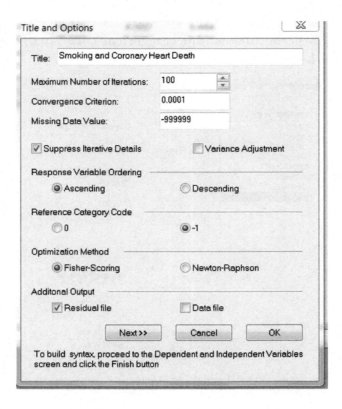

Then click **Next** to open the **Distributions and Links** dialog box. In this window choose **Poisson** as **Distribution type** and **LOG** as **Link** (since LOG is the only link function available for the Poisson distribution it is automatically selected once the Poisson distribution has been chosen):

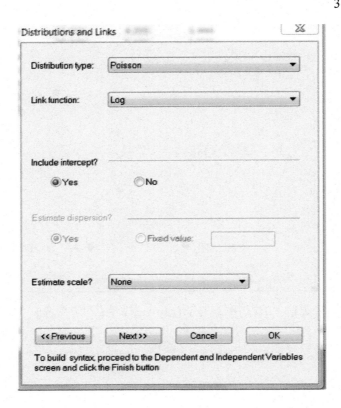

Click on **Next** to open the **Dependent and Independent Variables** dialog box. In this window select DEATHS as dependent variables and SMOKE, AGECAT, AGESQ, and SMKAGE as independent variables (in this example all independent variables are treated as continuous variables; see next example for a case with both continuous and categorical covariates), and choose LNPYEARS as **Offset variable** as shown here

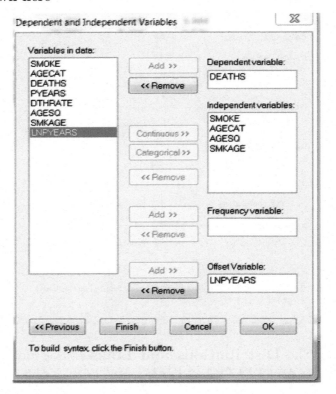

Click on **Finish**. Then a PRELIS syntax file **corheart.prl** is automatically generated:

```
GlimOptions Converge=0.0001 MaxIter=100 MissingCode=-999999
           Response=Ascending RefCatCode=-1 IterDetails=No
           Method=Fisher Output=Residuals;
Title=Smoking and Coronary Heart Death;
SY=corheart.LSF;
Distribution=POI;
Link=LOG;
Intcept=Yes;
Scale=None;
DepVar=DEATHS;
CoVars=SMOKE AGECAT AGESQ SMKAGE;
Offset=LNPYEARS;
```

This is a special type of **PRELIS** syntax valid only for GLM models. Note that each line must end with ;. The Glimoptions are explained in detail in the file **General Linear Modeling.pdf**. These options are the default options except for `Output=Residuals;`, so one can replace the lines

```
GlimOptions Converge=0.0001 MaxIter=100 MissingCode=-999999
           Response=Ascending RefCatCode=-1 IterDetails=No
           Method=Fisher Output=Residuals;
```

by the line

```
GlimOptions Output=Residuals;
```

The default options are sufficient for most cases so there is seldom any reason to change them.

Click on the Run Prelis button **P** to estimate the model. The output file gives the following results:

Estimated Regression Weights

Parameter	Estimate	Standard Error	z Value	P Value
intcept	-10.7939	0.4495	-24.0118	0.0000
SMOKE	1.4406	0.3717	3.8752	0.0001
AGECAT	2.3779	0.2078	11.4434	0.0000
AGESQ	-0.1979	0.0274	-7.2334	0.0000
SMKAGE	-0.3072	0.0970	-3.1689	0.0015

These results agree with those reported in Dobson and Barnett (2008, Table 9.2). They show that all regression weights are statistically significant.

The output also gives the following goodness-of-fit statistics

Goodness of Fit Statistics

Statistic	Value	DF	Ratio
Likelihood Ratio Chi-square	1.6510	5	0.3302
Pearson Chi-square	1.5665	5	0.3133
-2 Log Likelihood Function	-5455.2711		
Akaike Information Criterion	-5445.2711		
Schwarz Criterion	-5443.7581		

indicating that the fit of the model is good. Detailed information about the fit of the model is obtained in the file **corheart_RES.lsf**:

LISREL for Windows - [corheart_RES.LSF]

File Edit Data Transformation Statistics Graphs Multilevel SurveyGLIM View Window Help

	Observed	Fitted	Raw	Pearson	Deviance	Likelihd	SPearson	SDevianc
1	32.000	29.555	2.445	0.450	0.444	0.708	0.712	0.703
2	104.000	106.822	-2.822	-0.273	-0.274	-0.367	-0.367	-0.368
3	206.000	208.361	-2.361	-0.164	-0.164	-0.247	-0.247	-0.248
4	186.000	182.588	3.412	0.252	0.252	0.339	0.339	0.338
5	102.000	102.673	-0.673	-0.066	-0.066	-0.152	-0.152	-0.152
6	2.000	3.412	-1.412	-0.764	-0.829	-0.962	-0.907	-0.983
7	12.000	11.540	0.460	0.135	0.135	0.173	0.174	0.173
8	28.000	24.749	3.251	0.653	0.640	0.815	0.826	0.809
9	28.000	30.235	-2.235	-0.407	-0.412	-0.504	-0.500	-0.507
10	31.000	31.064	-0.064	-0.011	-0.011	-0.022	-0.022	-0.022
11	0.000	0.000	0.000	1.567	1.651	2.711	2.626	2.737

which gives observed and fitted values and various residuals. The numbers in the last line and the last five columns are residual sums of squares. All together these results indicate that the model fits the data well and all parameters are statistically significant.

The effects of the covariates are usually measured by rate ratios $e^{\hat{\gamma}_j}$. For **SMOKE** this is 4.22 which means that risk of coronary heart death is about 4 times larger for smokers than for non-smokers after the effect of age is taken into account. However, the effect is attenuated as age increases.

3.3.2 Example: Awards

The previous example included a variable **AGECAT** which was treated as continuous. It could also be treated as categorical, in which case one can estimate the relative effect of each age category. The following example includes a real categorical variable and illustrates how this is handled.

Two hundred students in a high school reported the number of awards earned and their math score as well as the type of program (vocational, general, or academic) they were enrolled in. Can the number of awards a student will receive be predicted by his/her math score and the type of program? The data file is **awards.lsf**, where the variables are

awards number of awards earned

program = 1 for vocational; = 2 for general; = 3 for academic

math = math score

A data screening reveals the following information about the three variables:

```
Total Sample Size(N) =     200
```

Univariate Distributions for Ordinal Variables

awards	Frequency	Percentage
0	124	62.0
1	49	24.5
2	13	6.5
3	9	4.5
4	2	1.0
5	2	1.0
6	1	0.5

program	Frequency	Percentage
1	45	22.5
2	105	52.5
3	50	25.0

Univariate Summary Statistics for Continuous Variables

Variable	Mean	St. Dev.	Skewness	Kurtosis	Minimum	Freq.	Maximum	Freq.
math	52.645	9.368	0.287	-0.649	33.000	1	75.000	2

It seems reasonable to assume that **awards** has a Poisson distribution.

To analyze the data by Poisson regression, one can use the Graphical User Interface (GUI) as illustrated in the previous example. However, one can also do as follows.

Open the file **corheart.prl** and save it as **awards1.prl** Then change this to be

```
GlimOptions;
Title=Awards by Students in High School;
SY=awards.LSF;
Distribution=POI;
Link=LOG;
Intcept=Yes;
Scale=None;
DepVar=awards;
CoVars=program$ math;
```

Note the $ sign after **program** on the line

```
CoVars=program$ math;
```

This indicates that the variable **program** is a categorical variable.

With categorical covariates, **LISREL** automatically generates $C - 1$ dummy variables, where C is the number of categories. One of the categories is specified as a reference category. By default **LISREL** chooses the last category as reference category but one can choose any category as a reference category. For example, to choose the first category as a reference category, include a line

`Refcats=1;`

The plural term `Refcats` is used because there can be more than one categorical covariate. For example, if there are two categorical variables and one wants to use the second category as reference category for the first variable and the first category as reference category for the second variable, one should write

`Refcats=2 1;`

To explain this mathematically, consider three students who choose program 1, 2, and 3, respectively (for example students 2, 12, and 1 in the data). Omitting the continuous variable `math`, the rows of the matrix $\mathbf{X}$ for these students are

$$\mathbf{X} = \begin{pmatrix} 1 & 1 & 0 & 0 \\ 1 & 0 & 1 & 0 \\ 1 & 0 & 0 & 1 \end{pmatrix}, \tag{3.19}$$

corresponding to the equations

$$\mu_1 = \alpha + \gamma_1$$
$$\mu_2 = \alpha + \gamma_2$$
$$\mu_3 = \alpha + \gamma_3$$

These are three linear equations in four unknowns. Obviously, there is no unique solution. In principle, one can choose three independent linear combinations of α, γ_1, γ_2, and γ_3 and estimate these. However, most programs use one of two alternative parameterizations.

One Zero: One of the γs is set to zero. If category 3 is chosen as reference category, $\gamma_3 = 0$ and

$$\alpha = \mu_3 \qquad\qquad\qquad \mu_1 = \alpha + \gamma_1$$
$$\gamma_1 = \mu_1 - \mu_3 \qquad\qquad\qquad \mu_2 = \alpha + \gamma_2$$
$$\gamma_2 = \mu_2 - \mu_3 \qquad\qquad\qquad \mu_3 = \alpha$$

This means that

$$\mathbf{X} = \begin{pmatrix} 1 & 1 & 0 \\ 1 & 0 & 1 \\ 1 & 0 & 0 \end{pmatrix},$$

If category 1 is chosen as reference category, $\gamma_1 = 0$ and

$$\alpha = \mu_1 \qquad\qquad\qquad \mu_1 = \alpha$$
$$\gamma_2 = \mu_2 - \mu_1 \qquad\qquad\qquad \mu_2 = \alpha + \gamma_2$$
$$\gamma_3 = \mu_3 - \mu_1 \qquad\qquad\qquad \mu_3 = \alpha + \gamma_3$$

This means that

$$\mathbf{X} = \begin{pmatrix} 1 & 0 & 0 \\ 1 & 1 & 0 \\ 1 & 0 & 1 \end{pmatrix},$$

These two choices of reference categories will result in different α's and therefore different estimates of α depending on which category is chosen as a reference category.

Sum Zero: The sum of the γ's are set to zero. If category 3 is chosen as reference category, $\gamma_3 = -\gamma_1 - \gamma_2$ and

$$\alpha = (1/3)(\mu_1 + \mu_2 + \mu_3) = \bar{\mu} \qquad\qquad \mu_1 = \alpha + \gamma_1$$

$$\gamma_1 = \mu_1 - \bar{\mu} \qquad\qquad\qquad \mu_2 = \alpha + \gamma_2$$

$$\gamma_2 = \mu_2 - \bar{\mu} \qquad\qquad\qquad \mu_3 = \alpha - \gamma_1 - \gamma_2$$

This means that

$$\mathbf{X} = \begin{pmatrix} 1 & 1 & 0 \\ 1 & 0 & 1 \\ 1 & -1 & -1 \end{pmatrix},$$

If category 1 is chosen as reference category, $\gamma_1 = -\gamma_2 - \gamma_3$ and

$$\alpha = (1/3)(\mu_1 + \mu_2 + \mu_3) = \bar{\mu} \qquad\qquad \mu_1 = \alpha - \gamma_2 - \gamma_3$$

$$\gamma_2 = \mu_2 - \bar{\mu} \qquad\qquad\qquad \mu_2 = \alpha + \gamma_2$$

$$\gamma_3 = \mu_3 - \bar{\mu} \qquad\qquad\qquad \mu_3 = \alpha + \gamma_3$$

This means that

$$\mathbf{X} = \begin{pmatrix} 1 & -1 & -1 \\ 1 & 1 & 0 \\ 1 & 0 & 1 \end{pmatrix},$$

These two choices of reference categories will result in the same α and therefore the same estimate of α regardless of which category is chosen as a reference category. Also, the estimate of γ_2 will be the same regardless of which category is chosen as reference category.

By default LISREL uses **Sum Zero** corresponding to the GlimOption `RefCatCode=-1`. To use **One Zero**, one must put `RefCatCode=0` on the `GlimOptions` line.

These alternatives are illustrated in the four files **awards1.prl**, **awards2.prl**, **awards3.prl**, and **awards4.prl**, where

- **awards1.prl** uses **Sum Zero** (default) with reference category 3 (default)

- **awards2.prl** uses **Sum Zero** (default) with reference category 1

- **awards3.prl** uses **One Zero** with reference category 3 (default)

- **awards4.prl** uses **One Zero** with reference category 1

Note that the matrix $\mathbf{X}$ is automatically generated from the data file **awards.lsf** and the specification of the `RefCatCode` on the `GlimOptions` line and the `Refcats` value. To see which dummy variable coding was used, put `Output=All` on the `GlimOptions` line. This will produce an **lsf** file called **awards3_RAW.lsf** containing the entire $\mathbf{X}$ matrix.

The estimated parameters for each of these alternative parameterizations are given here

- **awards1.prl** uses **Sum Zero** (default) with reference category 3 (default)

Parameter	Estimate	Standard Error	z Value	P Value
intcept	-4.7626	0.6126	-7.7745	0.0000
program1	-0.4846	0.2457	-1.9721	0.0486
program2	0.5993	0.1722	3.4798	0.0005
math	0.0702	0.0106	6.6187	0.0000

- **awards2.prl** uses **Sum Zero** (default) with reference category 1

Parameter	Estimate	Standard Error	z Value	P Value
intcept	-4.7626	0.6126	-7.7745	0.0000
program2	0.5993	0.1722	3.4798	0.0005
program3	-0.1147	0.2274	-0.5045	0.6139
math	0.0702	0.0106	6.6187	0.0000

- **awards3.prl** uses **One Zero** with reference category 3 (default)

Parameter	Estimate	Standard Error	z Value	P Value
intcept	-4.8773	0.6282	-7.7642	0.0000
program1	-0.3698	0.4411	-0.8384	0.4018
program2	0.7140	0.3200	2.2313	0.0257
math	0.0702	0.0106	6.6186	0.0000

- **awards4.prl** uses **One Zero** with reference category 1

Parameter	Estimate	Standard Error	z Value	P Value
intcept	-5.2471	0.6584	-7.9690	0.0000
program2	1.0839	0.3582	3.0255	0.0025
program3	0.3698	0.4411	0.8385	0.4018
math	0.0702	0.0106	6.6187	0.0000

The conclusion from all this is that `math` and `program2` are significant predictors of the number of awards a student will receive. Recall that the data is fictitious.

3.4   The Binomial-Logit/Probit Model

In this section we reconsider the model for binomial counts as given in Section 2.4.4 in Chapter 2.

Let $y_1, \cdots, y_N$ be independent variables corresponding to the number of successes out of n_i trials in N different subgroups, where

$$f(y_i) = \binom{n_i}{y_i} \pi_i^{y_i} (1 - \pi_i)^{n_i - y_i} , \quad y_i = 0, 1, \cdots, n_i ,$$

in subgroup i. Then the log-likelihood function is

$$\ln L = \sum_{i=1}^{N} \left[y_i \ln(\frac{\pi_i}{1 - \pi_i}) + n_i \ln(1 - \pi_i) + \ln \left(\begin{array}{c} n_i \\ y_i \end{array} \right) \right] . \tag{3.20}$$

The data may be viewed as frequencies for N binomial distributions:

	Subgroup			
	1	2	$\cdots$	N
Successes	y_1	y_2	$\cdots$	y_N
Failures	$n_1 - y_1$	$n_2 - y_2$	$\cdots$	$n_N - y_N$
Totals	n_1	n_2	$\cdots$	n_N

Data of this form is very common in the literature. For example, if residents of a city are asked whether they are satisfied with the public transportation system in that city, the responses for subpopulations are (fictitious data) may be

Subgroup	Response		Total
	Yes	No	
Male, low income	59	47	106
Male, high income	43	83	126
Female, low income	70	75	145
Female, high income	47	64	111

Here the subgroups are defined by the rows rather than by the columns.

The proportion of successes is $p_i = y_i/n_i$ in each subgroup. Since $E(y_i) = n_i\pi_i$, $E(p_i) = \pi_i$. The model for the response probabilities as functions of the explanatory variables is

$$\ln(\frac{\pi_i(\mathbf{x})}{1 - \pi_i(\mathbf{x})}) = \alpha + \boldsymbol{\gamma}'\mathbf{x}_i . \tag{3.21}$$

In the probit case this is

$$\Phi^{-1}[(\pi_i(\mathbf{x})] = \alpha + \boldsymbol{\gamma}'\mathbf{x}_i . \tag{3.22}$$

Alternatively, these equations can be written as

$$\pi_i(\mathbf{x}) = \frac{e^{\alpha+\boldsymbol{\gamma}'\mathbf{x}_i}}{1 + e^{\alpha+\boldsymbol{\gamma}'\mathbf{x}_i}} . \tag{3.23}$$

$$\pi_i(\mathbf{x}) = \Phi(\alpha + \boldsymbol{\gamma}'\mathbf{x}_i) . \tag{3.24}$$

The subgroups are often defined as combinations of categorical variables and the data is often presented as a contingency table. This is illustrated in the following example.

3.4.1 Example: Death Penalty Verdicts Revisited

Recall Table 2.9 showing the number of death penalty verdicts for cases involving multiple murders in Florida between 1976 and 1987, classified by the race of the defendant and the race of the victim. Does the probability of a death penalty depend on the race of the defendant and/or the race of the victim?

In Section 2.4.4 in Chapter 2 we analyzed these data by a regression model where the dependent variable is binary (death penalty yes/no) and the regression function was logistic or probit. Here we regard the dependent variable as a binomial variable (number of death penalty verdicts out of the total number of verdicts).

For the purpose here we use the data in the following form

```
DPV VR DR TOTALS
 53  0  0   467
 11  0  1    48
  1  1  0    16
  4  1  1   143
```

interpreting the DPV as binomial observations of the totals given in TOTALS. Note that this data file has only 4 rows; the data file used in Section 2.4.4 had 8 rows. Importing this to **lsf** gives the file **dpvbin.lsf**.

Before we estimate any model for the data we can calculate the observed probabilities by dividing the number of death penalty verdicts (DPV) by the total number of verdicts (TOTALS). These are shown in Table 3.6:

Table 3.6: Observed Probabilities of Death Penalty Verdicts

		Victim's Race			
		White		Black	
		Defendant's Race		Defendant's Race	
		White	Black	White	Black
Death	Yes	0.128	0.229	0.000	0.028
Penalty	No	0.872	0.771	1.000	0.972

We can now estimate the binomial logit model by the file **dpvbin1.prl**:

```
GlimOptions
Title=Binomial Logit Model for Death Penalty Verdicts;
SY=dpvbin.LSF;
Distribution=BIN;
Link=LOGIT;
Intcept=Yes;
Scale=None;
DepVar=DPV;
CoVars=VR DR;
NTrials=TOTALS;
```

The estimated regression results appear in the output as

```
                 Estimated Regression Weights

                          Standard
Parameter      Estimate    Error     z Value    P Value
---------      --------   --------   -------    -------
intcept        -2.0595     0.1458   -14.1208    0.0000
VR             -2.4044     0.6006    -4.0033    0.0001
DR              0.8678     0.3671     2.3641     0.0181
```

The results are almost the same as given in Section 2.4.4 in Chapter 2.

The value of $-2 \ln L$ is given in the output as

```
-2 Log Likelihood Function                    418.9565
```

To use the Probit link function instead of the logistic, just replace the line

```
Link=LOGIT;
```

by

```
Link=PROBIT;
```

see file **dpvbin2.prl**. This gives the following estimates

```
                 Estimated Regression Weights

                          Standard
Parameter      Estimate    Error     z Value    P Value
---------      --------   --------   -------    -------
intcept        -1.2102     0.0762   -15.8802    0.0000
VR             -1.2004     0.2836    -4.2332    0.0000
DR              0.4827     0.2092     2.3074     0.0210
```

As pointed out in Section 2.4.4 the parameter estimates differ by a scale factor. The value of $-2 \ln L$ for the Probit model is given in the output as

```
-2 Log Likelihood Function                    418.8440
```

which is slightly smaller than the previous value.

One can also use the complimentary log-log link function (CLL), see file **dpvbin3.prl**.

These three alternative ways of estimating the model constitute three different models for the same data. The models have the same number of parameters, so the one with the smallest value of $-2 \ln L$ is the one that fits the data best. This is the Probit model in this case.

We can calculate the probabilities for each cell as follows. For the logit link function, let

$$a_i = -2.060 - 2.404 \text{VR} + 0.868 \text{DR}$$

Then

$$\hat{\pi}_i = \frac{e^{a_i}}{1 + e^{a_i}}$$

This gives the expected probabilities shown together with the observed probabilities in Table 3.7:

According to these results for the logit link function, the probability of a death penalty verdict is 0.011 if a white kills a black and 0.233 if a black kills a white. Thus, the latter probability is 21 times larger than the former.

Table 3.7: Observed and Expected Probabilities for Death Penalty Verdicts

		Victim's Race			
		White		Black	
		Defendant's Race		Defendant's Race	
		White	Black	White	Black
Death	Obs	0.128	0.229	0.000	0.028
Penalty	Exp	0.113	0.233	0.011	0.027

3.5 Log-linear Models

Consider a categorical variable with C categories. The multinomial distribution is a generalization of the binomial distribution to more than two categories. The multinomial distribution is the distribution of y_1 occurrences of category 1, y_2 occurrences of category 2, ..., y_C occurrences of category C, where the sum of all occurrences is fixed at N. Let π_i, be the probability of category i, with $\pi_i > 0$ and $\pi_1 + \pi_2 + \cdots + \pi_C = 1$. The frequency distribution of the multinomial distribution is

$$f(\mathbf{y}; \boldsymbol{\pi}, N) = \frac{N!}{y_1! y_2! \cdots y_C!} \pi_1^{y_1} \pi_2^{y_2} \cdots \pi_C^{y_C}, \quad y_1 + y_2 + \cdots y_C = N , \qquad (3.25)$$

where $\mathbf{y}' = (y_1, y_2, \ldots, y_C)$ and $\boldsymbol{\pi}' = (\pi_1, \pi_2, \ldots, \pi_C)$. The multinomial distribution is not a member of exponential family. However, it can be shown to be the same as the joint distribution of C independent Poisson variables conditional upon their sum N. It is therefore used with a log link function. If this kind of model is used to analyze a contingency table, the model is often called a *log-linear model*. Log-linear models were made popular in books by Bishop et al. (1975), Goodman (1978), Haberman (1974), and Agresti (1990).

The log-linear model extends the analysis of two-way contingency tables to the analysis of relationships between categorical variables in multi-way contingency tables. The variables are treated equal,*i.e.*, there is no distinction between dependent and independent variables.

Here we consider an example of a two-way contingency table.

Let y_{ij} be the frequency of the cell in row i and column j of a two-way contingency table with I rows and J columns with $N = \sum \sum y_{ij}$ fixed. If the y_{ij} are independent Poisson variables with $E(y_{ij}) = \mu_{ij}$ then their sum is a Posson variable with $E(N) = \sum \sum \mu_{ij} = \mu$. Hence, the conditional distribution of the y_{ij}'s, given their sum is the multinomial distribution

$$f(\mathbf{y}|N) = N! \prod \prod \pi_{ij}^{y_{ij}} / y_{ij}! , \qquad (3.26)$$

where $\pi_{ij} = \mu_{ij}/\mu$ is the probability of cell (ij). Since

$$E(y_{ij}) = \mu_{ij} = N\pi_{ij} , \qquad (3.27)$$

we can use the log link function for the Poisson distribution:

$$\ln(\mu_{ij}) = \ln(N) + \ln(\pi_{ij}) , \qquad (3.28)$$

which is the same as (3.9) except that the offset is the same for all ij. A particular model is one where rows and columns are independent so that

$$\ln(\mu_{ij}) = \ln(N) + \ln(\pi_i) + \ln(\pi_j) . \tag{3.29}$$

In analogy with Analysis of Variance, this model is often written as

$$\ln(\mu_{ij}) = \mu + \alpha_i + \beta_j , \tag{3.30}$$

where the α's are row effects and the β's are column effects. If the independence between row and column effects does not hold, the model can be extended to include interaction effects $(\alpha\beta)_{ij}$:

$$\ln(\mu_{ij}) = \mu + \alpha_i + \beta_j + (\alpha\beta)_{ij} , \tag{3.31}$$

3.5.1 Example: Malignant Melanoma

Roberts et al. (1981) published data on a form of skin cancer called malignant melanoma classified by type of tumor

1. Hutchinson's melanotic freckle(HMF)

2. Superficial spreading melanoma(SSM)

3. Nodular(NOD)

4. Indeterminate(IND)

and tumor site

1. Head or Neck(HNK)

2. Trunk(TNK)

3. Extremities(EXT)

The observed frequences are given in Table 3.8.

Table 3.8: Melanoma: Frequences classified by tumor type and site. Originally published in Dobson and Barnett (2008, Table 9.4). Published with kind permission of © Taylor and Francis Group, LLC 2008. All Rights Reserved

Tumor Type	Tumor Site			Total
	HNK	TNK	EXT	
HMF	22	2	10	34
SSM	16	54	115	185
NOD	19	33	73	125
IND	11	17	28	56
Total	68	106	226	400

Is there any association between tumor type and site?

To analyze the data create the following data file **melanoma.lsf**:

Count	Type	Site
22,000	1,000	1,000
2,000	1,000	2,000
10,000	1,000	3,000
16,000	2,000	1,000
54,000	2,000	2,000
115,000	2,000	3,000
19,000	3,000	1,000
33,000	3,000	2,000
73,000	3,000	3,000
11,000	4,000	1,000
17,000	4,000	2,000
28,000	4,000	3,000

To test the hypothesis of independence of tumor type and site is equivalent to test the hypothesis $(\alpha\beta)_{ij} = 0$, for all i and j. To do this, one can use the following input file **melanoma1.prl**

```
GlimOptions RefCatCode=0 Output=All;
Title=melanoma;
SY=melanoma.LSF;
Distribution=POI;
Link=LOG;
Intercept=Yes;
Scale=None;
DepVar=Count;
CoVars=Type$ Site$;
```

Here we use the **One Zero** parameterization implied by `RefCatCode=0`. To get the first category as reference category for both `Type` and `Site` add the line

```
RefCats=1 1;
```

This means that $\alpha_1 = 0$ and $\beta_1 = 0$ in (3.30), *i.e.*, the effects of tumor type is measured relative to HMF and the effects of tumor site is measured relative to HNK. The estimated parameters are α_2, α_3, α_4 and β_2, β_3. In the output these are labeled `Type2 Type3 Type4` and `Site2 Site3`, respectively. The `Output=All` on the `GlimOptions` line is used to request both a residual file as well as a data file where the dummy variables are included. From the output we find that

Likelihood Ratio Chi-square	51.7950	6
Pearson Chi-square	65.8129	6

and

Estimated Regression Weights

Parameter	Estimate	Standard Error	z Value	P Value
intcept	1.7544	0.2040	8.6000	0.0000
Type2	1.6940	0.1866	9.0787	0.0000
Type3	1.3020	0.1934	6.7313	0.0000
Type4	0.4990	0.2174	2.2951	0.0217
Site2	0.4439	0.1554	2.8573	0.0043
Site3	1.2010	0.1383	8.6834	0.0000

These results agree with those reported in Dobson and Barnett (2008, Table 9.10). Both chi-squares are too large for a χ^2 with 6 degrees of freedom. An inspection of the residual file **melanoma1_res.lsf**

	Observed	Fitted	Raw	Pearson	Deviance	Likelihd	SPearson	SDevianc
1	22,000	5,780	16,220	6,747	5,135	6,387	7,742	5,893
2	2,000	9,010	-7,010	-2,335	-2,828	-3,264	-2,848	-3,449
3	10,000	19,210	-9,210	-2,101	-2,316	-3,470	-3,331	-3,671
4	16,000	31,450	-15,450	-2,755	-3,045	-4,324	-4,125	-4,559
5	54,000	49,025	4,975	0,711	0,699	1,123	1,130	1,112
6	115,000	104,525	10,475	1,025	1,008	2,111	2,119	2,085
7	19,000	21,250	-2,250	-0,488	-0,497	-0,653	-0,646	-0,658
8	33,000	33,125	-0,125	-0,022	-0,022	-0,031	-0,031	-0,031
9	73,000	70,625	2,375	0,283	0,281	0,516	0,517	0,514
10	11,000	9,520	1,480	0,480	0,468	0,558	0,568	0,554
11	17,000	14,840	2,160	0,561	0,548	0,695	0,705	0,689
12	28,000	31,640	-3,640	-0,647	-0,660	-1,066	-1,058	-1,079
13	0,000	0,000	0,000	65,813	51,795	90,524	104,542	89,109

reveals, in particular, that the Hutchinson's melanomic freckle is more common in the head and neck than can be expected if type and site were independent (observed value 22 and expected value 5.78).

To investigate if there are interactions between the categories of type and site, one can use the dummy variables directly. The dummy variables used to obtain the results in **melanoma1.out** appear in the file **melanoma1_RAW.lsf**. To set the width to 8 and the number of decimals to 0, go to **Edit** and **Format**.

	Count	Type2	Type3	Type4	Site2	Site3
1	22.	0.	0.	0.	0.	0.
2	2	0.	0.	0.	1.	0.
3	10.	0.	0.	0.	0.	1.
4	16.	1.	0.	0.	0.	0.
5	54.	1.	0.	0.	1.	0.
6	115.	1.	0.	0.	0.	1.
7	19.	0.	1.	0.	0.	0.
8	33.	0.	1.	0.	1.	0.
9	73.	0.	1.	0.	0.	1.
10	11.	0.	0.	1.	0.	0.
11	17.	0.	0.	1.	1.	0.
12	28.	0.	0.	1.	0.	1.

To add interaction terms in the model, multiply the dummy variable for each type and site and extend the **lsf** file to include these six variables. This can be done with the **Compute** dialog box in the **Transformation** menu. The resulting file is **melanoma_RAW.lsf**:

	Count	Type2	Type3	Type4	Site2	Site3	Typ2Sit2	Typ2Sit3	Typ3Sit2	Typ3Sit3	Typ4Sit2	Typ4Sit3
1	22.	0.	0.	0.	0.	0.	0.	0.	0.	0.	0.	0.
2	2	0.	0.	0.	1.	0.	0.	0.	0.	0.	0.	0.
3	10	0.	0.	0.	0.	1.	0.	0.	0.	0.	0.	0.
4	16.	1.	0.	0.	0.	0.	0.	0.	0.	0.	0.	0.
5	54.	1.	0.	0.	1.	0.	1.	0.	0.	0.	0.	0.
6	115	1.	0.	0.	0.	1.	0.	1.	0.	0.	0.	0.
7	19	0.	1.	0.	0.	0.	0.	0.	0.	0.	0.	0.
8	33	0.	1.	0.	1.	0.	0.	0.	1.	0.	0.	0.
9	73	0.	1.	0.	0.	1.	0.	0.	0.	1.	0.	0.
10	11	0.	0.	1.	0.	0.	0.	0.	0.	0.	0.	0.
11	17	0.	0.	1.	1.	0.	0.	0.	0.	0.	1.	0.
12	28	0.	0.	1.	0.	1.	0.	0.	0.	0.	0.	1.

To estimate a model with all interaction terms included, use **melanoma2.prl**:

```
GlimOptions;
Title=Melanoma with Interaction Terms;
SY=melanoma_RAW.LSF;
Distribution=POI;
Link=LOG;
Intercept=Yes;
Scale=None;
DepVar=Count;
CoVars=Type2 Type3 Type4 Site2 Site3
       Typ2Sit2 Typ2Sit3 Typ3Sit2 Typ3Sit3 Typ4Sit2 Typ4Sit3;
```

This is a saturated model with chi-square and degrees of freedom equal to 0. The estimates are given as

Estimated Regression Weights

| | | Standard | | |
Parameter	Estimate	Error	z Value	P Value
intcept	3.0910	0.2132	14.4983	0.0000
Type2	-0.3185	0.3286	-0.9692	0.3324
Type3	-0.1466	0.3132	-0.4681	0.6397
Type4	-0.6931	0.3693	-1.8771	0.0605
Site2	-2.3979	0.7378	-3.2499	0.0012
Site3	-0.7885	0.3814	-2.0674	0.0387
Typ2Sit2	3.6143	0.7908	4.5702	0.0000
Typ2Sit3	2.7608	0.4655	5.9314	0.0000
Typ3Sit2	2.9500	0.7920	3.7245	0.0002
Typ3Sit3	2.1345	0.4602	4.6381	0.0000
Typ4Sit2	2.8332	0.8332	3.4006	0.0007
Typ4Sit3	1.7228	0.5216	3.3028	0.0010

It is seen that all interaction terms are significant but now the main effects of the categories for Type are no longer significant at the 5% level.

The results for three models are summarized in Table 3.9.

3.6 Nominal Logistic Regression

Consider a dependent nominal variable with C unordered categories. Let y_j be the number of observations in category j and π_j the corresponding probability, $j = 1, 2, \ldots, C$. The frequency distribution is given by 3.25. Since $\pi_1 + \pi_2 + \cdots + \pi_C = 1$ one of the categories is chosen as a reference category. Suppose this is the first category. Then the model for nominal logistic regression is

$$\ln(\pi_j/\pi_1) = \alpha_j + \gamma_{j1}x_1 + \gamma_{j2}x_2 + \cdots + \gamma_{jq}x_q; , \tag{3.32}$$

$$j = 2, 3, \ldots, C ,$$

Table 3.9: Parameter Estimates (Standard Errors) for Three Models for the Melanoma Data

Term	Model Interaction	Additive	Intercept Only
Intercept	3.091(0.213)	1.754(0.204)	3.507(0.050)
SSM	-0.318(0.329)	1.694(0.187)	
NOD	-0.147(0.313)	1.302(0.193)	
IND	-0.693(0.369)	0.499(0.217)	
TNK	-2.398(0.739)	0.444(0.155)	
EXT	-0.788(0.381)	1.201(0.138)	
SSM & TNK	3.614(0.792)		
SSM & EXT	2.761(0.465)		
NOD & TNK	2.950(0.793)		
NOD & EXT	2.134(0.460)		
IND & TNK	2.833(0.834)		
IND & EXT	1.723(0.522)		
$-2\ln L$	-2300.449	-2248.654	-2005.246
χ_P^2	0	65.813	348.740
χ_D^2	0	51.795	295.203

where α_j is an intercept and γ_{j1}, γ_{j2},...,γ_{jq} is a set of parameters (regression weights) to be estimated. Note that there is one "regression equation" for each category j except the first. These "regression lines" can differ in intercepts and slopes as illustrated in Figure 3.2 for the case of 5 categories and a single continuous covariate x.

Figure 3.2: Nominal Logistic Regression: Four Regression Lines on a Log-odds Scale

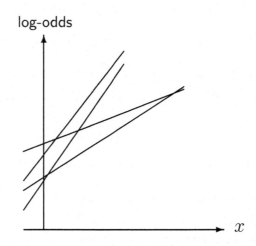

Solving (3.32) for the π's gives

$$\pi_1 = \frac{1}{1 + \sum e^{\alpha_j + \gamma_{j1}x_1 + \gamma_{j2}x_2 + \cdots + \gamma_{jq}x_q}} \ , \tag{3.33}$$

$$\pi_j = \frac{e^{\alpha_j + \gamma_{j1}x_1 + \gamma_{j2}x_2 + \cdots + \gamma_{jq}x_q}}{1 + \sum e^{\alpha_j + \gamma_{j1}x_1 + \gamma_{j2}x_2 + \cdots + \gamma_{jq}x_q}} \ i = 2, 3, \ldots, C \ , \tag{3.34}$$

where the sum is from $j = 2$ to C.

Let $(y_{ij}, x_{i1}, x_{i2}, \ldots, x_{iq})$, for $i =, 1, 2, \ldots, N$ be N independent observations of $y, x_1, x_2, \ldots, x_q$, where y_{ij} is the number of observations in category j of observation i. The log-likelihood of the ith observation is

$$\ln(L_i) = c + \sum_j y_{ij} \ln(\pi_j) , \tag{3.35}$$

where c is a constant which is a function of observations only. Omitting this constant, the log likelihood for all observations is

$$\ln(L) = \sum_i \ln(L_i) = \sum_i \sum_j y_{ij} \ln(\pi_j) = \sum_j y_j \ln(\pi_j) , \tag{3.36}$$

where $y_j = \sum_i y_{ij}$ is the total number of observations in category j. The log-likelihood (3.36) is to be maximized with respect to the parameter vector $\boldsymbol{\theta}$ consisting of $(\alpha_j, \gamma_{j1}, \gamma_{j2}, \ldots, \gamma_{jq})$, $j = 2, 3, \ldots, C$. Altogether there are $(C-1)(q+1)$ parameters.

Let $\hat{\boldsymbol{\theta}}$ be the value of $\boldsymbol{\theta}$ that maximizes $\ln(L)$ and let $\hat{\pi}_j$ be the value π_j evaluated at $\hat{\boldsymbol{\theta}}$. The values of $\hat{\pi}_j$ are computed from equations (3.33) and (3.34), for $j = 1, 2, \ldots, C$. Then the maximum value of $\ln(L)$ is

$$\ln(L_0) = c + \sum y_j \ln(\hat{\pi}_j) . \tag{3.37}$$

For a saturated model all π_j are free parameters subject only to the condition that $\pi_j > 0$ and $\sum \pi_j = 1$. Then the estimate of π_j is $p_j = y_j/N$ and if all $y_j > 0$, the value of $\ln(L)$ for a saturated model is

$$\ln(L_1) = \sum y_j \ln(p_j) . \tag{3.38}$$

To test the model, one can use the likelihood ratio statistic χ^2_{LR}, also called the deviance statistic D,

$$\chi^2_{\mathrm{LR}} = 2 \sum [(y_j \ln(p_j/\hat{\pi}_j)] , \tag{3.39}$$

or the Pearson chi-square statistic χ^2_{P} (also called the goodness-of-fit chi-square)[3]:

$$\chi^2_{\mathrm{P}} = \sum [(y_j - N\pi_j)^2/(N\pi_j)] = N \sum [(p_j - \pi_j)^2/\pi_j] . \tag{3.40}$$

These chi-square statistics should be used with $N - (C-1)(q+1)$ degrees of freedom.

Typically the data for multinomial logistic regression comes in two different forms:

individual data with one row $(y_{ij}, x_{i1}, x_{i2}, \ldots, x_{iq})$, for each individual $i = 1, 2, \ldots, N$.

contingency table for $q + 1$ variables where one of them is a dependent nominal variable.

We illustrate each of these cases in the following two examples.

3.6.1 Example: Program Choices 1

Students in a high school reported their score on a writing test, their socioeconomic status (ses) as well as the type of program (vocational, general, or academic) they were enrolled in. This example investigates if the program choices can be predicted by the writing score and ses. The data file is **programchoice.lsf**, where the variables are

---

[3]Some books use the notation X^2 instead of χ^2_{LR} and G^2 instead of χ^2_{P}.

program $= 1$ for vocational; $= 2$ for general; $= 3$ for academic

write $=$ writing score

ses $= 1$ for low; $= 2$ for middle; $= 3$ for high

The dependent variable is **program** which is a nominal variable with three categories. The covariates are **write** and **ses**, where **write** is a continuous variable and **ses** is a categorical variable.

A data screening reveals the following information about the three variables:

```
Total Sample Size(N) =    200
```

```
Univariate Distributions for Ordinal Variables

      ses Frequency Percentage Bar Chart
     low        47        23.5
    midd        95        47.5
    high        58        29.0

  program Frequency Percentage Bar Chart
    gene        45        22.5
    acad       105        52.5
    voca        50        25.0
```

```
Univariate Summary Statistics for Continuous Variables

Variable     Mean  St. Dev.  Skewness  Kurtosis  Minimum Freq.  Maximum Freq.
--------     ----  --------  -------   --------  ------- -----  ------- -----
   write   52.775     9.479   -0.482    -0.750   31.000     4   67.000     7
```

The bivariate distribution of **program** and **ses** is shown in the following graph

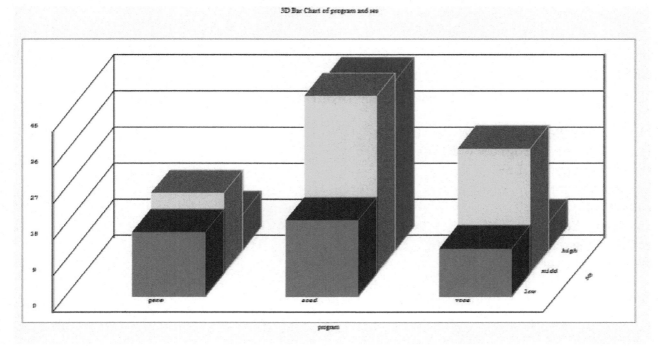

3D Bar Chart of program and ses

It is seen that **ses = middle** is the most common category for all program choices but **ses = high** is dominant among those students who chose the academic program.

The bivariate distribution of **write** and **program** is shown in the following graph

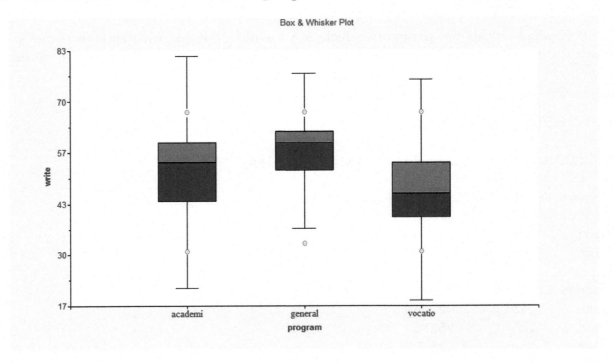

where it can be seen that the distribution is negatively skewed particularly for the general and academic group.

To estimate a multinomial logistic model, we use academic (the second category) as reference category for **program** and high (the last category) as reference category for **ses**. Let $x_1 = $ **write**, $x_2 = 1$ if **ses=1** and $x_3 = 1$ if **ses=2**. The model to be estimated is

$$\ln(\pi_1/\pi_2) = \alpha_1 + \gamma_{11}x_1 + \gamma_{12}x_2 + \gamma_{13}x_3$$

$$\ln(\pi_3/\pi_2) = \alpha_3 + \gamma_{31}x_1 + \gamma_{32}x_2 + \gamma_{33}x_3$$

Note that x_2 and x_3 are **zero-one** coded dummy variables which are automatically created by **LISREL**. They represent the same variable **ses**.

This model can be estimated using the file **programchoice1.prl**:

```
GlimOptions RefCatCode=0;
Title=Students Program Choices;
SY=programchoice.LSF;
Distribution=MUL;
Link=LOGIT;
Intercept=Yes;
DepVar=program Refcat=2;
CoVars=write ses$;
```

The output gives the following parameter estimates

Estimated Regression Weights

Parameter	Estimate	Standard Error	z Value	P Value
intcept 1	1.6894	1.2269	1.3769	0.1685
intcept 3	4.2355	1.2047	3.5159	0.0004
write 1	-0.0579	0.0214	-2.7056	0.0068
write 3	-0.1136	0.0222	-5.1127	0.0000
ses1 1	1.1628	0.5142	2.2614	0.0237
ses1 3	0.9827	0.5956	1.6500	0.0989
ses2 1	0.6295	0.4650	1.3538	0.1758
ses2 3	1.2741	0.5111	2.4927	0.0127

where the parameters are given in the order α_1, α_3, γ_{11}, γ_{31}, γ_{12}, and γ_{32}.

The interpretation of these results are

- One unit increase in **write** decreases the relative odds of general vs academic by 0.058.

- One unit increase in **write** decreases the relative odds of vocational vs academic by 0.114.

- A change in **ses** from low to high increases the relative odds of general vs academic by 1.163

- A change in **ses** from low to high increases the relative odds of vocational vs academic by 0.983

- A change in **ses** from middle to high increases the relative odds of general vs academic by 0.630

- A change in **ses** from middle to high increases the relative odds of vocational vs academic by 1.274

The output file **programchoice1.out** also give the following information about the fit of the model

Likelihood Ratio Chi-square	359.9524	192
Pearson Chi-square	410.2703	192
-2 Log Likelihood Function	359.9635	

Although no P-value is given in the output, a chi-square of 359.95 with 192 degrees of freedom is highly significant[4]. Some programs, including SPSS give another chi-square, namely one which compares the fit of the estimated model with the fit of the Intercept-Only model. To obtain this chi-square, first run the Intercept-Only model by omitting the line

CoVars=write ses$;

in **programchoice1.prl**, see file **programchoice2.prl**. This give the following fit measures

Likelihood Ratio Chi-square	408.1823	198
Pearson Chi-square	399.9992	198
-2 Log Likelihood Function	408.1933	

---

[4]Use the normal approximation $z = \sqrt{2\chi^2} - \sqrt{2df} = \sqrt{719.9} - \sqrt{384} = 7.23$.

Then subtract the corresponding values for the initial model from these values as shown in here

Model	$-2\ln(L)$	χ^2_{LR}	χ^2_P	df
Empty	408.19	408.18	399.99	198
Fitted	359.96	359.95	410.27	192
	48.23	48.23	-10-28	6

The chi-square value of 48.23 with 6 degrees of freedom tells us that the fitted model fits significantly better than the Intercept-Only (the empty) model which is the same as to say that not all γ coefficients are zero (which is also clear from their P-values). It does not, however, tell us whether the model fits the data adequately or not.

3.6.2   Example: Program Choices 2

A more common situation is that the data is in the form of contingency table where one of the variables is a dependent variable which is nominal and all the covariates are categorical.

Suppose we group the 200 writing scores in the previous example into three groups labeled wlow, wmid, and whig. The data will then be in the form of a $3 \times 3 \times 3$ contingency table as shown in Table 3.10.

Table 3.10: Number of Program Choices by SES and Writing

	Program								
	General			Academic			Vocational		
SES	wlow	wmid	whig	wlow	wmid	whig	wlow	wmid	whig
low	1	13	2	5	13	2	2	4	3
middle	4	9	6	2	32	10	5	20	17
high	5	6	1	11	17	3	2	4	1

For the analysis with LISREL these data are organized as in **programchoice1.lsf** where we have ordered the rows in descending order according **count**.

count	program	ses	write
32.	2.	2.	2.
20.	3.	2.	2.
17.	3.	2.	3.
17.	2.	3.	2.
13.	2.	1.	2.
13.	1.	1.	2.
11.	2.	3.	1.
10.	2.	2.	3.
9.	1.	2.	2.
6.	1.	3.	2.
6.	1.	2.	3.
5.	2.	1.	1.
5.	3.	2.	1.
5.	1.	3.	1.
4.	1.	2.	1.
4.	3.	1.	2.
4.	3.	3.	2.
3.	3.	1.	3.
3.	2.	3.	3.
2.	2.	1.	3.
2.	2.	2.	1.
2.	3.	1.	1.
2.	3.	3.	1.
2.	1.	1.	3.
1.	3.	3.	3.
1.	1.	1.	1.
1.	1.	3.	3.

To estimate the same model as in the previous example, one can use the following file **programchoice3.prl**

```
GlimOptions
Title=Students Program Choices;
SY=programchoice1.LSF;
Distribution=MUL;
Link=LOGIT;
Intcept=Yes;
DepVar=program Refcat=2;
CoVars=ses$ write$;
Freq=count;
```

The resulting estimates are

Parameter	Estimate	Standard Error	z Value	P Value
intcept 1	-0.7540	0.5568	-1.3541	0.1757
intcept 3	-0.8021	0.5576	-1.4384	0.1503
ses1 1	0.7660	0.4874	1.5717	0.1160
ses1 3	0.7216	0.5929	1.2172	0.2235
ses2 1	0.1295	0.4584	0.2824	0.7776
ses2 3	1.3683	0.4958	2.7597	0.0058
write1 1	-0.0464	0.6071	-0.0764	0.9391
write1 3	-0.5269	0.5672	-0.9290	0.3529
write2 1	-0.3292	0.4878	-0.6748	0.4998
write2 3	-0.9666	0.4182	-2.3116	0.0208

Note that most of the estimates are non-significant. This is a consequence of the loss of power due to the grouping of the **write** variable.

3.7   Ordinal Logistic Regression

If the categories of the dependent variables are ordered in a natural way, this order can be taken into account in the model and analysis. For examples the categories may be ordered in terms of importance from not important to very important, in terms of agreement from disagree very much to agree very much, in terms of likeness from dislike very much to like very much, or in terms ranked preferences.

Consider an ordinal variable with C ordered categories. Let $\pi_j > 0$ be the probability of category j, where

$$\pi_1 + \pi_2 + \cdots + \pi_C = 1 .$$

There are various alternatives to model these probabilities as functions of the covariates $\mathbf{x}$, the most common one being the cumulative logit model or proportional odds model, which we consider here. This is

$$logit(\pi_1 + \pi_2 + \cdots + \pi_j) = \ln \frac{\pi_1 + \pi_2 + \cdots + \pi_j}{\pi_{j+1} + \pi_{j+2} + \cdots + \pi_C} = \alpha_j + \gamma_1 x_1 + \gamma_2 x_2 + \cdots + \gamma_q x_q , \quad (3.41)$$

$$j = 1, 2, \ldots, C - 1 .$$

Note that the intercepts α_j depends on the category j but the γ's do not. This is the essential difference between ordinal and nominal logistic regression. Equation (3.41) can also be written

$$\ln \frac{Pr(y \leq j)}{Pr(y > j)} = \alpha_j + \gamma_1 x_1 + \gamma_2 x_2 + \cdots + \gamma_q x_q . \quad (3.42)$$

. Note that this is equivalent to

$$\ln \frac{Pr(y > j)}{Pr(y \leq j)} = -\alpha_j - \gamma_1 x_1 - \gamma_2 x_2 + \cdots - \gamma_q x_q . \quad (3.43)$$

.

The logodds of a response in category j or less is a linear function of the covariates $\mathbf{x}$. The effect of one unit increase in x_k on this odds ratio is e^{γ_k}, the same for all j.

This is illustrated in Figure 3.3 for the case of five categories and a single continuous covariate. One can solve equation (3.41) for the π's recursively. Let

$$\lambda_j = e^{\alpha_j + \gamma_1 x_1 + \gamma_2 x_2 + \cdots + \gamma_q x_q} , j = 1, 2, \ldots, C - 1 . \quad (3.44)$$

Then

$$\pi_1 = \frac{\lambda_1}{1 + \lambda_1} \quad (3.45)$$

$$\pi_2 = \frac{\lambda_2}{1 + \lambda_2} - \pi_1 \quad (3.46)$$

$$\pi_3 = \frac{\lambda_3}{1 + \lambda_3} - \pi_1 - \pi_2 \quad (3.47)$$

$$\pi_j = \frac{\lambda_j}{1 + \lambda_j} - \pi_1 - \pi_2 - \cdots - \pi_{j-1} , j = 4, 5, \ldots, C - 1 \quad (3.48)$$

$$\pi_C = 1 - \pi_1 - \pi_2 - \cdots - \pi_{C-1} \quad (3.49)$$

Figure 3.3: Nominal Logistic Regression: Four Regression Lines on a Log-odds Scale

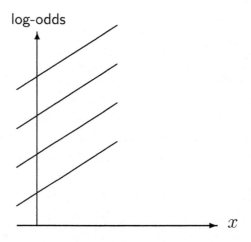

One can estimate a model for ordinal logistic regression in the same way as for the nominal logistic regression. The total number of parameters are $C - 1 + q + 1 = C + q$ which is much smaller than for the nominal logistic regression model.

3.7.1   Example: Mental Health

Fourty patients were interviewed about their mental impairment, here labelled as **Wellness** and categorized as

Impaired coded as 1

Moderate coded as 2

Mild coded as 3

Well coded as 4

The covariates are x_1, an index ranging from 0 to 9 measuring the number and severity of life events such as child birth, new job, divorce, or death in the family, and x_2, a binary measurement of socioeconomic status **Ses** coded 0 for low and 1 for high. The data file created from Table 9.8 in Agresti (1990) is **Wellness.lsf**, where **Wellness** is recorded in descending order.

The dependent variable is **Wellness** which is an ordered categorical variable. Using category **Impaired** as reference category, the ordinal logistic regression model to be estimated is

$$\frac{\pi_1}{\pi_2 + \pi_3 + \pi_4} = \alpha_1 + \gamma_1 x_1 + \gamma_2 x_2$$

$$\frac{\pi_1 + \pi_2}{\pi_3 + \pi_4} = \alpha_2 + \gamma_1 x_1 + \gamma_2 x_2$$

$$\frac{\pi_1 + \pi_2 + \pi_3}{\pi_4} = \alpha_3 + \gamma_1 x_1 + \gamma_2 x_2$$

Both x_1 and x_2 can be treated as continuous variables.

This model can be estimated by using the following syntax file **wellness1.prl**:

```
GlimOptions Response=Descending RefCatCode=0;
Title=Mental Impairment;
SY=wellness.LSF;
Distribution=MUL;
Link=OLOGIT;
DepVar=Wellness;
CoVars=Events Ses;
```

Note that the link function for ordinal logistic regression is OLOGIT. Response=Descending on the Glimoptions line tells the LISREL that the dependent variable Wellness is in descending order in the data file **wellness.lsf**. The last category wellness=1, *i.e.*, Impaired, is used as reference category by default.

The output gives the following parameter estimates

Estimated Regression Weights

Parameter	Estimate	Standard Error	z Value	P Value
Alpha1	-0.2819	0.6231	-0.4525	0.6509
Alpha2	1.2128	0.6511	1.8626	0.0625
Alpha3	2.2094	0.7171	3.0810	0.0021
Events	-0.3189	0.1194	-2.6699	0.0076
Ses	1.1112	0.6143	1.8090	0.0704

The estimated coefficient -0.319 for **Events** means that the odds ratio of low vs high mental impairment decreases with higher scores on the life events index, which is the same as to say that the chance of higher levels of **Wellness** increases with higher scores on the life events index. The estimated coefficient 1.111 means that the odds ratio of low vs high mental impairment increase at the high level of **Ses** which is the same as to say that high levels of mental impairment are associated with low **Ses**.

Consider the estimated probability $Pr(y > 2)$ as a function of x_1 and x_2. This can be calculated as

$$Pr(y > 2) = e^{-\hat{\alpha}_2 - \hat{\gamma}_1 x_1 - \hat{\gamma}_2 x_2} Pr(y \le 2) \tag{3.50}$$

$$= \frac{1}{\hat{\lambda}_2}(\hat{\pi}_1 + \hat{\pi}_2) \tag{3.51}$$

$$= \frac{1}{\hat{\lambda}_2} \frac{\hat{\lambda}_2}{1 + \hat{\lambda}_2} \tag{3.52}$$

$$= \frac{1}{1 + \hat{\lambda}_2} \tag{3.53}$$

$$= \frac{1}{1 + e^{\hat{\alpha}_2 + \hat{\gamma}_1 x_1 + \hat{\gamma}_2 x_2}} \tag{3.54}$$

$$= \frac{1}{1 + e^{1.2128 - 0.31891 x_1 + 1.1112 x_2}} \tag{3.55}$$

Plotting (3.55) as a function of x_1 for the two values of x_2 gives the plot shown in Figure 3.4, which clearly shows that the probability of a response **Mild** or **Well** increases with increasing scores on the Event index for both low and high Ses.

Figure 3.4: Probability $Pr(y > 2)$ for High and Low Ses

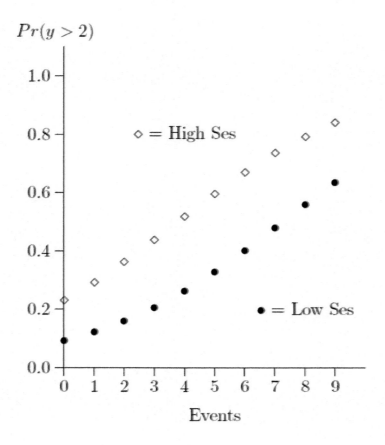

3.7.2 Example: Car Preferences

In a study of motor vehicle safety, Mc Fadden et al. (2000) give data about preferences for various features in cars, such as power steering and air conditioning. A subset of the data is shown in Table 3.11.

From these data, we constructed the **LISREL** system data file **carpref.lsf**:

	Men	Agegrp	Response	Count
1	0.000	1.000	1.000	26.000
2	0.000	1.000	2.000	12.000
3	0.000	1.000	3.000	7.000
4	0.000	2.000	1.000	9.000
5	0.000	2.000	2.000	21.000
6	0.000	2.000	3.000	15.000
7	0.000	3.000	1.000	5.000
8	0.000	3.000	2.000	14.000
9	0.000	3.000	3.000	41.000
10	1.000	1.000	1.000	40.000
11	1.000	1.000	2.000	17.000
12	1.000	1.000	3.000	8.000
13	1.000	2.000	1.000	17.000
14	1.000	2.000	2.000	15.000
15	1.000	2.000	3.000	12.000
16	1.000	3.000	1.000	8.000
17	1.000	3.000	2.000	15.000
18	1.000	3.000	3.000	18.000

Table 3.11: Importance of Air Conditioning and Power Steering in Cars. Originally published in Dobson and Barnett (2008, Table 8.1). Published with kind permission of © Taylor and Francis Group, LLC 2008. All Rights Reserved

Gender	Age	Response		
		Little Importance	Important	Very Important
Woman	18–23	26(58%)	12(27%)	7(16%)
	24–40	9(20%)	21(47%)	15(33%)
	>40	5(8%)	14(23%)	41(68%)
Men	18–23	40(62%)	17(26%)	8(12%)
	24–40	17(39%)	15(34%)	12(27%)
	>40	8(20%)	15(37%)	18(44%)

The dependent variable is **Response**, which is an ordinal variable, and the covariates are **Men** which is treated as a continuous variable coded with 1 for men and 0 for women and **Agegrp** which is a nominal variable with three categories. Using the last category (3 = Very Important) of **Response** as a reference category and the first category (1 = 18-25) as reference category for **Agegrp**, the ordinal logistic regression model with the proportional odds parameterization is

$$\ln(\frac{\pi_1}{\pi_2 + \pi_3}) = \alpha_1 + \gamma_1 x_1 + \gamma_2 x_2 + \gamma_3 x_3 , \tag{3.56}$$

$$\ln(\frac{\pi_1 + \pi_2}{\pi_3}) = \alpha_2 + \gamma_1 x_1 + \gamma_2 x_2 + \gamma_3 x_3 , \tag{3.57}$$

where x_1 is **Men**, x_2 is a dummy variable equal to 1 for age = 24-40, and x_3 is a dummy variable equal to 1 for age > 40. Note that x_2 and x_3 are automatically created by **LISREL** from the values of the **Agegrp** variable.

This model can be estimated using the following syntax file (**carpref1.prl**):

```
GlimOptions RefCatCode=0 Output=Residuals;
Title=car preferences;
SY=carpref.lsf;
Distribution=MUL;
Link=OLOGIT;
DepVar=Response;
CoVars=Men Agegrp$;
Refcats=1;
Freq=Count;
```

                    Estimated Regression Weights

                        Standard
    Parameter   Estimate   Error   z Value   P Value
    ---------   --------   --------   -------   -------
    Alpha1   0.0435   0.2303   0.1890   0.8501
    Alpha2   1.6550   0.2536   6.5259   0.0000

Men	0.5762	0.2261	2.5484	0.0108
Agegrp2	-1.1471	0.2773	-4.1371	0.0000
Agegrp3	-2.2325	0.2904	-7.6871	0.0000

These estimates suggest that

- For each age group, the odds ratios $\pi_1/(\pi_2 + \pi_3)$ and $(\pi_1 + \pi_2)/\pi_3$ are increased by the factor $e^{0.576}$ for men relative to women, i.e., air conditioning and power steering are more important for women than for men.

- The odds ratio $\pi_1/(\pi_2 + \pi_3)$ is decreased by the factor $e^{1.147}$ for age group 24–40 relative to the age group 18–23 and the odds ratio $(\pi_1 + \pi_2)/\pi_3$ is decreased by the factor $e^{2.233}$ for age group > 40 relative to the age group 18–23, i.e., air conditioning and power steering is more important for the younger age group than for the older.

From these estimates of the regression weights and the equations (3.45)–(3.49), one can compute the estimates of the cell probabilities π for each cell and estimates of the expected frequencies. The values are given in the file **carpref1_RES.LSF** (the estimated cell probabilities appear in the column **Prob** and the expected frequencies appear in the column **Fitted**; the column **Raw** is the difference between the observed and fitted values):

carpref1_RES.LSF

	Observed	Prob.	Fitted	Raw
1	26,000	0,511	22,990	3,010
2	12,000	0,329	14,791	-2,791
3	7,000	0,160	7,220	-0,220
4	9,000	0,249	11,208	-2,208
5	21,000	0,375	16,886	4,114
6	15,000	0,376	16,906	-1,906
7	5,000	0,101	6,045	-1,045
8	14,000	0,259	15,526	-1,526
9	41,000	0,640	38,429	2,571
10	40,000	0,650	42,261	-2,261
11	17,000	0,253	16,435	0,565
12	8,000	0,097	6,304	1,696
13	17,000	0,371	16,330	0,670
14	15,000	0,376	16,550	-1,550
15	12,000	0,253	11,120	0,880
16	8,000	0,166	6,815	1,185
17	15,000	0,333	13,672	1,328
18	18,000	0,500	20,513	-2,513

Chapter 4

Multilevel Analysis

4.1 Basic Concepts and Issues in Multilevel Analysis

4.1.1 Multilevel Data and Multilevel Analysis

In multilevel data some observational units are nested within other observational units. For example, students are grouped in classes, classes are grouped in schools, and so on. Multilevel models, also called hierarchical linear models, and their associated techniques take the inherent data structure into account in the data analysis.

The analysis of multilevel data has become very popular in recent years and special purpose computer programs have been developed for such analysis, the most widely used ones are HLM and MLWin, see, e.g., Bryk & Raudenbush (1992) and Goldstein (2003). However, multilevel models may be estimated directly by using any of the two multilevel modules in LISREL. In some cases, multilevel models may also be viewed as structural equation models (SEM) and estimated by any of the methods available in LISREL.

There are two multilevel modules in LISREL: One *simple* module which can handle two- and three-level models and one *advanced* module which can handle models with data up to five levels. The simple module is reached by selecting **Multilevel Models** in the **Statistics** menu. The advanced module is reached by selecting the **Multilevel Menu** in the **LSF Window**. This menu includes more advanced features such as **Multilevel Generalized Linear Models**. In this chapter we use the simple module for all examples except the example **Social Mobility** in Section 4.7.

4.1.2 Examples of Multilevel Data

Educational Research Students are nested within classes within schools

Cross-national Studies Individuals are nested within national units

Survey Research Individuals are nested within communities within regions

Organizational Research Individuals are nested within departments within organizations

Family Research Family members are nested within families

Methodological Research Respondents are nested within interviewers

Longitudinal Studies Measurements over individuals are nested within individuals

© Springer International Publishing Switzerland 2016
K.G. Jöreskog et al., *Multivariate Analysis with LISREL*, Springer Series in Statistics,
DOI 10.1007/978-3-319-33153-9_4

4.1.3   Terms Used for Two-level Models

Some of the terms used in multilevel analysis are

Level-2	Level-1
level-2 units	level-1 units
macro units	micro units
primary units	secondary units
clusters	elementary units
groups	observations

Typical examples of units for two-level models are

Level-2	Level-1
schools	teachers
classes	students
neighborhoods	families
firms	employees
jawbones	teeth
families	children
doctors	patients
subjects	measurements
interviewers	respondents
judges	suspects
groups	individuals

Here are some examples of statements which involve level-1 (in lower case) and level-2 variables (in upper case) as well as their interrelationships.

- Student score on a reading test (x) affects his/her score on a math test (y). The variables x and y are level-1 variables. If they are observed on level-1 units nested within higher level units, one can use multilevel analysis to estimate within and between variances of x and y and within and between regression lines. This model is specified in Section 4.3 and illustrated in Example 4.3.1.

- The organizational climate (Z) affects the company's profits (Y). If the observational units are companies, this does not require multilevel analysis.

- Religious norms (Z) in social networks affects individuals' opinion (y) about contraception. This is an example of cross-level analysis where the level-2 variable Z affects a variable y observed on level-1 units. This model is specified in Section 4.4.

- Teacher efficacy (Z) affects student motivation (y) after controlling for student aptitude (x). This is another example of cross-level analysis as specified in Section 4.4.

4.1.4   Multilevel Analysis vs Linear Regression

Traditionally, fixed parameter linear regression models have been used for the analysis of multilevel data. Statistical inference is based on the assumptions of linearity, normality, homo-scedasticity and independence. It has been shown by Aitkin and Longford (1986) that the aggregation of variables over individual observations may lead to misleading results. Both the aggregation of individual variables to a higher level of observation and the disaggregation of higher order variables to an individual level have been discredited (Bryk & Raudenbush, 1992). It has been pointed out by Holt, Scott and Ewings (1980) that serious inferential errors may result from the analysis of complex survey data, if it is assumed that the data have been obtained under a simple random sampling scheme.

4.1.5   Other Terminology

Multilevel data structures are pervasive. As Kreft and de Leeuw (1998) note: "Once you know that hierarchies exist, you see them everywhere". It was not until the 1980s and 1990s that techniques to properly analyze multilevel data became widely available. These techniques use maximum likelihood procedures to estimate random coefficients and are often referred to as "multilevel random coefficient models" (MRCM) in contrast to "hierarchical linear models" (HLM).

4.1.6   Populations and Subgroups

Many populations of subjects are organized in subgroups: gender, educational level, type of job etc. Statistical analyses can take advantage of this organization to improve the accuracy of statistical predictions. In education research, a problem is that students are assigned to particular classrooms.

- The educational environment in one classroom can be quite different from that in others: characteristics of the teacher and the interactions among the students vary.

- When students participate in an experiment, the effect of the treatment may depend on the classroom to which a student is assigned.

- Failure to account for classroom subgroups may distort the effectiveness of the treatment.

4.1.7   The Interaction Question

Interaction exists when the benefit of an intervention, or the effect of a treatment, depends on particular levels of other variables.

Simple Interaction: Differential performance is attributable to levels of a single variable. For example, students in classes with a female teacher perform better than students with a male teacher, on the same curriculum.

Complex Interaction: A variety of contextual or background variables are associated with performance. Differences in behavior cannot be attributed to any single variable. For example, on a common curriculum, students in certain classrooms perform differently than in others, but no specific variable accounts for the differential outcome.

4.2   Within and Between Group Variation

4.2.1   Univariate Analysis

The basic setup for two-level models is as follows. There are N level-2 units labeled $i = 1, 2, \ldots, N$ and n_i level-1 units labeled $j = 1, 2, \ldots, n_i$ within level-2 unit i. The n_i may differ between different level-2 units. Let y_{ij} be a random variable observed on level-1 unit j in level-2 unit i. The simplest two-level is

$$y_{ij} = \alpha + u_i + e_{ij} , \tag{4.1}$$

where α is a fixed parameter and u_i and e_{ij} are independent random variables with mean zero and variances τ^2 and σ^2, respectively. Then it follows that

$$E(y_{ij}) = \alpha , \quad Var(y_{ij}) = Var(u_i) + Var(e_{ij}) = \tau^2 + \sigma^2 \tag{4.2}$$

The quantities α, τ^2 and σ^2 are fixed parameters to be estimated. α is called the overall mean or grand mean. τ^2 is called the between group variance and σ^2 is called the within group variance. In the literature this model is also referred to as the Random ANOVA Model, or the Random Intercept Model, or sometimes as the Empty Model. We shall call it the Intercept-Only Model.

To understand the terminology that **LISREL** uses to specify the model, consider the right hand side of (4.1). This consists of two parts: the fixed part α and the random part $u_i + e_{ij}$, where u_i is a level-2 variable and e_{ij} is a level-1 variable. Every term in (4.1) which does not have a coefficient in front of it is called an **Intcept** which is short for Intercept. Thus, all three terms on the right side of (4.1) are Intcept.

4.2.2   Example: Netherlands Schools, Univariate Case

To illustrate this and some other models we use a dataset analyzed extensively by Snijders and Bosker (1999). The data set contain various test scores on 2287 students in 131 schools in The Netherlands. The students were in 8th grade and approximately 11 years old. The variables used here is a subset of the variables used by Snijders and Bosker (1999). The test scores are two IQ tests called IQ_verb and IQ_perf and one arithmetic test and one language test. The two latter tests are measured twice, so they are called aritpret and aritpost and langpret and langpost. Presumably these are pretests and posttests. The first ten lines of the data file **nlschools.lsf** are

nlschools.lsf	School	Student	IQ_verb	IQ_perf	aritpret	aritpost	langpret	langpost
1	1,000	1,000	15,000	12,330	14,000	24,000	36,000	46,000
2	1,000	2,000	14,500	10,000	12,000	19,000	36,000	45,000
3	1,000	3,000	9,500	11,000	10,000	24,000	33,000	33,000
4	1,000	4,000	11,000	10,000	13,000	26,000	29,000	46,000
5	1,000	5,000	8,000	6,666	8,000	9,000	19,000	20,000
6	1,000	6,000	9,500	9,000	8,000	13,000	22,000	30,000
7	1,000	7,000	9,500	10,330	7,000	13,000	20,000	30,000
8	1,000	8,000	13,000	14,330	17,000	30,000	44,000	57,000
9	1,000	9,000	9,500	8,666	10,000	23,000	34,000	36,000
10	1,000	10,000	11,000	15,000	14,000	22,000	31,000	36,000

To specify the multilevel model in **LISREL**, one can use either the **LISREL**'s Graphical Users Interface GUI or one can type an input file using a special multilevel syntax. In the first case the syntax file will be written by the GUI. For this first example we use the GUI. For all other examples we will use syntax files which are either provided or to be written by the user.

The variable `langpost` is the response variable y_{ij} and `School` is the level-2 identifier. For all other examples we will use syntax files which are either provided or to be written by the user.

While **nlschools.lsf** is visible select **Multilevel Analysis** in the **Statistics** menu. Then go through the four screens as described here. The first screen is the **Title and Options** screen. Fill in a title, leave everything else as is and click **Next**.

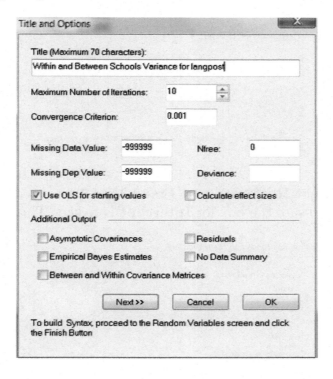

The next screen is the **Identification Variables** screen. Select `School` as the level-2 identifier and click **Next**.

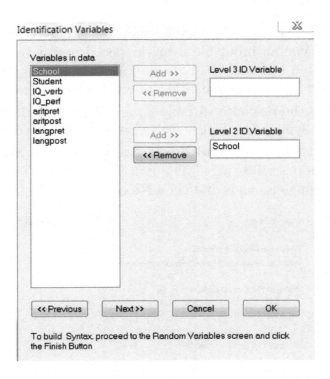

The next screen is the **Select Response and Fixed Variables** screen. Select `langpost` as the response variable and make sure the box for **Intercept** is ticked. Then click **Next**.

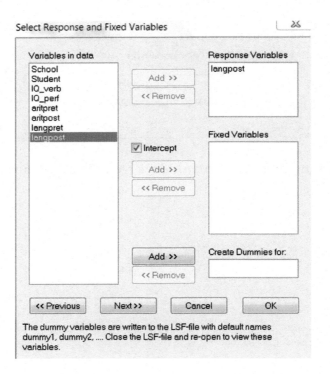

The last screen is the **Random Variables** screen. In this case, just make sure that both tick boxes in the column **Intercept** are filled in. Then click **Finish**.

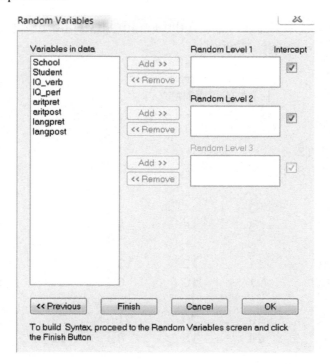

After clicking **Finish**, LISREL will generate the following input file **nlschools.prl**:

```
OPTIONS OLS=YES CONVERGE=0.001000 MAXITER=10 OUTPUT=STANDARD ;
TITLE=Within and Between Schools Variance for langpost;
SY=nlschools.lsf;
ID2=School;
RESPONSE=langpost;
FIXED=intcept;
RANDOM1=intcept;
RANDOM2=intcept;
```

This is a special type of syntax valid only for Multilevel models. Note that each line must end with ;. The Options are explained in detail in the file **Multilevel (Hierarchical Linear) Modeling Guide.pdf**. To read or print this file click on **Help** and select **LISREL User & Reference Guides** and select **Multilevel (Hierarchical Linear) Modeling Guide.**

The options generated in the file **nlschools.prl** are the default options, so one can replace the line

```
OPTIONS OLS=YES CONVERGE=0.001000 MAXITER=10 OUTPUT=STANDARD ;
```

by the line

```
OPTIONS;
```

The default options are sufficient for most examples so there is seldom any reason to change them. For the examples in this book, there are just a few options we need to use and we will learn them as we move along. Click on the Run **PRELIS** button **P** to estimate the model.

Each time one uses the data file **nlschools.lsf** with a different model, LISREL will overwrite the input file **nlschools.prl**, and, as a consequence also the output file **nlschools.out**. So it is suggested that you save **nlschools.prl** to **nlschools1.prl** before you run it.

The output file **nlschools1.out** first gives a data summary containing a numbered list of schools and the number of students in each school.

```
                    +--------------+
                    | DATA SUMMARY |
                    +--------------+


   NUMBER OF LEVEL 2 UNITS :      131
   NUMBER OF LEVEL 1 UNITS :     2287


   N2  :        1         2         3         4         5         6         7         8
   N1  :       25         7         5        15         8         8        24        17


   N2  :        9        10        11        12        13        14        15        16
   N1  :       24        18        21        10         6        14        23        28


   N2  :       17        18        19        20        21        22        23        24
   N1  :       35        13        14        15         8        11         9        21


   N2  :       25        26        27        28        29        30        31        32
   N1  :       31        30        24        22        20        21        22        11


   N2  :       33        34        35        36        37        38        39        40
   N1  :       26        20        16        17        14        21        24        21


   N2  :       41        42        43        44        45        46        47        48
   N1  :       10        15        15        18        17        20        23         4


   N2  :       49        50        51        52        53        54        55        56
   N1  :       10        17         9        20        17        18         7        30


   N2  :       57        58        59        60        61        62        63        64
   N1  :       19         9        12        10         4        20        26        13


   N2  :       65        66        67        68        69        70        71        72
   N1  :       21        16        18        20        24        22        29        21


   N2  :       73        74        75        76        77        78        79        80
   N1  :       25        23        17        33        12        31        23        31


   N2  :       81        82        83        84        85        86        87        88
   N1  :       24        25        26        14        23        17         6         9


   N2  :       89        90        91        92        93        94        95        96
   N1  :       31        28        12        17         8        12        20        27


   N2  :       97        98        99       100       101       102       103       104
   N1  :       20         7        30        24        11        23        16        25
```

```
N2  :       105      106      107      108      109      110      111      112
N1  :        21       12        7        8       24       14       24       13

N2  :       113      114      115      116      117      118      119      120
N1  :         8       18       21       21        9       13       10       15

N2  :       121      122      123      124      125      126      127      128
N1  :        12       23       11        8       13       12       24       15

N2  :       129      130      131
N1  :        11       10        7
==============================================================================
```

One can see that there are 131 schools and the number of students per school varies from 4 to 35. Since this DATA SUMMARY is the same for every model based on **nlschools.lsf**, one can omit this part of the output by using the option line

```
OPTIONS SUMMARY=NONE
```

PRELIS will estimate the parameters by maximum likelihood using an iterative procedure, see Section 4.4. The model used here is a special case of a more general model presented in Section 4.4, so we postpone the specification of the likelihood function until that section.

The output file **nlschools1.out** gives starting values for the iterative procedure after the line

```
ITERATION NUMBER      1
```

These starting values are generally of no interest, so one should just skip them and continue to browse the output file until one finds the lines

```
-----------------------------------------
CONVERGENCE REACHED IN    6 ITERATIONS
-----------------------------------------
```

The parameter estimates are found after the line

```
ITERATION NUMBER         6
```

```
                      +-----------------------+
                      | FIXED PART OF MODEL   |
                      +-----------------------+


---------------------------------------------------------------------------
   COEFFICIENTS          BETA-HAT      STD.ERR.      Z-VALUE      PR > |Z|
---------------------------------------------------------------------------

     intcept             40.36409       0.42637      94.67021      0.00000
```

```
                          +-----------------------+
                          | RANDOM PART OF MODEL   |
                          +-----------------------+
```

```
    ----------------------------------------------------------------------
    LEVEL 2                     TAU-HAT      STD.ERR.      Z-VALUE    PR > |Z|
    ----------------------------------------------------------------------
    intcept /intcept            19.42856     2.93256       6.62511    0.00000


    ----------------------------------------------------------------------
    LEVEL 1                     TAU-HAT      STD.ERR.      Z-VALUE    PR > |Z|
    ----------------------------------------------------------------------
    intcept /intcept            64.56783     1.96604      32.84151    0.00000
```

To interpret the output in terms of the model (4.1) and (4.2), the estimate $\hat{\alpha}$ is called BETA-HAT and the variances $\hat{\tau}^2$ and $\hat{\sigma}^2$ are both called TAU-HAT. They are distinguished by the LEVEL 2 and LEVEL 1. Thus, the estimate of the overall mean is $\hat{\alpha} = 40.36$, the estimated between schools variance is $\hat{\tau}^2 = 19.43$, and the estimated within schools variance is $\hat{\sigma}^2 = 64.57$. These estimates agree with the results reported by Snijders and Bosker (1999, Table 4.1). The PRELIS output also give standard errors, z-values, and P-values for these estimates.

Articles on multilevel analysis sometimes report a quantity called *intraclass correlation* defined as

$$\hat{\rho} = \frac{\hat{\tau}^2}{\hat{\tau}^2 + \hat{\sigma}^2} = \frac{19.42}{83.99} = 0.23 \tag{4.3}$$

This means that the between schools variance accounts for 23% of the total variance of langpost. Thus, most of the variance of langpost is within schools.

For a small number of level-2 units N, one can calculate an estimate of σ^2 as follows. Define

$$M = n_1 + n_2 + \cdots + n_N , \tag{4.4}$$

$$\bar{y}_i = (1/n_i) \sum_{j=1}^{n_i} y_{ij} , \quad s_i^2 = [1/(n_i - 1)] \sum_{j=1}^{n_i} (y_{ij} - \bar{y}_i)^2 \tag{4.5}$$

and

$$s_w^2 = [(1/(M - N)] \sum_i^N (n_i - 1)s_i^2 = [(1/(M - N)] \sum_i^N \sum_{j=1}^{n_i} (y_{ij} - \bar{y}_i)^2 \tag{4.6}$$

Then the following result hold

$$E(s_w^2) = \sigma^2 , \tag{4.7}$$

so that s_w^2 can be used to estimate σ^2.

To obtain an explicit estimate of $\hat{\tau}^2$ is a bit more complicated but Snijders and Bosker (1999, page 19) suggest to use the following formulas:

$$\bar{n} = N/M \tag{4.8}$$

$$s_n^2 = [1/(N - 1)] \sum_{i=1}^N (n_i - \bar{n})^2 \tag{4.9}$$

$$\tilde{n} = \bar{n} - s_n^2/N\bar{n} \tag{4.10}$$

$$s_b^2 = \frac{\sum_{i=1}^{N} n_i(\bar{y}_i - \bar{y})^2}{\tilde{n}(N-1)} \tag{4.11}$$

$$E(s_b^2) = \tau^2 + \frac{\sigma^2}{\tilde{n}} \tag{4.12}$$

$$\hat{\tau}^2 = s_b^2 - \frac{s_w^2}{\tilde{n}} \tag{4.13}$$

4.2.3   Multivariate Analysis

The model (4.1) can be generalized to the multivariate case. Let $\mathbf{y}_{ij}$ be a vector of observed response variables in level-1 unit j within level-2 unit i. Then (4.1) generalizes to

$$\mathbf{y}_{ij} = \boldsymbol{\alpha} + \mathbf{u}_i + \mathbf{e}_{ij} , \tag{4.14}$$

where $\boldsymbol{\alpha}$ is the grand mean vector, $\mathbf{u}_i$ and $\mathbf{e}_{ij}$ are independent random vectors with covariance matrices $\boldsymbol{\Sigma}_{\mathrm{b}}$ and $\boldsymbol{\Sigma}_{\mathrm{w}}$, respectively. The main objective is to estimate the between group covariance matrix $\boldsymbol{\Sigma}_{\mathrm{b}}$ and the within group covariance matrix $\boldsymbol{\Sigma}_{\mathrm{w}}$.

4.2.4   Example: Netherlands Schools, Multivariate Case

Let $\mathbf{y}_{ij}$ be the vector of observed response variables on `IQ_verb`, `IQ_perf`, `aritpret`, `aritpost`, `langpret`, and `langpost`. The multivariate model (4.14) can be estimated using a three-level model, where the scores on the six variables are treated as level-1 units, students as level-2 units and schools as level-3 units. The input file **nlschools2.prl** for this is

```
OPTIONS SUMMARY=NONE COVBW=YES;
TITLE=Within and Between Covariance Matrix for 6 Test Scores;
SY=nlschools.lsf;
ID3=School;
ID2=Student;
RESPONSE=IQ_verb IQ_perf aritpret aritpost langpret langpost;
FIXED=intcept;
RANDOM1=intcept;
RANDOM2=intcept;
RANDOM3=intcept;
```

where `COVBW=YES` has been added on the `Options` line to produce the estimates of the between and within group covariance matrices $\hat{\boldsymbol{\Sigma}}_{\mathrm{b}}$ and $\hat{\boldsymbol{\Sigma}}_{\mathrm{w}}$.

The estimated grand mean vector $\boldsymbol{\alpha}$ is

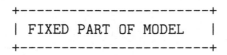

```
+-----------------------+
| FIXED PART OF MODEL   |
+-----------------------+
```

COEFFICIENTS	BETA-HAT	STD.ERR.	Z-VALUE	PR > |Z|
intcept1	11.75097	0.07663	153.33900	0.00000
intcept2	10.98571	0.06392	171.85698	0.00000
intcept3	11.73904	0.15190	77.28386	0.00000
intcept4	18.95263	0.33508	56.56127	0.00000
intcept5	33.88039	0.26608	127.33025	0.00000
intcept6	40.35982	0.41861	96.41444	0.00000

and the between and within covariance matrices appear in the output as

LEVEL 3 COVARIANCE MATRIX

	intcept1	intcept2	intcept3	intcept4	intcept5
intcept1	0.52351				
intcept2	0.26196	0.25167			
intcept3	0.87099	0.45958	2.37204		
intcept4	1.93179	1.24817	4.19631	12.53172	
intcept5	1.61381	0.94915	3.21832	6.93619	6.73197
intcept6	2.50844	1.58392	4.93494	14.26373	9.03740

	intcept6
intcept6	18.66449

LEVEL 2 COVARIANCE MATRIX

	intcept1	intcept2	intcept3	intcept4	intcept5
intcept1	3.83022				
intcept2	1.71733	4.62199			
intcept3	2.49665	3.26362	9.82812		
intcept4	5.59173	6.74055	10.94488	32.27615	
intcept5	7.36608	5.61583	9.49776	19.02252	38.96517
intcept6	9.27566	6.88912	12.31897	28.66060	35.96826

	intcept6
intcept6	64.69375

It is seen that the within schools variance is much larger than the between schools variance for all variables. The between and within schools covariance matrices are also saved as the files **nlschools2_between.cov** and **nlschools2_within.cov**. In these files, the estimated between schools covariance matrix $\hat{\Sigma}_b$ is

```
0.52351
0.26196        0.25167
0.87099        0.45958        2.37204
1.93179        1.24817        4.19631        12.53172
1.61381        0.94915        3.21832         6.93619        6.73197
2.50844        1.58392        4.93494        14.26373        9.03740        18.66449
```

and the estimated within schools covariance matrix $\hat{\Sigma}_{\mathrm{w}}$ is

```
3.83022
1.71733       4.62199
2.49665       3.26362       9.82812
5.59173       6.74055      10.94488      32.27615
7.36608       5.61583       9.49776      19.02252      38.96517
9.27566       6.88912      12.31897      28.66060      35.96826      64.69375
```

These matrices can be used in multi-group structural equation modeling as described in Chapter 10. The sample sizes associated with these files are 17 and 2287, respectively. These numbers appear in the output file as

```
            +--------------+
            | DATA SUMMARY |
            +--------------+


NUMBER OF LEVEL 3 UNITS :      131
NUMBER OF LEVEL 2 UNITS :     2287
NUMBER OF LEVEL 1 UNITS :    13722



   Adjusted between cluster sample size=      17
   Within cluster sample size=     2287
```

Here there are 2287 level-2 units (students) and 6 level-1 units for each level-2 unit (the six test scores). So the total number of level-1 units is $2287 \times 6 = 13722$.

4.3   The Basic Two-Level Model

The setup for this model is the same as in Section 4.2.1 but in addition to the response variable y_{ij} an explanatory (covariate) x_{ij} is also observed on each level-1 unit. This may well be several variables but to simplify the presentation here we assume that it is a single variable. The x-values are considered as fixed values just as in ordinary regression models. The model is

$$y_{ij} = a_i + b_i x_{ij} + e_{ij} , \tag{4.15}$$

with

$$i = 1, 2, \ldots, N \ \text{ level 2 units} \tag{4.16}$$

$$j = 1, 2, \ldots, n_i \ \text{ level 1 units} \tag{4.17}$$

If the N level-2 units constitute a random sample, the intercept a_i and the slope b_i are random variables, and it is assumed that

$$a_i = \alpha + u_i \tag{4.18}$$

$$b_i = \beta + v_i \tag{4.19}$$

with

$$\begin{pmatrix} u_i \\ v_i \end{pmatrix} \sim N(\mathbf{0}, \mathbf{\Phi}) \tag{4.20}$$

$$e_{ij} \sim N(0, \sigma_e^2) \qquad (4.21)$$

The error term e_{ij} is assumed to be independent of a_i and b_i, so that (4.15) may be regarded as a regression equation for each level-2 unit.

In principle, the intercepts a_i and slopes b_i could all be different across level-2 units. It is of interest to know if the intercepts and/or the slopes are equal across level-2 units. Four possible cases are illustrated in Figure 4.1.

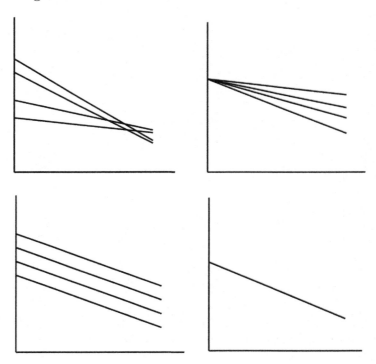

Figure 4.1: Four Cases of Intercepts and Slopes

The objective is to estimate the mean vector (α, β) and covariance matrix $\boldsymbol{\Phi}$ of (a_i, b_i) as well as the within group error variance σ_e^2. The parameters α, β, ϕ_{11}, ϕ_{21}, and ϕ_{22} are level-2 parameters and σ_e^2 is a level-1 parameter.

Substituting (4.18) and (4.19) into (4.15) gives

$$y_{ij} = (\alpha + \beta x_{ij}) + (u_i + v_i x_{ij} + e_{ij}), \qquad (4.22)$$

where the parentheses have been added for the following explanation needed to understand the input specification for **LISREL**. The part $\alpha + \beta x_{ij}$ is called the fixed part of the model and the part $u_i + v_i x_{ij} + e_{ij}$ is called the random part of the model. To estimate the fixed part, one specifies the multiplier of α and β, where the invisible multiplier 1 of α is an intercept and the multiplier of β is the x-variable. The random part consists of one level-2 part $u_i + v_i x_{ij}$ and one level-1 part e_{ij}. To estimate the level-2 random part, one specifies the multiplier of u_i which is the intercept 1 and the multiplier of v_i which is the x-variable. The random level-1 part is an intercept.

There are two special cases of model (4.15) of particular interest:

- Special Case A: The Intercept-Only Model This is the model with no x-variables considered previously in Section 4.2.1 which can also be written.

$$y_{ij} = \alpha + u_i + e_{ij}. \qquad (4.23)$$

- Special Case B: Random Intercept Model. This model has $\phi_{22} = 0$ (and hence $\phi_{21} = 0$) and can be written as

$$y_{ij} = \alpha + u_i + \beta x_{ij} + e_{ij} \, . \tag{4.24}$$

4.3.1   Example: Math on Reading with Career-Revisited

We considered this example in Section 2.1.15, where we used conditional regression, see Section 2.1.12. With this approach we obtained a separate regression equation for each of the nine career options. Here we will use multilevel analysis to analyze the same data and specifically focus on the distinction between the two approaches. The motivation for the example was given in Section 2.1.15 and will not be repeated here.

Following the general formulation in Section 4.3, the setup for multilevel analysis is

- $i = 1, 2, \ldots, 9$ are the 9 different careers,

- $j = 1, 2, \ldots, n_i$ are the students pursuing career i,

- y_{ij} is the math score of student j in career group i,

- x_{ij} is the reading score of student j in career group i.

The data file **Math on Reading with Career.lsf** is the same as before. By running ordinary regression of y on x, one obtains the overall regression equation:

$$y = 38.0 + 0.64x + z \tag{4.25}$$

Extracting the results from Section 2.1.15, the regression estimates from conditional regression are

	Career								
	Tr	B	M	I	R	P	S	T	U
a_i	38.03	36.15	40.51	36.09	39.53	34.43	37.44	41.44	35.29
b_i	0.66	0.71	0.38	0.74	0.63	0.83	0.65	0.68	0.65

To estimate the multilevel model 4.15, use the following input file **mathread1.prl**:

```
OPTIONS ALL;
TITLE=Math on Reading by Career;
SY='Math on Reading by Career.lsf';
ID2=Career;
RESPONSE=Math;
FIXED=intcept Reading;
RANDOM1=intcept;
RANDOM2=intcept Reading;
```

Here the `OUTPUT=ALL` on the `OPTIONS` line is needed to obtain a_i and b_i and residuals e_{ij}. In the literature these are called empirical Bayes estimates as they are estimated as the means in a posterior distribution.

The line

```
ID2=Career;
```

specifies that the level-2 identifier is `Career`. No level-1 identifier needs to be specified; the program figures out all the level-1 identifier values $1, 2, \ldots, n_i$ for each value of the level-2 identifier. The line

```
RESPONSE=Math;
```

specifies that the response (dependent) variable is Math. The line

```
FIXED=intcept Reading;
```

defines $\alpha + \beta x_{ij}$ as the fixed part of the model, which means that α and β will be estimated. Similarly, the line

```
RANDOM2=intcept Reading;
```

defines the level-2 random part $u_i + v_i x_{ij}$, which means that ϕ_{11}, ϕ_{21}, and ϕ_{22} will be estimated, i.e., the variances and covariance of a_i and b_i. Finally, the line

```
RANDOM1=intcept;
```

specifies that the level-1 error variance σ_e^2 should be estimated.

Take a look at the output file **mathread1.out**. The

```
+--------------+
| DATA SUMMARY |
+--------------+


NUMBER OF LEVEL 2 UNITS :        9
NUMBER OF LEVEL 1 UNITS :      147


N2 :        1         2         3         4         5         6         7         8
N1 :       12        23        19        24        22         9        12        17


N2 :        9
N1 :        9
========================================================================
```

shows that there are 147 students altogether and the number of students per career varies from 9 to 24.

Convergence is obtained after 5 iterations and the results are

```
+-----------------------+
| FIXED PART OF MODEL   |
+-----------------------+
```

COEFFICIENTS	BETA-HAT	STD.ERR.	Z-VALUE	PR > |Z|
intcept	37.68531	0.76864	49.02825	0.00000
Reading	0.65545	0.04066	16.12127	0.00000

```
+-----------------------+
| RANDOM PART OF MODEL  |
+-----------------------+
```

LEVEL 2	TAU-HAT	STD.ERR.	Z-VALUE	PR > |Z|
intcept /intcept	0.91092	2.32327	0.39208	0.69500
Reading /intcept	0.00319	0.10399	0.03068	0.97552
Reading /Reading	0.00494	0.00669	0.73802	0.46050

LEVEL 1	TAU-HAT	STD.ERR.	Z-VALUE	PR > |Z|
intcept /intcept	13.55506	1.67523	8.09146	0.00000

The fixed part of the model gives the estimates $\hat{\alpha} = 37.69$ and $\hat{\beta} = 0.66$ both of which are highly significant. The estimates of ϕ_{11}, ϕ_{21}, and ϕ_{22} are 0.91, 0.003, and 0.005, respectively but none of these are significant. This means that the variation of intercepts and slopes are not statistically significant. So the conclusion is that all nine regression lines are the same. This corresponds to the lower right figure in Figure 4.1.

In general, it can be of interest to estimate each a_i and b_i and we use this example to illustrate how this can be done. One can obtain estimates of these as follows. First obtain estimates of the error terms u_i and v_i for each level-2 unit i. They are given in the output file **mathread1.ba2**:

```
1    1    0.10707        0.61606        intcept
1    2    0.72738E-02    0.26697E-02    Reading
2    1   -0.35773        0.56775        intcept
2    2    0.63561E-02    0.13663E-02    Reading
3    1   -0.32494        0.62351        intcept
3    2   -0.11737        0.16012E-02    Reading
4    1   -0.41321        0.49411        intcept
4    2    0.18879E-01    0.18271E-02    Reading
5    1    0.52911        0.57515        intcept
5    2    0.30752E-01    0.20819E-02    Reading
6    1   -0.17362        0.66952        intcept
6    2    0.19155E-01    0.24943E-02    Reading
```

7	1	-0.69131E-01	0.68260	intcept
7	2	-0.68889E-02	0.18396E-02	Reading
8	1	1.2686	0.60906	intcept
8	2	0.92458E-01	0.17909E-02	Reading
9	1	-0.56617	0.69489	intcept
9	2	-0.50620E-01	0.19943E-02	Reading

The first line gives the estimate of $u_1 = 0.10707$ and its standard error 0.61606. The next line gives the estimate of $v_1 = 0.007727$ and its standard error 0.00267. The next two lines gives estimated u_2 and v_2, and so on. The fact that most standard errors are larger than the estimates is in line with previous results about a_i and b_i: All these error terms are essentially zero.

Using the estimated u_i and v_i and the parameter estimates of α and β, one can obtain estimates of a_i and b_i for each level-2 unit i from equations (4.18) and (4.19). For $i = 1$ we get

$$a_1 = \alpha + u_1 = 37.68531 + 0.10707 = 37.792$$

$$b_1 = \beta + v_1 = 0.65545 + 0.00727 = 0.663 \ ,$$

and for $i = 9$ we get

$$a_9 = \alpha + u_9 = 37.68531 - 0.56618 = 37.119$$

$$b_9 = \beta + v_9 = 0.65545 - 0.0506 = 0.604 \ .$$

The complete results are given in the following table

	Career								
	Tr	B	M	I	R	P	S	T	U
a_i	37.79	37.33	37.36	37.27	38.21	37.51	37.62	38.95	37.12
b_i	0.66	0.66	0.54	0.67	0.69	0.68	0.65	0.75	0.60

This can be compared with the results obtained previously with conditional regression. It is seen that both intercept and slope estimates are different between conditional regression and multilevel analysis. Why is this? It is because the two approaches are based on different assumptions. Conditional regression assumes that there is a random sample for each level-2 unit and estimates the regression intercepts and slopes as fixed parameters in each level-2 population. Multilevel analysis, on the other hand, assumes that the level-2 units come from a population of level-2 units, so that the intercepts and slopes are random variables. The empirical Bayes estimates of these are estimates of their expected values in the population.

The estimates of a_i and b_i can be used to obtain the equation

$$E(y \mid x) = a_i + b_i x \tag{4.26}$$

for predicting y from x for each level-2 unit i.

It is sometime of interest to examine the fit of the model in detail. To do so, one can obtain estimates of e_{ij} for each level-1 unit ij. These are called *residuals* and are obtained in the output file **mathread1.res**. The numbers are calculated by

$$\hat{y}_{ij} = \hat{\alpha} + \hat{u}_i + (\hat{\beta} + \hat{v}_i)x_{ij} \quad \hat{e}_{ij} = y_{ij} - \hat{y}_{ij}$$

For $i = 1$ and $i = 9$ and some j-values they are summarized in the following table

Case	i	j	y_{ij}	$\hat{y}_{ij}$	$\hat{e}_{ij}$
1	1	1	48.710	47.892	0.8177
2	1	2	43.490	41.987	1.5026
3	1	3	44.080	47.733	-3.6533
4	1	4	47.500	53.035	-5.5351
.	.	.	.	.	.
11	1	11	42.230	43.883	-1.6528
12	1	12	56.150	49.185	6.9654
.	.	.	.	.	.
139	9	1	49.300	52.560	-3.2605
140	9	2	61.370	60.907	0.4628
141	9	3	46.970	48.859	-1.8889
.	.	.	.	.	.
145	9	7	47.840	46.458	1.3823
146	9	8	59.420	60.254	-0.8339
147	9	9	44.800	44.698	0.1023

4.4   Two-Level Model with Cross-Level Interaction

As in Section 4.3, we consider the regression of y on x for each level-2 unit:

$$y_{ij} = a_i + b_i x_{ij} + e_{ij} \qquad (4.27)$$

$$i = 1, 2, \ldots, N \ \text{ level 2 units} \qquad (4.28)$$

$$j = 1, 2, \ldots, n_i \ \text{ level 1 units} \qquad (4.29)$$

Suppose the intercept a_i and slope b_i depend on a variable z_i observed on each level-2 unit i. As a first approximation, these relationships are modeled as linear, so that

$$a_i = \alpha + \gamma_a z_i + u_i \ , \qquad (4.30)$$

and

$$b_i = \beta + \gamma_b z_i + v_i \ , \qquad (4.31)$$

where

$$\begin{pmatrix} u_i \\ v_i \end{pmatrix} \sim N(\mathbf{0}, \mathbf{\Phi}) \ . \qquad (4.32)$$

As before we assume that

$$e_{ij} \sim N(0, \sigma_e^2) \ , \qquad (4.33)$$

where σ_e^2 is the same across all level-2 units. This assumption can be relaxed as shown in Section 4.6. Furthermore, it is assumed that e_{ij} is independent of u_i and v_i.

Substituting a_i and b_i into (4.27) gives

$$y_{ij} = (\alpha + \beta x_{ij} + \gamma_a z_i + \gamma_b x_{ij} z_i) + (u_i + v_i x_{ij} + e_{ij}) \ . \qquad (4.34)$$

The part $\alpha + \beta x_{ij} + \gamma_a z_i + \gamma_b x_{ij} z_i$ is called the *fixed* part of the model as it contains the fixed parameters to be estimated, and the part $u_i + v_i x_{ij} + e_{ij}$ is called the *random* part of the model as it contains the random variables whose variances and covariances should be estimated

Equation (4.34) can be written in matrix form as

$$\mathbf{y}_i = \mathbf{X}_{1i}\boldsymbol{\beta} + \mathbf{X}_{2i}\mathbf{b}_i + \mathbf{e}_i , \quad i = 1, 2, \ldots, N , \tag{4.35}$$

where $\mathbf{y}_i = (y_{i1}, y_{i2}, \ldots, y_{n_i})'$, $\mathbf{b}_i = (u_i, v_i)'$, $\boldsymbol{\beta} = (\alpha, \beta, \gamma_a, \gamma_b)'$ $\mathbf{e}_i = (e_{i1}, e_{i2}, \ldots, e_{n_i})'$, and

$$\mathbf{X}_{1i} = \begin{bmatrix} 1 & x_{i1} & z_i & x_{i1}z_i \\ 1 & x_{i2} & z_i & x_{i2}z_i \\ \vdots & \vdots & & \vdots & \vdots \\ 1 & x_{in_i} & z_i & x_{in_i}z_i \end{bmatrix} \quad \mathbf{X}_{2i} = \begin{bmatrix} 1 & x_{i1} \\ 1 & x_{i2} \\ \vdots & \vdots \\ 1 & x_{in_i} \end{bmatrix} .$$

Note that the first two columns of $\mathbf{X}_{1i}$ are the same as the two columns of $\mathbf{X}_{2i}$.

The mean vector and covariance matrix of $\mathbf{y}_i$ are

$$\boldsymbol{\mu}_i = E(\mathbf{y}_i) = \mathbf{X}_{1i}\boldsymbol{\beta} , \tag{4.36}$$

$$\boldsymbol{\Sigma}_i = Cov(\mathbf{y}_i) = \mathbf{X}_{2i}\boldsymbol{\Phi}\mathbf{X}_{2i}' + \sigma_e^2\mathbf{I}_{n_i} , \tag{4.37}$$

where $\boldsymbol{\Phi}$ is the covariance matrix of $\mathbf{b}$.

The normality of the random variables in the model implies that $\mathbf{y}_i$ is normally distributed. The likelihood of $\mathbf{y}_i$ is

$$L_i = (2\pi)^{-\frac{1}{2}} \mid \boldsymbol{\Sigma}_i \mid^{-\frac{1}{2}} exp\{-\frac{1}{2}(\mathbf{y}_i - \boldsymbol{\mu}_i)'\boldsymbol{\Sigma}_i^{-1}(\mathbf{y}_i - \boldsymbol{\mu}_i)\} , \tag{4.38}$$

so that

$$-2\ln L_i = \ln(2\pi) + \ln \mid \boldsymbol{\Sigma}_i \mid + (\mathbf{y}_i - \boldsymbol{\mu}_i)'\boldsymbol{\Sigma}_i^{-1}(\mathbf{y}_i - \boldsymbol{\mu}_i) . \tag{4.39}$$

The *deviance* $-2\ln L$ of all observation y_{ij} is

$$D = -2\ln L = \sum_{i=1}^{N} -2\ln L_i \tag{4.40}$$

D is to be minimized with respect to the parameters $\boldsymbol{\beta}$, $\boldsymbol{\Phi}$, and σ_e^2. See in the appendix (Chapter 17).

4.5   Likelihood, Deviance, and Chi-Square

Consider a model M_0 with t_0 independent parameters and an alternative model M_1 with $t_1 > t_0$ independent parameters. It is assumed that model M_0 is a special case of model M_1 such that the parameters in M_0 are restricting the parameters in M_1 in some way. The most common situation is when some of the parameters in M_1 are fixed to 0 or some other value.

Maximizing L is equivalent to minimizing D. Let

$$L_0 = maxL \text{ under } M_0 \tag{4.41}$$

and

$$L_1 = maxL \text{ under } M_1 \tag{4.42}$$

Then

$$L_1 \geq L_0 \implies \log L_1 \geq \log L_0 \implies D_1 \leq D_0 \tag{4.43}$$

Chi-square is defined as

$$\chi^2 = D_0 - D_1 = 2\log L_1 - 2\log L_0 = 2\log \frac{L_1}{L_0} = 2\log(LR) \tag{4.44}$$

with

$$\text{Degrees of freedom } d = t_1 - t_0 . \tag{4.45}$$

The model described in equations (4.27)–(4.34) can be generalized in various ways. For example, equation (4.27) may be non-linear in x_{ij} or there can be more than one x_{ij}-variable, and there can be more than one variable z_i in the relationships describing a_i and b_i. The multivariate case of more than one y_{ij}-variable is considered in Section 4.9.

4.5.1   Example: Math Achievement and Socioeconomic Status

The file **mathach.lsf** contains six variables, namely **mathach, ses, school, sector, meanses** and **cses**. These data include information on 7185 students from 160 high schools in United States: 90 public and 70 Catholic. This is one of the highlighted examples in Bryk and Raudenbush (1992 p. 60), which is often regarded as a benchmark example. The variables are

mathach a standardized measure of student mathematics achievement.

ses student socioeconomic status, a composite measure of parental education, parental occupation and income.

school a school identifier

sector =1 for a catholic school, and = 0, for a public school

meanses the average of the student ses values within each school.

cses = **ses** - **meanses**

Bryk and Raudenbush (1992 p. 61) list the following research questions:

- How much do U.S. high schools vary in their mean mathematics achievement?

- Do schools with high socioeconomic status also have high math achievement?

- Is the strength of association between socioeconomic status and math achievement similar across schools? Or is socioeconomic status a better predictor of math achievement in some schools than others?

- How do public and Catholic schools compare in terms of mean math achievement and in terms of the strength of the ses-math achievement relationship after controlling for the mean socioeconomic status?

To answer these questions we will consider four models.

In the formulas we will use the notation in Section 4.4 and write

- $y_{ij} =$ **mathach** of student j in school i

- $x_{ij} =$ **cses** = **ses** - **meanses**, i.e., the deviation of **ses** of student j in school i from the mean of **ses** in school i

There are two covariates observed at the school level:

- $z_{1i} =$ **meanses**

- $z_{2i} =$ **sector**

Model 1: Intercept-Only Model

$$y_{ij} = a_i + e_{ij}$$

$$a_i = \alpha + u_i$$

$$y_{ij} = \alpha + u_i + e_{ij}$$

where

- $a_i =$ mean math achievement in school i

- $\alpha =$ mean math achievement in the population of schools

This model can be estimated using the following multilevel syntax **mathach1.prl**

```
OPTIONS;
TITLE=Mathach Model 1;
SY=mathach.lsf;
ID2=school;
RESPONSE=mathach;
FIXED=intcept;
RANDOM1=intcept;
RANDOM2=intcept;
```

From the output file **mathach1.out** we see that all three parameter estimates are highly significant. The parameter estimates are

$$\hat{\alpha} = 12.64, \ \hat{\sigma}^2(u_i) = 8.55, \ \hat{\sigma}^2(e_{ij}) = 39.15$$

The between school variance of math achievement is 8.55 and the within school variance is 39.15. This gives an intraclass correlation of $8.55/(8.55+39.15) = 0.18$. Thus, 18% of the variance of math achievement is due to between school variation.

Model 2: Regression with Means as Outcomes This model investigates the extent to which the between schools variance can be accounted for by **meanses**. Hence, we regress a_i on z_{1i}. The model is

$$y_{ij} = a_i + e_{ij}$$

$$a_i = \alpha + \gamma_1 z_{1i} + u_i$$

$$y_{ij} = (\alpha + \gamma_1 z_{1i}) + (u_i + e_{ij})$$

where

- a_i = mean math achievement in school i

- α = regression intercept

- γ_1 = regression slope

- u_i = error variance in the regression

To estimate this model one can add `SUMMARY=NONE` on the `OPTIONS` line since the data summary is already given in the output file **mathach1.out**. We also add `meanses` on the `FIXED` line, see file **mathach2.prl**.

The resulting parameter estimates are

$$\hat{\alpha} = 12.65, \; \hat{\gamma}_1 = 5.86$$
$$\hat{\sigma}^2(u_i) = 2.59, \; \hat{\sigma}^2(e_{ij}) = 39.16$$

which are all highly significant. An increase of one unit on meanses increases the school mean of math achievement by 5.86 units. The estimated regression variance is 2.59 which can compared with the total between school variance of 8.55 obtained for Model 1. This gives $R^2 = 1 - \frac{2.59}{8.55} = 0.70$. Thus, 70% of the between school variance of math achievement is explained by the school mean socioeconomic status.

Model 3: Random Regression Model

In this model we consider the within schools regression of y_{ij} (mathach) on x_{ij} (cses):

$$y_{ij} = a_i + b_i x_{ij} + e_{ij}$$

and estimate mean vector and covariance matrix of the intercept a_i and slope b_i. The model is the same as in Section 4.3:

$$a_i = \alpha + u_i$$
$$b_i = \beta + v_i$$
$$y_{ij} = (\alpha + \beta x_{ij}) + (u_i + v_i x_{ij} + e_{ij})$$

where

- a_i = random intercept in school i

- b_i = random slope in school i

- α = mean intercept in the population of schools

- β = mean slope in the population of schools

Because of the way x_{ij} is defined, a_i can be interpreted as the mean math achievement in school i.

This model can be estimated using the following multilevel syntax file (file **mathach3.prl**)

```
OPTIONS SUMMARY=NONE;
TITLE=Mathach Model 3;
SY=mathach.lsf;
ID2=school;
RESPONSE=mathach;
FIXED=intcept cses;
RANDOM1=intcept;
RANDOM2=intcept cses;
```

As shown in file **mathach3.out** all parameters are highly significant except the covariance between a_i and b_i which is estimated as 0.05. Hence, the intercept and slope can be assumed to be uncorrelated. The other parameter estimates are

$$\hat{\alpha} = 12.65, \ \hat{\beta} = 2.19,$$
$$\hat{\sigma}^2(u_i) = 8.62, \ \hat{\sigma}^2(v_i) = 0.68, \ \hat{\sigma}^2(e_{ij}) = 36.70$$

These results can be interpreted as follows. Both the intercept, *i.e.*, the school mean math achievement, and the slope, *i.e.*, rate of increase of math achievement with ses, varies considerably across schools. The situation corresponds to the upper left figure in Figure 4.1. The average school mean is 12.65 and the variance is 8.62 across schools. The average within school slope is 2.19 with a variance across schools of 0.68.

Model 4: Intercepts and Slopes as Outcomes

In Model 4 we investigate if the variances of the intercepts and slopes in the within schools regression of mathach on ses, as obtained in Model 3 can be explained by meanses and sector. To do so, we write the model equations as

$$y_{ij} = a_i + b_i x_{ij} + e_{ij} \tag{4.46}$$

$$a_i = \alpha + \gamma_{11} z_{i1} + \gamma_{12} z_{i2} + u_i \tag{4.47}$$

$$b_i = \beta + \gamma_{21} z_{i1} + \gamma_{22} z_{i2} + v_i \tag{4.48}$$

Equation (4.47) is the regression of the intercept a_i on meanses and sector, here called the intercept regression, and (4.48) is the regression of the slope b_i on meanses and sector, here called the slope regression.

Substituting the (4.47) and (4.48) into (4.46) gives

$$
\begin{aligned}
y_{ij} &= [\alpha + \beta x_{ij} + \gamma_{11} z_{i1} + \gamma_{12} z_{i2} \\
&+ \ \gamma_{21} x_{ij} z_{i1} + \gamma_{22} x_{ij} z_{i2}] \\
&+ \ [u_i + v_i x_{ij} + e_{ij}]
\end{aligned} \tag{4.49}
$$

where

- a_i, b_i = random intercept and slope in the within schools regression of mathach on cses in school i

- α, β = intercepts in the intercept and slope regressions

- γ_{11}, γ_{12} = effect of meanses and sector on the intercept

- γ_{21}, γ_{22} = effect of meanses and sector on the slope

To estimate this model use the following multilevel syntax file, **mathach4.prl**

```
OPTIONS SUMMARY=NONE;
TITLE=Mathach Model 4;
SY=mathach.lsf;
ID2=school;
RESPONSE=mathach;
FIXED=intcept cses meanses sector cses*meanses cses*sector;
RANDOM1=intcept;
RANDOM2=intcept cses;
```

Note that this includes the products **cses*meanses** and **cses*sector** corresponding to $x_{ij}z_{i1}$ and $x_{ij}z_{i2}$ in (4.49).

The output files for each of the models considered in this example gives the deviance $-2\ln L$ and the number of estimated parameters. For Model 4 this is

```
DEVIANCE= -2*LOG(LIKELIHOOD) =      46496.4338826074
  NUMBER OF FREE PARAMETERS =                10
```

The deviances and number of parameters and their differences as chi-squares and degrees of freedom are given in Table 4.1.

Model	Deviance	Parameters	Chi-Square	Degrees of Freedom
1	47115.81	3		
2	46959.11	4	156.70	1
3	46710.98	6	248.13	2
4	46496.43	10	214.55	4

Table 4.1: Deviances and Chi-Squares for Models 1 – 4

All four models are not in a nested sequence. Model 1 is a special case of Model 2 when $\gamma_1 = 0$. Also Model 3 is a special case of Model 4 when $\gamma_{11} = 0$, $\gamma_{12} = 0$, $\gamma_{21} = 0$, $\gamma_{22} = 0$. So the chi-squares 156.70 can be used to conclude that Model 2 fits significantly much better than Model 1 and the chi-square 214.55 can be used to conclude that Model 4 fits significantly much better than Model 3. However, Model 2 is not a special case of Model 3. So 248.13 cannot be used as a chi-square in a similar way. Because there is no saturated model, there is no deviance for a saturated model that can be used to test the fit of a model in an absolute sense.

The output file **mathach4.out** gives the following results

```
+-----------------------+
| FIXED PART OF MODEL   |
+-----------------------+
```

```
------------------------------------------------------------
COEFFICIENTS      BETA-HAT  STD.ERR.    Z-VALUE    PR > |Z|
------------------------------------------------------------
intcept          12.11360   0.19692   61.51554    0.00000
cses              2.93936   0.15350   19.14949    0.00000
meanses           5.33795   0.36569   14.59696    0.00000
sector            1.21694   0.30338    4.01132    0.00006
meanses *cses     1.04237   0.29604    3.52108    0.00043
sector  *cses    -1.64387   0.23736   -6.92564    0.00000
```

```
             +-----------------------+
             | RANDOM PART OF MODEL   |
             +-----------------------+
```

```
------------------------------------------------------------
LEVEL 2           TAU-HAT   STD.ERR.    Z-VALUE    PR > |Z|
------------------------------------------------------------
intcept/intcept   2.31890   0.35535    6.52574    0.00000
cses   /intcept   0.18810   0.19584    0.96048    0.33681
cses   /cses      0.06524   0.20764    0.31418    0.75338
```

```
------------------------------------------------------------
LEVEL 1           TAU-HAT   STD.ERR.    Z-VALUE    PR > |Z|
------------------------------------------------------------
intcept/intcept  36.72115   0.62593   58.66614    0.00000
```

From the fixed part of the model we extract the parameter estimates rounded to 2 decimals as $\hat{\alpha} = 12.11$, $\hat{\beta} = 2.94$, $\hat{\gamma}_{11} = 5.34$, $\hat{\gamma}_{12} = 1.22$, $\hat{\gamma}_{21} = 1.04$, and $\hat{\gamma}_{22} = -1.64$.

Since $\hat{\gamma}_{11} = 5.34$ is significant, we see that meanses is positively related to school mean math achievement. This holds for both public and catholic schools. Also, since $\hat{\gamma}_{12} = 1.22$ is significant, we see that catholic schools have significantly higher mean math achievement than public schools even after controlling for meanses.

For the slope relationships both $\hat{\gamma}_{21} = 1.04$ and $\hat{\gamma}_{22} = -1.64$ are significant. So we conclude that there is a tendency for schools with high meanses to have larger slopes than schools with low meanses. Furthermore, catholic schools have significantly weaker SES slopes than public schools.

How strong are the intercept and slope relationships (4.47) and (4.48)? We can estimate the strength of these linear relationships by comparing the estimated variances of u_i and v_i with the estimated variances of a_i and b_i obtained for Model 3. For the intercept equation we obtain $R^2 = 1 - 2.32/8.62 = 0.73$ and for the slope equation we obtain $R^2 = 1 - 0.065/0.68 = 0.90$. Thus, the slope can be predicted very accurately from meanses and sector.

These results are further clarified as follows. Selecting values 0 and 1 for z_2 and values -0.4, 0, and 0.4 for z_1, we can calculate the expected values of a_i and b_i from equations (4.47) and (4.48) These expected values are shown in Table 4.2

Using the values of a_i and b_i in Table 4.2, we can calculate and draw the regression lines for each combination of z_1 and z_2. These regression lines are shown in Figure 4.2.

Table 4.2: Expected Values of a_i and b_i for Selected Values of z_1 and z_2

z_2	z_1	$E(a_i)$	$E(b_i)$
0	-0.4	6.78	2.52
0	0.0	12.11	2.94
0	0.4	17.44	3.36
1	-0.4	7.99	0.88
1	0.0	13.32	1.30
1	0.4	18.65	1.72

Figure 4.2: Within and Between Schools Regressions for Public (left) and Catholic (right) Schools for Different Values of Meanses

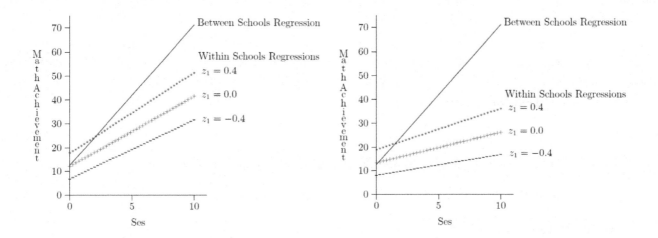

4.6   Multilevel Analysis of Repeated Measurements

One typical application of the multilevel model in Section 4.4 is in the analysis of repeated measurements data where each unit of observation is followed over a number of successive time points (occasions). For example, in psychology one studies the development of cognitive abilities in children, in economics one studies the growth of nations or sales of companies, and in sociology one studies the change in crime rates across communities. Such data is often called longitudinal but we shall use this term in a more general sense than repeated measurements where the occasions occur at regular intervals.

The time intervals may be daily, weekly, monthly, quarterly or annually. The unit of observation may be individuals, businesses, regions, or countries. "Regardless of the subject area or the time interval, social and behavioral scientists have a keen interest in describing and explaining the time trajectories of their variables." (Bollen & Curran (2006, p. 1)

We give several examples here with a variation of ingredients. These examples can also be considered as *Latent Curve Models*, see Bollen and Curran (2006). Section 9.3 of this book shows how these type of models can be estimated using **SIMPLIS** or **LISREL** syntax. Here we illustrate how longitudinal data with missing values can be estimated using multilevel analysis.

4.6.1   Example: Treatment of Prostate Cancer

A medical doctor offered all his patients diagnosed with prostate cancer a treatment aimed at reducing the cancer activity in the prostate. The severity of prostate cancer is often assessed by a plasma component known as prostate specific antigen (PSA), an enzyme that is elevated in the presence of prostate cancer. The PSA level was measured regularly every three months. The data contains five repeated measurements of PSA. The age of the patient is also included in the data. Not every patient accepted the offer at the first occasion and several patients chose to enter the program after the first occasion. Some patients, who accepted the initial offer, are absent at some later occasions for various reasons. Thus, there are missing values in the data.

The aim of this study is to answer the following questions: What is the average initial PSA value? Do all patients have the same initial PSA value? Is there an overall effect of treatment. Is there a decline of PSA values over time, and, if so, what is the average rate of decline? Do all patients have the same rate of decline? Does the individual initial PSA value and/or the rate of decline depend on the patient's age?

This is a typical example of repeated measurements data, the analysis of which is sometimes done within the framework of multilevel analysis. It represents the simplest type of a two-level model.

The data file is **psa.lsf** where the variables are

$y_{ij} = \text{PSA}$

$x_{ij} = \text{Months} = \text{Number of months since first time visit to the doctor}$

$z_i = \text{Age} = \text{Age of the patient at first time visit to the doctor}$

The variable Patient is used as the level-2 identifier.

We used this data in Chapter 1 to explore the characteristics of the data with particular focus on missing values. In that chapter we used the data in longitudinal form, see the display of the first few lines given on page 20. Here we must use the data in multilevel form where the time of the occasions is included as a variable. The distinction between data in multilevel form and data in longitudinal form will be explained in Section 9.3.

The first few lines of the data file **psa.lsf** are shown here

	Patient	Months	PSA	Age
1	1	0.	30.	69.
2	1	3.	28.	69.
3	1	6.	27.	69.
4	1	9.	25.	69.
5	1	12.	20.	69.
6	2	0.	28.	58.
7	2	3.	27.	58.
8	2	6.	21.	58.
9	2	9.	19.	58.
10	2	12.	19.	58.
11	3	0.	27.	53.
12	3	3.	22.	53.
13	3	6.	18.	53.
14	3	9.	18.	53.
15	3	12.	15.	53.
16	4	0.	25.	61.
17	4	3.	25.	61.
18	4	6.	20.	61.
19	4	9.	20.	61.
20	4	12.	19.	61.
21	5	0.	34.	63.
22	5	3.	30.	63.

In this kind of data it is inevitable that there are missing values. For example, a patient may be on vacation or ill or unable to come to the doctor for any reason at some occasion or a patient may die and therefore will not come to the doctor after a certain occasion. For example, a quick perusal of the data shows that

- Patients 9 and 10 are missing at 3 months

- Patient 15 is missing at 3 and 6 months

- Patient 16 is missing at 0, 3, and 12 months

Such missing data are automatically handled if the time of record is included in the data. In this case this is the variable Months.

We begin by analyzing the data according to model (4.15), *i.e.*, we leave Age out. This can be done using the PRELIS syntax file **psa1.prl**:

```
OPTIONS;
TITLE=Treatment of Prostate Cancer;
SY=psa.LSF;
ID2=Patient;
RESPONSE=PSA;
FIXED=intcept Months;
RANDOM1=intcept;
RANDOM2=intcept Months;
```

The model converges after 6 iterations and gives the following results:

```
+-----------------------+
| FIXED PART OF MODEL   |
+-----------------------+
```

COEFFICIENTS	BETA-HAT	STD.ERR.	Z-VALUE	PR > |Z|
intcept	31.93378	0.57101	55.92473	0.00000
Months	-0.74214	0.01861	-39.86912	0.00000

```
+-----------------------+
| RANDOM PART OF MODEL  |
+-----------------------+
```

LEVEL 2	TAU-HAT	STD.ERR.	Z-VALUE	PR > |Z|
intcept /intcept	30.89912	4.61240	6.69914	0.00000
Months /intcept	0.30243	0.10758	2.81119	0.00494
Months /Months	0.00392	0.00539	0.72798	0.46663

LEVEL 1	TAU-HAT	STD.ERR.	Z-VALUE	PR > |Z|
intcept /intcept	2.28753	0.20830	10.98199	0.00000

This means that the average initial PSA value is 31.93 and PSA decreases by 0.74 per quarter on average or about 0.25 per month. The variation of the intercept is highly significant, so the initial PSA value varies considerably across patients. The variation of the slope is not statistically significant, so that the rate of decrease is essentially the same for all patients. So this example corresponds to the lower left figure in Figure 4.1. The error variance σ_e^2 is estimated as 2.29.

To see if the intercept a_i and the slope b_i can be predicted by Age, we use model (4.34). Then we need to use the variable $z_i = $ Age as well as the product variable $x_{ij}z_i$, i.e., the product of Months and Age. The latter variable is not included in the data file **psa.lsf**. However, this is not necessary since it can be introduced in the input file directly as shown in the line

```
FIXED=intcept Months Age Months*Age;
```

Otherwise, the input is the same as in the previous analysis, see file **psa2.prl**.

The result for the fixed part of the model is

```
+----------------------+
| FIXED PART OF MODEL  |
+----------------------+
```

COEFFICIENTS	BETA-HAT	STD.ERR.	Z-VALUE	PR > |Z|
intcept	15.89819	3.73784	4.25331	0.00002
Months	-0.60000	0.13558	-4.42535	0.00001
Age	0.28920	0.06676	4.33182	0.00001
Months *Age	-0.00255	0.00242	-1.05524	0.29131

which means that the effect of Age on the initial PSA value is highly significant. The initial PSA value increases by 0.29 per year of Age. However, the effect of Age on the rate of decrease of PSA is not significant, as indicated by the P-value of the product variable. The latter result was anticipated as we saw in the first analysis that the slope has almost no variance.

The output files also gives the value the deviance $-2\ln L$ and the number of parameters estimated. For the first model this is given as

```
DEVIANCE= -2*LOG(LIKELIHOOD) =      2008.60066876969
NUMBER OF FREE PARAMETERS =              6
```

and for the second model this is

```
DEVIANCE= -2*LOG(LIKELIHOOD) =      1988.75995647571
NUMBER OF FREE PARAMETERS =              8
```

As shown in Section 4.5, the deviance difference 19.84 is a chi-square with 2 degrees of freedom for testing the hypothesis that both γ_a and γ_b are 0. Clearly, this hypothesis is rejected, which is in agreement with what we have already found, namely that the estimate of γ_a is highly significant.

4.6.2 Example: Learning Curves of Air Traffic Controllers

The data used in this example are described by Kanfer and Ackerman (1989)[1]. The data consists of information for 141 U.S. Air Force enlisted personnel who carried out a computerized air traffic controller task developed by Kanfer and Ackerman.

The subjects were instructed to accept planes into their hold pattern and land them safely and efficiently on one of four runways, varying in length and compass directions, according to rules governing plane movements and landing requirements. For each subject, the success of a series of between three and six 10-minute trials was recorded. The measurement employed was the number of correct landings per trial.

The Armed Services Vocational Battery (ASVB) was also administered to each subject. A global measure of cognitive ability, obtained from the sum of scores on the 10 subscales, is included in the data.

The data for this example can be found in the **kanfer1.lsf** file. The variable labels and first few records of this data file are shown below.

kanfer1.lsf	control	time	measure	ability
1	1.000	1.000	24.000	142.160
2	1.000	2.000	27.000	142.160
3	1.000	3.000	30.000	142.160
4	1.000	4.000	32.000	142.160
5	1.000	5.000	38.000	142.160
6	1.000	6.000	41.000	142.160
7	2.000	1.000	2.000	-7.630
8	2.000	2.000	3.000	-7.630
9	2.000	3.000	9.000	-7.630
10	2.000	4.000	13.000	-7.630
11	2.000	5.000	13.000	-7.630
12	2.000	6.000	14.000	-7.630
13	3.000	1.000	12.000	-67.430

where the variables are

control controller $i = 1, 2, \ldots, 141$ (the level-2 identifier)

time trial 1, 2, 3, 4, 5, 6 (the x_{ij}-variable)

measure the performance measure (the y_{ij}-variable)

ability the cognitive ability measure(the z_i-variable)

A quick browse at the end of the data file shows that controllers 138, 139, and 140 only completed 5 trials, and controller 141 only completed 3 trials. Thus there are missing values also in this data.

A box plot of measure against time is shown in Figure 4.3. This suggests that the growth curve may be non-linear so we will consider a quadratic and cubic growth curve as well as a linear.

---

[1] Permission for SSI to use the copyrighted raw data was provided by R. Kanfer and P.L. Ackerman. The data are from experiments reported in Kanfer, R., and Ackerman, P.L. (1989). The data remain the copyrighted property of Ruth Kanfer and Phillip L. Ackerman. Further publication or further dissemination of these data is not permitted without the expressed consent of the copyright owners.

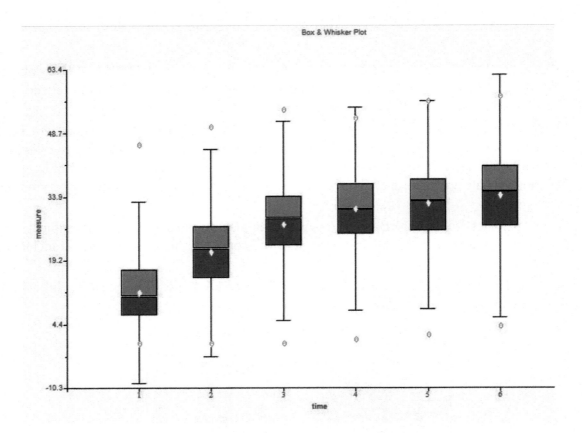

Figure 4.3: Box-Whisker Plot of measure against time

Therefore we will consider three models. To emphasize that x_{ij} is a measure of time in these examples we shall write T_{it} instead of x_{ij}, where T_{it} is time at occasion t, $t = 1, 2, \ldots, 6$. In this example, $T_{it} = t$, for all i, which is the variable **time** in the data. The three models are

Linear Model: $y_{it} = a_i + b_{i1}T_{it} + e_{it}$

Quadratic Model: $y_{it} = a_i + b_{i1}T_{it} + b_{i2}T_{it}^2 + e_{it}$

Cubic Model: $y_{it} = a_i + b_{i1}T_{it} + b_{i2}T_{it}^2 + b_{i3}T_{it}^3 + e_{it}$

For the quadratic model we need to add the variable $\texttt{time2} = \texttt{time}^2$ as shown here

	control	time	measure	ability	time2	time3
1	1.000	1.000	24.000	142.160	1.000	1.000
2	1.000	2.000	27.000	142.160	4.000	8.000
3	1.000	3.000	30.000	142.160	9.000	27.000
4	1.000	4.000	32.000	142.160	16.000	125.000
5	1.000	5.000	38.000	142.160	25.000	216.000
6	1.000	6.000	41.000	142.160	36.000	216.000
7	2.000	1.000	2.000	-7.630	1.000	1.000
8	2.000	2.000	3.000	-7.630	4.000	8.000
9	2.000	3.000	9.000	-7.630	9.000	27.000
10	2.000	4.000	13.000	-7.630	16.000	125.000
11	2.000	5.000	13.000	-7.630	25.000	216.000
12	2.000	6.000	14.000	-7.630	36.000	216.000
13	3.000	1.000	12.000	-67.430	1.000	1.000

For the cubic model we need to add the variable $\texttt{time3} = \texttt{time}^3$ in addition.

We will also investigate if the random effects $a_i, b_{i1}, b_{i2}, b_{i3}$ can be predicted by the ability measure z_i. For this purpose, we need also the products of z_i and T_{it}, T_{it}^2, T_{it}^3, as shown here

	control	time	measure	ability	time2	time3	timeab	time2ab	time3ab
1	1.000	1.000	24.000	142.160	1.000	1.000	142.160	142.160	142.160
2	1.000	2.000	27.000	142.160	4.000	8.000	284.320	568.640	1137.280
3	1.000	3.000	30.000	142.160	9.000	27.000	426.480	1279.440	3838.320
4	1.000	4.000	32.000	142.160	16.000	125.000	568.640	2274.560	17770.000
5	1.000	5.000	38.000	142.160	25.000	216.000	710.800	3554.000	30706.560
6	1.000	6.000	41.000	142.160	36.000	216.000	852.960	5117.760	30706.560
7	2.000	1.000	2.000	-7.630	1.000	1.000	-7.630	-7.630	-7.630
8	2.000	2.000	3.000	-7.630	4.000	8.000	-15.260	-30.520	-61.040
9	2.000	3.000	9.000	-7.630	9.000	27.000	-22.890	-68.670	-206.010
10	2.000	4.000	13.000	-7.630	16.000	125.000	-30.520	-122.080	-953.750
11	2.000	5.000	13.000	-7.630	25.000	216.000	-38.150	-190.750	-1648.080
12	2.000	6.000	14.000	-7.630	36.000	216.000	-45.780	-274.680	-1648.080
13	3.000	1.000	12.000	-67.430	1.000	1.000	-67.430	-67.430	-67.430

To estimate the linear model use file **kanfer1.prl**:

```
OPTIONS;
TITLE=Analysis of Air Traffic Controller Data: Linear Growth Curve;
SY=kanfer1.lsf;
ID2=control;
RESPONSE=measure;
FIXED=intcept time;
RANDOM1=intcept;
RANDOM2=intcept time;
```

For the quadratic model add `time2` on the `FIXED` line and on the `RANDOM2` line and change `kanfer1.lsf` to `kanfer2.lsf`, see file **kanfer2.prl**. For the cubic model add `time3` on the `FIXED` line and on the `RANDOM2` line and change `kanfer2.lsf` to `kanfer3.lsf`, see file **kanfer3.prl**.

The output file **kanfer1.out** gives the following data summary, including the beginning and ending part and omitting the middle part

```
                              +--------------+
                              | DATA SUMMARY |
                              +--------------+


      NUMBER OF LEVEL 2 UNITS :           141
      NUMBER OF LEVEL 1 UNITS :           840


      N2  :       1        2        3        4        5        6        7        8
      N1  :       6        6        6        6        6        6        6        6


      N2  :       9       10       11       12       13       14       15       16
      N1  :       6        6        6        6        6        6        6        6


      N2  :     129      130      131      132      133      134      135      136
      N1  :       6        6        6        6        6        6        6        6

             :
             :
             :

      N2  :     137      138      139      140      141
      N1  :       6        5        5        5        3
    ==========================================================================
```

showing that controllers 138, 139, and 140 completed only 5 trials and controller 141 completed only 4 trials.

The deviances, numbers of parameters, chi-squares, and degrees of freedom for the three models are shown in Table 4.3.

Table 4.3: Deviances and Chi-Squares for Three Growth Curve Models

Model	Deviance	Parameters	Chi-Square	Degrees of Freedom
Linear	5492.00	6		
Quadratic	5095.47	10	396.53	4
Cubic	5051.41	15	44.06	5

This suggests that the quadratic model fits very much better than the linear model. Also, since $\chi^2 = 44.06$ with 5 degrees of freedom has a P-value less than 0.001, the cubic model fits significantly better than the quadratic model. However, the cubic model is non-admissible since the estimated variance of b_{i3} is negative, see file **kanfer3.out**. In such cases it is recommended to choose another functional form for the growth curve or a more parsimonious model by restricting the parameters in the cubic model. There are various ways to do this but we shall not pursue this matter in detail here.

Alternative functional forms for growth curves may be a **Piecewise Linear Growth Curve**, for example

$$y_{it} = a_i + b_{i1}t + e_{it}, \quad t = 1, 2, 3 \tag{4.50}$$

$$y_{it} = a_i + b_{i2}t + e_{it}, \quad t = 4, 5, 6 , \tag{4.51}$$

an **Exponential Growth Curve**, see Du Toit and Cudeck (2001)

$$y_{it} = a_i + b_i(1 - e^{-\gamma t}) + e_{it} \tag{4.52}$$

or a **Logistic Growth Curve**

$$y_{it} = \frac{a_i}{1 + e^{(b_i - c_i t)}} + e_{it} \tag{4.53}$$

Obtaining the Deviance for a Saturated Model

A problem with the chi-squares in Table 4.3 is that they don't tell whether the model fits the data or not. The reason for this is that the deviance for a saturated model is not automatically computed by the program. Here we will show how this deviance can be obtained and how goodness-of-fit chi-squares can be obtained for each of the three models.

For this purpose we need to generate six dummy variables $d_{it}, i = 1, 2, \ldots, 6$ such that $d_{it} = 1$ if $T_{it} = t$ and $d_{it} = 0$ otherwise. First save a copy of **kanfer1.lsf** as **kanferwithdummies.lsf**. The dummy variables can the be obtained by running the following multilevel syntax file **kanfer1withdummies.prl**

```
OPTIONS;
TITLE=Air Controller Data: Generating Dummies for time
SY=kanferwithdummies.lsf;
ID2=control;
FIXED=intcept;
DUMMY=time;
RANDOM1=intcept;
RANDOM2=intcept;
```

Here we use the intercept-only model but the model is irrelevant although some model must be specified. The important line is the line

```
DUMMY=time;
```

There is no need to read the output file **kanferwithdummies.out**. The only purpose of this run is to generate the file **kanferwithdummies.lsf** containing the six dummy variables. To see this file close it and then open it. The beginning if this file looks like this

kanfer0.lsf

	control	time	measure	ability	dummy1	dummy2	dummy3	dummy4	dummy5	dummy6
1	1.000	1.000	24.000	142.160	1.000	0.000	0.000	0.000	0.000	0.000
2	1.000	2.000	27.000	142.160	0.000	1.000	0.000	0.000	0.000	0.000
3	1.000	3.000	30.000	142.160	0.000	0.000	1.000	0.000	0.000	0.000
4	1.000	4.000	32.000	142.160	0.000	0.000	0.000	1.000	0.000	0.000
5	1.000	5.000	38.000	142.160	0.000	0.000	0.000	0.000	1.000	0.000
6	1.000	6.000	41.000	142.160	0.000	0.000	0.000	0.000	0.000	1.000
7	2.000	1.000	2.000	-7.630	1.000	0.000	0.000	0.000	0.000	0.000
8	2.000	2.000	3.000	-7.630	0.000	1.000	0.000	0.000	0.000	0.000
9	2.000	3.000	9.000	-7.630	0.000	0.000	1.000	0.000	0.000	0.000
10	2.000	4.000	13.000	-7.630	0.000	0.000	0.000	1.000	0.000	0.000
11	2.000	5.000	13.000	-7.630	0.000	0.000	0.000	0.000	1.000	0.000
12	2.000	6.000	14.000	-7.630	0.000	0.000	0.000	0.000	0.000	1.000
13	3.000	1.000	12.000	-67.430	1.000	0.000	0.000	0.000	0.000	0.000

The deviance for a saturated model can now be obtained by running the following multilevel syntax file **kanfer0.prl**:

```
OPTIONS;
TITLE=Air Controller Data: Estimating a Saturated Model;
SY=kanferwithdummies.lsf;
ID2=control;
RESPONSE=measure;
FIXED=dummy1 dummy2 dummy3 dummy4 dummy5 dummy6;
RANDOM2=dummy1 dummy2 dummy3 dummy4 dummy5 dummy6;
```

The output file **kanfer0.out** now gives the required deviance for a saturated model as

```
DEVIANCE= -2*LOG(LIKELIHOOD) =     5001.02145301858
NUMBER OF FREE PARAMETERS =               27
```

These values should now be added on the OPTIONS line in each of the input files **kanfer1.prl**, **kanfer2.prl**, and **kanfer3.prl**, as shown here

```
OPTIONS NFREE=27 DEVIANCE=5001.0215;
```

One can now rerun each of the input files **kanfer1.prl**, **kanfer2.prl**, and **kanfer3.prl**.

For the linear model, the following goodness-of.fit measures are obtained at the end of the output file **kanfer1.out**.

```
|-------------------------------------------------|
| Chi-square Statistic for Testing the Fit of     |
| the Current Model versus an Alternative Model   |
|-------------------------------------------------|

 -2Log(L)=        5491.9964 with        6  Free Parameters (Current Model)
 -2Log(L)=        5001.0215 with       27  Free Parameters (Alternative Model)
  Chi-Square=      490.9749, df=       21, p-value=  0.00000

    ROOT MEAN SQUARE ERROR OF APPROXIMATION
    Steiger-Lind : RMSEA = SQRT[(CHISQ-DF)/(N*DF)]
    where N, the number of level 2 units=       141

    Point Estimate     : 0.3984
    95.0 Percent C.I.  : ( 0.3624 ; 0.4352)
```

For the quadratic model the corresponding information is

```
 -2Log(L)=        5095.4659 with       10  Free Parameters (Current Model)
 -2Log(L)=        5001.0215 with       27  Free Parameters (Alternative Model)
  Chi-Square=       94.4444, df=       17, p-value=  0.00000

    ROOT MEAN SQUARE ERROR OF APPROXIMATION
    Steiger-Lind : RMSEA = SQRT[(CHISQ-DF)/(N*DF)]
```

```
        where N, the number of level 2 units=      141


        Point Estimate        : 0.1797
        95.0 Percent C.I.   : ( 0.1384 ; 0.2226)
```

and for the cubic model the corresponding information is

```
        -2Log(L)=         5051.4109 with      15  Free Parameters (Current Model)
        -2Log(L)=         5001.0215 with      27  Free Parameters (Alternative Model)
         Chi-Square=        50.3894, df=       12, p-value=  0.00000


        ROOT MEAN SQUARE ERROR OF APPROXIMATION
        Steiger-Lind : RMSEA = SQRT[(CHISQ-DF)/(N*DF)]
        where N, the number of level 2 units=      141


        Point Estimate        : 0.1506
        95.0 Percent C.I.   : ( 0.1005 ; 0.2028)
```

From these results it is clear that none of the three models fit the data well. However, to find a suitable and meaningful model that fits the data well is a considerably difficult task.

Now we return to the quadratic model and proceed to investigate if a_i, b_{i1}, and b_{i2} depend on z_i. This can be done by the following input file **kanfer4.prl**:

```
OPTIONS SUMMARY=NONE;
TITLE=Analysis of Air Traffic Controller Data;
SY=kanfer3.lsf;
ID2=control;
RESPONSE=measure;
FIXED=intcept time time2 timeab time2ab;
RANDOM1=intcept;
RANDOM2=intcept time time2;
```

The output gives the following results

```
                    +-----------------------+
                    | FIXED PART OF MODEL   |
                    +-----------------------+
```

COEFFICIENTS	BETA-HAT	STD.ERR.	Z-VALUE	PR > |Z|
intcept	1.68243	0.87326	1.92660	0.05403
time	11.48570	0.53234	21.57595	0.00000
time2	-1.03520	0.06430	-16.09960	0.00000
timeab	0.01537	0.00342	4.49032	0.00001
time2ab	-0.00184	0.00046	-3.96788	0.00007

```
+-----------------------+
|   -2 LOG-LIKELIHOOD   |
+-----------------------+
```

DEVIANCE= -2*LOG(LIKELIHOOD) =   5079.75182623279
NUMBER OF FREE PARAMETERS =   12


```
+-----------------------+
| RANDOM PART OF MODEL  |
+-----------------------+
```

LEVEL 2		TAU-HAT	STD.ERR.	Z-VALUE	PR > |Z|
intcept	/intcept	76.97237	12.97701	5.93144	0.00000
time	/intcept	-31.64105	7.07515	-4.47214	0.00001
time	/time	26.80489	4.84242	5.53543	0.00000
time2	/intcept	3.09478	0.82789	3.73813	0.00019
time2	/time	-2.84206	0.57736	-4.92253	0.00000
time2	/time2	0.32391	0.07156	4.52632	0.00001

LEVEL 1		TAU-HAT	STD.ERR.	Z-VALUE	PR > |Z|
intcept	/intcept	9.49461	0.65580	14.47793	0.00000

This suggest that both the linear and the quadratic coefficient depend on Ability.

4.6.3   Example: Growth Curves for the Weight of Mice

In the two preceding examples we have seen that LISREL handles missing values automatically as long as the time at each occasion is recorded in the data file. In the psa example in Section 4.6.1 the missing values were distributed seemingly at random over the level-2 units (patients). In the Kanfer example in Section 4.6.2 the level-2 units (controllers) were ordered so that only the last four level-2 units had missing values. In the following example the missing values are systematic in the sense that almost half of the level-2 units (mice) are missing on one occasion.

The data set contains repeated measurements on 82 striped mice and was obtained from the Department of Zoology at the University of Pretoria, South Africa. A number of male and female mice were released in an outdoor enclosure with nest boxes and sufficient food and water. They were allowed to multiply freely. Occurrence of birth was recorded daily and newborn mice were weighed weekly, from the end of the second week after birth until physical maturity was reached. The data set consists of the weights of 42 male and 40 female mice. For male mice, 9 repeated weight measurements are available and for the female mice 8 repeated measurements.

The data file is **mouse.lsf**, where gender is coded +1 for male mice and -1 for female mice. A box-whisker plot of weight against time is shown in Figure 4.4. This shows that the growth

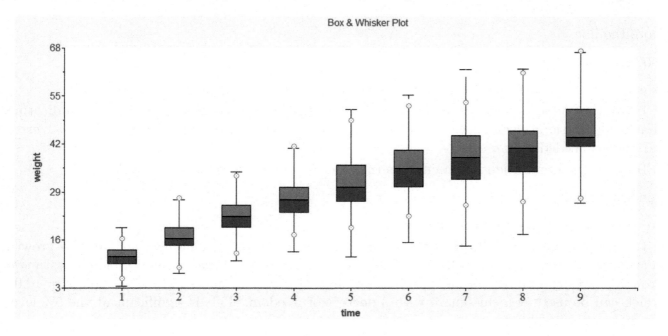

Figure 4.4: Box-Whisker Plot of weight against time

curve may not be linear and that the variance of the level-1 error e_{ij} in (4.27) is not constant but increasing with time. We will therefore consider a quadratic growth curve with a level-1 error variance proportional to time. This model is

$$y_{it} = a_i + b_{i1}t + b_{i2}t^2 + e_{it} \quad t = 1, 2, \ldots, 9 \tag{4.54}$$

where

$$a_i = \alpha + v_i , \tag{4.55}$$

$$b_{i1} = \beta_1 + u_{i1} , \tag{4.56}$$

$$b_{i2} = \beta_2 + u_{i2} , \tag{4.57}$$

and

$$\begin{pmatrix} a_i \\ b_{i1} \\ b_{i2} \end{pmatrix} \sim N \left[\begin{pmatrix} \alpha \\ \beta_1 \\ \beta_2 \end{pmatrix} , \begin{pmatrix} \phi_{11} & & \\ \phi_{21} & \phi_{22} & \\ \phi_{31} & \phi_{32} & \phi_{33} \end{pmatrix} \right] , \tag{4.58}$$

$$e_{it} \sim N(\sigma_e^2 t) . \tag{4.59}$$

This model can be estimated using the following multilevel syntax file (**mouse1.prl**):

```
OPTIONS;
TITLE=Weight of Mice Data: Quadratic Growth Curve.Complex Error Variance;
SY=mouse.lsf;
```

```
ID2=mouse;
RESPONSE=weight;
FIXED=intcept time timesq;
RANDOM1=time;
RANDOM2=intcept time timesq;
```

Note the line

```
RANDOM1=time;
```

which makes the variance of e_{it} proportional to t.

The output file **mouse1.out** shows that all 10 parameters are statistically significant, indicating that a quadratic growth curve is needed and also that the assumption about the level-1 error variance is needed.

To see if the quadratic growth curve is the same for male and female mice, add

```
gender gender*time gender*timesq;
```

on the FIXED line, see file **mouse2.pr2**.

The output file **mouse2.out** gives insufficient evidence for concluding that the two growth curves are different. A test of the model with a common growth curve against the model with two growth curves is obtained as the difference between the two deviances $3615.27 - 3605.26 = 10.01$ which can be used as a chi-square with 3 degrees of freedom. This is significant at the 5% level but not at the 1% level.

4.6.4   Example: Growth Curves for Weight of Chicks on Four Diets

We will consider yet another example of repeated measurements, this time in an experimental design setting. This example illustrates how to test contrasts between different experimental effects similar to analysis of variance (AOV).

The data file **chicks.lsf** contains the weights (in grams) of chicks on four different protein diets. Measurements were taken on days 0 2, 4, ..., 20, and 21, thus all together 12 times, see Crowder and Hand (1990, Example 5.3). Also this data contain missing values in the sense that some weight measurements are missing at certain times for some of the chicks. One would like to know if any of the four diets are superior to the others in terms of weight growth of the chicks.

The first 15 rows of the data file **chicks.lsf** looks like this

	Chick	Weight	Diet1	Diet2	Diet3	Diet4	Day	Day_Sq
1	1.	42.	1.	0.	0.	0.	0.	0.
2	1.	51.	1.	0.	0.	0.	2.	4.
3	1.	59.	1.	0.	0.	0.	4.	16.
4	1.	64.	1.	0.	0.	0.	6.	36.
5	1.	76.	1.	0.	0.	0.	8.	64.
6	1.	93.	1.	0.	0.	0.	10.	100.
7	1.	106.	1.	0.	0.	0.	12.	144.
8	1.	125.	1.	0.	0.	0.	14.	196.
9	1.	149.	1.	0.	0.	0.	16.	256.
10	1.	171.	1.	0.	0.	0.	18.	324.
11	1.	199.	1.	0.	0.	0.	20.	400.
12	1.	205.	1.	0.	0.	0.	21.	441.
13	2.	40.	1.	0.	0.	0.	0.	0.
14	2.	49.	1.	0.	0.	0.	2.	4.
15	2.	58.	1.	0.	0.	0.	4.	16.

The first variable `Chick` is an identification number for the chick, variable 2 is the `Weight` and variables 3 to 6 labeled `Diet1`, `Diet`, `Diet3` and `Diet4`, respectively, are dummy variables corresponding to the four diets. These are coded as follows

	Diet1	Diet2	Diet3	Diet4
Protein diet 1	1	0	0	0
Protein diet 2	0	1	0	0
Protein diet 3	0	0	1	0
Protein diet 4	0	0	0	1

The last 2 variables in the data set (columns 7 and 8) are `Day` and `Day_sq`, where `Day` is the number of days since the first measurement was taken and `Day_sq = Day x Day`.

A Box-whisker plot of Weight against Day is shown in Figure 4.5. indicating that the growth

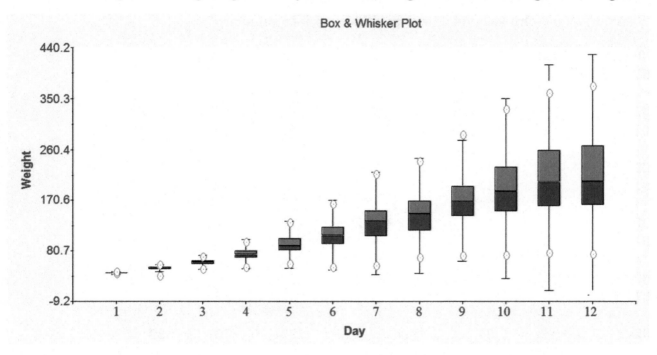

Figure 4.5: Box-Whisker Plot og Weight against Day

curve may be quadratic and that the within chick error variance increase over time. This suggests a model as follows.

$$y_{it} = \beta_1(Diet1)_{it} + \beta_2(Diet2)_{it} + \beta_3(Diet3)_{it} + \beta_4(Diet4)_{it} +$$
$$a_i + b_{1i}(Day)_{it} + b_{2i}(Day\_sq)_{it} + (Day)_{it}e_{it} \tag{4.60}$$

where

$$a_i = u_i$$
$$b_{1i} = \beta_5 + u_{1i} \tag{4.61}$$
$$b_{2i} = \beta_6 + u_{2i}.$$

An intercept term is not included in the fixed part of the model, since its inclusion together with the four dummy variables will cause a singular design matrix, and thus one of the coefficients will be non-estimable.

This model can be estimated using the following multilevel syntax file **chicks1.prl**:

```
OPTIONS MAXITER=20;
TITLE=Weight of Chicks on 4 Protein Diets;
SY=CHICKS.LSF;
ID1=Day;
ID2=Chick;
RESPONSE=Weight;
FIXED= Diet1  Diet2  Diet3  Diet4  Day  Day_Sq;
RANDOM1=Day;
RANDOM2=intcept Day Day_Sq;
```

MAXITER=20 has been added on the Options line to ascertain convergence. Since the variation in the weight measurements increases over time, the level-1 error term is expressed as Day $\times e_{it}$, where the variance σ_e^2 of e_{it} will be estimated.

The output file **chicks1.out** gives the following data summary

```
NUMBER OF LEVEL 2 UNITS :          50
NUMBER OF LEVEL 1 UNITS :         578
```

N2 :	1	2	3	4	5	6	7	8
N1 :	12	12	12	12	12	12	12	11

N2 :	9	10	11	12	13	14	15	16
N1 :	12	12	12	12	12	12	8	7

N2 :	17	18	19	20	21	22	23	24
N1 :	12	2	12	12	12	12	12	12

N2 :	25	26	27	28	29	30	31	32
N1 :	12	12	12	12	12	12	12	12

N2 :	33	34	35	36	37	38	39	40
N1 :	12	12	12	12	12	12	12	12

N2 :	41	42	43	44	45	46	47	48
N1 :	12	12	12	10	12	12	12	12

N2 :	49	50
N1 :	12	12

which shows that most of the chicks have complete weight measurements at 12 occasions, but chicks 8, 15, 16, 18, and 44 have less than 12 weight measurements. A notable case is chick 18 which has only 2 weight measurements.

The output file shows further the following estimates obtained after 13 iterations:

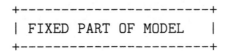

```
+-----------------------+
| FIXED PART OF MODEL   |
+-----------------------+
```

COEFFICIENTS	BETA-HAT	STD.ERR.	Z-VALUE	PR > |Z|
Diet1	34.76194	1.11540	31.16547	0.00000
Diet2	37.47333	1.29962	28.83412	0.00000
Diet3	39.09500	1.29962	30.08193	0.00000
Diet4	40.29288	1.29992	30.99654	0.00000
Day	5.49498	0.42191	13.02406	0.00000
Day_Sq	0.14664	0.03172	4.62273	0.00000

```
+-----------------------+
| RANDOM PART OF MODEL  |
+-----------------------+
```

LEVEL 2		TAU-HAT	STD.ERR.	Z-VALUE	PR > |Z|
intcept	/intcept	41.94086	9.79004	4.28403	0.00002
Day	/intcept	-15.70037	3.90172	-4.02396	0.00006
Day	/Day	7.77129	1.76392	4.40570	0.00001
Day_Sq	/intcept	0.32469	0.22965	1.41384	0.15741
Day_Sq	/Day	-0.31089	0.10712	-2.90225	0.00370
Day_Sq	/Day_Sq	0.04603	0.00995	4.62416	0.00000

LEVEL 1		TAU-HAT	STD.ERR.	Z-VALUE	PR > |Z|
Day	/Day	0.41172	0.02986	13.78701	0.00000

Since almost all parameters are significant both the assumption of a quadratic growth curve and the assumption of a linearly increasing level-1 error variance are reasonable. Also, it is clear that the coefficients a_i, b_{i1} and b_{i2} of the quadratic growth curve vary considerably across chicks as indicated by the fact that their variances 41.94, 7.77, and 0.046 are highly significant.

The fixed part of the model gives the following estimates for the effect of the four diets: $\beta_1 = 34.76$, $\beta_2 = 37.47$, $\beta_3 = 39.10$, and $\beta_4 = 40.29$. Thus, it appears that diet 4 is the most effective diet. To see if the differences between the diet effects are significant one can use contrasts. One can test the diets for each pair of diets or test all diets simultaneously. A recommended procedure is to do the simultaneous test first. This correspond to test the hypothesis

$$\beta_1 = \beta_2 = \beta_3 = \beta_4$$

If this hypothesis is rejected one can test the hypothesis $\beta_i = \beta_j$ for each pair i and j with $i \neq j$. To test the joint hypothesis, one should type a **.ctr** file **chicks1.ctr** as follows

```
3
1  0  0 -1  0 0
0  1  0 -1  0 0
0  0  1 -1  0 0
```

The first line gives the number of hypotheses to be tested. This is 3. The other lines has six numbers corresponding to the fixed effects $\beta_1, \ldots, \beta_6$. So the first contrast corresponds to $\beta_1 - \beta_4$, the second contrast corresponds to $\beta_2 - \beta_4$ and the third contrast corresponds to $\beta_3 - \beta_4$.

To test these hypotheses one at a time on should type a **.ctr** file as follows, file **chicks2.ctr**:

```
1
1 -1  0  0  0 0
1
1  0 -1  0  0 0
1
1  0  0 -1  0 0
1
0  1 -1  0  0 0
1
0  1  0 -1  0 0
1
0  0  1 -1  0 0
```

This corresponds to test the six hypotheses $\beta_1 - \beta_2 = 0$, $\beta_1 - \beta_3 = 0$, $\beta_1 - \beta_4 = 0$, $\beta_2 - \beta_3 = 0$, $\beta_2 - \beta_4 = 0$, $\beta_3 - \beta_4 = 0$. To test the joint hypothesis add the line

```
contrast=chicks1.ctr;
```

as the last line of the multilevel syntax file as shown in file **chicks2.prl** To test the six hypotheses one at a time, use the file **chicks2.ctr** instead, see file **chicks3.prl**.

The results of the tests appear at the end of the output file. For **chicks1.ctr** the results are

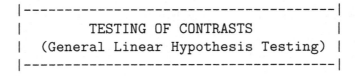

```
 TEST OF CONTRAST SET NO.    :          1
 ---------------------------------------------------------------
 Coefficients          Beta-hat      Contrasts
 ---------------------------------------------------------------
     1   Diet1          34.76194      1.000     0.000     0.000
     2   Diet2          37.47333      0.000     1.000     0.000
     3   Diet3          39.09500      0.000     0.000     1.000
     4   Diet4          40.29288     -1.000    -1.000    -1.000
     5   Day             5.49498      0.000     0.000     0.000
     6   Day_Sq          0.14664      0.000     0.000     0.000


 CHISQ STATISTIC           :      28.352
 DEGREES OF FREEDOM        :           3
 EXCEEDENCE PROBABILITY    :       0.000
```

The test is chi-square statistic with 3 degrees of freedom. The hypothesis is rejected. The conclusion is that not all β's are equal.

For **chicks3.prl** the results are

```
|---------------------------------------|
|          TESTING OF CONTRASTS         |
|   (General Linear Hypothesis Testing) |
|---------------------------------------|
```

TEST OF CONTRAST SET NO. :   1

Coefficients	Beta-hat	Contrasts
1 Diet1	34.76194	1.000
2 Diet2	37.47333	-1.000
3 Diet3	39.09500	0.000
4 Diet4	40.29288	0.000
5 Day	5.49498	0.000
6 Day_Sq	0.14664	0.000

CHISQ STATISTIC   :   5.546
DEGREES OF FREEDOM   :   1
EXCEEDENCE PROBABILITY   :   0.019

TEST OF CONTRAST SET NO. :   2

Coefficients	Beta-hat	Contrasts
1 Diet1	34.76194	1.000
2 Diet2	37.47333	0.000
3 Diet3	39.09500	-1.000
4 Diet4	40.29288	0.000
5 Day	5.49498	0.000
6 Day_Sq	0.14664	0.000

CHISQ STATISTIC   :   14.164
DEGREES OF FREEDOM   :   1
EXCEEDENCE PROBABILITY   :   0.000

TEST OF CONTRAST SET NO. :   3

Coefficients	Beta-hat	Contrasts
1 Diet1	34.76194	1.000
2 Diet2	37.47333	0.000
3 Diet3	39.09500	0.000
4 Diet4	40.29288	-1.000
5 Day	5.49498	0.000
6 Day_Sq	0.14664	0.000

CHISQ STATISTIC   :   23.074
DEGREES OF FREEDOM   :   1
EXCEEDENCE PROBABILITY   :   0.000

```
TEST OF CONTRAST SET NO.  :          4
-------------------------------------------------
Coefficients          Beta-hat     Contrasts
-------------------------------------------------

    1  Diet1          34.76194       0.000
    2  Diet2          37.47333       1.000
    3  Diet3          39.09500      -1.000
    4  Diet4          40.29288       0.000
    5  Day             5.49498       0.000
    6  Day_Sq          0.14664       0.000

CHISQ STATISTIC             :       1.492
DEGREES OF FREEDOM          :          1
EXCEEDENCE PROBABILITY      :       0.222

TEST OF CONTRAST SET NO.  :          5
-------------------------------------------------
Coefficients          Beta-hat     Contrasts
-------------------------------------------------

    1  Diet1          34.76194       0.000
    2  Diet2          37.47333       1.000
    3  Diet3          39.09500       0.000
    4  Diet4          40.29288      -1.000
    5  Day             5.49498       0.000
    6  Day_Sq          0.14664       0.000

CHISQ STATISTIC             :       4.509
DEGREES OF FREEDOM          :          1
EXCEEDENCE PROBABILITY      :       0.034

TEST OF CONTRAST SET NO.  :          6
-------------------------------------------------
Coefficients          Beta-hat     Contrasts
-------------------------------------------------

    1  Diet1          34.76194       0.000
    2  Diet2          37.47333       0.000
    3  Diet3          39.09500       1.000
    4  Diet4          40.29288      -1.000
    5  Day             5.49498       0.000
    6  Day_Sq          0.14664       0.000

CHISQ STATISTIC             :       0.814
DEGREES OF FREEDOM          :          1
EXCEEDENCE PROBABILITY      :       0.367
```

If we use the 5% significance level, it appears as if Diet 2, Diet 3 and Diet 4 are all significantly better than Diet 1, Diet 4 is significantly better than Diet 2 but the difference between Diet 4 and Diet 3 is not significant. Thus, although Diet 4 appears to be best, is not certain that it is uniformly better than all the other diets.

4.7   Multilevel Generalized Linear Models

In all previous examples of multilevel analysis in this chapter, it was assumed that the response variable is continuous. More precisely, it was assumed that it is normally distributed. Just as in Chapter 3, this assumption can be relaxed so that the distribution can be any distribution in the exponential family and the response variable can be linked to the covariates by a link function. In fact, any of the distributions and link functions in Table 3.4 of Chapter 3 can be used also with multilevel data. This is called Multilevel Generalized Linear Models and is well documented in the pdf file **Multilevel Generalized Linear Modeling Guide.pdf** which can be reached by clicking on **Help** and selecting **LISREL User & Reference Guides**. This document gives several examples. Here we consider just one example. This is an example with a categorical response variable with two categories.

4.7.1   Example: Social Mobility

The data file **occupations.lsf** used in this example is based on a 30% random sample of the data discussed in Biblarz and Raftery (1993). The data set contains six variables, F_Career, Son_ID, S_Choice, Famstruc, and Race. The first variable, F_Career, defines seventeen level-2 units, each being the father's occupation. Son_ID is used to number the sons within each level-2 unit. The response variable S_Choice is a dichotomous outcome variable such that S_Choice = 1 if son's current occupation is different from his father's occupation and = 2, otherwise. The variable Famstruc is family structure, coded 0 in the case of an intact family background, and 1 if the family background was not intact. The fifth variable, Race, is coded 0 in the case of white respondents and 1 in the case of black respondents.

The father's occupation, F_Career, is coded as

Code	Description
17	Professional, Self-Employed
16	Professional, Salaried
15	Manager
14	Salesman-Non retail
13	Proprietor
12	Clerk
11	Salesman-Retail
10	Craftsman-Manufacturing
9	Craftsman-Other
8	Craftsman-Construction
7	Service Worker
6	Operative-Non manufacturing
5	Operative-Manufacturing
4	Laborer-Manufacturing
3	Laborer-Non manufacturing
2	Farmer/Farm Manager
1	Farm Laborer

We would like to investigate the following:

- Is the ratio of the probabilities of different to same occupation consistent over the 17 occupations?

- Is this ratio influenced by race and/or the intactness of the family?

We begin by an exploratory analysis. The following bar chart of S_Choice shows

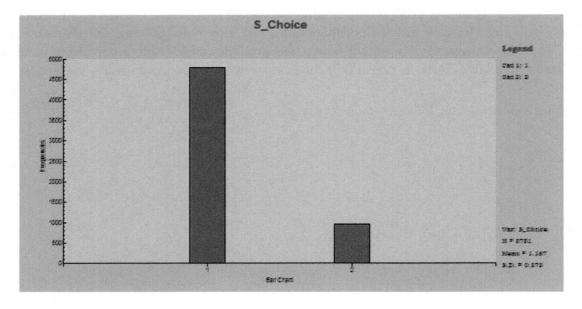

shows that approximately 17% of the 5751 sons chose the same occupation as their father.

To see if this percentage varies with FarmStruc and Race we use a 3-dimensional bar chart as shown in the next two graphs.

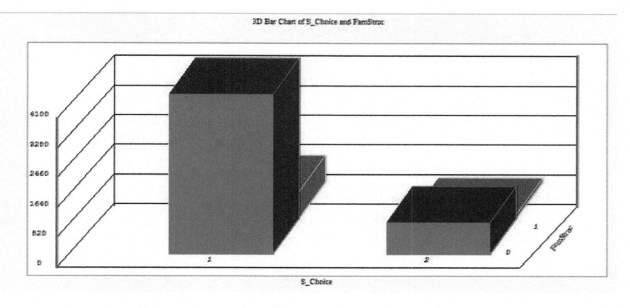

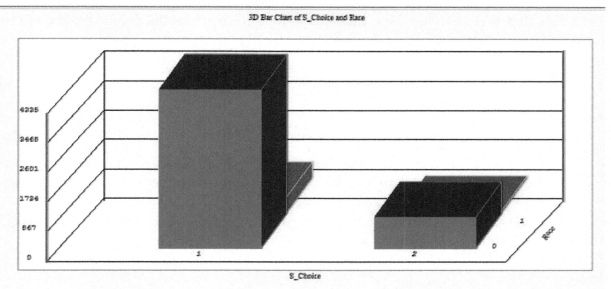

These suggest that the distribution of S_Choice is very similar for the two categories of FarmStruc and Race.

To see if the distribution of S_Choice varies across occupations we use a line plot of S_Choice on F_Career as shown in the next graph

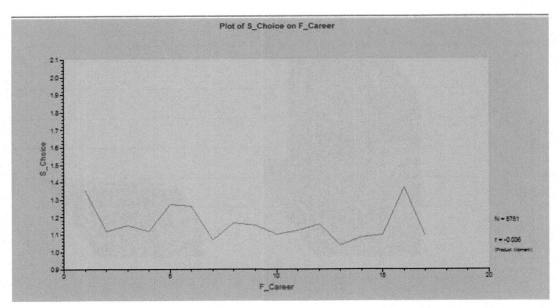

This suggests that there may be some variation of S_Choice across the different occupations. To see if these variations are significant we use a multilevel logistic response model.

This model is

$$P(\text{same}_{ij}|\text{occupation}_i) = \frac{1}{1 + \exp(-\hat{y}_{ij})};\,, \tag{4.62}$$

where

$$y_{ij} = a_i + b_{i1}\text{FamStruc}_{ij} + b_{i2}\text{Race}_{ij}\,. \tag{4.63}$$

Note that there is no error term e_{ij} on the right side of this equation. The probability on the left side of equation (4.62) is the probability that a son j whose father has occupation i chooses the same occupation as his father.

Equation (4.62) corresponds to a multilevel logistic regression model where the coefficients of the regression equation a_1, b_{i1}, and b_{i2} vary across the fathers occupations $i = 1, 2, \ldots, 17$. These coefficients are regarded as random variables in a fictitious population of occupations, for which we define

$$a_i = \alpha + u_i\,, \tag{4.64}$$

$$b_{i1} = \beta_1 + v_{i1}\,, \tag{4.65}$$

$$b_{i2} = \beta_2 + v_{i2}\,, \tag{4.66}$$

where u_1, v_{i1}, and v_{i2} have zero means. So the population logistic regression is

$$y = \alpha + \beta_1\text{FamStruc} + \beta_2\text{Race} \tag{4.67}$$

The covariance matrix of u_1, v_{i1}, and v_{i2} which is the same as the covariance matrix of a_1, $bi1$, and b_{i2} will be estimated.

The model can be estimated using the following multilevel syntax file **occupations1.prl**:

```
MGlimOptions Converge=0.0001 MaxIter=100
          Method=Quad NQUADPTS=10;
Title=Fathers and Sons Occupations;
SY=occupations.lsf;
```

```
ID2=F_Career;
Distribution=BER;
Link=LOGIT;
Intercept=Yes;
DepVar=S_Choice;
CoVars=FamStruc Race;
RANDOM2=intcept FamStruc Race;
```

Because of the random variables $bi1$, and b_{i2} the likelihood values cannot be obtained analytically but have to be evaluated by numerical integration. The is specified by the options

```
Method=Quad NQUADPTS=10;
```

on the `MGlimOptions` line, indicating that adaptive quadrature[2] with 10 quadrature points will be used.

The output file **occupations1.out** gives the following result for the fixed parameters α, β_1, and β_2:

Estimated regression weights

Parameter	Estimate	Standard Error	z Value	P Value
intcept	-1.7850	0.1774	-10.0625	0.0000
FamStruc	-0.2813	0.1919	-1.4660	0.1426
Race	-0.0882	0.2147	-0.4109	0.6811

The output file **occupations1.out** also gives odds ratios corresponding to these estimates and 95% confidence interval for these:

Odds Ratio and 95% Odds Ratio Confidence Intervals

Parameter	Estimate	Odds Ratio	Bounds Lower	Upper
intcept	-1.7850	0.1678	0.1185	0.2376
FamStruc	-0.2813	0.7548	0.5182	1.0994
Race	-0.0882	0.9156	0.6011	1.3945

---

[2]In applied mathematics, adaptive quadrature is a process in which the integral of a function f(x) is approximated using static quadrature rules on adaptively refined subintervals of the integration domain. Generally, adaptive algorithms are just as efficient and effective as traditional algorithms for "well behaved" integrands, but are also effective for "badly behaved" integrands for which traditional algorithms fail *Wikipedia*.

For the random level-2 parameters a_1, $bi1$, and b_{i2} the result is

Estimated level 2 variances and covariances

Parameter	Estimate	Standard Error	z Value	P Value
intcept/intcept	0.4833	0.1836	2.6316	0.0085
FamStruc/intcept	0.0922	0.1137	0.8108	0.4175
FamStruc/FamStruc	0.1344	0.1423	0.9445	0.3449
Race/intcept	-0.1712	0.1413	-1.2119	0.2256
Race/FamStruc	0.0851	0.1103	0.7714	0.4405
Race/Race	0.1796	0.1567	1.1460	0.2518

Of the random level-2 parameters a_1, $bi1$, and b_{i2} it is only the intercept a_i that has a significant variance at the 5% level. Therefore, the conclusion is that neither **FamStruc** nor **Race** has any significant effect on the probability that the a son chooses the same occupation as the father. Statistically this probability may vary across all occupations. We shall therefore take a closer look.

In addition to the normal output file **occupations1.out**, two other output files **occupations1.ba2** are obtained. These contain empirical Bayes estimates of u_i, v_{i1}, and v_{i2}, for $i = 1, 2, \ldots, 17$. Furthermore the file **occupations1.res** contains residuals for all i and j. Estimated values of u_i, v_{1i} and v_{2i} for $i = 1, 2, \ldots, 17$ are obtained from the third column in the file **occupations1.ba2**. The fourth column gives estimates of their variances. The estimated values of u_i, v_{1i} and v_{2i} are given in Table 4.4

Table 4.4: Empirical Bayes Estimates of u_i, v_{1i} and v_{2i}

i	u_i	v_{i1}	v_{i2}
1	1.179082	0.427537	-0.212921
2	-0.157454	-0.164961	-0.081050
3	0.058747	0.162261	0.131351
4	-0.147479	-0.140632	-0.061467
5	0.803852	0.199324	-0.237776
6	0.730259	0.486378	0.090533
7	-1.034789	0.318305	0.887542
8	0.215513	-0.242085	-0.361833
9	0.114379	-0.188651	-0.252997
10	-0.357393	-0.183170	0.010635
11	-0.108741	-0.210814	-0.153245
12	0.048615	0.316977	0.293378
13	-1.212837	-0.280799	0.379641
14	-0.549663	-0.185186	0.113675
15	-0.408998	-0.310846	-0.089939
16	1.232535	0.161523	-0.510632
17	-0.405625	-0.165225	0.055040

Predicted values of y_{ij} can be obtained from

$$\hat{y}_{ij} = (\hat{\alpha} + u_i) + (\hat{\beta}_1 + v_{i1})\text{FamStruc} + (\hat{\beta}_2 + v_{i2})\text{Race}. \tag{4.68}$$

From the fixed parameters of the output file **occupations1.out** we obtain $\hat{\alpha} = -1.7850$, $\hat{\beta}_1 = -0.2813$, and $\hat{\beta}_2 = -0.0882$. Thus, the predicted value of y_{ij} is

$$\hat{y}_{ij} = (-1.7850 + u_i) + (-0.2813 + v_{1i})\text{FamStruc} + (-0.0882 + v_{2i})\text{Race}, \qquad (4.69)$$

where `FamStruc` is 1 for an intact family background, and 2 otherwise, and `Race` is 1 for whites and 2 for blacks.

Then the predicted $\hat{y}_{ij}$ are obained from (4.69) for $i = 1, 2, \ldots, 17$. The $\hat{y}_{ij}$-values can be used to calculate the predicted probability that a son's current occupation will be the same as that of the father. This probability is

$$P(\text{same}_{ij}|\text{occupation}_i) = \frac{1}{1 + \exp(-\hat{y}_{ij})}.$$

These probabilities are plotted in Figure 4.6.

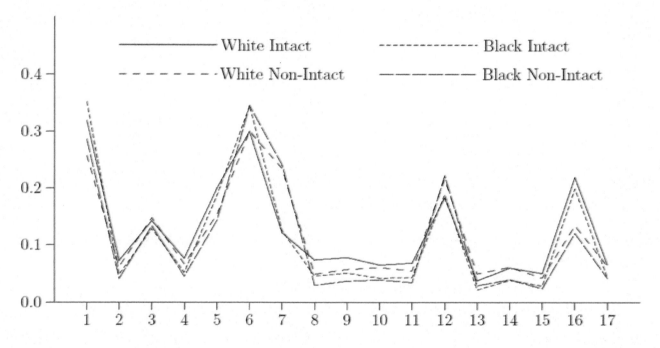

Figure 4.6: Probabilities of Same Occupation as Father

From the figures one can clearly see that the family structure and the race do not have any strong effect on the probability of choosing the same occupation as father but that this probability varies across occupations. The probability is small for occupations 2, 4, 8-11, 13-15, and 17 and large for occupations 1, 6, 12, and 16.

4.8 The Basic Three-Level Model

Consider the situation where a response variable y may depend on a set of p predictors $x_1, x_2, \ldots, x_p$. The general level-3 model is defined as

$$\mathbf{y}_{ijk} = \mathbf{x}'_{(f)ijk}\boldsymbol{\beta} + \mathbf{x}'_{(3)ijk}\mathbf{u}_i + \mathbf{x}'_{(2)ijk}\mathbf{v}_{ij} + \mathbf{e}_{ijk}, \qquad (4.70)$$

where

- $i = 1, 2, \ldots, N$ denotes level-3 units (*e.g.*, educational departments),

- $j = 1, 2, \ldots, n_i$ denotes level-2 units (*e.g.*, schools), and

- $k = 1, 2, \ldots, n_{ij}$ denotes level-1 units (*e.g.*, pupils).

- $\mathbf{x}'_{(f)ijk} : 1 \times s$ is a typical row of the design matrix of the fixed part of the model, the elements being a subset of the p predictors.

- $\mathbf{x}'_{(3)ijk} : 1 \times q$ is a typical row of the design matrix for the random part at level 3, the elements being a subset of the p predictors.

- $\mathbf{x}'_{(2)ijk} : 1 \times m$ is a typical row of the design matrix for the random part at level 2, the elements being a subset of the p predictors.

- $\boldsymbol{\beta} : s \times 1$ is a vector of fixed, but unknown parameters to be estimated.

- $\mathbf{u}_i : q \times 1$ are random effects at level 3

- $\mathbf{v}_{ij} : m \times 1$ are random effects at level 2

It is assumed that $\mathbf{u}_1, \mathbf{u}_2, \ldots, \mathbf{u}_N$ are independently and identically distributed with mean $\mathbf{0}$ and covariance matrix $\boldsymbol{\Phi}_{(3)}$. It is further assumed that $\mathbf{v}_{i1}, \mathbf{v}_{i2}, \ldots, \mathbf{v}_{in_i}$ are *i.i.d.* with mean $\mathbf{0}$ and covariance matrix $\boldsymbol{\Phi}_{(2)}$, while $\mathbf{e}_{ij1}, \mathbf{e}_{ij2}, \ldots, \mathbf{e}_{ijn_{ij}}$ are *i.i.d.* with mean $\mathbf{0}$ and covariance matrix $\boldsymbol{\Phi}_{(1)}$. Finally, it is assumed that $\mathbf{u}_i$, $\mathbf{v}_{ij}$, and $\mathbf{e}_{ijk}$ are independent.

4.8.1 Example: CPC Survey Data

The data for this example comes from the March 1995 Current Population Survey in USA. There are two groups of respondents:

Educational sector	Respondents with professional specialty in the educational sector
Construction sector	Operators, constructors, and laborers in the construction sector

The variable group is coded 0 for the Construction Sector and 1 for the Education Sector. Other demographic variables and their values are:

gender 0 = female; 1 = male

age Age in single years

marital 1 = married; 0 = other

hours Hours worked during last week at all jobs

citizen 1 for native Americans, 0 for all foreign born respondents

LnInc The natural logarithm of the personal income during 1994

degree 1 for respondents with master's degrees, professional school degree, or doctoral degree; 0 otherwise

The Region and State codes are

New England region (Region = 1)	11	Maine
	12	New Hampshire
	13	Vermont
	14	Massachusetts
	15	Rhode Island
	16	Connecticut
Middle Atlantic region (Region=2)	21	New York
	22	New Jersey
	23	Pennsylvania
East North Central region (Region=3)	31	Ohio
	32	Indiana
	33	Illinois
	34	Michigan
	35	Wisconsin
West North Central region (Region=4)	41	Minnesota
	42	Iowa
	43	Missouri
	44	North Dakota
	45	South Dakota
	46	Nebraska
	47	Kansas
South Atlantic region (Region=5)	51	Delaware
	52	Maryland
	53	District of Columbia
	54	Virginia
	55	West Virginia
	56	North Carolina
	57	South Carolina
	58	Georgia
	59	Florida
East South Central region (Region=6)	61	Kentucky
	62	Tennessee
	63	Alabama
	64	Mississippi
West South Central region (Region=7)	71	Arkansas
	72	Louisiana
	73	Oklahoma
	74	Texas
Mountain region (Region=8)	81	Montana
	82	Idaho
	83	Wyoming
	84	Colorado
	85	New Mexico
	86	Arizona
	87	Utah
	88	Nevada
Pacific region (Region=9)	91	Washington
	92	Oregon
	93	California
	94	Alaska
	95	Hawaii

The first 10 rows of the data file **cpc1995.lsf** is

cpc1995.lsf	region	state	age	gender	marital	hours	citizen	degree	group	LnInc
1	1.000	11.000	59.000	0.000	0.000	40.000	1.000	0.000	1.000	10.114
2	1.000	11.000	56.000	1.000	1.000	40.000	1.000	0.000	1.000	10.342
3	1.000	11.000	64.000	0.000	1.000	12.000	1.000	0.000	1.000	10.429
4	1.000	11.000	30.000	1.000	1.000	40.000	1.000	0.000	0.000	9.629
5	1.000	11.000	27.000	0.000	1.000	40.000	1.000	0.000	1.000	9.976
6	1.000	11.000	49.000	0.000	1.000	40.000	1.000	1.000	1.000	10.685
7	1.000	11.000	41.000	1.000	1.000	40.000	1.000	0.000	0.000	10.604
8	1.000	11.000	36.000	0.000	1.000	40.000	1.000	0.000	1.000	10.012
9	1.000	11.000	46.000	0.000	1.000	36.000	1.000	0.000	0.000	10.190
10	1.000	11.000	39.000	0.000	1.000	55.000	1.000	0.000	1.000	10.023

A 3-level variance components model can be estimated using the following multilevel syntax file **cpc1995.prl**:

```
OPTIONS;
TITLE=Analysis of CPC 1995 Survey;
SY=cpc1995.lsf;
ID3=region;
ID2=state;
RESPONSE=LnInc;
FIXED=intcept age gender marital hours citizen degree group;
RANDOM1=intcept;
RANDOM2=intcept;
RANDOM3=intcept;
```

The output file **cpc1995.out** give the following data summary

```
NUMBER OF LEVEL 3 UNITS :        9
NUMBER OF LEVEL 2 UNITS :       51
NUMBER OF LEVEL 1 UNITS :     6062


ID3 :       1        2        3        4        5        6        7        8
N2  :       6        3        5        7        9        4        4        8
N1  :     545      862      785      521     1095      291      598      704


ID3 :       9
N2  :       5
N1  :     661
```

There are 6062 individuals and the nine regions have between 291 and 1095 respondents, nested within states. The smallest number of level-2 units within a level-3 unit was 3, for the middle Atlantic region which included only New York, New Jersey, and Pennsylvania.

The fixed effects are estimated as

```
--------------------------------------------------------------------------------
  COEFFICIENTS          BETA-HAT      STD.ERR.      Z-VALUE        PR > |Z|
--------------------------------------------------------------------------------
  intcept               8.19488       0.06867      119.34530       0.00000
  age                   0.01636       0.00101       16.17417       0.00000
  gender                0.23710       0.02853        8.31169       0.00000
  marital               0.08456       0.02243        3.76975       0.00016
  hours                 0.01344       0.00065       20.63542       0.00000
  citizen               0.28652       0.03449        8.30714       0.00000
  degree                0.41226       0.02846       14.48697       0.00000
  group                 0.19798       0.03135        6.31519       0.00000
```

All the fixed effects are highly significant. The coefficient for the variable `Intcept`, representing the mean log income, was 8.19488. Since the response variable is the natural logarithm of a respondent's annual income, this number translates to a mean income of

$$\exp(8.19488 + 21(0.01636) + 40(0.01344)) = \$8,743$$

for a respondent from the construction sector who is 21 years of age, working 40 hours per week, unmarried, without a higher degree, and not a USA citizen.

Although the size of the coefficients is quite small, it should be kept in mind that the natural logarithm of income is used as response variable. The relatively large positive coefficients for `gender` (0.23710), `citizen` (0.28652), and `degree` (0.41226) indicate that males, citizens of the USA, and respondents with a high education level tend to earn more when other variables are held constant.

A comparison of two respondents with different demographic profiles follows.

Respondent 1	Respondent 2
age=30	age=30
hours=40	hours=40
group=1	group=1
marital=0	marital=0
gender=0	gender=1
citizen=0	citizen=1
degree=0	degree=1

The first respondent's expected income is calculated as **Expected income** =

$$\exp[8.19488 + 30(0.01636) + 40(0.01344) + 0.19798] =$$

$$\exp[9.42126] = \$12,348$$

The expected income of the second respondent is **Expected income** =

$$\exp[8.19488 + 30(0.01636) + 40(0.01344) + 0.19798 + 0.23710 +$$

$$0.28652 + 0.41226] = \exp[10.35714] = \$31,481$$

The random part of the model is

```
              +-----------------------+
              | RANDOM PART OF MODEL  |
              +-----------------------+
```

LEVEL 3	TAU-HAT	STD.ERR.	Z-VALUE	PR > |Z|
intcept /intcept	0.00783	0.00472	1.65774	0.09737

LEVEL 2	TAU-HAT	STD.ERR.	Z-VALUE	PR > |Z|
intcept /intcept	0.00522	0.00250	2.08936	0.03668

LEVEL 1	TAU-HAT	STD.ERR.	Z-VALUE	PR > |Z|
intcept /intcept	0.60688	0.01107	54.83990	0.00000

from which we see that log income varies most over the respondents (level-1 units), and least over the states (level-2 units) since the variances at these levels are estimated as 0.60688 and 0.00522, respectively.

4.9 Multivariate Multilevel Analysis

In all previous examples there was one single response variable. However, multilevel analysis have been extended such that several jointly dependent response or outcome variables can be handled.

The simplest and most commonly used generalization to multivariate response variables is the Multivariate Random Intercept Model, which generalizes equation (4.24) to:

$$\mathbf{y}_{ij} = \mathbf{a}_i + \mathbf{X}_{ij}\boldsymbol{\beta} + \mathbf{e}_{ij} \; , \tag{4.71}$$

where

$$\mathbf{a}_i = \boldsymbol{\alpha} + \mathbf{u}_i \; , \tag{4.72}$$

and $\mathbf{X}_{ij}$ is a design matrix. Substituting (4.72) into (4.71) gives

$$\mathbf{y}_{ij} = \boldsymbol{\alpha} + \mathbf{X}_{ij}\boldsymbol{\beta} + \mathbf{u}_i + \mathbf{e}_{ij} \; , \tag{4.73}$$

If there are p response variables and k fixed parameters in $\boldsymbol{\beta}$, $\mathbf{y}_{ij}$, $\mathbf{u}_i$ and $\mathbf{e}_{ij}$ are of order $p \times 1$ and $\mathbf{X}_{ij}$ is of order $p \times k$. Equation (4.73) can also be written

$$\mathbf{y}_{ij} = \mathbf{X}_{ij}^{\star}\boldsymbol{\beta}^{\star} + \mathbf{u}_i + \mathbf{e}_{ij} \; , \tag{4.74}$$

where $\mathbf{X}_{ij}^{\star}$ is X_{ij} with a first column of 1's and

$$\boldsymbol{\beta}^{\star} = \begin{pmatrix} \alpha \\ \beta \end{pmatrix} \; . \tag{4.75}$$

This formulation can be illustrated with the following example. Consider $p = 3$ response variables, y_1, y_2, and y_3 made on each of 4 students within N schools. This data set can be schematically represented for school i as follows

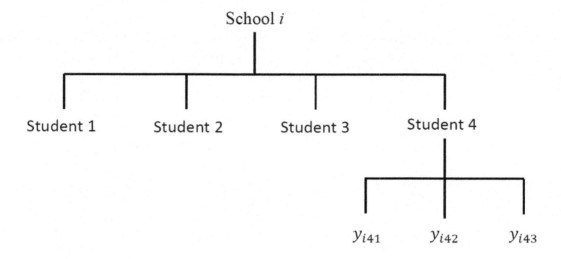

For the i-th school, we can write

$$\mathbf{y}_i = \begin{pmatrix} \mathbf{y}_{i1} \\ \mathbf{y}_{i2} \\ \mathbf{y}_{i3} \\ \mathbf{y}_{i4} \end{pmatrix},$$

where for student 4 within school i

$$\mathbf{y}'_{i4} = \begin{pmatrix} y_{i41} & y_{i42} & y_{i43} \end{pmatrix}.$$

A model which allows for between- and within-schools variation in the response variables is the following simple variance component model

$$\mathbf{y}_{i1} = \mathbf{u}_i + \mathbf{e}_{i1}$$
$$\mathbf{y}_{i2} = \mathbf{u}_i + \mathbf{e}_{i2}$$
$$\mathbf{y}_{i3} = \mathbf{u}_i + \mathbf{e}_{i3}$$
$$\mathbf{y}_{i4} = \mathbf{u}_i + \mathbf{e}_{i4}.$$

In general, if there are N schools and n_i students in school i,

$$\mathbf{y}_{ij} = \mathbf{u}_i + \mathbf{e}_{ij}, \quad i = 1, 2, \ldots N \ \ j = 1, 2, \ldots, n_i,$$

and let

$$\mathbf{y}'_i = (\mathbf{y}'_{i1}, \mathbf{y}'_{i2}, \ldots, \mathbf{y}'_{in_i}).$$

It is assumed that $\mathbf{u}_1, \mathbf{u}_2, \ldots \mathbf{u}_N$ are i.i.d. $N(0, \boldsymbol{\Sigma}_B)$, that $\mathbf{e}_{i1}, \ldots, \mathbf{e}_{in_i}$ are i.i.d $N(0, \boldsymbol{\Sigma}_W)$, and that

$$Cov(\mathbf{u}_i, \mathbf{e}_{ij}) = 0, \ i = 1, \ldots, N; \ j = 1, 2, \ldots n_i.$$

From these distributional assumptions it follows that

$$\boldsymbol{\Sigma}_i = Cov(\mathbf{y}_i) = \begin{pmatrix} \boldsymbol{\Sigma}_B + \boldsymbol{\Sigma}_W & \boldsymbol{\Sigma}_B & \boldsymbol{\Sigma}_B & \boldsymbol{\Sigma}_B \\ \boldsymbol{\Sigma}_B & \boldsymbol{\Sigma}_B + \boldsymbol{\Sigma}_W & \boldsymbol{\Sigma}_B & \boldsymbol{\Sigma}_B \\ \boldsymbol{\Sigma}_B & \boldsymbol{\Sigma}_B & \boldsymbol{\Sigma}_B + \boldsymbol{\Sigma}_W & \boldsymbol{\Sigma}_B \\ \boldsymbol{\Sigma}_B & \boldsymbol{\Sigma}_B & \boldsymbol{\Sigma}_B & \boldsymbol{\Sigma}_B + \boldsymbol{\Sigma}_W \end{pmatrix}$$

or

$$\mathbf{\Sigma}_i = Cov(\mathbf{y}_i) = \mathbf{11}' \otimes \mathbf{\Sigma}_B + \mathbf{I} \otimes \mathbf{\Sigma}_W \ . \tag{4.76}$$

It also follows that

$$E(\mathbf{y}_i) = \mathbf{0}.$$

In practice, the latter assumption $E(\mathbf{y}_i) = \mathbf{0}$ is seldom realistic, since the response variables usually do not have zero means but one can add a fixed component $\boldsymbol{\alpha}_i + \mathbf{X}_{ij}\boldsymbol{\beta}$ to the model $\mathbf{y}_{ij} = \mathbf{u}_i + \mathbf{e}_{ij}$, so that this becomes equal to (4.73).

Thus, $\mathbf{y}_{ij} \sim N(\boldsymbol{\mu}_{ij}, \mathbf{\Sigma}_i)$, where $\boldsymbol{\mu}_{ij} = \boldsymbol{\alpha}_i + \mathbf{X}_{ij}\boldsymbol{\beta}$ and $\mathbf{\Sigma}_i$ is given by (4.76). So, apart from a constant,

$$-2\ln L_i = \ln | \mathbf{\Sigma}_i | + \sum_j^{n_i} (\mathbf{y}_{ij} - \boldsymbol{\mu}_{ij})' \mathbf{\Sigma}_i^{-1}(\mathbf{y}_{ij} - \boldsymbol{\mu}_{ij}) \ . \tag{4.77}$$

If the $\mathbf{y}_{ij}$ for $i = 1, 2, \ldots, N$ $j = 1, 2, \ldots, n_i$ are independent, the *deviance* $-2\ln L$ of all $\mathbf{y}_{ij}$ is

$$D = -2\ln L = \sum_{i=1}^{N} -2\ln L_i \tag{4.78}$$

D is to be minimized with respect to the parameters $\boldsymbol{\alpha}$, $\boldsymbol{\beta}$, $\mathbf{\Sigma}_B$, and $\mathbf{\Sigma}_W$.

In Multivariate Multilevel Models one uses the schools as the level-3 identification variable, the students as the level-2 identification variable, and the response variables as level-1 variables. From these specifications and the values of any fixed x-variables, if any, LISREL automatically creates the design matrices $\mathbf{X}_{ij}$. This will be clear from the following example.

4.9.1   Example: Analysis of the Junior School Project Data (JSP)

The data set used in this example forms part of the data library of the Multilevel Project at the University of London and comes from the Junior School Project (Mortimore, *et al.*, 1988).

Mathematics and English tests were administered in three consecutive years to more than 1000 students from 50 primary schools, which were randomly selected from primary schools maintained by the Inner London Education Authority.

The following variables are available in the data file **JSP.LSF**:

school	School code (1 to 50)
student	Student ID (1 to 1402)
gender	Gender (boy=1; girl=0)
ravens	Ravens test score in year 1(score 4–36)
math1	Score on mathematics test in year 1 (score 1–40)
math2	Score on mathematics test in year 2 (score 1–40)
math3	Score on mathematics test in year 3 (score 1–40)
eng1	Score on language test in year 1 (score 0–98)
eng2	Score on language test in year 2 (score 0–98)
eng3	Score on language test in year 3 (score 0–98)

Two models will be discussed

- A variance component model giving the between and within schools covariance matrices of the six test scores.

- A model using **gender** and **ravens** as explanatory variables.

The school number (**school**) is used as the level-3 identification variable, with Student ID (**student**) as the level-2 identification variable. The level-1 units are the scores obtained by a student for the mathematics and language tests, as represented by **math1** to **math3** and **eng1** to **eng3**. As with most repeated measurements data, missing values occur, in the sense that not every student has a score on all three English or Math tests. These missing values are coded -9 in this data set.

A simple data screening reveals that the numbers of missing values in the response variables are

```
Number of Missing Values per Variable
    math1       math2       math3       eng1       eng2       eng3
      38          63          239         38         63         239
```

The aim of this analysis is to examine the variation in test scores over schools and over students. One of the main benefits of analyzing different responses simultaneously in one multivariate analysis is that the way in which measurements relate to the explanatory variables can be directly explored.

As the residual covariance matrices for the second and third levels of the hierarchy are also obtained from this analysis, differences between coefficients of explanatory variables for different responses can be studied.

Finally, each respondent does not have to be measured on each response, as is the case for the data set we consider here, where scores for all three years are not available for all students. In the case of missing responses, a multilevel multivariate analysis of the responses that are available for respondents can be used to provide information in the estimation of those that are missing.

Variance Component Model

The variance component model can be estimated using the following multilevel syntax file (**jsp1.prl**):

```
OPTIONS COVBW=YES;
TITLE=First Analysis of JSP Data;
SY=JSP.lsf;
ID3=school;
ID2=student;
RESPONSE=math1 math2 math3 eng1 eng2 eng3;
FIXED=intcept;
RANDOM1=intcept;
RANDOM2=intcept;
RANDOM3=intcept;
MISSING_DEP=-9;
```

The COVBV=YES on the OPTIONS line is used to save the estimates of Σ_B and Σ_W in files for further use. These files are called **JSP1_between.cov** and **JSP1_within.cov**, respectively.

The last line MISING_DEP=-9 is used to specify the missing value code as -9.

In the output file **jsp1.out**, the following data summary is obtained

```
                          +--------------+
                          | DATA SUMMARY |
                          +--------------+


NUMBER OF LEVEL 3 UNITS :         49
NUMBER OF LEVEL 2 UNITS :       1192
NUMBER OF LEVEL 1 UNITS :       6472


ID3 :        1        2        3        4        5        6        7        8
N2  :       34       13       21       24       29       24       15       31
N1  :      184       72       96      144      166      120       78      174


ID3 :        9       10       11       12       13       14       15       16
N2  :       22       14       12       28       25       15       24       19
N1  :      130       48       70      154      138       86      106      106


ID3 :       17       18       19       20       21       22       23       24
N2  :        7       20       17       15       32       16       20       21
N1  :       40      110       88       84      184       92      116      126


ID3 :       25       26       27       28       29       30       31       32
N2  :       18       29       27       18       25       37       36       26
N1  :       94      156      146       96      124      182      214      148


ID3 :       33       34       35       36       37       38       39       40
N2  :       46       34       19       32       22       14       16       13
N1  :      262      184      106      178      118       78       90       68


ID3 :       41       42       43       44       45       46       47       48
N2  :       13       47        6       17       19       37       72       25
N1  :       70      262       32       88      102      204      412      144


ID3 :       49
N2  :       46
N1  :      202
```

This shows that there are 49 schools with a total of 1192 students. If all students had complete data on all 6 response variables there would be a total of $1192 \times 6 = 7152$ level-1 observations. However, since some observations are missing there are only 6472 level-1 observations.

The number of students per school varies between 6 and 72. The number of observations per school varies from 32 and 412.

Based on these numbers the output file gives the following information, where the number 24 is approximately the average number of students per school. Later the numbers 24 and 1192 will be used as between schools and within schools sample sizes, respectively.

```
Adjusted between cluster sample size=       24
Within cluster sample size=      1192
```

All parameter estimates are highly significant. The fixed parameters are given in the output as

```
COEFFICIENTS        BETA-HAT       STD.ERR.      Z-VALUE        PR > |Z|
intcept1            24.90370       0.33546       74.23792       0.00000
intcept2            24.87234       0.40108       62.01311       0.00000
intcept3            30.04909       0.37761       79.57736       0.00000
intcept4            47.15338       1.32158       35.67945       0.00000
intcept5            64.96594       1.22017       53.24352       0.00000
intcept6            40.71988       1.38296       29.44399       0.00000
```

Since there are six response variables there are six fixed parameters, the α's in (4.73). Note that the term $\mathbf{X}_{ij}\boldsymbol{\beta}$ is not included in this model. In the part of the output given above, the first three intercepts correspond to the intercepts of the Math scores and the last three intercepts correspond to the intercepts of the English scores.

The value of $-2\ln L$ and the number of estimated parameters are given as

```
+-----------------------+
|   -2 LOG-LIKELIHOOD   |
+-----------------------+


DEVIANCE= -2*LOG(LIKELIHOOD) =    45991.2578958991
NUMBER OF FREE PARAMETERS =        48
```

The number 48 equals the total number of independent parameters in $\boldsymbol{\alpha}$, $\boldsymbol{\Sigma}_{\mathrm{B}}$, and $\boldsymbol{\Sigma}_{\mathrm{W}}$, which are 6, 21, and 21, respectively.

The estimated between schools covariance matrix $\hat{\boldsymbol{\Sigma}}_{\mathrm{B}}$ is given as

```
        LEVEL 3 COVARIANCE MATRIX
         intcept1  intcept2  intcept3  intcept4  intcept5  intcept6
intcept1  3.31028
intcept2  2.29203    5.23047
intcept3  2.36877    3.13485   4.79640
intcept4  9.95162    9.49602   9.95900  59.32680
intcept5  9.93290   11.41833  11.59906  42.49228  53.08595
intcept6 10.16830   10.71620  13.71315  45.00410  51.61419  71.08796
```

The estimated within schools covariance matrix $\hat{\Sigma}_W$ is given as

```
    LEVEL 2 COVARIANCE MATRIX
              intcept1    intcept2    intcept3  intcept4  intcept5  intcept6
intcept1      47.17671
intcept2      38.63798    55.45831
intcept3      31.21276    36.54802    41.33874
intcept4     109.03175   109.44780    88.00030  549.41232
intcept5      88.73168    94.99391    79.52971  388.36174  409.27297
intcept6      94.37662    99.80683    86.38268  382.76743  317.45309  425.49404
```

We see that the variation within schools, *i.e.*, over students is much higher than the variation between schools. We also see that the variation is larger for English test (response variables 4, 5, and 6) than for the Math test (response variables 1, 2, and 3), but this may just be a reflection of the larger range of scores for the English test.

Explanatory Model

To see if the variables **gender** and **ravens** have any effect on the response variables, we use the model

$$\mathbf{y}_{ij} = \boldsymbol{\alpha} + \text{gender}_{ij}\boldsymbol{\beta}_1 + \text{ravens}_{ij}\boldsymbol{\beta}_2 + \mathbf{e}_{ij} \,, \tag{4.79}$$

where $\boldsymbol{\beta}_1$ and $\boldsymbol{\beta}_2$ are vectors of order 6×1. This correspond to taking $\mathbf{X}_{ij}$ and $\boldsymbol{\beta}$ in (4.73) as

$$\mathbf{X}_{ij} = \begin{pmatrix} \mathbf{X}_{1ij} & \mathbf{X}_{2ij} \end{pmatrix} \quad \boldsymbol{\beta} = \begin{pmatrix} \boldsymbol{\beta}_1 \\ \boldsymbol{\beta}_2 \end{pmatrix} \,, \tag{4.80}$$

where $\mathbf{X}_{1ij} = \text{gender}_{ij}\mathbf{I}$ and $\mathbf{X}_{2ij} = \text{ravens}_{ij}\mathbf{I}$.

To estimate this model, just add on the **FIXED** line so that the input file becomes **jsp2.prl**:

```
OPTIONS;
TITLE=Second Analysis of JSP Data;
SY=JSP.lsf;
ID3=school;
ID2=student;
RESPONSE=math1 math2 math3 eng1 eng2 eng3;
FIXED=intcept gender ravens;
RANDOM1=intcept;
RANDOM2=intcept;
RANDOM3=intcept;
MISSING_DEP=-9;
```

The output file **jsp2.out** gives the fixed effects as

COEFFICIENTS	BETA-HAT	STD.ERR.	Z-VALUE	PR > |Z|
intcept1	7.68926	0.80046	9.60607	0.00000
intcept2	6.48252	0.90269	7.18136	0.00000
intcept3	14.84541	0.84586	17.55061	0.00000
intcept4	2.64745	2.90410	0.91163	0.36197
intcept5	28.95062	2.57309	11.25128	0.00000
intcept6	-2.58204	2.74794	-0.93963	0.34741
gender1	-0.48513	0.33815	-1.43466	0.15138
gender2	-0.79986	0.37167	-2.15205	0.03139
gender3	-0.45790	0.34277	-1.33588	0.18159
gender4	-10.57862	1.20875	-8.75172	0.00000
gender5	-9.21809	1.06819	-8.62960	0.00000
gender6	-7.02657	1.10494	-6.35926	0.00000
ravens1	0.69639	0.02943	23.66040	0.00000
ravens2	0.74945	0.03248	23.07353	0.00000
ravens3	0.61505	0.03029	20.30821	0.00000
ravens4	1.97944	0.10546	18.77035	0.00000
ravens5	1.61390	0.09328	17.30225	0.00000
ravens6	1.86293	0.09754	19.09957	0.00000

Each column correspond to α, β_1, and β_2 in that order. In each group of six values, the first three correspond to the Math scores and the last three values to the English scores. It is seen that **gender** does not have an effect on the Math scores but on the English scores. Girls are better than boys in English. **ravens** has a positive effect on both Math and English. Students with better scores on **ravens** are also better in both Math and English.

Chapter 5

Principal Components (PCA)

Many multivariate statistical methods are based on the idea of forming linear combinations of variables for various purposes. For example, in regression one forms linear combinations of predictors which have maximum correlation with a specified response variable. An extension of this is canonical correlation which maximizes the correlation between two linear combinations. A third example is discriminant analysis which forms linear combinations of variables to have maximum between group variance.

Principal components are linear combinations with maximum variance. It is best characterized as a sequential procedure for forming uncorrelated linear combinations of random variables with maximum variances. It is often confused with factor analysis but, as explained in Section 5.2 and in Chapter 6, unlike principal components, factor analysis is a model which can be tested. Principal components are often viewed as a dimensionality reduction problem or as a matrix approximation.

Principal components analysis (PCA) is described in chapters in several books on multivariate statistical analysis, for example, Anderson (1984), Johnson and Wichern (2002), Mardia, Kent, and Bibby (1980), and Sharma (1996). The most cited reference on principal components is the book by Joliffe (1986) which gives a comprehensive account of PCA with its extensions to biplots and correspondence analysis and its application in time series analysis. PCA is also closely related to partial least squares (PLS), see, *i.e.*, Vinci *et al* (2010). In this chapter we focus only on the features available in LISREL and, in particular, on how to do PCA with LISREL.

Section 5.1 defines principal components of a covariance matrix. Section 5.2 gives a discussion of the distinction between principal components and factor analysis. Section 5.3 defines principal components of a data matrix. Principal components have several interesting least squares properties. These aspects of principal components are considered in Sections 5.2 and 5.3.

5.1 Principal Components of a Covariance Matrix

The 2×2 Case

Consider two random variables x_1 and x_2 with covariance matrix

$$\Sigma = \begin{pmatrix} \sigma_{11} & \\ \sigma_{21} & \sigma_{22} \end{pmatrix}, \tag{5.1}$$

© Springer International Publishing Switzerland 2016
K.G. Jöreskog et al., *Multivariate Analysis with LISREL*, Springer Series in Statistics,
DOI 10.1007/978-3-319-33153-9_5

and let

$$c = a_1 x_1 + a_2 x_2 \tag{5.2}$$

be a linear combination of x_1 and x_2. How large can the variance of c be by choice of a_1 and a_2? This question is meaningless, for we can make the variance of c indefinitely large by making a_1 or a_2 sufficiently large. To make some sense of the question we must norm a_1 and a_2 in some way. The usual normalization rule is

$$a_1^2 + a_2^2 = 1 \; . \tag{5.3}$$

If a_1 and a_2 is so normalized, we say that c is a normalized linear combination. So the question is now: How large can the variance of a normalized linear combination c be?

The variance of c is

$$Var(c) = a_1^2 \sigma_{11} + a_2^2 \sigma_{22} + 2a_1 a_2 \sigma_{21} \; . \tag{5.4}$$

Let $a_2 = \pm \sqrt{(1 - a_1^2)}$. Then

$$Var(c) = a_1^2 \sigma_{11} + (1 - a_1^2)\sigma_{22} \pm 2a_1 \sqrt{(1 - a_1^2)}\sigma_{21} \; . \tag{5.5}$$

We can then let a_1 vary from -1 to +1 and find the largest value of $Var(c)$.

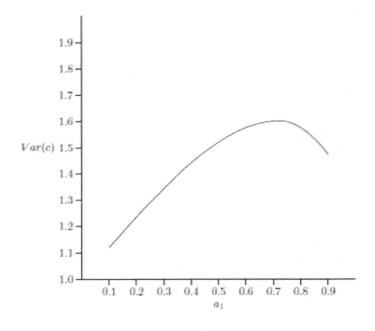

Figure 5.1: Plot of $Var(c)$ against a_1

Example 1

Let

$$\Sigma = \begin{pmatrix} 1 & \\ .6 & 1 \end{pmatrix} , \tag{5.6}$$

Taking the $+$ sign in (5.5) and plotting $Var(c)$ against a_1 gives the plot in Figure 5.1 showing a maximum of $Var(c)$ of 1.6 at approximately 0.7. More precisely, $a_1 = \frac{1}{\sqrt{2}} \approx 0.707107$, corresponding to the normalized linear combination

$$c = \frac{1}{\sqrt{2}}x_1 + \frac{1}{\sqrt{2}}x_2 \ . \tag{5.7}$$

Equation (5.7) defines the first principal component c_1 of Σ. There is also a second principal component c_2:

$$c_2 = -\frac{1}{\sqrt{2}}x_1 + \frac{1}{\sqrt{2}}x_2 \ , \tag{5.8}$$

with variance $Var(c_2) = 0.4$. It is easily verified that c_1 and c_2 are uncorrelated. The second principal component is defined as the normalized linear combination which has maximum variance among all normed linear combinations which are uncorrelated with c_1.

Example 2

Suppose we multiply x_1 by 2 while leaving x_2 unchanged, so that

$$\Sigma = \begin{pmatrix} 4 & \\ 1.2 & 1 \end{pmatrix} , \tag{5.9}$$

The two principal components of this Σ are (rounded to two decimals)

$$c_1 = 0.94x_1 + 0.33x_2 \tag{5.10}$$

$$c_2 = -0.33x_1 + 0.94x_2 \tag{5.11}$$

with variances 4.42 and 0.58, respectively. There is no clear relationships between the principal components of the two covariance matrices. The only noted difference is that for the second covariance matrix the first principal component has a larger coefficient for the variable x_1 which has the largest variance. This is a weakness of principal components: they are not scale invariant, i.e., if one change the unit of measurement in the x-variables one will get totally different principal components.

Principal components are best applied to data where the x-variables have at least approximately the same variances or where they are all measured in the same units of measurement. It is often applied to correlation matrices.

The $p \times p$ Case

The 2×2 case can be generalized to the $p \times p$ case as follows. Let $\mathbf{x}' = (x_1, x_2, \ldots, x_p)$ be p random variables with a positive definite covariance matrix Σ and consider the normed linear combination[1]

$$c = a_1 x_1 + a_2 x_2 + \cdots + a_p x_p = \mathbf{a}'\mathbf{x} \ , \tag{5.12}$$

where

$$a_1^2 + a_2^2 + \cdots + a_p^2 = \mathbf{a}'\mathbf{a} = 1 \ . \tag{5.13}$$

---

[1]To follow the development in this section we use some results from basic matrix algebra outlined in the appendix (Chapter 11). For further results, see e.g. Anderson (1984).

The variance of c is

$$Var(c) = \mathbf{a}'\mathbf{\Sigma}\mathbf{a} \ . \tag{5.14}$$

To determine the largest variance of c we should maximize $Var(c)$ subject to (5.13). It is clear that $\mathbf{a}$ can only be determined up to its sign, for changing the sign of $\mathbf{a}$ changes the sign of c but leaves the variance of c unchanged. To maximize $Var(c)$ we use the equation

$$f(\mathbf{a}, \gamma) = \mathbf{a}'\mathbf{\Sigma}\mathbf{a} - \gamma(\mathbf{a}'\mathbf{a} - 1) \ , \tag{5.15}$$

where γ is a Lagrange multiplier. For a maximum the first-order condition $\partial f / \partial \mathbf{a} = 0$ and $\partial f / \partial \gamma = 0$ must be satisfied. The last condition is the same as (5.13) and the first condition requires that

$$\mathbf{\Sigma}\mathbf{a} = \gamma \mathbf{a} \ , \tag{5.16}$$

which shows that γ must be an eigenvalue of $\mathbf{\Sigma}$ and $\mathbf{a}$ must be a corresponding eigenvector. Premultiplying (5.16) by $\mathbf{a}'$ gives

$$\mathbf{a}'\mathbf{\Sigma}\mathbf{a} = \gamma \ , \tag{5.17}$$

which shows that γ equals the variance we want to maximize. Therefore γ must be the largest eigenvalue γ_1 of $\mathbf{\Sigma}$ and $\mathbf{a}$ must be the eigenvector $\mathbf{a}_1$ corresponding to the largest eigenvalue γ_1. With γ_1 and $\mathbf{a}_1$ determined in this way, we define the first principal component of $\mathbf{\Sigma}$ as:

$$c_1 = \mathbf{a}'_1 \mathbf{x}; \tag{5.18}$$

with variance $Var(c_1) = \gamma_1$.

Next, we want to determine another normed linear combination

$$c = a_1 x_1 + a_2 x_2 + \cdots + a_p x_p = \mathbf{a}'\mathbf{x} \ , \tag{5.19}$$

uncorrelated with c_1 which have maximum variance. For c and c_1 to be uncorrelated, requires that

$$Cov(c, c_1) = \mathbf{a}'\mathbf{\Sigma}\mathbf{a}_1 = \gamma_1 \mathbf{a}'\mathbf{a}_1 = 0 \ . \tag{5.20}$$

We find that $\mathbf{a}$ is the eigenvector $\mathbf{a}_2$ corresponding to the second largest eigenvalue γ_2 of $\mathbf{\Sigma}$. So the second principal component of $\mathbf{\Sigma}$ is

$$c_2 = \mathbf{a}'_2 \mathbf{x}; \tag{5.21}$$

with variance $Var(c_2) = \gamma_2$.

In the same way we can determine the third principal component c_3 as the normed linear combination which has maximum variance γ_3 and is uncorrelated with c_1 and c_2. This process can be continued until all p principal components have been defined.

For simplicity of the presentation we assume that all eigenvalues of $\mathbf{\Sigma}$ are distinct, so that

$$\gamma_1 > \gamma_2 > \ldots > \gamma_p > 0 \ .$$

Then there will be p principal components $c_1, c_2 \ldots c_p$ with successively smaller variances and where $c_i = \mathbf{a}'_i \mathbf{x}$ with $\mathbf{a}_i$ equal to the eigenvector of $\mathbf{\Sigma}$ corresponding to γ_i. Each $\mathbf{a}_i$ is determined only up to its sign.

Collecting these results for $i = 1, 2, \ldots p$ into one matrix equation

$$\mathbf{c} = \mathbf{A}'\mathbf{x} \ , \tag{5.22}$$

where $\mathbf{A}$ is a matrix with columns $\mathbf{a}_1, \mathbf{a}_2, \ldots, \mathbf{a}_p$. By construction $\mathbf{A}$ satisfies

$$\mathbf{A}'\mathbf{A} = \mathbf{I} = \mathbf{A}\mathbf{A}' . \tag{5.23}$$

Pre-multiplying (5.22) by $\mathbf{A}$ and using the last part of (5.23) gives

$$\mathbf{x} = \mathbf{A}\mathbf{c} , \tag{5.24}$$

or

$$x_i = \sum_{j=1}^{p} a_{ij}c_j , \tag{5.25}$$

which expresses the original variables x_i in terms of the principal components c_j. Since the principal components $c_1, c_2, \ldots, c_p$ are all uncorrelated, it follows from (5.25) that

$$\sigma_{ii} = Var(x_i) = \sum_{j=1}^{p} a_{ij}^2 \gamma_j . \tag{5.26}$$

Generally, we have

$$\mathbf{\Sigma} = Cov(\mathbf{x}) = \mathbf{A}Cov(\mathbf{c})\mathbf{A}' = \mathbf{A}\mathbf{\Gamma}\mathbf{A}' , \tag{5.27}$$

where $\mathbf{\Gamma}$ is a diagonal matrix with the γ's along the diagonal. Pre-multiplying (5.27) by $\mathbf{A}'$ and post-multiplying by $\mathbf{A}$ and using (5.23) gives

$$\mathbf{A}'\mathbf{\Sigma}\mathbf{A} = \mathbf{\Gamma} , \tag{5.28}$$

which can be expressed as $\mathbf{A}$ is a matrix which transforms $\mathbf{\Sigma}$ to diagonal form. This is the same as to say that $\mathbf{A}$ transforms $\mathbf{x}$ to a set of uncorrelated variables.

In practice $\mathbf{\Sigma}$ is unknown and one uses the sample covariance matrix or correlation matrix instead.

5.1.1   Example: Five Meteorological Variables

Mardia, Kent, and Bibby (1980, p. 248) give data on 11 yearly measurements on five meteorological variables. Various types of analysis of these data are discussed by Dagnelie (1975). The variables are

$x_1 =$ rainfall in November and December (in millimeters)

$x_2 =$ average July temperature (in degrees Celsius)

$x_3 =$ rainfall in July (in millimeters)

$x_4 =$ radiation in July (in millimeters of alcohol)

$x_5 =$ average harvest yield (in quintals per hectare)

The covariance matrix is given in Table 5.1.

We illustrate how the principal components can be obtained for the covariance matrix using a LISREL command file (see file **meteor1b.lis**):

Table 5.1: Covariance Matrix of Five Meteorological Variables

x_1	x_2	x_3	x_4	x_5
1973.298				
-4.921	1.637			
799.564	-29.279	1346.859		
-2439.351	217.198	-6822.728	52914.656	
-57.214	1.735	-62.080	361.803	4.496

```
!Principal Components of 5 meteorological variables
DA NI=5 NO=11
LA
X1 X2 X3 X4 X5
CM
   1973.298
     -4.921      1.637
    799.564    -29.279    1346.859
  -2439.351    217.198   -6822.728   52914.656
    -57.214      1.735     -62.080     361.803       4.496
PC
OU
```

The results are given in the output as follows.

```
Eigenvalues and Eigenvectors
```

	PC_1	PC_2	PC_3	PC_4	PC_5
	--------	--------	--------	--------	--------
Eigenvalue	53927.94	1999.96	311.26	1.26	0.52
StandError	22994.95	852.79	132.72	0.54	0.22
% Variance	95.89	3.56	0.55	0.00	0.00
Cum. % Var	95.89	99.44	100.00	100.00	100.00
	--------	--------	--------	--------	--------
X1	-0.048	0.954	-0.296	0.013	-0.013
X2	0.004	0.003	-0.008	0.508	0.861
X3	-0.129	0.288	0.949	0.016	-0.001
X4	0.990	0.084	0.109	-0.005	-0.001
X5	0.007	-0.021	-0.008	0.861	-0.508

```
Correlations between Variables. and Principal Components
```

	PC_1	PC_2	PC_3	PC_4	PC_5
	--------	--------	--------	--------	--------
X1	-0.254	0.960	-0.118	0.000	0.000
X2	0.738	0.091	-0.115	0.446	0.486
X3	-0.818	0.351	0.456	0.000	0.000
X4	1.000	0.016	0.008	0.000	0.000
X5	0.750	-0.443	-0.065	0.456	-0.173

Variance Contributions

	PC_1	PC_2	PC_3	PC_4	PC_5
X1	0.064	0.922	0.014	0.000	0.000
X2	0.544	0.008	0.013	0.199	0.236
X3	0.669	0.123	0.208	0.000	0.000
X4	1.000	0.000	0.000	0.000	0.000
X5	0.562	0.196	0.004	0.208	0.030

The first line gives the eigenvalues of the covariance matrix. These are the variances of the principal components. The second line gives the estimated standard errors of the eigenvalues. If PCA is based on a random sample of size N with a covariance matrix $\mathbf{S}$, the eigenvalues $\hat{\gamma}_1, \hat{\gamma}_2, \ldots, \hat{\gamma}_p$ of $\mathbf{S}$ are random variables. The asymptotic variance of $\hat{\gamma}_i$ is $(2/N)\gamma_i^2$. So the estimated standard error of $\hat{\gamma}_i$ is $\sqrt{(2/N)}\hat{\gamma}_i$. In this example the estimated standard errors are unreliable since the sample size is only 11. The third line gives the eigenvalues in percentage of the total variance and the fourth line gives these percentages cumulatively. The next five columns give the eigenvectors of the covariance matrix normalized so that their sum of squares is 1 and such that the largest value is positive. These eigenvectors are the coefficients (weights) in the linear combinations defining the principal components.

The next set of five columns shows the correlations between the observed variables and the principal components and the last set of five columns gives the variance contribution of each principal component to the variance of each observed variable. If all principal components are computed, as is the case here, these variance contributions sum to 1 row-wise.

The first two principal components account for 99.44% of the variance. The first component is dominated by the radiation variable x_4. The second component is a rainfall variable dominated by x_1 and x_3. It is often difficult to get a clear interpretation of the principal components.

To analyze the correlation matrix instead of the covariance matrix of the five meteorological variables, just add MA=KM on the DA command in **meteor1b.lis**, see file **meteor2b.lis**. As shown here this gives a very different result and interpretation.

Eigenvalues and Eigenvectors

	PC_1	PC_2	PC_3	PC_4	PC_5
Eigenvalue	3.40	1.02	0.29	0.16	0.13
StandError	1.45	0.43	0.13	0.07	0.06
% Variance	67.97	20.31	5.90	3.21	2.60
Cum. % Var	67.97	88.29	94.18	97.40	100.00
X1	-0.293	0.809	-0.231	0.173	-0.419
X2	0.423	0.482	0.677	-0.338	0.123
X3	-0.500	0.040	0.505	0.585	0.389
X4	0.483	0.286	-0.433	0.365	0.604
X5	0.502	-0.170	0.214	0.617	-0.541

```
Correlations between Variables and Principal Components
```

	PC_1	PC_2	PC_3	PC_4	PC_5
X1	-0.540	0.816	-0.125	0.070	-0.151
X2	0.780	0.486	0.368	-0.136	0.044
X3	-0.921	0.040	0.274	0.234	0.140
X4	0.890	0.288	-0.235	0.146	0.218
X5	0.926	-0.171	0.116	0.247	-0.195

Here we see that the first two principal components account only for 88.29% of the total variance and the first principal component seems to be a contrast x_2, x_4, x_5 and x_1, x_3.

Principal Components of Nine Psychological Variables

To analyze more than just a few variables it is not practical to give the covariance matrix directly in the input file. Instead one should read the covariance matrix from a file. However, if raw data is available in a **lsf** file or in a text file it is much easier to use **PRELIS** syntax. We illustrate this by analyzing first the covariance matrix and then the correlation matrix for the nine psychological variables for the Pasteur school sample, see Chapter 1. We created the **lsf** file in that chapter and saved in the file **pasteur_npv.lsf**.

To estimate the principal components of the covariance matrix of the nine psychological variables for the Pasteur school, use the following simple three-line **PRELIS** syntax file **pcnpv1.prl**:

```
SY=pasteur_npv.lsf
PC
OU MA=CM
```

The output reveals the following

```
Eigenvalues and Eigenvectors
```

	PC_1	PC_2	PC_3	PC_4	PC_5	PC_6
Eigenvalue	1457.41	587.13	259.49	102.66	63.99	26.54
StandError	165.02	66.48	29.38	11.62	7.25	3.01
% Variance	57.60	23.21	10.26	4.06	2.53	1.05
Cum. % Var	57.60	80.81	91.06	95.12	97.65	98.70
VISPERC	0.057	-0.038	0.035	0.487	-0.119	0.850
CUBES	0.018	-0.052	0.031	0.173	0.048	0.023
LOZENGES	0.069	-0.072	0.068	0.679	0.621	-0.333
PARCOMP	0.019	0.026	-0.008	0.175	-0.239	-0.042
SENCOMP	0.027	0.010	0.008	0.225	-0.471	-0.256
WORDMEAN	0.046	0.036	0.033	0.411	-0.559	-0.313
ADDITION	0.326	0.853	-0.395	0.047	0.073	0.018
COUNTDOT	0.278	0.317	0.902	-0.089	0.007	0.010
SCCAPS	0.897	-0.401	-0.145	-0.107	-0.022	-0.014

Eigenvalues and Eigenvectors

	PC_7	PC_8	PC_9
Eigenvalue	19.50	9.55	3.91
StandError	2.21	1.08	0.44
% Variance	0.77	0.38	0.15
Cum. % Var	99.47	99.85	100.00
VISPERC	-0.102	0.006	-0.096
CUBES	0.979	0.069	0.024
LOZENGES	-0.154	0.058	-0.013
PARCOMP	-0.062	0.322	0.895
SENCOMP	-0.048	0.687	-0.433
WORDMEAN	0.008	-0.645	-0.013
ADDITION	0.039	0.008	-0.024
COUNTDOT	-0.003	0.016	0.007
SCCAPS	-0.012	-0.008	0.008

This illustrates an example of how the results of a principal component analysis can be almost useless. Note that the principal components are approximately equal to the variables. The first principal component is essentially equal to the variable with the largest variance, which is SCCAPS. The second principal component is essentially equal to the variable with the second largest variance which is ADDITION, and so on until the ninth principal component which is essentially equal to the variable with the smallest variance which is PARCOMP.

Since the variances of the variables vary considerably across variables it is better to analyze the correlation matrix instead of the covariance matrix. To do so, just change MA=CM on the OU line to MA=KM, see file **pcnpv2.prl**. This gives the following principal components.

Eigenvalues and Eigenvectors

	PC_1	PC_2	PC_3	PC_4	PC_5	PC_6
Eigenvalue	3.10	1.55	1.45	0.72	0.61	0.56
StandError	0.35	0.18	0.16	0.08	0.07	0.06
% Variance	34.40	17.25	16.15	7.99	6.79	6.25
Cum. % Var	34.40	51.65	67.80	75.79	82.58	88.83
VISPERC	0.380	0.301	-0.114	-0.428	0.004	-0.062
CUBES	0.168	0.511	-0.175	0.711	-0.233	0.317
LOZENGES	0.238	0.546	0.018	-0.433	-0.262	0.010
PARCOMP	0.448	-0.280	-0.200	-0.085	-0.046	0.135
SENCOMP	0.429	-0.302	-0.265	0.185	0.176	-0.121
WORDMEAN	0.470	-0.183	-0.179	0.086	-0.081	-0.098
ADDITION	0.197	-0.289	0.549	-0.090	-0.286	0.659
COUNTDOT	0.229	-0.006	0.576	0.239	-0.341	-0.642
SCCAPS	0.272	0.252	0.425	0.100	0.800	0.082

Eigenvalues and Eigenvectors

	PC_7	PC_8	PC_9
Eigenvalue	0.48	0.31	0.21
StandError	0.05	0.04	0.02
% Variance	5.33	3.47	2.37
Cum. % Var	94.16	97.63	100.00
VISPERC	0.722	0.104	0.181
CUBES	0.127	-0.061	0.009
LOZENGES	-0.614	-0.076	0.080
PARCOMP	0.022	-0.662	-0.463
SENCOMP	-0.214	-0.077	0.724
WORDMEAN	-0.154	0.706	-0.410
ADDITION	0.042	0.123	0.191
COUNTDOT	0.084	-0.149	-0.004
SCCAPS	-0.084	-0.013	-0.132

When principal component analysis (PCA) is used in this context a major issue is how many components to retain and interpret.

Various *ad hoc* procedures have been suggested in the literature. One of them is to plot $\hat{\gamma}_i$ against i for $i = 1, 2, \ldots, p$ and look for an elbow in the screeplot as shown in Figure 5.2. In this example, this might suggest to retain three principal components.

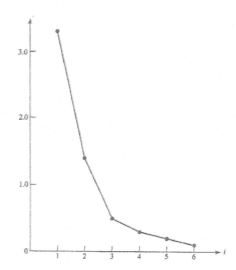

Figure 5.2: A Hypothetical Scree Plot

For the principal components of the nine psychological variables estimated for the Pasteur school, the screeplot is shown in Figure 5.3.

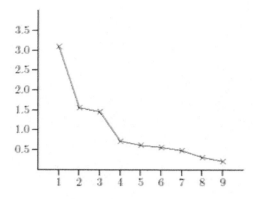

Figure 5.3: Scree Plot for Nine Psychological Variables

The main idea is to retain only those principal components which are statistically different, *i.e.*, if the k largest eigenvalues are statistically different and the $p - k$ smallest eigenvalues are not statistically different then one should retain k principal components.

A test of the hypothesis that $\gamma_{k+1} = \gamma_{k+2} = \cdots = \gamma_p$ has been developed, see *i.e.*,Anderson (1984, p. 475). Experience suggests that this test often leads to too many components especially in large samples. Therefore this test is not included in **LISREL**.

A very simple rule[2] is the following. Let k be the number of eigenvalues of the correlation matrix which are greater than 1. This rule does not seem to have any rationale but it can be rationalized as follows. In a correlation matrix all variables have variance 1. In terms of explaining variance it is only reasonable to keep those principal components which have a variance greater than 1 since the others explain less variance than any variable. This simple rule works remarkably well in practice. To apply this rule in the case of analyzing a covariance matrix, one must first obtain the eigenvalues of the correlation matrix to determine k and then extract the first k principal components of the covariance matrix.

Applying this rule to the principal component analysis of the correlation matrix of the Pasteur school sample, suggests that we should retain 3 components. To estimate only the first 3 principal components add NC=3 on the PC line, see file **pcnpv3.prl**. The resulting output is

```
Eigenvalues and Eigenvectors
                 PC_1        PC_2        PC_3

                --------    --------    --------

Eigenvalue       3.10        1.55        1.45
StandError       0.35        0.18        0.16
% Variance      34.40       17.25       16.15
Cum. % Var      34.40       51.65       67.80

                --------    --------    --------

   VISPERC       0.380       0.301      -0.114
     CUBES       0.168       0.511      -0.175
  LOZENGES       0.238       0.546       0.018
   PARCOMP       0.448      -0.280      -0.200
```

[2]This is called Kaiser's little jiffy and was originally suggested in the context of factor analysis in Henry Kaiser's dissertation

SENCOMP	0.429	-0.302	-0.265
WORDMEAN	0.470	-0.183	-0.179
ADDITION	0.197	-0.289	0.549
COUNTDOT	0.229	-0.006	0.576
SCCAPS	0.272	0.252	0.425

Correlations between Variables and Principal Components

	PC_1	PC_2	PC_3
	------	-------	-------
VISPERC	0.668	0.375	-0.137
CUBES	0.296	0.637	-0.211
LOZENGES	0.419	0.681	0.021
PARCOMP	0.788	-0.348	-0.241
SENCOMP	0.755	-0.376	-0.319
WORDMEAN	0.827	-0.228	-0.216
ADDITION	0.347	-0.360	0.661
COUNTDOT	0.404	-0.007	0.695
SCCAPS	0.479	0.314	0.513

Principal components seldom have a clear interpretation other than in terms of explained variance. One should not interpret the eigenvectors or the correlations as factor loadings. As explained in the next section principal components analysis and factor analysis are different types of analysis with different aims.

5.2   Principal Components vs Factor Analysis

Principal components are often confused with factor analysis. This section clarifies the distinction between the two types of analysis.

Consider equation (5.25) and suppose that for some $k < p$, $\gamma_{k+1}, \gamma_{k+2}, \ldots, \gamma_p$ are so small that we want to ignore the effects of $c_{k+1}, c_{k+2}, \ldots, c_p$. We can then write

$$x_i = \sum_{j=1}^{k} a_{ij} c_j + z_i \,, \tag{5.29}$$

where

$$z_i = \sum_{j=k+1}^{p} a_{ij} c_j \,. \tag{5.30}$$

Viewed as a regression equation (5.29) is the regression of x_i on a set of uncorrelated variables $c_1, c_2 \ldots, c_k$ with error term z_i. In matrix form this can be written

$$\mathbf{x} = \mathbf{A}_k \mathbf{c} + \mathbf{z} \,, \tag{5.31}$$

where $\mathbf{A}_k$ is the matrix formed by the first k columns of $\mathbf{A}$.

Equations (5.29) and (5.31) resemble equations (6.1) and (6.4) in Chapter 6. We reproduce these equations here as

$$x_i = \lambda_{i1}\xi_1 + \lambda_{i2}\xi_2 + \cdots + \lambda_{ik}\xi_k + \delta_i \ , i = 1, 2, \ldots, p \ , \tag{5.32}$$

and

$$\mathbf{x} = \mathbf{\Lambda}\boldsymbol{\xi} + \boldsymbol{\delta} \ , \tag{5.33}$$

In the factor analysis model defined by (5.32) $\xi_1, \xi_2, \ldots, \xi_k$ are underlying latent factors *assumed* to account for the correlations between the observed variables $x_1, x_2, \ldots, x_p$, whereas the principal components $c_1, c_2 \ldots, c_k$ in (5.29) are supposed to account for maximum variance. This leads to the following fundamental distinction.

In equation (5.32), δ_i and δ_j in are uncorrelated for $i \neq j$, whereas z_i and z_j in (5.29) are correlated for $i \neq j$. The last part of this statement follows because z_i and z_j are linear functions of $c_{k+1}, c_{k+2}, \ldots, c_p$, so z_i and z_j have $c_{k+1}, c_{k+2}, \ldots, c_p$ as common factors.

Using the matrix forms (5.31) and (5.32) we see that

$$Cov(\mathbf{z}) = \mathbf{A}_2\mathbf{\Gamma}_2\mathbf{A}_2' \ , \tag{5.34}$$

where $\mathbf{\Gamma}_2$ is a diagonal matrix with the $p - k$ smallest eigenvalues of $\mathbf{\Sigma}$ and $\mathbf{A}_2$ consists of the corresponding eigenvectors. Thus, $Cov(\mathbf{z})$ is not diagonal, whereas $Cov(\boldsymbol{\delta})$ is diagonal by definition.

Numeric Illustration

To illustrate the distinction between principal components and factor analysis we use the following covariance matrix $\mathbf{\Sigma}$ (which is a correlation matrix here):

$$\mathbf{\Sigma} = \begin{pmatrix} 1.000 & & & & & \\ 0.720 & 1.000 & & & & \\ 0.378 & 0.336 & 1.000 & & & \\ 0.324 & 0.288 & 0.420 & 1.000 & & \\ 0.270 & 0.240 & 0.350 & 0.300 & 1.000 & \\ 0.270 & 0.240 & 0.126 & 0.108 & 0.090 & 1.000 \end{pmatrix} \tag{5.35}$$

We regard this as a population covariance matrix $\mathbf{\Sigma}$ rather than a sample covariance matrix $\mathbf{S}$ because, as shown in Chapter 6, it has been specially constructed so as to satisfy a factor analysis model with two factors.

To obtain the principal components of $\mathbf{\Sigma}$ one can use the following LISREL syntax file **pcex1b.lis**:

```
Example of Principal Components
DA NI=6 NO=100
CM
1
.720 1
.378 .336 1
.324 .288 .420 1
.270 .240 .350 .300 1
.270 .240 .126 .108 .090 1
PC
OU
```

Here the DA line specifies the number of variables NI and the sample size. In this case, the sample size NO is irrelevant but some value must be specified. The CM tells LISREL to read a covariance matrix Σ. Only the lower half including the diagonal should be included. Note that 1.000 can be written as 1 and 0.378 can be written as .378. PC is the command for Principal Components. In this case all principal components will be computed. To compute only the first 2 principal components, say, use the line PC NC=2.

Alternatively, one can use the following SIMPLIS syntax file **pcex1a.spl**:

```
Example of Principal Components
Observed Variables: x1-x6
Covariance Matrix
1
.720 1
.378 .336 1
.324 .288 .420 1
.270 .240 .350 .300 1
.270 .240 .126 .108 .090 1
Principal Components
End of Problem
```

Both input files give the same results. The principal components are given as

```
Principal Component Analysis
```

Eigenvalues and Eigenvectors

	PC_1	PC_2	PC_3	PC_4	PC_5	PC_6
Eigenvalue	2.57	1.06	0.82	0.71	0.57	0.28
StandError	0.36	0.15	0.12	0.10	0.08	0.04
% Variance	42.83	17.66	13.63	11.77	9.49	4.62
Cum. % Var	42.83	60.49	74.12	85.89	95.38	100.00

It is seen that the first two principal components account for only 60.49 % of the total variance. The next six lines gives the columns of eigenvectors. Each column has a sum of squares equal to 1.

x1	0.506	0.299	-0.342	0.076	-0.044	0.728
x2	0.484	0.330	-0.417	0.101	-0.072	-0.683
x3	0.426	-0.347	0.088	-0.235	0.796	-0.038
x4	0.390	-0.390	0.131	-0.595	-0.569	-0.021
x5	0.344	-0.440	0.288	0.755	-0.186	-0.013
x6	0.242	0.579	0.775	-0.064	0.016	-0.033

The next six lines gives the correlations ρ_{ij} between the x-variables and the principal components:

Correlations between Variables and Principal Components

	PC_1	PC_2	PC_3	PC_4	PC_5	PC_6
x1	0.811	0.308	-0.310	0.064	-0.034	0.383
x2	0.776	0.340	-0.377	0.085	-0.054	-0.360
x3	0.683	-0.357	0.080	-0.197	0.601	-0.020
x4	0.624	-0.401	0.119	-0.500	-0.430	-0.011
x5	0.552	-0.453	0.260	0.635	-0.140	-0.007
x6	0.387	0.596	0.701	-0.054	0.012	-0.018

Each ρ_{ij} is calculated as

$$\rho_{ij} = \frac{a_{ij}}{\sqrt{\sigma_{ii}\gamma_j}} \, . \tag{5.36}$$

The next six lines give the variance contribution v_{ij} of each principal component c_j to each variable x_i:

Variance Contributions

	PC_1	PC_2	PC_3	PC_4	PC_5	PC_6
x1	0.657	0.095	0.096	0.004	0.001	0.147
x2	0.602	0.116	0.142	0.007	0.003	0.129
x3	0.466	0.128	0.006	0.039	0.361	0.000
x4	0.390	0.161	0.014	0.250	0.185	0.000
x5	0.304	0.205	0.068	0.403	0.020	0.000
x6	0.150	0.355	0.491	0.003	0.000	0.000

These variance contributions are calculated as

$$v_{ij} = a_{ij}^2 \frac{\gamma_j}{\sigma_{ii}} \, . \tag{5.37}$$

They sum to 1 row-wise, see equation (5.26).

Suppose we ignore the last four principal components and represent $\boldsymbol{\Sigma}$ as in equation (5.31) with $\mathbf{A}_k$ as the first two eigenvectors. Then

$$Cov(\mathbf{z}) = \boldsymbol{\Sigma} - \mathbf{A}_k\boldsymbol{\Gamma}_k\mathbf{A}_k' = \mathbf{A}_2\boldsymbol{\Gamma}_2\mathbf{A}_2' \, , \tag{5.38}$$

where $\boldsymbol{\Gamma}_2$ is a diagonal matrix with the 4 smallest eigenvalues and $\mathbf{A}_2$ consists of the corresponding eigenvectors. The result is

$$Cov(\mathbf{z}) = \begin{pmatrix} 0.247 & & & & & \\ -0.014 & 0.283 & & & & \\ -0.066 & -0.073 & 0.406 & & & \\ -0.060 & -0.061 & -0.151 & 0.448 & & \\ -0.038 & -0.034 & -0.189 & -0.227 & 0.491 & \\ -0.228 & -0.264 & 0.074 & 0.105 & 0.146 & 0.495 \end{pmatrix} \tag{5.39}$$

The total explained variance is

$$6 - 0.247 - 0.283 - 0.406 - 0.448 - 0.491 - 0.495 = 0.363$$

As will be shown in Chapter 6 a factor analysis of $\boldsymbol{\Sigma}$ gives a diagonal matrix $Cov(\boldsymbol{\delta})$:

$$Cov(\boldsymbol{\delta}) = diag(0.19, 0.36, 0.51, 0.64, 0.75, 0.91) . \tag{5.40}$$

The total explained variance in factor analysis is

$$6 - 0.19 - 0.36 - 0.51 - 0.64 - 0.75 - 0.91 = 0.264$$

What this example shows is that factor analysis explains the correlations whereas principal components account for more variance. This holds in general.

5.3   Principal Components of a Data Matrix

Suppose $\mathbf{X}$ is a data matrix of order $N \times p$ with measurements on p variables for N cases (e.g.,individuals or time points). We refer to this as a sample.

To begin with we assume that the variables are mean centered so that the means of the variables are zero. The sample covariance matrix[3] is $\mathbf{S} = (1/N)\mathbf{X}'\mathbf{X}$. The sample principal components $\hat{\mathbf{c}} = \hat{\mathbf{A}}'\mathbf{x}$ of $\mathbf{S}$ are defined in analogy with the principal components of $\boldsymbol{\Sigma}$. So $\mathbf{S} = \hat{\mathbf{A}}\hat{\boldsymbol{\Gamma}}\hat{\mathbf{A}}'$, with $\hat{\mathbf{A}}$ and $\hat{\boldsymbol{\Gamma}}$ defined in analogy with $\mathbf{A}$ and $\boldsymbol{\Gamma}$ in (5.27). For simplicity we assume that all eigenvalues $\hat{\gamma}_1, \hat{\gamma}_2, \ldots, \hat{\gamma}_p$ of $\mathbf{S}$ are positive.

Let $\mathbf{x}_i'$ be the ith row of $\mathbf{X}$. Applying equation (5.22) to $\mathbf{x}_i'$ gives

$$\hat{\mathbf{c}}_i' = \mathbf{x}_i'\hat{\mathbf{A}} . \tag{5.41}$$

This gives the *principal component scores* for case i in the sample. Combining these equations for $i = 1, 2, \ldots, N$ gives the matrix equation

$$\hat{\mathbf{C}} = \mathbf{X}\hat{\mathbf{A}} , \tag{5.42}$$

where $\hat{\mathbf{C}}$ is a $N \times p$ matrix of principal component scores.

Post-multiplying (5.42) by $\hat{\mathbf{A}}'$ and using (5.23) gives

$$\mathbf{X} = \hat{\mathbf{C}}\hat{\mathbf{A}}' . \tag{5.43}$$

Using (5.23) once more, it follows from (5.42) that

$$(1/N)\hat{\mathbf{C}}'\hat{\mathbf{C}} = (1/N)\hat{\mathbf{A}}'\mathbf{X}'\mathbf{X}\hat{\mathbf{A}} = \hat{\mathbf{A}}'\mathbf{S}\hat{\mathbf{A}} = \hat{\boldsymbol{\Gamma}} , \tag{5.44}$$

which shows that the covariance matrix of the principal component scores is $\hat{\boldsymbol{\Gamma}}$, *i.e.*, the principal component scores are uncorrelated and their variances are $\hat{\gamma}_1, \hat{\gamma}_2, \ldots, \hat{\gamma}_p$.

Consider $\hat{\mathbf{C}}$ as a data matrix of N cases on p variables. Since the variables have successively smaller and smaller variances one can decide to retain only the first k variables for some $k < p$ and approximate $\mathbf{X}$ by

$$\mathbf{X}^\star = \hat{\mathbf{C}}\hat{\mathbf{A}}_k , \tag{5.45}$$

where $\hat{\mathbf{A}}_k$ consist of the first k columns of $\hat{\mathbf{A}}$. $\mathbf{X}^\star$ is a rank k least squares approximation to $\mathbf{X}$ in the sense that

$$\sum_i \sum_j (x_{ij} - x_{ij}^\star)^2 \tag{5.46}$$

---

[3]Actually it does not matter whether we divide by N or not.

has a minimum. Although the minimum is unique, the matrix $\mathbf{X}^\star$ is not unique for it can be post-multiplied by any orthogonal matrix $\mathbf{T}$ of order $k \times k$.

Geometrically, the matrix $\mathbf{X}$ represents N points in p-dimensional space and $\mathbf{X}^\star$ represents the projection of these N points onto a k-dimensional space. We illustrate this for a very small example with $N = 8$, $p = 2$ and $k = 1$.

The following mean centered data matrix on two variables is given

$$\mathbf{X} = \begin{bmatrix} -8 & -1 \\ 6 & 10 \\ -2 & -10 \\ 8 & 1 \\ 0 & 3 \\ -6 & -6 \\ 0 & -3 \\ 2 & 6 \end{bmatrix} \tag{5.47}$$

So

$$\mathbf{X}'\mathbf{X} = \begin{pmatrix} 208 & \\ 144 & 292 \end{pmatrix} \tag{5.48}$$

with eigenvalues $\gamma_1 = 400$ and $\gamma_2 = 100$ and eigenvectors

$$\mathbf{A} = \begin{bmatrix} 0.6 & 0.8 \\ 0.8 & -0.6 \end{bmatrix} \tag{5.49}$$

Geometrically the matrix $\mathbf{X}$ represents the 8 points marked $\times$ in Figure 5.4 and the first principal component $c_1 = 0.6x_1 + 0.8x_2$ is the line in the figure. This line has direction cosines 0.6 and 0.8, which means that the cosine of the angle between the line and x_1-axis is 0.6 and the cosine of the angle between the line and x_2-axis is 0.8. Each of the 8 points are projected onto the line which correspond to the lines from the point to the projected point on the line marked $\star$. These lines are perpendicular to the principal component line.

From (5.42) we get principal component sores as

$$\mathbf{C} = \mathbf{X}\mathbf{A} = \begin{bmatrix} -8 & -1 \\ 6 & 10 \\ -2 & -10 \\ 8 & 1 \\ 0 & 3 \\ -6 & -6 \\ 0 & -3 \\ 2 & 6 \end{bmatrix} \begin{bmatrix} 0.6 & 0.8 \\ 0.8 & -0.6 \end{bmatrix} = \begin{bmatrix} -5.6 & 5.8 \\ 11.6 & 1.2 \\ -9.2 & -4.4 \\ 5.6 & -5.8 \\ 2.4 & 1.8 \\ -8.4 & 1.2 \\ -2.4 & -1.8 \\ 6.0 & -2.0 \end{bmatrix} \tag{5.50}$$

A rank 1 approximation to $\mathbf{X}$ is obtained by taking the first column of $\mathbf{C}$.

5.3.1   Example: PCA of Nine Psychological Variables

Continuing the example of the nine psychological variables for the Pasteur school and using three principal components, one can obtain the principal component scores using the following PRELIS syntax file **pcnpv4.prl**:

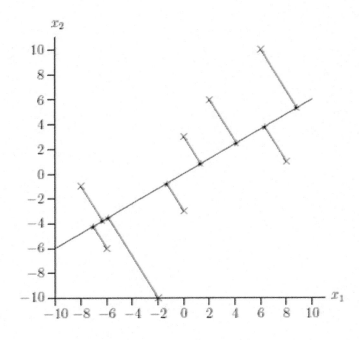

Figure 5.4: Onto First Principal Component

```
SY=pasteur_npv.lsf
PC NC=3 PS
OU MA=KM
```

The principle components scores will appear in text form in the file **pcnpv4.psc**. One can add a first line with variable names as shown here for the first few cases

PC_1	PC_2	PC_3
0.502	0.832	0.024
0.366	2.487	1.858
-3.468	0.748	-0.863
0.339	0.842	-1.475
0.403	-0.132	0.462
-0.343	1.907	2.417
0.893	-1.900	-0.202
-1.091	-0.927	-1.549
1.949	0.926	-0.097
-0.511	-1.504	-1.345
-1.557	-0.548	-0.295

and then import this to **pcnpv4.lsf** as shown in Chapter 1. Defining the variables as continuous one can compute the covariance matrix of the principal component scores and verify that they are uncorrelated with variances 3.10, 1.55, and 1.45.

5.3.2   Example: Stock Market Prices

Johnson and Wichern (2002) gave an exercise and provided data on a diskette for five chemical and oil companies on the New York Stock Market Exchange. The companies are Allied Chemical, Du Pont, Union Carbide, Exxon, and Texaco and the data consists of the percentage weekly rate of returns for 100 weeks. The data matrix is **smp.lsf**.

To use Kaiser's little Jiffy we first estimate the principal components of the correlation matrix using the PRELIS syntax file **smp1.prl**:

```
sy=smp.lsf
pc
ou ma=km
```

This gives the following eigenvalues

	PC_1	PC_2	PC_3	PC_4	PC_5
Eigenvalue	2.86	0.81	0.54	0.45	0.34
StandError	0.40	0.11	0.08	0.06	0.05
% Variance	57.13	16.18	10.80	9.03	6.86
Cum. % Var	57.13	73.31	84.11	93.14	100.00

So Kaiser's rule suggests using only one principal component. Next we use the covariance matrix to estimate the first principal component and the corresponding principal component scores. The PRELIS syntax for this is **smp2.prl**:

```
sy=smp.lsf
pc nc=1 ps
ou ma=cm
```

The principal component scores appear in the text file **smp2.psc**. Add the variable name Wrrt (for Weekly rate of returns) on the first line. Then import this as shown in Chapter 1 and save it as **coilindx.lsf**, say,. For now this file contains only one variable Wrrt which equals the estimated principal component scores. We can then compute a price index for chemical and oil companies here called Coilindx, using the formula

$$p_t = p_{t-1} + p_{t-1}(\delta_t/100) \,, \tag{5.51}$$

where δ_t is Wrrt with $\delta_0 = 0$ and p_t is Coilindx with $p_0 = 100$. Recall that δ_t is the weekly rate of returns in percent, so the actul increase in the price index is $p_{t-1}(\delta_t/100)$. The first few lines of the file **coilindx.lsf** are

	Week	Wrr	Coilindx
1	1.000	0.413	100.413
2	2.000	0.108	100.521
3	3.000	5.366	105.915
4	4.000	1.703	107.719
5	5.000	-1.743	105.842
6	6.000	3.942	110.014
7	7.000	-1.664	108.183
8	8.000	1.335	109.628
9	9.000	-1.222	108.288
10	10.000	-1.928	106.200
11	11.000	1.019	107.282
12	12.000	-2.837	104.239
13	13.000	3.643	108.036

The keyword TIME can be used to obtain a variable Week with values 1, 2, ...,100. A line plot of Coilindx against Week gives the following graph

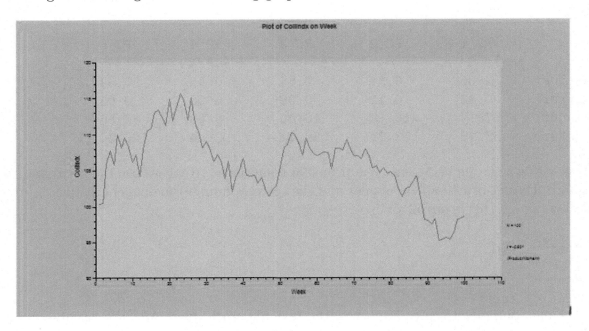

Figure 5.5: Line Plot of Stock Market Chemical Index

Chapter 6

Exploratory Factor Analysis (EFA)

Although its roots can be traced back to the work of Francis Galton, it is generally considered that factor analysis began with the celebrated article by Spearman (1904). In the first half of the 20th century factor analysis was mainly developed by psychologists for the purpose of identifying mental abilities by means of psychological testing. Various theories of mental abilities and various procedures for analyzing the correlations among psychological tests emerged. The most prominent factor analysts in the first half of the 20th century seem to be Godfrey Thomson, Cyril Burt, Raymond Cattell, Karl Holzinger, Louis Thurstone and Louis Guttman. A later generation of psychological factor analysts that played important roles are Ledyard Tucker, Henry Kaiser, and Chester Harris. For the early history of factor analysis, see the articles by Bartholomew (1995) and Hägglund (2001), the text books by Harman (1967) and Mulaik (1972) and the review article by Mulaik (1986).

Factor analysis was found to be quite useful in the early stages of experimentation and test development. Thurstone's (1938) primary mental abilities, French's (1951) factors in aptitude and achievement tests, Guilford's (1956) structure of intelligence, and the structure of personal characteristics of Romney and Bynner (1992) are good examples of this.

In the 1950's there seem to be two schools of factor analysis: the psychometric school and the statistical school. The psychometric school regarded the battery of tests as a selection from a large domain of tests that could be developed for the same psychological phenomenon and focused on the factors in this domain. By contrast, the statistical school regarded the number of tests as fixed and focused on the inference from the individuals being tested to a hypothetical population of individuals. The distinction between the two perspectives is particularly contrasted with regard to the number of factors. In the psychometric perspective it was assumed that there are a small number of major factors and possibly a large number of minor factors, whereas in the statistical perspective the number of factors is assumed to be small relative to the number of tests.

Whereas the factor analysis literature in the first half of the 20th century was dominated by psychologists, the literature of the second half of the century was dominated by statisticians. In fact, there has been an enormous development of the statistical methodology for factor analysis in the last 50 years. This has been accompanied by an equally enormous development of computational methods for factor analysis. During this period the applications of factor analysis spread from psychology to many other disciplines, e.g., international relations, economics, sociology, communications, taxonomy, biology, physiology, medicine, geology, and meteorology.

Section 6.3 covers exploratory factor analysis for continuous variables. Factor analysis of ordinal variables is covered in Section 6.4.

© Springer International Publishing Switzerland 2016

K.G. Jöreskog et al., *Multivariate Analysis with LISREL*, Springer Series in Statistics,

DOI 10.1007/978-3-319-33153-9_6

6.1   The Factor Analysis Model and Its Estimation

The basic idea of factor analysis is the following. For a given set of manifest response variables $x_1, \ldots, x_p$ one wants to find a set of underlying latent factors $\xi_1, \ldots, \xi_k$, fewer in number than the observed variables. These latent factors are supposed to account for the inter-correlations of the response variables in the sense that when the factors are held constant, there should no longer remain any correlations between the response variables. If both the observed response variables and the latent factors are measured in deviations from their means, this leads to the linear factor analysis model:

$$x_i = \lambda_{i1}\xi_1 + \lambda_{i2}\xi_2 + \cdots + \lambda_{ik}\xi_k + \delta_i \, , \ \ i = 1, 2, \ldots, p \, , \tag{6.1}$$

where δ_i, the unique part of x_i, is uncorrelated with $\xi_1, \xi_2, \ldots, \xi_k$ and with δ_j for $j \neq i$. In this classical factor analysis model (6.1) the factors $\xi_1, \xi_2, \ldots, \xi_k$ are assumed to be standardized to zero means and unit standard deviation. Since they are also assumed to be uncorrelated, it follows that

$$Var(x_i) = \lambda_{i1}^2 + \lambda_{i2}^2 + \cdots + \lambda_{ik}^2 + Var(\delta_i) \, . \tag{6.2}$$

Since δ_i and δ_j are uncorrelated for $j \neq i$, it also follows that

$$Cov(x_i, x_j) = \lambda_{i1}\lambda_{j1} + \lambda_{i2}\lambda_{j2} + \cdots + \lambda_{ik}\lambda_{jk} \, , \ \ j \neq i \, . \tag{6.3}$$

In matrix notation (6.1) is

$$\mathbf{x} = \boldsymbol{\Lambda}\boldsymbol{\xi} + \boldsymbol{\delta} \, , \tag{6.4}$$

where $\mathbf{x} = (x_1, x_2, \ldots, x_p)'$, $\boldsymbol{\xi} = (\xi_1, \xi_2, \ldots, \xi_k)'$ and $\boldsymbol{\delta} = (\delta_1, \delta_2, \ldots, \delta_p)'$ are random vectors and $\boldsymbol{\Lambda}$ is a matrix of order $p \times k$ of parameters called factor loadings. The objective of factor analysis is to determine the number of factors k and the factor loadings λ_{ij}. The covariance matrix $\boldsymbol{\Sigma}$ of $\mathbf{x}$ is

$$\boldsymbol{\Sigma} = \boldsymbol{\Lambda}\boldsymbol{\Lambda}' + \boldsymbol{\Psi} \, , \tag{6.5}$$

where $\boldsymbol{\Psi} = Cov(\boldsymbol{\delta})$ is a diagonal matrix.

Exploratory factor analysis is illustrated in Figure 6.1 for the case of 6 manifest variables and 2 factors.

There is a one-way arrow (path) from each factor to all the manifest variables. This corresponds to a factor matrix

$$\boldsymbol{\Lambda} = \begin{pmatrix} \lambda_{11} & \lambda_{12} \\ \lambda_{21} & \lambda_{22} \\ \lambda_{31} & \lambda_{32} \\ \lambda_{41} & \lambda_{42} \\ \lambda_{51} & \lambda_{52} \\ \lambda_{61} & \lambda_{62} \end{pmatrix} \, , \tag{6.6}$$

and there is no two-way arrow between the two factors meaning that they are uncorrelated. Path diagrams are further explained in Chapter 7.

Through the 1950's factor analysis was characterized by a set of procedures for analyzing the correlation matrix $\mathbf{R}$ of the manifest variables. Four problems of factor analysis emerged:

- Number of factors

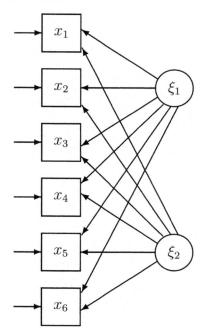

Figure 6.1: Path Diagram of Exploratory Factor Analysis Model

- Communalities

- Factor extraction

- Factor rotation

These four problems are highly interrelated.

Number of Factors

There is no unique way to determine the number of factors k. This is best done by the investigator who knows what the variables (are supposed to) measure. Then the number of factors can be specified *a priori* at least tentatively. Many procedures have been suggested in the literature to determine the number of factors analytically. One of them is to continue to extract factors until there no longer are at least three large loadings in each column of the rotated solution. The question was what one should mean by large. Another *ad hoc* rule is "Kaiser's little jiffy" which says that the number of factors should be equal to the number of eigenvalues of the correlation matrix $\mathbf{R}$ which are greater than 1[1]. A similar *ad hoc* rule, also based on the eigenvalues of $\mathbf{R}$, is Cattell's (1966) scree-plot, where the eigenvalues are plotted against their rank and the number of factors is indicated by the 'elbow' of the curve. Other procedures for deciding on the number of factors are based on statistical fit. For example, with the maximum likelihood method, one can test the hypothesis that $k = k_0$ where k_0 is a specified number of factors. This can be developed into a sequential procedure to estimate the number of factors.

---

[1]Henry Kaiser gave this rule in his dissertation 1956, see Kaiser (1970). Actually, this is the first part of the little jiffy; the second part says that the factors should be rotated by Varimax.

Communalities

The communality problem is as follows. What numbers should be put in the diagonal of the correlation matrix $\mathbf{R}$ to make this approximately equal to $\mathbf{\Lambda\Lambda}'$, where $\mathbf{\Lambda}$ is a $p \times k$ matrix of factor loadings? Such numbers are called communalities and the correlation matrix with communalities in the diagonal is denoted $\mathbf{R}_c$. Thus, the communalities should be chosen such that $\mathbf{R}_c$ is positive semi-definite and of rank k, (for definitions, see the appendix in Chapter 11) . Then $\mathbf{\Lambda}$ could be determined such that

$$\mathbf{R}_c \approx \mathbf{\Lambda\Lambda}' . \tag{6.7}$$

The problem of communalities was involved in much discussion of factor analysis in the 1950's. For example, in Harman's (1967) book a whole chapter is devoted to this problem. The communality problem is highly related to the factor extraction problem. The most important papers on these problems are those by Guttman (1944, 1954a, 1956, 1957).

Guttman (1956) showed that the squared multiple correlation R_i^2 in the regression of the ith manifest variable on all the other manifest variables is a lower bound for the communality c_i^2 of the ith variable:

$$c_i^2 \geq R_i^2 , \quad i = 1, 2, \ldots, p . \tag{6.8}$$

Factor Extraction

Once the communalities have been determined, there are various methods for determining $\mathbf{\Lambda}$ in (6.7). The most common method in the early literature is one which chooses the columns of $\mathbf{\Lambda}$ proportional to the eigenvectors of $\mathbf{R}_c$ corresponding to the k largest eigenvalues. This is in fact a least squares solution in the sense that it minimizes

$$tr(\mathbf{R}_c - \mathbf{\Lambda\Lambda}')^2 . \tag{6.9}$$

Rotation

The matrix $\mathbf{\Lambda}$ in (6.7) is only determined up to an orthogonal rotation. If $\mathbf{\Lambda}$ in (6.7) is replaced by $\mathbf{\Lambda}^\star = \mathbf{\Lambda U}$, where $\mathbf{U}$ is an arbitrary orthogonal matrix of order $k \times k$, then

$$\mathbf{\Lambda}^\star \mathbf{\Lambda}^{\star\prime} = \mathbf{\Lambda\Lambda}' .$$

This is a special case of a more general formula for factor transformation given in Section 6.1 and further discussion of the rotation problem is postponed until that section.

An Iterative Solution

The least squares solution in (6.9) lends itself to an iterative solution. After $\mathbf{\Lambda}$ has been determined, the communalities can be re-estimated as the sum of squares of each row in $\mathbf{\Lambda}$. Putting these new communalities in the diagonal of $\mathbf{R}$ gives a new matrix $\mathbf{R}_c$ from which a new $\mathbf{\Lambda}$ can be obtained. This process can be repeated. In this process it can happen that one or more of the communalities exceed 1, so called *Heywood cases*. Such Heywood cases occurred quite often in practice and caused considerable problems.

This kind of factor analysis solution, called Minres by Harman (1967) was commonly discussed in the literature in the 1950's and 1960's but few writers seemed to realize that this is the solution to a more general least squares problem, namely one which minimizes the *Unweighted Least Squares* (ULS) fit function

$$F_{\mathsf{ULS}}(\mathbf{\Lambda}, \mathbf{\Psi}) = \frac{1}{2} tr[(\mathbf{R} - \mathbf{\Lambda}\mathbf{\Lambda}' - \mathbf{\Psi})^2] , \qquad (6.10)$$

with respect to both $\mathbf{\Lambda}$ and $\mathbf{\Psi}$, where $\mathbf{\Psi}$ is a diagonal matrix of unique variances. Jöreskog (1977) showed that the ULS solution can be obtained by minimizing the concentrated fit function

$$f_{\mathsf{ULS}}(\mathbf{\Psi}) = \min_{\mathbf{\Lambda}} F_{\mathsf{ULS}}(\mathbf{\Lambda}, \mathbf{\Psi}) . \qquad (6.11)$$

Maximum Likelihood Factor Analysis

In the first half of the 20th century factor analysts did not care much about where the data came from. No distributional assumptions were made. One just computed correlations and factor analyzed these. An exception is the paper by Lawley (1940) who made the assumption that the data represents a random sample from a multivariate normal distribution with covariance matrix

$$\mathbf{\Sigma} = \mathbf{\Lambda}\mathbf{\Lambda}' + \mathbf{\Psi} . \qquad (6.12)$$

Assuming that the mean vector of the manifest variables is unconstrained, he showed that logarithm of the likelihood function is

$$\log L = -\frac{1}{2} N[\log \|\mathbf{\Sigma}\| + tr(\mathbf{S}\mathbf{\Sigma}^{-1})] ,$$

where N is the sample size, see the appendix in Chapter 16. This leads to the problem of minimizing the fit function, see *e.g.*, Jöreskog (1967),

$$F_{\mathsf{ML}}(\mathbf{\Lambda}, \mathbf{\Psi}) = \log \|\mathbf{\Sigma}\| + tr(\mathbf{S}\mathbf{\Sigma}^{-1}) - \log \|\mathbf{S}\| - p , \qquad (6.13)$$

At the minimum F_{ML} equals $-2/N$ times the logarithm of the likelihood ratio for testing the hypothesis that $\mathbf{\Sigma}$ is of the form (6.12) against the alternative that $\mathbf{\Sigma}$ is an unconstrained positive definite matrix, see *e.g.*, Jöreskog (1967). Hence, F_{ML} is always non-negative and N times the minimum value of F_{ML} can be used as a test statistic for testing the model. See also Section 16.1.3 in the appendix (Chapter 16).

The assumption of normality is used in much of the factor analysis literature in the second half of 20th century and the maximum likelihood method has become the most common method of factor analysis. However, as late as the mid 60's there was no good way for computing the estimates. An exception is Howe (1955) who developed a Gauss-Seidel algorithm. Browne (1968) discussed the computational problems of factor analysis and compared several analytic techniques. Jöreskog (1967) approached the computational problem by focusing on the concentrated fit function

$$f_{\mathsf{ML}}(\mathbf{\Psi}) = \min_{\mathbf{\Lambda}} F_{\mathsf{ML}}(\mathbf{\Lambda}, \mathbf{\Psi}) , \qquad (6.14)$$

which could be minimized numerically. If one or more of the ψ_i gets close to zero, this procedure becomes unstable, a problem that can be circumvented by reparameterizing

$$\theta_i = \ln \psi_i , \quad \psi_i = +e^{\theta_i} , \qquad (6.15)$$

see Jöreskog (1977). This leads to a very fast and efficient algorithm Heywood cases can still occur in the sense that $\hat{\psi}_i$ becomes practically 0 when $\theta_i \to -\infty$. Other fast algorithms have been developed by Jennrich and Robinson (1969) and by Jennrich (1986).

In practice, the assumption of multivariate normality seldom holds but the ML method has been found to be very robust to departures from normality at least as far as parameter estimates is concerned, see *e.g.*, Boomsma and Hoogland (2001). However, certain adjustments are needed to obtain correct standard errors and test statistics under non-normality, see Satorra and Bentler (1988). A more general treatment that avoids the assumption of multivariate normality is outlined in Section 16.1.5.

Generalized Least Squares

A generalized least squares fit function

$$F_{\mathsf{GLS}}(\boldsymbol{\Lambda}, \boldsymbol{\Psi}) = \frac{1}{2} tr[(\mathbf{I} - \mathbf{S}^{-1}\boldsymbol{\Sigma})^2] \qquad (6.16)$$

was formulated by Jöreskog and Goldberger (1972) and was further developed by Browne (1977). The GLS estimates can also be obtained by minimizing the corresponding concentrated fit function (6.14), see Jöreskog (1977).

Weighted Least Squares

The three fit functions for ULS, GLS, and ML described earlier can be seen to be special cases of one general family of fit functions for weighted least squares:

$$F_V(\boldsymbol{\Lambda}, \boldsymbol{\Psi}) = \frac{1}{2} tr[(\mathbf{S} - \boldsymbol{\Sigma})\mathbf{V}]^2 \qquad (6.17)$$

where $\mathbf{V}$ is any positive definite weight matrix. The special cases are

$$\mathsf{ULS}: \quad \mathbf{V} = \mathbf{I} \qquad (6.18)$$

$$\mathsf{GLS}: \quad \mathbf{V} = \mathbf{S}^{-1} \qquad (6.19)$$

$$\mathsf{ML}: \quad \mathbf{V} = \hat{\boldsymbol{\Sigma}}^{-1} \qquad (6.20)$$

In particular, it can be shown that the ML estimates can be obtained as iteratively reweighted least squares, see *e.g.*, Browne (1977).

Factor Transformation

As stated previously, when $k > 1$, the factor loadings in $\boldsymbol{\Lambda}$ are not uniquely defined. Geometrically the factor loadings may be viewed as p points in a k-dimensional space. In this space the points are fixed but their coordinates can be referred to different factor axes. If the factor axes are orthogonal we say we have an *orthogonal solution*; if they are oblique we say that we have an *oblique solution* where the cosine of the angles between the factor axes are interpreted as correlations between the factors. In statistical terminology, an orthogonal solution corresponds to *uncorrelated factors* and an oblique solution corresponds to *correlated factors*. One can also have solutions in which some factors are uncorrelated and some are correlated, see *e.g.*, Jöreskog (1969).

To facilitate the interpretation of the factors one makes an orthogonal or oblique rotation of the factor axes. This rotation is usually guided by Thurstone's principle of simple structure which essentially states that only a small fraction of the loadings in each row and column should be large. Geometrically, this means that the factor axes pass through or near as many points as possible.

In early times the rotation was done by hand which could be very tedious. A real breakthrough came when Carroll (1953) and Kaiser (1958) developed procedures for analytic rotation which could be performed automatically by computer. See Browne (2001) for a review of these and other analytical procedures. A further breakthrough came when Archer and Jennrich (1973) and Jennrich (1973) developed methods for estimating standard errors of analytically rotated factor loadings, see also Jennrich (2007). This made it possible for researchers to use statistical criteria such as z-values to judge whether a factor loading is statistically non-zero.

Previously it was assumed that the factors $\xi_1, \ldots, \xi_k$ in (6.1) are uncorrelated and have variances 1. These assumptions can be relaxed and the factors may be correlated and they need not have variance 1. If $\boldsymbol{\xi}$ has covariance matrix $\boldsymbol{\Phi}$, the covariance matrix of $\mathbf{x}$ is

$$\boldsymbol{\Sigma} = \boldsymbol{\Lambda}\boldsymbol{\Phi}\boldsymbol{\Lambda}' + \boldsymbol{\Psi} \, . \tag{6.21}$$

Let $\mathbf{T}$ be an arbitrary non-singular matrix of order $k \times k$ and let

$$\boldsymbol{\xi}^* = \mathbf{T}\boldsymbol{\xi} \quad \boldsymbol{\Lambda}^* = \boldsymbol{\Lambda}\mathbf{T}^{-1} \quad \boldsymbol{\Phi}^* = \mathbf{T}\boldsymbol{\Phi}\mathbf{T}' \, .$$

Then

$$\boldsymbol{\Lambda}^*\boldsymbol{\xi}^* \equiv \boldsymbol{\Lambda}\boldsymbol{\xi} \quad \boldsymbol{\Lambda}^*\boldsymbol{\Phi}^*\boldsymbol{\Lambda}^{*'} \equiv \boldsymbol{\Lambda}\boldsymbol{\Phi}\boldsymbol{\Lambda}'$$

Since $\mathbf{T}$ has k^2 independent elements, this shows that at least k^2 independent conditions must be imposed on $\boldsymbol{\Lambda}$ and/or $\boldsymbol{\Phi}$ to make these identified.

Factor analysis is typically done in two steps. In the first step, one obtains an arbitrary orthogonal solution in which $\boldsymbol{\Phi} = \mathbf{I}$ in (6.21). In the second step, this is rotated orthogonally or obliquely to achieve a simple structure. For the rotated factors to have unit variance, $\mathbf{T}$ must satisfy

$$diag(\mathbf{T}\mathbf{T}') = \mathbf{I} \, , \tag{6.22}$$

for an oblique solution and

$$\mathbf{T}\mathbf{T}' = \mathbf{I} \, , \tag{6.23}$$

for an orthogonal solution.

Reference Variables Solution

If $\boldsymbol{\Lambda}$ has rank k, one can choose a transformation $\mathbf{T}$ such that there will be k rows of $\boldsymbol{\Lambda}$ that form an identity matrix. For simplicity of exposition, we assume that the variables in $\mathbf{x}$ have been ordered so that the first k rows of $\boldsymbol{\Lambda}$ form an identity matrix. The variables that correspond to the identity matrix are called reference variables and this solution is called a reference variables solution. In practice, it is best to choose as reference variables those variables that are the best indicators of each factor.

Partitioning $\mathbf{x}$ into two parts $\mathbf{x}_1(k \times 1)$ and $\mathbf{x}_2(q \times 1)$, where $q = p - k$, and $\boldsymbol{\delta}$ similarly into $\boldsymbol{\delta}_1(k \times 1)$ and $\boldsymbol{\delta}_2(q \times 1)$, (6.4) can be written

$$\mathbf{x}_1 = \boldsymbol{\xi} + \boldsymbol{\delta}_1 \tag{6.24}$$

$$\mathbf{x}_2 = \boldsymbol{\Lambda}_2\boldsymbol{\xi} + \boldsymbol{\delta}_2 , \tag{6.25}$$

where $\boldsymbol{\Lambda}_2(q \times k)$ consists of the last $q = p - k$ rows of $\boldsymbol{\Lambda}$. The matrix $\boldsymbol{\Lambda}_2$ may, but need not, contain *a priori* specified elements. We say that the model is *unrestricted* when $\boldsymbol{\Lambda}_2$ is entirely unspecified and that the model is *restricted* when $\boldsymbol{\Lambda}_2$ contains *a priori* specified elements. For a more general discussion of restricted and unrestricted solutions, see Jöreskog (1969).

Solving (6.24) for $\boldsymbol{\xi}$ and substituting this into (6.25) gives

$$\mathbf{x}_2 = \boldsymbol{\Lambda}_2\mathbf{x}_1 + \mathbf{u} , \tag{6.26}$$

where $\mathbf{u} = \boldsymbol{\delta}_2 - \boldsymbol{\Lambda}_2\boldsymbol{\delta}_1$. Each equation in (6.26) is of the form (2.52) but it is not a regression equation because $\mathbf{u}$ is correlated with $\mathbf{x}_1$, since $\boldsymbol{\delta}_1$ is correlated with $\mathbf{x}_1$.

Hägglund (1982) showed that instrumental variables can be obtained as follows. Let

$$x_i = \boldsymbol{\lambda}'_i\mathbf{x}_1 + u_i , \tag{6.27}$$

be the i-th equation in (2.24), where $\boldsymbol{\lambda}'_i$ is the i-th row of $\boldsymbol{\Lambda}_2$, and let $\mathbf{x}_{(i)}(q - 1 \times 1)$ be a vector of the remaining variables in $\mathbf{x}_2$. Then u_i is uncorrelated with $\mathbf{x}_{(i)}$ so that $\mathbf{x}_{(i)}$ can be used as instrumental variables for estimating (6.27). Provided $q \geq k + 1$, this can be done for each $i = 1, 2, \ldots, q$.

The factors in a reference variables solution are neither standardized nor uncorrelated. After $\boldsymbol{\Lambda}$ has been estimated, the covariance matrix $\boldsymbol{\Phi}$ and the error covariance matrix $\boldsymbol{\Theta}$ can be estimated directly (non-iteratively) by unweighted or generalized least squares because the covariance structure in (6.21) is linear in $\boldsymbol{\Phi}$ and $\boldsymbol{\Theta}$, see Browne (1977). Since most people prefer to interpret factors that are standardized, the solution is rescaled to standardized factors in the output.

Some different rotations are illustrated in the following figures

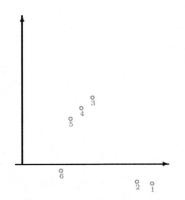
Factor Rotation: Unrotated Solution

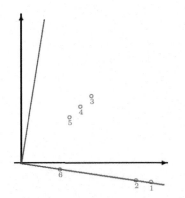
Factor Rotation: Orthogonal Solution

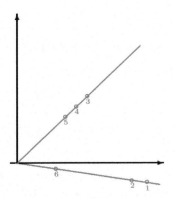
Factor Rotation: Oblique Solution

The left figure illustrates an arbitrary unrotated solution. If one rotates the coordinate axes such that one axis passes through three points while maintaining orthogonal axes, *i.e.*, the 90 degree angle between the axes, one obtains the middle figure. If one keep this axis and rotates the other axis such that it passes through the other three points, one obtains an oblique solution where the cosine of the angle between the axis correspond to the correlation between the factors.

6.2   A Population Example

6.2.1   Example: A Numeric Illustration

To illustrate the basic idea of exploratory factor analysis we will use the same covariance matrix Σ (which is a correlation matrix in this example) as was used in Section 5.2:

$$\Sigma = \begin{pmatrix} 1.000 \\ 0.720 & 1.000 \\ 0.378 & 0.336 & 1.000 \\ 0.324 & 0.288 & 0.420 & 1.000 \\ 0.270 & 0.240 & 0.350 & 0.300 & 1.000 \\ 0.270 & 0.240 & 0.126 & 0.108 & 0.090 & 1.000 \end{pmatrix} \tag{6.28}$$

This covariance matrix was used in Chapter 5 to illustrate the fundamental difference between principal component analysis (PCA) and exploratory factor analysis (EFA). We regard this as a population covariance matrix Σ rather than a sample covariance matrix S because it has been specially constructed so as to satisfy a factor analysis model with two factors. But LISREL does not know how Σ was constructed. So the question is: Can LISREL determine the number of factors and find the simple structure that exists in this covariance matrix.

To do exploratory factor analysis (EFA) one can use either SIMPLIS syntax or LISREL syntax. The SIMPLIS command file **efaex1a.spl** is:

```
Exploratory Factor Analysis
Observed Variables: X1 - X6
Correlation Matrix
1
.720 1
.378 .336 1
.324 .288 .420 1
.270 .240 .350 .300 1
.270 .240 .126 .108 .090 1
Sample Size 1000
Factor Analysis
End of Problem
```

The corresponding LISREL command file **efaex1b.lis** is:

```
Exploratory Factor Analysis
da ni=6 no=1000
la
X1 X2 X3 X4 X5 X6
cm
1
.720 1
.378 .336 1
.324 .288 .420 1
.270 .240 .350 .300 1
.270 .240 .126 .108 .090 1
fa
ou
```

Note that none of these input file specifies the number of factors. Both files give the same output:

```
Maximum Likelihood Factor Analysis
```

```
Decision Table for Number of Factors
```

Factors	Chi2	df	P	DChi2	Ddf	PD	RMSEA
0	1387.99	15	0.000				0.303
1	177.62	9	0.000	1210.37	6	0.000	0.137
2	0.00	4	1.000	177.62	5	0.000	0.000

This decision table will be explained in detail in the next example. For now we need only note that chi-square is 0 for 2 factors. So 2 factors fit the data perfectly, *i.e.*, 2 factors account for all the correlations exactly. The output then gives three different rotations for the two-factor solution, first an unrotated solution:

```
Unrotated Factor Loadings
```

	Factor 1	Factor 2	Unique Var
X1	0.889	-0.138	0.190
X2	0.791	-0.122	0.360
X3	0.501	0.489	0.510
X4	0.429	0.419	0.640
X5	0.358	0.349	0.750
X6	0.296	-0.046	0.910

then a Varimax rotation:

Varimax-Rotated Factor Loadings

	Factor 1	Factor 2	Unique Var
X1	0.854	0.285	0.190
X2	0.759	0.253	0.360
X3	0.221	0.664	0.510
X4	0.190	0.569	0.640
X5	0.158	0.474	0.750
X6	0.285	0.095	0.910

then a Promax rotation with a factor correlation of 0.541:

Promax-Rotated Factor Loadings

	Factor 1	Factor 2	Unique Var
X1	0.867	0.059	0.190
X2	0.771	0.052	0.360
X3	0.014	0.692	0.510
X4	0.012	0.593	0.640
X5	0.010	0.494	0.750
X6	0.289	0.019	0.910

Factor Correlations

	Factor 1	Factor 2
Factor 1	1.000	
Factor 2	0.541	1.000

and finally a reference variables rotation with a factor correlation of 0.6 (standard errors and z-values are eliminated since they are irrelevant in this example):

Reference Variables Factor Loadings Estimated by TSLS

	Factor 1	Factor 2	Unique Var
X1	0.900	0.000	0.190
X2	0.800	0.000	0.360
X3	0.000	0.700	0.510
X4	0.000	0.600	0.640
X5	0.000	0.500	0.750
X6	0.300	0.000	0.910

Factor Correlations

	Factor 1	Factor 2
Factor 1	1.000	
Factor 2	0.600	1.000

The principles that **LISREL** uses for the rotations are explained in detail in the next example.

It is clear that the factor solution is

$$\mathbf{\Lambda} = \begin{pmatrix} .9 & 0 \\ .8 & 0 \\ 0 & .7 \\ 0 & .6 \\ 0 & .5 \\ .3 & 0 \end{pmatrix} \quad \mathbf{\Phi} = \begin{pmatrix} 1 & \\ .6 & 1 \end{pmatrix} \quad \mathbf{\Psi} = diag(.19, .36, .51, .64, .75, .91) \tag{6.29}$$

The reader can verify that

$$\mathbf{\Sigma} = \mathbf{\Lambda}\mathbf{\Phi}\mathbf{\Lambda}' + \mathbf{\Psi} \tag{6.30}$$

This numerical illustration shows that **LISREL** can determine the correct number of factors and the simple structure.

6.3   EFA with Continuous Variables

6.3.1   Example: EFA of Nine Psychological Variables (NPV)

To analyze more than just a few variables it is not practical to give the covariance matrix directly in the input file. Instead one could read the covariance matrix from a file. However, if raw data is available in a **lsf** file or in a text file it is much easier to use this instead with **SIMPLIS** or **LISREL** syntax. We illustrate this by analyzing first the covariance matrix and then the correlation matrix for the nine psychological variables for the Pasteur school sample, see Chapter 1. We created the **lsf** file in that chapter and saved in the file **pasteur_npv.lsf**.

To estimate the factor analysis model of the covariance matrix of the nine psychological variables for the Pasteur school, use the following simple **PRELIS** syntax file **efanpv1.prl**:

```
Exploratory Factor Analysis of Nine Psychological Variables
sy=pasteur_npv.lsf
fa
ou ma=cm
```

One can specify the number of factors by adding **NF=3**, say, on the **FA** line. If **NF=3** is omitted, **LISREL** will determine the number of factors automatically. Furthermore if raw data is provided, as is the case here, one can obtain factor scores by adding **FS** on the **FA** line. The **MA=CM** on the **OU** line is important. It means that the sample covariance matrix should be computed and the factor analysis should be based on this. To analyze the sample correlations (product-moment, Pearson) put **MA=KM** instead.

We did not specify the number of factors in the input file **efanpv1.prl** as we want to see if **LISREL** can determine a suitable number of factors automatically.

The output gives the following information about the number of factors

```
Decision Table for Number of Factors
Factors     Chi2    df      P       DChi2 Ddf     PD      RMSEA
-------     ----    --      -       ----- ---     --      -----
   0      438.00   36    0.000                            0.268
   1      144.78   27    0.000    293.21   9    0.000     0.167
   2       69.48   19    0.000     75.30   8    0.000     0.131
   3       18.63   12    0.098     50.85   7    0.000     0.060
```

Normal Maximum Likelihood Factor Analysis with 3 Factors

The number of factors has been determined as the number of eigenvalues of
the correlation matrix which is greater than 1 (Kaiser's little jiffy).

Minimum Fit Function Chi-Square with 12 Degrees of Freedom = 18.63

To explain the procedure that **LISREL** uses to determine the number of factors we define
Chi2 df P DChi2 Ddf PD RMSEA as the corresponding quantities c_k, d_k, P_k, Δc_k, Δd_k, $P_{\Delta c}$,
and ρ_k, as follows, where

$$c_k = [n - (2p+5)/6 - 2k/3][\ln|\hat{\boldsymbol{\Sigma}}| - \ln|\mathbf{S}|] \,, \tag{6.31}$$
$$k = 0, 1, \ldots, k_{max}$$
$$d_k = [(p-k)^2 - (p+k)]/2 \,, k = 0, 1, \ldots, k_{max} \tag{6.32}$$
$$P_k = Pr\{\chi^2_{d_k} > c_k\} \,, k = 0, 1, \ldots, k_{max} \tag{6.33}$$
$$\Delta c_k = c_k - c_{k-1} \,, k = 1, 2, \ldots, k_{max} \tag{6.34}$$
$$\Delta d_k = d_k - d_{k-1} \,, k = 1, 2, \ldots, k_{max} \tag{6.35}$$
$$P_{\Delta c} = Pr\{\chi^2_{\Delta d_k} > \Delta c_k\} \,, k = 1, 2, \ldots, k_{max} \tag{6.36}$$
$$\rho_k = \sqrt{[(c_k - d_k)/nd_k]} \,, k = 0, 1, \ldots, k_{max} \tag{6.37}$$

Here c_k is the chi-square statistic for testing the fit of k factors, see Lawley and Maxwell (1971,
pp. 35–36). If the model holds and the variables have a multivariate normal distribution, this is
distributed in large samples as χ^2 with d_k degrees of freedom.[2] The P-value of this test is P_k,
i.e., the probability that a random χ^2 with d_k degrees of freedom exceeds the chi-square value
actually obtained. As suggested by Browne & Cudeck (1993) it is better to regard these quantities
as approximate measures of fit rather than as test statistics. Δc_k measures how much better the
fit is with k factors than with $k-1$ factors. Δd_k and $P_{\Delta c}$ are the corresponding degrees of freedom
and P-value. ρ_k is Steiger's (1990) *Root Mean Squared Error of Approximation* (**RMSEA**) which
is a measure of population error per degree of freedom, see Browne & Cudeck (1993)

LISREL investigates these quantities for $k = 1, 2, \ldots, k_{max}$ and determines the smallest acceptable k with the following decision procedure: If $P_k > .10$, k factors are accepted. Otherwise, if
$P_{\Delta c} > .10$, $k - 1$ factors are accepted. Otherwise, if $\rho_k < .05$, k factors are accepted. If none of
these conditions are satisfied, k is increased by 1.

The first criterion, $P_k > .10$, guarantees that one stops at k if the overall fit is good. The
second criterion, $P_{\Delta c} > .10$, guarantees that one will not increase the number of factors unless
the improvement in fit is statistically significant at the 10% level. The third criterion, $\rho_k < .05$, is
the Browne–Cudeck guideline (Browne and Cudeck, 1993, p. 144). This guarantees that one does

---

[2]For $k = 0$ this is a test of the hypothesis that the variables are uncorrelated. If this hypothesis cannot be
rejected, it is meaningless to do a factor analysis.

not get too many factors in large samples. This procedure may not give a satisfactory answer to the number of factors in all respects, but at least there will not be a tendency to overfit, *i.e.*, to take too many factors.

For the values here the decision will be $k = 3$ factors, because for $k = 2$, P_k and $P_{\Delta c}$ are too small and ρ_k is too large, but for $k = 3$, P_k is acceptable.

LISREL will also use Kaiser's Little Jiffy to decide the number of factors. This is an **ad hoc** rule which says that the number of factors should be determined as the number of eigenvalues of the correlation matrix which is greater than 1. When these two automatic procedures disagree, **LISREL** will use the latter procedure. In this example both procedures will reach the decision $k = 3$.

Once the number of factors has been determined to 3, the problem is to estimate the factor loadings in $\mathbf{\Lambda}$ in the model (6.4) where

$$\mathbf{\Lambda} = \begin{pmatrix} \lambda_{11} & \lambda_{12} & \lambda_{13} \\ \lambda_{21} & \lambda_{22} & \lambda_{23} \\ \lambda_{31} & \lambda_{32} & \lambda_{33} \\ \lambda_{41} & \lambda_{42} & \lambda_{43} \\ \lambda_{51} & \lambda_{52} & \lambda_{53} \\ \lambda_{61} & \lambda_{62} & \lambda_{63} \\ \lambda_{71} & \lambda_{72} & \lambda_{73} \\ \lambda_{81} & \lambda_{82} & \lambda_{83} \\ \lambda_{91} & \lambda_{92} & \lambda_{93} \end{pmatrix} , \tag{6.38}$$

and the unique variances in the diagonal matrix $\mathbf{\Psi}$. The number of parameters to estimate is $27 + 9 = 36$ but not all of the 27 factor loadings are independent since $\mathbf{\Lambda}$ can be rotated orthogonally. The number of independent factor loadings is therefore 27 - 3 = 24 since an orthogonal matrix of order 3×3 has 3 independent elements, see the appendix in Chapter 11. So the degrees of freedom will be 45 - 24 - 9 = 12. The general formula for the degrees of freedom in EFA is

$$d_k = \frac{1}{2}p(p+1) - pk - p + \frac{1}{2}k(k-1) = \frac{1}{2}[(p-k)^2 - (p+k)] , \tag{6.39}$$

where p is the number of manifest variables and k is the number of factors.

Following the old factor analysis tradition, it might be tempting to analyze the correlation matrix of the nine psychological variables instead of the covariance matrix. However, in principle, analyzing a correlation matrix by maximum likelihood (ML) is problematic in several ways as pointed out by Cudeck (1989). See also Lawley and Maxwell (1971, p. 59). The analysis of the covariance matrix rather than the correlation matrix, as is done here, has no disadvantage since **LISREL** gives the solutions in both standardized and unstandardized form as the output file **efanpv1.out** shows. Here we present only the standardized solutions.

Varimax-Rotated Factor Loadings

	Factor 1	Factor 2	Factor 3	Unique Var
VISPERC	0.338	0.131	0.609	0.498
CUBES	0.066	-0.083	0.482	0.756
LOZENGES	0.015	0.125	0.702	0.492
PARCOMP	0.790	0.192	0.115	0.326
SENCOMP	0.897	0.027	0.046	0.193
WORDMEAN	0.784	0.171	0.225	0.305
ADDITION	0.130	0.832	-0.186	0.257
COUNTDOT	0.092	0.511	0.127	0.714
SCCAPS	0.098	0.396	0.354	0.708

Promax-Rotated Factor Loadings

	Factor 1	Factor 2	Factor 3	Unique Var
VISPERC	0.607	0.227	-0.043	0.498
CUBES	0.503	-0.001	-0.205	0.756
LOZENGES	0.757	-0.134	-0.038	0.492
PARCOMP	0.010	0.786	0.091	0.326
SENCOMP	-0.089	0.944	-0.072	0.193
WORDMEAN	0.128	0.764	0.045	0.305
ADDITION	-0.176	0.013	0.886	0.257
COUNTDOT	0.149	-0.023	0.486	0.714
SCCAPS	0.385	-0.034	0.314	0.708

Factor Correlations

	Factor 1	Factor 2	Factor 3
Factor 1	1.000		
Factor 2	0.358	1.000	
Factor 3	0.247	0.291	1.000

Reference Variables Factor Loadings

	Factor 1	Factor 2	Factor 3	Unique Var
VISPERC	0.605	0.323	0.016	0.498
	(0.19)	(0.11)	(0.17)	
	3.225	2.992	0.094	
CUBES	0.449	0.082	-0.170	0.756
	(0.17)	(0.11)	(0.18)	
	2.577	0.725	-0.942	
LOZENGES	0.713	0.000	0.000	0.492
PARCOMP	0.111	0.766	0.155	0.326
	(0.12)	(0.09)	(0.12)	
	0.953	8.698	1.304	
SENCOMP	0.000	0.899	0.000	0.193
WORDMEAN	0.213	0.763	0.115	0.305
	(0.11)	(0.09)	(0.12)	
	1.884	8.653	0.963	
ADDITION	0.000	0.000	0.862	0.257
COUNTDOT	0.235	0.014	0.486	0.714
	(0.15)	(0.12)	(0.22)	
	1.570	0.113	2.163	
SCCAPS	0.430	0.040	0.331	0.708
	(0.16)	(0.12)	(0.18)	
	2.618	0.337	1.821	

The first solution is the varimax solution of Kaiser (1958). This is an orthogonal solution, *i.e.*, the factors are uncorrelated. The second solution is the promax solution of Hendrickson and White (1964). This is an oblique solution, *i.e.*, the factors are correlated. The varimax and the promax solutions are transformations of the unrotated solution and as such they are still maximum likelihood solutions. The third solution is the reference variables solution as described earlier. The reference variables are chosen as those variables in the promax solution that have the largest factor loadings in each column. This gives LOZENGES, SENCOMP, and ADDITION as reference variables. The advantage of the reference variables solution is that standard errors can be obtained for all the variables except for the reference variables. This makes it easy to determine which loadings are statistically significant or not. The standard errors are given in parentheses below the loading estimate and the z-values are given below the standard errors. A simple rule to follow is to judge a factor loading statistically significant if its z-value is larger than 2 in magnitude. On the basis of the reference variables solution one can formulate an hypothesis for confirmatory factor analysis by specifying that all non-significant loadings are zero. This hypothesis should be tested on an independent sample.

To analyze the correlation matrix instead of the covariance matrix, replace MA=CM by MA=KM on the OU line, see file **efanpv2.prl**. To obtain factor scores add NF=3 FS on the FA line, see file **efanpv3.prl**. The factor scores are given in the text file **efanpv3.fsc**. The first few lines of this file are given here

```
-1.238      0.577      -0.810
 0.581     -0.907      -0.161
-0.566     -1.671      -1.297
 0.641      0.362      -1.383
-0.645      0.250      -0.247
 0.449     -1.205       0.255
-1.684      1.302       0.366
-0.393      0.284      -1.059
-0.546      0.969       0.122
-1.760      0.737      -0.379
-0.776     -0.439      -0.606
 0.257      0.471      -0.822
 0.669     -0.068       1.124
-0.455      0.424       0.446
```

This can be imported into a **lsf** file as showed in Chapter 1.

Exploratory factor analysis can also be performed with data containing missing values both with continuous and ordinal variables. A full discussion of these issues is given in Chapter 7.

6.4 EFA with Ordinal Varaibles

Observations on an ordinal variable represent responses to a set of ordered categories, such as a five-category Likert scale. It is only assumed that a person who selected one category has more of a characteristic than if he/she had chosen a lower category, but we do not know how much more. Ordinal variables are not continuous variables and should not be treated as if they are. It is common practice to treat scores 1, 2, 3, ... assigned to categories as if they have metric properties but this is wrong. Ordinal variables do not have origins or units of measurements. Means, variances, and covariances of ordinal variables have no meaning. The only information we have are counts of cases in response vectors. To use ordinal variables in structural equation models requires other techniques than those that are traditionally employed with continuous variables.

Several methods have been developed for exploratory factor analysis of ordinal variables. Here we consider the methods described in *e.g.*, Jöreskog and Moustaki (2001), also described in Section 16.2.

To estimate the ordinal factor analysis model by maximum likelihood uses a likelihood function which is quite different than the one used for continuous variables. The log-likelihood is the sum of individual log-likelihoods for each case in the data and each such individual log-likelihood is an integral which must be evaluated numerically, see Section 16.2.1.

6.4.1   EFA of Binary Test Items

A special case of an ordinal variable is a dichotomous (binary) variable. This situation is very common in achievement tests where each item is scored 1 for a correct answer and 0 otherwise. For classical test theory see, *i.e.*, Lord & Novick (1968) and for more modern approaches see, *i.e.*, Bartholomew and Knott (1999).

6.4.2   Example: Analysis of LSAT6 Items

Bock and Lieberman (1970) and Christoffersson (1975) published the data in Table 6.1 giving observed frequencies for the 32 response patterns arising from five items (11 through 15) of Section 6 of the Law School Admissions Test (LSAT). All items are dichotomous. The sample is a subsample of 1000 from a larger sample of those who took the test. This small dataset is often used as a benchmark problem, see Muthen (1978) and Bartholomew & Knott (1999). It is seen

Table 6.1: Observed frequencies of response patterns for five items of LSAT6

Index	\multicolumn{5}{c}{Response pattern}	Frequency
	1	2	3	4	5	
1	0	0	0	0	0	3
2	0	0	0	0	1	6
3	0	0	0	1	0	2
4	0	0	0	1	1	11
5	0	0	1	0	0	1
6	0	0	1	0	1	1
7	0	0	1	1	0	3
8	0	0	1	1	1	4
9	0	1	0	0	0	1
10	0	1	0	0	1	8
11	0	1	0	1	0	0
12	0	1	0	1	1	16
13	0	1	1	0	0	0
14	0	1	1	0	1	3
15	0	1	1	1	0	2
16	0	1	1	1	1	15
17	1	0	0	0	0	10
18	1	0	0	0	1	29
19	1	0	0	1	0	14
20	1	0	0	1	1	81
21	1	0	1	0	0	3
22	1	0	1	0	1	28
23	1	0	1	1	0	15
24	1	0	1	1	1	80
25	1	1	0	0	0	16
26	1	1	0	0	1	56
27	1	1	0	1	0	21
28	1	1	0	1	1	173
29	1	1	1	0	0	11
30	1	1	1	0	1	61
31	1	1	1	1	0	28
32	1	1	1	1	1	298
			Total			1000

that 298 persons out of 1000 answered all items correct and only 3 persons failed on all five items. 173 persons had all but item 3 correct. This might indicate that item 3 is more difficult than the other four. Also note that two out of 32 response patterns did not occur in the subsample.

The data file used here is **LSAT6.lsf**:

	X1	X2	X3	X4	X5	FREQ
1	0.	0.	0.	0.	0.	3
2	0	0.	0.	0.	1.	6.
3	0	0.	0.	1.	0.	2.
4	0.	0.	0.	1.	1.	11.
5	0	0.	1.	0.	0.	1.
6	0.	0.	1.	0.	1.	1.
7	0	0.	1.	1.	0.	3.
8	0	0.	1.	1.	1.	4.
9	0	1.	0.	0.	0.	1.
10	0	1.	0.	0.	1.	8.
11	0.	1.	0.	1.	1.	16.
12	0	1.	1.	0.	1.	3.
13	0	1.	1.	1.	0.	2.
14	0.	1.	1.	1.	1.	15.
15	1	0.	0.	0.	0.	10.
16	1	0.	0.	0.	1.	29.
17	1	0.	0.	1.	0.	14.
18	1	0.	0.	1.	1.	81.
19	1	0.	1.	0.	0.	3.
20	1	0.	1.	0.	1.	28.
21	1	0.	1.	1.	0.	15.
22	1	0.	1.	1.	1.	80.
23	1	1.	0.	0.	0.	16.
24	1	1.	0.	0.	1.	56.
25	1	1.	0.	1.	0.	21.
26	1	1.	0.	1.	1.	173.
27	1	1.	1.	0.	0.	11.
28	1	1.	1.	0.	1.	61.
29	1	1.	1.	1.	0.	28.
30	1	1.	1.	1.	1.	298.

The variables X1 - X5 are ordinal by default but it is important to define the variable FREQ as a weight variable. To do this select **Data** and **Weight Cases**. Then select the variable FREQ and click on **Add** and **OK**. Then save the **lsf** file.

To estimate the one-factor model by Maximum Likelihood use following **PRELIS** syntax file **LSAT6.PRL**:

```
Factor Analysis of LSAT6 Data by ML
SY=LSAT6.LSF
OFA NOR
OU
```

The general **PRELIS** command for Factor Analysis of Ordinal Variables is

```
OFA NF=k NOR
```

or

```
OFA NF=k POM
```

where **k** is the number of factors 1 or 2 with 1 as default, and where **NOR** correspond to the PROBIT link function and **POM** correspond to the LOGIT link function.

The output file **LSAT6.OUT** gives

```
Univariate Marginal Parameters

Variable      Mean St. Dev.    Thresholds
--------      ---- --------    ----------
      X1     0.000   1.000     -1.433
      X2     0.000   1.000     -0.550
      X3     0.000   1.000     -0.133
      X4     0.000   1.000     -0.716
      X5     0.000   1.000     -1.126
```

The mean, standard deviation, and thresholds refer to the underlying variables used to estimate the tetrachoric correlations. All thresholds are negative because there are more correct answers than incorrect answers on each item as shown in the output

```
Univariate Distributions for Ordinal Variables

    X1 Frequency Percentage
0       76          7.6
1      924         92.4

    X2 Frequency Percentage
0      291         29.1
1      709         70.9

    X3 Frequency Percentage
0      447         44.7
1      553         55.3

    X4 Frequency Percentage
0      237         23.7
1      763         76.3

    X5 Frequency Percentage
0      130         13.0
1      870         87.0
```

The factor solution is given as

```
Unrotated Factor Loadings

           Factor 1 Unique Var
           -------- ----------
      X1     0.382     0.854
      X2     0.397     0.842
      X3     0.472     0.777
      X4     0.375     0.859
      X5     0.341     0.884
```

Three other files are also produced in this analysis: **LSAT6.NOR**, **BIVFITS.NOR** and **MULFITS.NOR**. The first of these gives estimates of unstandardized factor loadings and thresholds and their standard errors. The other two give detailed information about the fit of the model. We suggest that you examine these files in detail.

The file **MULFITS.NOR** gives the contributions of each observation to these chi-squares. This is useful to examine the fit in detail to see which observations contribute to the overall chi-square. The file **MULFITS.NOR** is:

Contributions to Chi-square

Number	Obs	Exp	LR	GF	Pattern				
1	298.	297.33	1.34	0.00	1	1	1	1	1
2	173.	173.10	-0.19	0.00	1	1	0	1	1
3	81.	76.51	9.23	0.26	1	0	0	1	1
4	80.	83.15	-6.18	0.12	1	0	1	1	1
5	61.	62.11	-2.20	0.02	1	1	1	0	1
6	56.	56.15	-0.31	0.00	1	1	0	0	1
7	29.	34.62	-10.28	0.91	1	0	0	0	1
8	28.	25.15	6.02	0.32	1	0	1	0	1
9	28.	28.68	-1.35	0.02	1	1	1	1	0
10	21.	25.76	-8.58	0.88	1	1	0	1	0
11	16.	11.38	10.91	1.88	1	1	0	0	0
12	16.	13.69	4.99	0.39	0	1	0	1	1
13	15.	11.55	7.83	1.03	1	0	1	1	0
14	15.	13.79	2.52	0.11	0	1	1	1	1
15	14.	15.73	-3.27	0.19	1	0	0	1	0
16	11.	8.52	5.63	0.72	1	1	1	0	0
17	11.	8.97	4.48	0.46	0	0	0	1	1
18	10.	9.47	1.09	0.03	1	0	0	0	0
19	8.	6.47	3.40	0.36	0	1	0	0	1
20	6.	5.73	0.56	0.01	0	0	0	0	1
21	4.	6.04	-3.30	0.69	0	0	1	1	1
22	3.	4.44	-2.35	0.47	0	1	1	0	1
23	3.	1.21	5.46	2.66	0	0	1	1	0
24	3.	4.74	-2.75	0.64	1	0	1	0	0
25	3.	2.13	2.07	0.36	0	0	0	0	0
26	2.	2.04	-0.07	0.00	0	1	1	1	0
27	2.	2.56	-0.99	0.12	0	0	0	1	0
28	1.	0.69	0.74	0.14	0	0	1	0	0
29	1.	1.82	-1.20	0.37	0	1	0	0	0
30	1.	2.65	-1.95	1.03	0	0	1	0	1
TOTAL			21.30	14.19					

The two chi-squares 21.30 and 14.19 can be used as approximate chi-squares with degrees of freedom (df) equal to the number of response patterns occuring in the sample minus 1 minus the number of estimated parameters. In this case $df = 30 - 1 - 10 = 19$. Thus, it appears that the fit of the model is quite good.

6.4.3 EFA of Polytomous Tests and Survey Items

Surveys are very often used in the political, social, marketing and other sciences to measure peoples attitudes to various questions or issues. Survey items are usually ordinal with number of categories varying from 4 to 10, say. LISREL can be used to analyze such variables with exploratory or confirmatory factor analysis. Here we give one example of this. Other examples are given in Chapter 7.

6.4.4   Example: Attitudes Toward Science and Technology

In the Eurobarometer Survey 1992, citizens of Great Britain were asked questions about Science and Technology. The questions are given below.

1. Science and technology are making our lives healthier, easier and more comfortable [COMFORT].

2. Scientific and technological research cannot play an important role in protecting the environment and repairing it [ENVIRON].

3. The application of science and new technology, will make work more interesting [WORK].

4. Thanks to science and technology, there will be more opportunities for the future generations [FUTURE].

5. New technology does not depend on basic scientific research [TECHNOL].

6. Scientific and technological research do not play an important role in industrial development [INDUSTRY].

7. The benefits of science are greater than any harmful effects it may have [BENEFIT].

Note that the items COMFORT, WORK, FUTURE, and BENEFIT have a positive question wording, whereas the items ENVIRON, TECHNOL, and INDUSTRY have a negative question wording.

The response alternatives were strongly disagree, disagree to some extent, agree to some extent, and, strongly agree. These were coded as 1, 2, 3, 4, respectively. The data file is **scitech.lsf**. Missing values have been deleted beforehand, see Bartholomew et al. (2002). This data set is used here to illustrate the case of no missing values. The sample size is 392.

Since there are 7 variables each with 4 categories, there are $4^7 = 16384$ possible response patterns. Obviously, since the sample size is only 392, not all of these can occur in the sample. As will be seen later, there are only 298 response patterns occuring in the sample. This suggests that not much information will be lost if one would dichotomize all the variables.

To perform an exploratory factor analysis of ordinal variables by the FIML method, open the file **scitech.lsf** and select **EFA of Ordinal Variables** in the **Statistics Menu**. This shows the following window

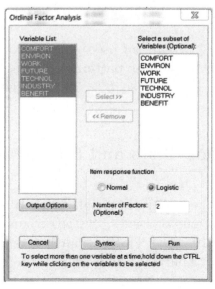

Then select the variables to be analyzed, tick the box **Logistic**[3], for example, and insert 2 for **Number of Factors**. Then click **Run**.

Instead of using the Graphics User Unterface GUI one can use the following simple PRELIS syntax file **sctitech.prl**

```
SY=scitech.lsf
OFA NF=2 POM
```

to perform exploratory factor analysis with ordinal variables.

The output file **scitech.out** gives the following unrotated and rotated standardized factor loadings:

Unrotated Factor Loadings

	Factor 1	Factor 2	Unique Var
COMFORT	0.764	0.000	0.416
ENVIRON	0.237	0.817	0.277
WORK	0.669	-0.404	0.388
FUTURE	0.847	-0.323	0.179
TECHNOL	0.234	0.832	0.252
INDUSTRY	0.462	0.716	0.274
BENEFIT	0.713	-0.225	0.441

Varimax-Rotated Factor Loadings

	Factor 1	Factor 2	Unique Var
COMFORT	0.719	0.259	0.416
ENVIRON	-0.054	0.849	0.277
WORK	0.767	-0.153	0.388
FUTURE	0.906	-0.016	0.179
TECHNOL	-0.062	0.862	0.252
INDUSTRY	0.192	0.830	0.274
BENEFIT	0.747	0.030	0.441

Promax-Rotated Factor Loadings

	Factor 1	Factor 2	Unique Var
COMFORT	0.710	0.240	0.416
ENVIRON	-0.086	0.852	0.277
WORK	0.774	-0.175	0.388
FUTURE	0.908	-0.041	0.179
TECHNOL	-0.094	0.866	0.252
INDUSTRY	0.161	0.826	0.274
BENEFIT	0.747	0.010	0.441

---

[3]One can choose between the normal link funtion called NOR or the logistic link function called POM; here we use POM, see Section 16.2 or see *e.g.*, Jöreskog and Moustaki (2001)

Factor Correlations

	Factor 1	Factor 2
Factor 1	1.000	
Factor 2	0.064	1.000

Reference Variables Factor Loadings

	Factor 1	Factor 2	Unique Var
COMFORT	0.738	0.273	0.416
ENVIRON	0.006	0.851	0.277
WORK	0.757	-0.140	0.388
FUTURE	0.906	0.000	0.179
TECHNOL	0.000	0.865	0.252
INDUSTRY	0.251	0.837	0.274
BENEFIT	0.750	0.043	0.441

Factor Correlations

	Factor 1	Factor 2
Factor 1	1.000	
Factor 2	-0.089	1.000

These rotations were explained in Section 6.3. The reference variables are chosen as those variables in the promax solution that have the largest factor loadings in each column. In this case the reference variables are FUTURE and TECHNOL.

The Promax-Rotated Factor Loadings and the Reference Variables Factor Loadings suggest that there are two nearly uncorrelated factors and that Factor 1 is a Positive factor, having large loadings on the positively worded items COMFORT, WORK, FUTURE, and BENEFIT and Factor 2 is a Negative factor having large loadings on the negatively worded items ENVIRON, TECHNOL, and INDUSTRY. We will consider a confirmatory factor analysis of the same data in the next chapter.

Three other files are also produced in this analysis: **SCITECH.POM**, **BIVFITS.POM** and **MULFITS.POM**. The first of these gives estimates of unstandardized factor loadings and thresholds and their standard errors. The other two give detailed information about the fit of the model. We suggest that you examine these files in detail.

Here we give the following short information about these output files.

The file **SCITECH.POM** gives technical information about the iterations and some further information as follows

Number of Possible Response Patterns = 16384

Number of Distinct Response Patterns = 298

Full Information Coverage Ratio = 0.018

Minimum Fit Function Value = 1.9029815146

-2ln L Under Model   = 5823.037

-2ln L Under Alternative = 4331.100

LR Chi-square with 263 Degrees of Freedom = 1491.94

GF Chi-square with 263 Degrees of Freedom = 80813.19

  The same file gives the following unstandardized parameters and standard errors

ORFIML Estimates for Logistic Response Function (POM)

Unstandardized Thresholds Alpha^(i)_a
```
  COMFORT    -5.015    -2.744     1.535
  ENVIRON    -3.429    -1.243     1.002
     WORK    -2.941    -0.905     2.279
   FUTURE    -4.998    -2.125     1.885
  TECHNOL    -4.164    -1.482     1.086
 INDUSTRY    -4.695    -2.507     0.461
  BENEFIT    -3.379    -1.006     1.705
```

Unstandardized Factor Loadings Beta_ij
```
  COMFORT     1.184     0.000
  ENVIRON     0.450     1.552
     WORK     1.074    -0.649
   FUTURE     2.002    -0.763
  TECHNOL     0.467     1.656
 INDUSTRY     0.882     1.366
  BENEFIT     1.073    -0.339
```

Standard Errors for Unstandardized Thresholds Alpha^(i)_a
```
  COMFORT     0.589     0.240     0.173
  ENVIRON     0.304     0.173     0.172
     WORK     0.251     0.147     0.227
   FUTURE     0.600     0.331     0.307
  TECHNOL     0.376     0.195     0.195
 INDUSTRY     0.444     0.243     0.168
  BENEFIT     0.290     0.143     0.179
```

```
Standard Errors for Unstandardized Factor Loadings Beta_ij
   COMFORT    0.187    0.000
   ENVIRON    0.222    0.228
      WORK    0.192    0.191
    FUTURE    0.434    0.298
   TECHNOL    0.253    0.272
  INDUSTRY    0.252    0.233
   BENEFIT    0.172    0.188
```

It is also of interest to examine the fit in detail. This is done in in the file **MULFITS.POM** which gives the LR and GF contributions of each response pattern. Since there are 298 distinct response patterns we cannot list all of them here. We give the most frequent response patterns and those that contribute over 1000 to the GF chi-square.

```
Contributions to Chi-square
Number     Obs    Exp       LR       GF     Pattern
     1     11.    3.88    22.94    13.09    3    3    3    3    3    3    3
     2      8.    1.67    25.08    24.04    3    3    2    3    3    3    3
     3      7.    2.51    14.35     8.02    4    4    3    3    4    4    3
     4      6.    5.18     1.75     0.13    3    4    3    3    4    4    3
     5      5.    1.14    14.79    13.08    3    4    3    4    4    4    4
     6      5.    2.42     7.24     2.74    3    2    3    3    3    3    3
     7      4.    1.04    10.79     8.46    3    3    2    3    3    3    2
     8      4.    0.74    13.53    14.44    3    3    2    2    3    3    2
     9      4.    0.47    17.18    26.73    3    2    3    3    2    2    2

    56      1.    0.00    15.36  2167.39    4    4    3    4    1    1    4
    64      1.    0.00    16.04  3041.54    3    2    2    4    1    4    1

    89      1.    0.00    16.99  4891.78    1    3    4    3    4    4    1

   243      1.    0.00    21.76 53132.89    4    2    1    1    1    3    4

   266      1.    0.00    17.24  5547.17    2    3    4    2    3    1    1
 TOTAL                  1491.94 80813.19
```

It is seen that the most common response pattern is to agree to all 7 items. But note that 5 response patterns each contributes more than 2000 to the total GF chi-square. If these are eliminated from the data the fit of the model would appear much better.

One can also evaluate the fit of the model to the univariate and bivariate marginals in the sample. The file **BIVFITS.POM** gives the contributions associated with each univariate and bivariate margins.

Chapter 7

Confirmatory Factor Analysis(CFA)

Exploratory factor analysis is used in the following situation. One has a set of tests or other variables and one would like to know how many factors are needed to account for their intercorrelations and what these factors are measuring. Both the number of factors and the meaning of the factors are unknown. The interpretation and the naming of the factors are usually done after analytic rotation.

In contrast, a confirmatory factor analysis begins by defining the latent variables one would like to measure. This is based on substantive theory and/or previous knowledge. One then constructs observable variables to measure these latent variables. This construction must follow certain rules of correspondence, see *e.g.*, Costner (1969). Thus, in a confirmatory factor analysis, the number of factors is known and equal to the number of latent variables. The confirmatory factor analysis is a model that should be estimated and tested.

Factor analysis needs not be strictly exploratory or strictly confirmatory. Most studies are to some extent both exploratory and confirmatory since they involve some variables of known and other variables of unknown composition. The former should be chosen with great care in order that as much information as possible about the latter may be extracted. It is highly desirable that a hypothesis which has been suggested by mainly exploratory procedures should subsequently be confirmed, or disproved, by obtaining new data and subjecting these to more rigorous statistical tests. Jöreskog and Lawley (1968) give an example where a model is developed on one sample and replicated on new fresh data, see also Kroonenberg and Lewis (1982). Cudeck and Browne (1983) discuss problems and methods for cross-validation.

Although estimation and testing of the CFA model, can easily be done by using **SIMPLIS** syntax, one can also use **LISREL** syntax. To understand the connection between the model and the syntax we present a general framework in matrix form, in which all models in this chapter will fit, and its connection to the **LISREL** syntax. This is done in Section 7.1.

A special case of a CFA model is a model with a single latent variable. Measurement models for a single latent variable are presented in Section 7.2. For general CFA models, we then distinguish between analysis of continuous and ordinal variables. CFA for continuous variables are covered in Section 7.3 and CFA for ordinal variables are covered in Section 7.4. Within each of these Sections we distinguish between cases where the data is complete and where the data contains missing values.

© Springer International Publishing Switzerland 2016
K.G. Jöreskog et al., *Multivariate Analysis with LISREL*, Springer Series in Statistics,
DOI 10.1007/978-3-319-33153-9_7

7.1   General Model Framework

The models in this chapter are of the following general form

$$\mathbf{x} = \boldsymbol{\tau}_x + \boldsymbol{\Lambda}_x \boldsymbol{\xi} + \boldsymbol{\delta} \ , \tag{7.1}$$

where $\boldsymbol{\xi}$ $(n \times 1)$ and $\boldsymbol{\delta}$ $(p \times 1)$ are independent random vectors with $\boldsymbol{\xi} \sim N(\boldsymbol{\kappa}, \boldsymbol{\Phi})$ and $\boldsymbol{\delta} \sim N(\mathbf{0}, \boldsymbol{\Theta}_\delta)$ and $\boldsymbol{\Lambda}_x$ is either a fixed matrix or a parameter matrix. In the context of confirmatory factor analysis, $\boldsymbol{\Lambda}_x$ is a matrix of factor loadings, often called the factor matrix, $\boldsymbol{\Phi}$ is a factor covariance or correlation matrix and $\boldsymbol{\Theta}_\delta$ is an error covariance matrix, which is usually assumed to be diagonal. In other contexts other terms may be used.

It follows that $\mathbf{x} \sim N(\boldsymbol{\mu}, \boldsymbol{\Sigma})$, where

$$\boldsymbol{\mu} = \boldsymbol{\tau}_x + \boldsymbol{\Lambda}_x \boldsymbol{\kappa} \ , \tag{7.2}$$

and covariance matrix

$$\boldsymbol{\Sigma} = \boldsymbol{\Lambda}_x \boldsymbol{\Phi} \boldsymbol{\Lambda}_x{'} + \boldsymbol{\Theta}_\delta \ . \tag{7.3}$$

Later the assumption of normality will be relaxed and the consequences of non-normality will be discussed.

The parameter matrices are $\boldsymbol{\tau}_x$, $\boldsymbol{\kappa}$, $\boldsymbol{\Lambda}_x$, $\boldsymbol{\Phi}$, and $\boldsymbol{\Theta}_\delta$, the elements of which may be

- *fixed parameters* that have been assigned specified values,

- *constrained parameters* that are unknown but equal to one or more other parameters, or equal to some specified function of other parameters and

- *free parameters* that are unknown and not constrained.

The independent parameters in the parameter matrices are collected in a parameter vector denoted $\boldsymbol{\theta}$. The objective is to estimate $\boldsymbol{\theta}$ from N observations of $\mathbf{x}$ which may consist of a raw data matrix $\mathbf{X}$ of order $N \times p$ or a sample mean vector $\bar{\mathbf{x}}$ and covariance matrix $\mathbf{S}$. This estimation is done by minimizing some fit function. The fit functions in LISREL are

ULS Unweighted Least Squares

GLS Generalized Least Squares

ML Maximum Likelihood

DWLS Diagonally Weighted Least Squares

WLS Weighted Least Squares

The first three were defined Chapter 6. A more general definition is given in the appendix (see Chapter 16) where also DWLS and WLS are defined. The ULS, GLS, and ML methods can be used with only the sample mean vector and covariance matrix whereas DWLS and WLS require raw data. The default fit function is ML, so unless otherwise specified, ML will be used.

In LISREL syntax $\text{NX} = p$ and $\text{NK} = n$ and the parameter matrices, their possible forms and default values are given in Table 7.1.

The meaning of the possible form values are as follows:

Table 7.1: Parameter Matrices in LISREL: Their Possible Forms and Default Values

Name	Math Symbol	LISREL Name	Order	Possible Forms	Default Form	Default Mode
TAU-X	τ_x	LX	NX×1	ZE	FU	FI
KAPPA	κ	LX	NX×1	ZE	FU	FI
LAMBDA-X	Λ_x	LX	NX×NK	ID,IZ,ZI,DI,FU	FU	FI
PHI	Φ	PH	NK×NK	ID,DI,SY,ST	SY	FR
THETA-DELTA	Θ_δ	TD	NX×NX	ZE,DI,SY	DI	FR

- ZE = $\mathbf{0}$ (zero matrix)

- ID = $\mathbf{I}$ (identity matrix)

- IZ =$(\ \mathbf{I}\ \ \mathbf{0}\ )$ or $\begin{pmatrix} \mathbf{I} \\ \mathbf{0} \end{pmatrix}$ (partitioned identity and zero)

- ZI = $(\ \mathbf{0}\ \ \mathbf{I}\ )$ or $\begin{pmatrix} \mathbf{0} \\ \mathbf{I} \end{pmatrix}$ (partitioned zero and identity)

- DI = a diagonal matrix

- SD = a full square matrix with fixed zeros in and above the diagonal and all elements under the diagonal free (refers to $\mathbf{B}$ only)

- SY = a symmetric matrix which is not diagonal

- ST = a symmetric matrix with fixed ones in the diagonal (a correlation matrix)

- FU = a rectangular or square non-symmetric matrix.

The default specification for Θ_δ needs a special explanation. In most cases Θ_δ is diagonal and this is the default. To specify correlated errors two other alternatives are available. Thus,

- Default means diagonal with free diagonal elements. Off-diagonal elements *cannot* be relaxed.

- SY means symmetric with free diagonal and *fixed* off-diagonal elements. Off-diagonal elements *can* be relaxed.

- SY,FR means that the whole symmetric matrix Θ_δ is free but off-diagonal elements can be fixed.

On the MO line (model line), one can make any number of specifications of the form

```
...  MN = AA,BB
```

where MN is a matrix name (column 3), AA is a matrix form (column 5) and BB is FR (free) or FI (fixed) (column 7). Either AA or BB may be omitted in which case the defaults of Table 7.1 are used. Any element in the parameter matrices can be specified to be fixed or free by listing the element on a FI or a FR line.

The most common special case of model (7.1) is when $\boldsymbol{\kappa} = \mathbf{0}$ and $\boldsymbol{\tau}_x$ is unconstrained. Then the mean vector $\boldsymbol{\mu}$ is unconstrained and the model can be estimated using only the covariance matrix $\mathbf{S}$. We assume this special case throughout this chapter. In this case, the parameter vector $\boldsymbol{\theta}$ consists of the independent elements in $\boldsymbol{\Lambda}_x$, $\boldsymbol{\Phi}$, and $\boldsymbol{\Theta}_\delta$. With the ML method the parameters $\boldsymbol{\theta}$ are estimated by minimizing the fit function with respect to $\boldsymbol{\theta}$, see the appendix in Chapter 16)

$$F(\boldsymbol{\theta}) = \log \|\boldsymbol{\Sigma}\| + tr(\mathbf{S}\boldsymbol{\Sigma}^{-1}) - \log \|\mathbf{S}\| - p \,. \tag{7.4}$$

7.2 Measurement Models

Broadly speaking, there are two basic problems that are important in the social and behavioral sciences. The first problem is concerned with the measurement properties—validities and reliabilities—of the measurement instruments. The second problem concerns the causal relationships among the variables and their relative explanatory power.

Most theories and models in the social and behavioral sciences are formulated in terms of theoretical or hypothetical concepts, or constructs, or latent variables, which are not directly measurable or observable. Examples of such constructs are prejudice, radicalism, alienation, conservatism, trust, self-esteem, discrimination, motivation, ability, and anomie. The measurement of a hypothetical construct is accomplished indirectly through one or more observable indicators, such as responses to questionnaire items, that are *assumed* to represent the construct adequately.

The purpose of a measurement model is to describe how well the observed indicators serve as a measurement instrument for the latent variables. The key concepts here are measurement, reliability, and validity. Measurement models often suggest ways in which the observed measurements can be improved.

Measurement models are important in the social and behavioral sciences when one tries to measure such abstractions as people's behavior, attitudes, feelings and motivations. Most measures employed for such purposes contain sizable measurement errors and the measurement models allow us to take these errors into account.

Most measurements used in the behavioral and social sciences contain sizable measurement errors, which, if not taken into account, can cause severe bias in results. Adequate modeling should take measurement errors into account whenever possible.

Measurement errors occur because of imperfections in measurement instruments (questionnaires, interviews, tests, etc) and measuring procedures (recording, coding, scaling, grouping, aggregation, etc).

7.2.1 The Congeneric Measurement Model

The most common type of measurement model is the one-factor congeneric measurement model; see Jöreskog (1971). A path diagram of this model is shown in Figure 7.1.

The equations corresponding to Figure 7.1 are written in matrix form as

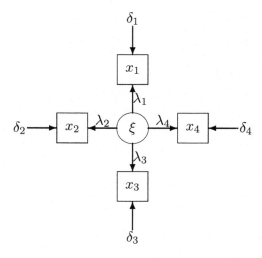

Figure 7.1: Path Diagram of The Congeneric Measurement Model

$$
\begin{pmatrix} x_1 \\ x_2 \\ x_3 \\ x_4 \end{pmatrix} = \begin{pmatrix} \lambda_1 \\ \lambda_2 \\ \lambda_3 \\ \lambda_4 \end{pmatrix} \xi + \begin{pmatrix} \delta_1 \\ \delta_2 \\ \delta_3 \\ \delta_4 \end{pmatrix} \tag{7.5}
$$

or

$$
\mathbf{x} = \boldsymbol{\Lambda}\,\xi + \boldsymbol{\delta} \ .
$$

The model (7.5) is empirically not directly verifiable since there are more unobserved variables than observed. However, with the assumption that the latent variable is standardized, the equations imply that the covariance matrix of the observed variables is of the form

$$
\boldsymbol{\Sigma} = \boldsymbol{\Lambda}\,\boldsymbol{\Lambda}' + \boldsymbol{\Theta} = \begin{pmatrix} \lambda_1^2 + \theta_{11} & & & \\ \lambda_2\lambda_1 & \lambda_2^2 + \theta_{22} & & \\ \lambda_3\lambda_1 & \lambda_3\lambda_2 & \lambda_3^2 + \theta_{33} & \\ \lambda_4\lambda_1 & \lambda_4\lambda_2 & \lambda_4\lambda_3 & \lambda_4^2 + \theta_{44} \end{pmatrix} \tag{7.6}
$$

In this equation, $\boldsymbol{\Theta}$ is a diagonal matrix with elements θ_{ii}, the variances of δ_i $(i = 1, 2, 3, 4)$.

The hypothesis that the population covariance matrix has this form is testable from a random sample of observations. In addition, the following sub-hypotheses are testable.

7.2.2   Congeneric, parallel, and tau-equivalent measures

The classical test theory of Lord and Novick (1968) uses the notation

$$
x_i = \tau_i + \delta_i \ , \tag{7.7}
$$

where τ_i is called the true score and δ_i the measurement error. The true score τ_i and the measurement error are assumed to be independent, so the variance of x_i is

$$
\sigma_{ii} = Var(\tau_i) + Var(\delta_i) = Var(\tau_i) + \theta_{ii} \ . \tag{7.8}
$$

The model (7.5) is called the *congeneric measurement* model. The measures x_1, x_2, ..., x_q are said to be *congeneric* if their true values $\tau_1, \tau_2, \ldots, \tau_q$ have all pair-wise correlations equal to unity. This is true for the model, since $\tau_i = \lambda_i \xi = x_i - \delta_i$ ($i = 1, 2, 3, 4$) and all τ's are linearly related and hence have unit correlation. The true variance of x_i is $Var(\tau_i) = \lambda_i^2$. The reliability of x_i is defined as the true score variance divided by the total variance,

$$\rho_{ii} = \frac{Var(\tau_i)}{\sigma_{ii}} = \frac{\lambda_i^2}{\sigma_{ii}} = \frac{\lambda_i^2}{\lambda_i^2 + \theta_{ii}} = 1 - \frac{\theta_{ii}}{\lambda_i^2 + \theta_{ii}} \ . \tag{7.9}$$

Strictly speaking, the error δ_i is considered to be the sum of two uncorrelated random components s_i and e_i, where s_i is a specific factor (specific to x_i) and e_i is the true measurement error. However, unless there are several replicate measures x_i with the same s_i, one cannot distinguish between these two components or separately estimate their variances. As a consequence, ρ_{ii} in (7.9) is a lower bound for the true reliability See Section 9 11 for how. to estimate the specific variance $Var(s_i)$ and the pure measurement error variance $Var(e_i)$.

Parallel measures have equal true score variances and equal error variances, *i.e.*,

$$\lambda_1^2 = \cdots = \lambda_4^2 \qquad \theta_{11} = \cdots = \theta_{44} \ .$$

Tau-equivalent measures have equal true score variances, but possibly different error variances. Tests of parallelism and tau-equivalence are demonstrated in the following example.

7.2.3   Example: Analysis of Reader Reliability in Essay Scoring

In an experiment to establish methods of obtaining reader reliability in essay scoring, 126 persons were given a three-part English Composition examination. Each part required the person to write an essay, and for each person, scores were obtained on the following: (1) the original part 1 essay, (2) a handwritten copy of the original part 1 essay, (3) a carbon copy of the handwritten copy in (2), and (4) the original part 2 essay. Scores were assigned by a group of readers using procedures designed to counterbalance certain experimental conditions. The investigator would like to know whether, on the basis of this sample of size 126, the four scores can be used interchangeably or whether scores on the copies (2) and (3) are less reliable than the originals (1) and (4).

The covariance matrix of the four measurements is given in Table 7.2.

Table 7.2: Essay Scoring Data: Covariance Matrix

	x_1	x_2	x_3	x_4
ORIGPRT1	25.0704			
WRITCOPY	12.4363	28.2021		
CARBCOPY	11.7257	9.2281	22.7390	
ORIGPRT2	20.7510	11.9732	12.0692	21.8707

Source: Votaw (1948).

The hypotheses to be tested are that the measurements are (1) parallel, (2) tau-equivalent, and (3) congeneric, respectively. All analyses use the ML fit function. Since the raw data is not available, we have to use the covariance matrix.

The SIMPLIS syntax is (see file **votaw1a.spl**):

```
Analysis of Reader Reliability in Essay Scoring; Votaw's Data
Congeneric model estimated by ML
Observed Variables: ORIGPRT1  WRITCOPY  CARBCOPY  ORIGPRT2
Covariance Matrix
25.0704
12.4363     28.2021
11.7257      9.2281     22.7390
20.7510     11.9732     12.0692     21.8707
Sample Size: 126
Latent Variable: Ability
Relationship
ORIGPRT1 - ORIGPRT2 = Ability
End of Problem
```

The corresponding LISREL syntax file is (see file **votaw1b.lis**):

```
Analysis of Reader Reliability in Essay Scoring; Votaw's Data
Congeneric model estimated by ML
DA NI=4 NO=126
LA
ORIGPRT1  WRITCOPY  CARBCOPY  ORIGPRT2
CM
25.0704
12.4363     28.2021
11.7257      9.2281     22.7390
20.7510     11.9732     12.0692     21.8707
MO NX=4 NK=1 LX=FR
LK
Ability
OU
```

The DA line specifies four observed variables and a sample size of 126; the MA default is assumed, so the covariance matrix will be analyzed. Labels for the input variables follow the LA command. The CM command indicates that a covariance matrix is to be input. Because an external file is not specified, the matrix follows in the command file. A format statement does not appear, so the input is in free format.

The MO command specifies four x-variables and one latent variable; the elements of λ are all free (LX=FR), and the latent variable is standardized by default. A label for the latent variable follows the LK command. No additional output is requested on the OU line.

To obtain the input for the hypothesis of tau-equivalence, insert the line

```
EQ LX(1) - LX(4)
```

before the **OU** line, see file **votaw2b.lis**. This specifies that the elements of $\boldsymbol{\lambda}$ should be equal.

The hypothesis of parallel measurements is specified by adding one more **EQ** line, see file **votaw3b.lis**:

```
EQ TD(1) - TD(4)
```

Although equality constraints can be specified also in **SIMPLIS** syntax, it is a rather tedious process. So we consider this here only for **LISREL** syntax.

In the results of this analysis, as summarized in Table 7.3, it is seen that the hypotheses (1) and (2) are untenable, but the hypothesis (3) is acceptable.

Table 7.3: Essay Scoring Data: Summary of Chi-Squares

	Hypothesis	df	χ^2	p
(1)	Parallel	8	109.996	0.0000
(2)	Tau-equivalent	5	40.747	0.0000
(3)	Congeneric	2	2.298	0.3169

The results under the hypothesis (3) are given in Table 7.4. The reliabilities in column 3 appear where the output says "squared multiple correlations for x-variables."

Inspecting the different λ's, it is evident that these are different even taking their respective standard errors of estimate into account. Comparing the reliabilities in the last column, one sees that they are high for scores (1) and (4) and low for scores (2) and (3). Thus, it appears that scores obtained from originals are more reliable than scores based on copies.

7.3 CFA with Continuous Variables

7.3.1 Continuous Variables without Missing Values

The principal idea of confirmatory factor analysis is illustrated in Figure 7.2 with $q = 6$ and $n = 2$. Compare Figure 7.2 with Figure 6.1. In CFA the hypothesis is that x_1, x_2, and x_3 measure ξ_1 only (and not ξ_2) and that x_4, x_5, and x_6 measure ξ_2 only (and not ξ_1). In Figure 7.2 this corresponds to the lack of 6 arrows from the ξ' to the x's compared to Figure 6.1. Another difference is that

Table 7.4: Essay Scoring Data: Results for Congeneric Model

i	$\hat{\lambda}_i$	s.e.$(\hat{\lambda}_i)$	$\hat{\rho}_{ii}$
1	4.573	0.361	0.834
2	2.676	0.452	0.254
3	2.651	0.400	0.309
4	4.535	0.325	0.940

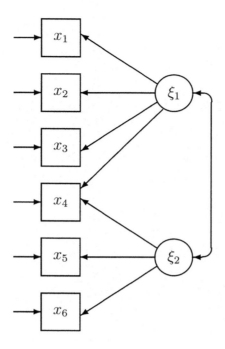

Figure 7.2: Path Diagram of Confirmatory Factor Analysis Model

in Figure 7.2 the two latent variables ξ_1 and ξ_2 are allowed to correlate which is indicated by the two-way arrow between ξ_1 and ξ_2.

In terms of the general framework in Section 7.1, this corresponds to

$$\mathbf{\Lambda}_x = \begin{pmatrix} \lambda_{11} & 0 \\ \lambda_{21} & 0 \\ \lambda_{31} & 0 \\ 0 & \lambda_{42} \\ 0 & \lambda_{52} \\ 0 & \lambda_{62} \end{pmatrix}, \tag{7.10}$$

$$\mathbf{\Phi} = \begin{pmatrix} 1 & \\ \mathbf{\Phi}_{21} & 1 \end{pmatrix} \tag{7.11}$$

and

$$\mathbf{\Theta}_\delta = diag(\theta_{11}^{(\delta)}, \theta_{22}^{(\delta)}, \theta_{33}^{(\delta)}, \theta_{44}^{(\delta)}, \theta_{55}^{(\delta)}, \theta_{66}^{(\delta)}) . \tag{7.12}$$

7.3.2   Example: CFA of Nine Psychological Variables

To illustrate confirmatory factor analysis we use the data on the nine psychological variables for the Grant White school introduced in Chapter 1. The data was saved in the file **grant-white_npv.lsf**. For simplicity, in this chapter, we will call this **npv.lsf** and use the short form **npv** in all files. Bear in mind that this refers to the Grant White school sample.

It is hypothesized that these variables have three correlated common factors: visual perception here called Visual, verbal ability here called Verbal and speed here called Speed such that the first three variables measure Visual, the next three measure Verbal, and the last three measure Speed. A path diagram of the model to be estimated is given in Figure 7.3.

7.3.3 Estimating the Model by Maximum Likelihood

With the **npv.lsf** file on hand, one can estimate the model by normal theory maximum likelihood[1]. One can use either SIMPLIS or LISREL syntax to estimate the model with LISREL. In many ways SIMPLIS syntax is easier to use and explain. The difference between SIMPLIS and LISREL syntax is that with SIMPLIS syntax one specifies the model by means of easily understood equations whereas with LISREL syntax one specifies the model by means of matrices. This is also manifested in the outputs where the estimated model is given in equation form or in matrix form, respectively.

SIMPLIS Syntax

The first SIMPLIS file is (see file **npv1a.spl**):

```
Estimation of the NPV Model by Maximum Likelihood
Raw Data from File npv.lsf
Latent Variables: Visual Verbal Speed
Relationships:
    VISPERC - LOZENGES = Visual
    PARCOMP - WORDMEAN = Verbal
    ADDITION - SCCAPS = Speed
Path Diagram
End of Problem
```

One can also include a line

```
Path Diagram
```

to display path diagrams with parameter estimates, standard errors, and t-values.

The output file is **npv1a.out** which gives the sample covariance matrix **S** as

```
        Covariance Matrix

              VISPERC      CUBES   LOZENGES    PARCOMP    SENCOMP   WORDMEAN
             --------   --------   --------   --------   --------   --------
   VISPERC     47.801
     CUBES     10.013     19.758
  LOZENGES     25.798     15.417     69.172
   PARCOMP      7.973      3.421      9.207     11.393
   SENCOMP      9.936      3.296     11.092     11.277     21.616
  WORDMEAN     17.425      6.876     22.954     19.167     25.321     63.163
  ADDITION     17.132      7.015     14.763     16.766     28.069     33.768
  COUNTDOT     44.651     15.675     41.659      7.357     19.311     20.213
    SCCAPS    124.657     40.803    114.763     39.309     61.230     79.993
```

---

[1]The term normal theory maximum likelihood is used to mean that the estimation of the model is based on the assumption that the variables have a multivariate normal distribution.

Covariance Matrix

	ADDITION	COUNTDOT	SCCAPS
ADDITION	565.593		
COUNTDOT	293.126	440.792	
SCCAPS	368.436	410.823	1371.618

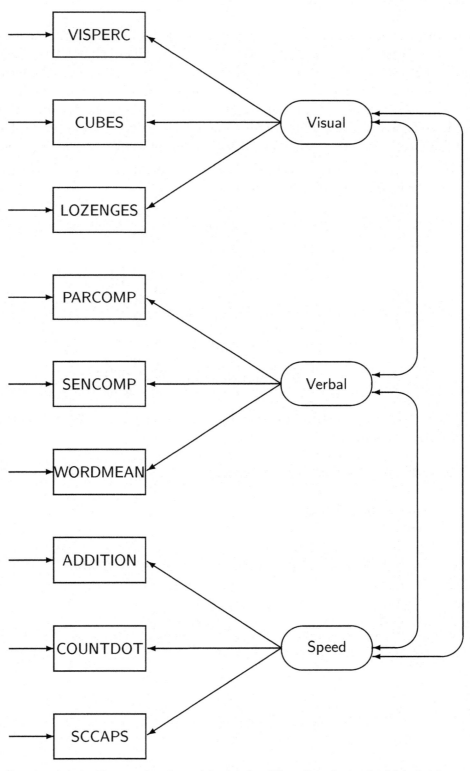

Figure 7.3: Confirmatory Factor Analysis Model for Nine Psychological Variables

After the covariance matrix the following lines are given

```
Total Variance = 2610.906 Generalized Variance = 0.106203D+17

Largest Eigenvalue = 1734.725 Smallest Eigenvalue = 3.665

Condition Number = 21.756
```

The total variance is the sum of the diagonal elements of $\mathbf{S}$ and the generalized variance is the determinant of $\mathbf{S}$ which equals the product of all the eigenvalues of $\mathbf{S}$. The largest and smallest eigenvalues of $\mathbf{S}$ are also given. These quantities are useful in principal components analysis. The condition number is the square root of the ratio of the largest and smallest eigenvalue. A large condition number indicates multicollinearity in the data. If the condition number is larger than 30, LISREL gives a warning. This might indicate that one or more variables are linear or nearly linear combinations of other variables. If this is intentional, the warning may be ignored.

Let $d_1, d_2, \cdots, d_9$ be the eigenvalues of $\mathbf{S}$ in descending order. Then one can express these quantities in terms of these eigenvalues as follows

$$\mathrm{Total\,Variance} = \sum_i d_i \tag{7.13}$$

$$\mathrm{Generalized\,Variance} = \prod_i d_i \tag{7.14}$$

$$\mathrm{Largest\,Eigenvalue} = d_1 \tag{7.15}$$

$$\mathrm{Smallest\,Eigenvalue} = d_9 \tag{7.16}$$

$$\mathrm{Condition\,Number} = \sqrt{\frac{d_1}{d_9}} \tag{7.17}$$

LISREL gives parameter estimates, standard errors, Z-values, P-values and R^2 for the measurement equations as follows

```
LISREL Estimates (Maximum Likelihood)

        Measurement Equations


  VISPERC = 4.678*Visual, Errorvar.= 25.915, R² = 0.458
Standerr  (0.624)                    (4.582)
Z-values  7.499                       5.656
P-values  0.000                       0.000
```

```
   CUBES = 2.296*Visual, Errorvar.= 14.487, R² = 0.267
Standerr  (0.408)                      (1.981)
Z-values   5.622                        7.313
P-values   0.000                        0.000


LOZENGES = 5.769*Visual, Errorvar.= 35.896, R² = 0.481
Standerr  (0.751)                      (6.660)
Z-values   7.684                        5.390
P-values   0.000                        0.000


 PARCOMP = 2.922*Verbal, Errorvar.= 2.857 , R² = 0.749
Standerr  (0.237)                      (0.589)
Z-values  12.312                        4.854
P-values   0.000                        0.000


 SENCOMP = 3.856*Verbal, Errorvar.= 6.749 , R² = 0.688
Standerr  (0.333)                      (1.165)
Z-values  11.590                        5.792
P-values   0.000                        0.000


WORDMEAN = 6.567*Verbal, Errorvar.= 20.034, R² = 0.683
Standerr  (0.569)                      (3.419)
Z-values  11.532                        5.859
P-values   0.000                        0.000


ADDITION = 15.676*Speed, Errorvar.= 319.868, R² = 0.434
Standerr  (2.012)                      (48.754)
Z-values   7.792                        6.561
P-values   0.000                        0.000


COUNTDOT = 16.709*Speed, Errorvar.= 161.588, R² = 0.633
Standerr  (1.752)                      (38.166)
Z-values   9.535                        4.234
P-values   0.000                        0.000


  SCCAPS = 25.956*Speed, Errorvar.= 697.900 , R² = 0.491
Standerr  (3.117)                      (116.524)
Z-values   8.328                        5.989
P-values   0.000                        0.000
```

By default **LISREL** standardizes the latent variables. This seems most reasonable since the latent variables are unobservable and have no definite scale. The correlations among the latent variables, with standard errors and Z- values are given as follows

```
            Correlation Matrix of Independent Variables


                   Visual       Verbal        Speed
                  --------     --------     --------
    Visual          1.000

    Verbal          0.541        1.000
                   (0.085)
                    6.355

     Speed          0.523        0.336        1.000
                   (0.094)      (0.091)
                    5.562        3.674
```

These estimates have been obtained by maximizing the likelihood function L under multivariate normality, see the appendix in Chapter 16. Therefore it is possible to give the log-likelihood values at the maximum of the likelihood function. It is common to report the value of $-2ln(L)$, sometimes called deviance, instead of L. **LISREL** gives the value $-2\ln(L)$ for the estimated model and for a saturated model. A saturated model is a model where the mean vector and covariance matrix of the multivariate normal distribution are unconstrained,

The log-likelihood values are given in the output as

```
                    Log-likelihood Values


                Estimated Model              Saturated Model
                ---------------              ---------------
Number of free parameters(t)     21                        45
-2ln(L)                     6707.266                  6655.724
AIC (Akaike, 1974)*         6749.266                  6745.724
BIC (Schwarz, 1978)*        6811.777                  6879.677
```

*LISREL uses AIC= 2t - 2ln(L) and BIC = tln(N)- 2ln(L)

LISREL also gives the values of AIC and BIC. These can be used for the problem of selecting the "best" model from several *a priori* specified models. One then chooses the model with the smallest AIC or BIC. The original papers of Akaike (1974) and Schwarz (1978) define AIC and BIC in terms of $\ln(L)$ but **LISREL** uses $-2\ln(L)$ and the formulas:

$$\text{AIC} = 2t - 2ln(L) \ , \tag{7.18}$$

$$\text{BIC} = tln(N) - 2ln(L) \ , \tag{7.19}$$

where t is the number of independent parameters in the model and N is the total sample size.

LISREL Syntax

The **LISREL** syntax file corresponding to **npv1a.spl** is **npv1b.lis**:

```
Estimation of the NPV Model by Maximum Likelihood
DA NI=9
RA=NPV.LSF
MO NX=9 NK=3
LK
Visual Verbal Speed
FR LX(1,1) LX(2,1) LX(3,1) LX(4,2) LX(5,2) LX(6,2) LX(7,3) LX(8,3) LX(9,3)
PD
OU
```

Every **LISREL** syntax file begins with a **DA** line with the number of variables in the data file specified. This needs not be the same as the number of x-variables in the model, as is the case here, but can be much larger and the variables included in the model can be selected by including an **SE** line.

The only parameter matrices in this model are $\mathbf{\Lambda}_x$, $\mathbf{\Phi}$, and $\mathbf{\Theta}_\delta$. On the **MO** line these are all default. All we need to do is to specify the free elements in $\mathbf{\Lambda}_x$. This is done by the **FR** line.

The output file **npv1b.out** gives the same results as **npv1a.out** but the estimates, standard errors, and Z-values, are given in matrix form giving $\mathbf{\Lambda}_x$, $\mathbf{\Phi}$, and $\mathbf{\Theta}_\delta$. In addition to these estimated matrices, **LISREL** also gives the parameter specification of these matrices. This is useful to check that **LISREL** has interpreted the model as intended. Here 0 means a fixed parameter and the independent parameters are numbered in sequence. Note that these parameter specifications do not specify the fixed values. In this case all fixed values in $\mathbf{\Lambda}_x$ are zero by default and the fixed diagonal element of $\mathbf{\Phi}$ are fixed by default. Other fixed values can be specified on **VA** lines or by reading a matrix as described in Section 9.3.

```
Parameter Specifications


     LAMBDA-X


              Visual      Verbal       Speed
  VISPERC          1           0           0
    CUBES          2           0           0
 LOZENGES          3           0           0
  PARCOMP          0           4           0
  SENCOMP          0           5           0
 WORDMEAN          0           6           0
 ADDITION          0           0           7
 COUNTDOT          0           0           8
   SCCAPS          0           0           9


        PHI


              Visual      Verbal       Speed
   Visual          0
   Verbal         10           0
    Speed         11          12           0
```

```
        THETA-DELTA

                VISPERC      CUBES    LOZENGES    PARCOMP    SENCOMP    WORDMEAN
                   13          14         15         16         17         18

        THETA-DELTA

                ADDITION    COUNTDOT    SCCAPS
                   19          20         21
```

This shows that the number of independent parameters is 21. It also shows exactly which these parameters are, *i.e.*,

$$
\begin{aligned}
\boldsymbol{\theta}' \;=\; & (\lambda_{11}^{(x)}, \lambda_{21}^{(x)}, \lambda_{31}^{(x)}, \lambda_{42}^{(x)}, \lambda_{52}^{(x)}, \lambda_{62}^{(x)}, \lambda_{73}^{(x)}, \lambda_{83}^{(x)}, \lambda_{93}^{(x)} \\
& \phi_{21}, \phi_{31}, \phi_{32}, \theta_{11}^{(\delta)}, \theta_{22}^{(\delta)}, \theta_{33}^{(\delta)}, \theta_{44}^{(\delta)}, \theta_{55}^{(\delta)}, \theta_{66}^{(\delta)}, \theta_{77}^{(\delta)}, \theta_{88}^{(\delta)}, \theta_{99}^{(\delta)})
\end{aligned}
\tag{7.20}
$$

The estimated parameter matrices are given in the output as

```
LISREL Estimates (Maximum Likelihood)

        LAMBDA-X

                 Visual     Verbal      Speed
                --------   --------   --------
    VISPERC       4.678       - -        - -
                 (0.622)
                  7.525

      CUBES       2.296       - -        - -
                 (0.407)
                  5.642

   LOZENGES       5.769       - -        - -
                 (0.748)
                  7.711

    PARCOMP        - -       2.922       - -
                            (0.236)
                             12.355

    SENCOMP        - -       3.856       - -
                            (0.332)
                             11.630

   WORDMEAN        - -       6.567       - -
                            (0.568)
                             11.572
```

```
ADDITION      - -        - -        15.676
                                    (2.005)
                                     7.819

COUNTDOT      - -        - -        16.709
                                    (1.746)
                                     9.568

SCCAPS        - -        - -        25.956
                                    (3.106)
                                     8.357
```

```
          PHI

          Visual      Verbal      Speed
Visual     1.000

Verbal     0.541       1.000
          (0.085)
           6.355

Speed      0.523       0.336       1.000
          (0.094)     (0.091)
           5.562       3.674
```

THETA-DELTA

VISPERC	CUBES	LOZENGES	PARCOMP	SENCOMP	WORDMEAN
25.915	14.487	35.896	2.857	6.749	20.034
(4.566)	(1.974)	(6.637)	(0.587)	(1.161)	(3.407)
5.675	7.339	5.409	4.870	5.812	5.880

THETA-DELTA

ADDITION	COUNTDOT	SCCAPS
319.868	161.588	697.900
(48.586)	(38.034)	(116.121)
6.584	4.248	6.010

Testing the Model

Various chi-square statistics are used for testing structural equation models. If normality holds and the model is fitted by the maximum likelihood (ML) method, one such chi-square statistic is obtained as N times the minimum of the ML fit function, where N is the sample size. An asymptotically equivalent chi-square statistic can be obtained from a general formula developed by Browne (1984) and using an asymptotic covariance matrix estimated under multivariate normality, see Section 16.1.5. These chi-square statistics are denoted C_1 and $C_{2\mathrm{NT}}$, respectively. They are valid under multivariate normality of the observed variables and if the model holds.

For this analysis, **LISREL** gives the two chi-square values C1 and C2_NT as

```
Degrees of Freedom For (C1)-(C2)                     24
Maximum Likelihood Ratio Chi-Square (C1)             51.542 (P = 0.0009)
Browne's (1984) ADF Chi-Square (C2_NT)               48.952 (P = 0.0019)
```

The degrees of freedom is the number non-duplicated elements of **S** minus the number of independent parameters in the model, in this case $45 - 21 = 24$.

These chi-squares indicate that the model does not fit the data. We will consider this issue in Section 7.3.4.

Robust Estimation

The analysis just described assumes that the variables have a multivariate normal distribution. This assumption is questionable in many cases. Although the maximum likelihood parameter estimates are considered to be robust against non-normality, their standard errors and chi-squares tends to be too large. It is therefore recommended to use the maximum likelihood method with robustified standard errors and chi-squares, which in **LISREL** is called *Robust Maximum Likelihood* (RML), see the appendix in Chapter 16.

Under certain conditions the standard errors of some parameters may be the same for ML and RML. In the literature this is called Asymptotic Robustness (AR), see Shapiro (1987), Browne and Shapiro (1988), Amemiya and Anderson (1990), or Satorra and Bentler (1990).

RML requires an estimate of the asymptotic covariance matrix (ACM) of the variances and covariances of the elements of **S** under non-normality. This in turn requires the computations of fourth-order moments of the data, see the appendix in Chapter 16.

As the term *asymptotic* indicates, RML requires a large sample. A recommendation is that

$$N \geq \frac{1}{2}k(k+1) \,, \tag{7.21}$$

where k is the number of observed variables in the model. In the context of CFA k is the number of x-variables.

To do RML in **LISREL**, include a line

```
Robust Estimation
```

anywhere between the second line and the last line in the **SIMPLIS** syntax, see file **npv2a.spl**. In **LISREL** syntax include a line

```
RO
```

see file **npv2b.lis**.

The output file gives the following information about the distribution of the variables.

Total Sample Size =   145

Univariate Summary Statistics for Continuous Variables

Variable	Mean	St. Dev.	Skewness	Kurtosis	Minimum	Freq.	Maximum	Freq.
VISPERC	29.579	6.914	-0.119	-0.046	11.000	1	51.000	1
CUBES	24.800	4.445	0.239	0.872	9.000	1	37.000	2
LOZENGES	15.966	8.317	0.623	-0.454	3.000	2	36.000	1
PARCOMP	9.952	3.375	0.405	0.252	1.000	1	19.000	1
SENCOMP	18.848	4.649	-0.550	0.221	4.000	1	28.000	1
WORDMEAN	17.283	7.947	0.729	0.233	2.000	1	41.000	1
ADDITION	90.179	23.782	0.163	-0.356	30.000	1	149.000	1
COUNTDOT	109.766	20.995	0.698	2.283	61.000	1	200.000	1
SCCAPS	191.779	37.035	0.200	0.515	112.000	1	333.000	1

This shows that the ranges of the variables are quite different, reflecting the fact that they are composed of different number of items. For example, PARCOMP has a range of 1 to 19, whereas SCCAPS has a range of 112 to 333. This is also reflected in the means and standard deviations.

LISREL also gives tests of univariate and multivariate skewness and kurtosis. These tests were introduced in chapter 1

Test of Univariate Normality for Continuous Variables

Variable	Skewness Z-Score	Skewness P-Value	Kurtosis Z-Score	Kurtosis P-Value	Skewness and Kurtosis Chi-Square	Skewness and Kurtosis P-Value
VISPERC	-0.604	0.546	0.045	0.964	0.367	0.833
CUBES	1.202	0.229	1.843	0.065	4.842	0.089
LOZENGES	2.958	0.003	-1.320	0.187	10.491	0.005
PARCOMP	1.995	0.046	0.761	0.447	4.559	0.102
SENCOMP	-2.646	0.008	0.693	0.489	7.483	0.024
WORDMEAN	3.385	0.001	0.720	0.472	11.977	0.003
ADDITION	0.826	0.409	-0.937	0.349	1.560	0.458
COUNTDOT	3.263	0.001	3.325	0.001	21.699	0.000
SCCAPS	1.008	0.313	1.273	0.203	2.638	0.267

Relative Multivariate Kurtosis = 1.072

Test of Multivariate Normality for Continuous Variables

Skewness Value	Skewness Z-Score	Skewness P-Value	Kurtosis Value	Kurtosis Z-Score	Kurtosis P-Value	Skewness and Kurtosis Chi-Square	Skewness and Kurtosis P-Value
11.733	5.426	0.000	106.098	3.023	0.003	38.579	0.000

It is seen that the hypothesis of zero skewness and kurtosis is rejected for LOZENGES, SENCOMP, WORDMEAN, and COUNTDOT.

The output file **npv2a.out** gives the same parameter estimates as before but different standard errors. As a consequence, also t-values and P-values are different. The parameter estimates and the two sets of standard errors are given in Table 7.5.

Table 7.5: Parameter Estimates, Normal Standard Errors, and Robust Standard Errors

Parameter Factor Loadings		Parameter Estimates	Standard Errors Normal	Robust
VISPERC on Visual	$\lambda_{11}^{(x)}$	4.678	0.622	0.693
CUBES on Visual	$\lambda_{21}^{(x)}$	2.296	0.407	0.375
LOZENGES on Visual	$\lambda_{31}^{(x)}$	5.769	0.748	0.725
PARCOMP on Verbal	$\lambda_{42}^{(x)}$	2.992	0.236	0.250
SENCOMP on Verbal	$\lambda_{52}^{(x)}$	3.856	0.332	0.331
WORDMEAN on Verbal	$\lambda_{62}^{(x)}$	6.567	0.568	0.573
ADDITION on Speed	$\lambda_{73}^{(x)}$	15.676	2.005	1.830
COUNTDOT on Speed	$\lambda_{83}^{(x)}$	16.709	1.746	1.775
SCCAPS on Speed	$\lambda_{93}^{(x)}$	25.956	3.106	3.077
Factor Correlations				
Verbal vs Visual	ϕ_{21}	0.541	0.085	0.094
Verbal vs Speed	ϕ_{31}	0.523	0.094	0.100
Verbal vs Speed	ϕ_{32}	0.336	0.091	0.115

If the observed variables are non-normal, one can use the asymptotic covariance matrix (ACM) estimated under non-normality to obtain another chi-square. This chi-square, often called the ADF (Asymptotically Distribution Free) chi-square statistic, is denoted $C_{2\mathrm{NNT}}$. It has been found in simulation studies that the ADF statistic does not work well because it is difficult to estimate the ACM accurately unless N is huge, see *e.g*, Curran, West, and Finch (1996).

Satorra and Bentler (1988) proposed another approximate chi-square statistic C_3, often called the SB chi-square statistic, which is C_1 multiplied by a scale factor which is estimated from the sample and involves estimates of the ACM both under normality and non-normality. The scale factor is estimated such that C_3 has an asymptotically correct mean even though it does not have an asymptotic chi-square distribution. In practice, C_3 is conceived of as a way of correcting C_1 for the effects of non-normality and C_3 is often used as it performs better than the ADF test $C_{2\mathrm{NT}}$ in **LISREL**, particularly if N is not very large, see e.g., Hu, Bentler, and Kano (1992).

Satorra and Bentler (1988) also mentioned the possibility of using a Satterthwaite (1941) type correction which adjusts C_1 such that the corrected value has the correct asymptotic mean and variance. This type of fit measure has not been much investigated, neither for continuous nor for ordinal variables. However, this type of chi-square fit statistic has been implemented in **LISREL**, where it is denoted C_4. The formulas for C_1–C_4 are given in Section 16.1.5 in the appendix (Chapter 16). For studies investigating the performance of C_3 and C_4, see Fouladi (2000), Savalei (2010) and Foldnes and Olsson (2015).

For our present example, C_1–C_4 appear in the output as

```
Degrees of Freedom For (C1)-(C3)                     24
Maximum Likelihood Ratio Chi-Square (C1)             51.542 (P = 0.0009)
Browne's (1984) ADF Chi-Square (C2_NT)               48.952 (P = 0.0019)
Browne's (1984) ADF Chi-Square (C2_NNT)              64.648 (P = 0.0000)
Satorra-Bentler (1988) Scaled Chi-square (C3)        50.061 (P = 0.0014)
```

```
Satorra-Bentler (1988) Adjusted Chi-square (C4)      35.134 (P = 0.0056)
Degrees of Freedom For C4                            16.844
```

C_1 and $C_{2\mathrm{NT}}$ are the same as before but with robust estimation LISREL also gives $C_{2\mathrm{NNT}}$, C_3 and C_4 so that one can see what the effect of non-normality is. In particular,the difference $C_{2\mathrm{NNT}} - C_{2\mathrm{NT}}$ can be viewed as an effect of non-normality. One can also regard the difference $C_1 - C_3$ as an effect of non-normality, see Foldnes and Olsson (2015).

Note that C_4 has its own degrees of freedom which is different from the model degrees of freedom. LISREL gives the degrees of freedom for C_4 as a fractional number and uses this fractional degrees of freedom to compute the P-value for C_4. The formulas for C_4 and its degrees of freedom are given in Section 16.1.5 in the appendix (see Chapter 16).

No matter which chi-square we use the conclusion is that the model does not fit well. In addition to these five chi-squares, LISREL gives many other fit measures in the output as follows

```
Estimated Non-centrality Parameter (NCP)            27.542
90 Percent Confidence Interval for NCP              (10.616 ; 52.208)


Minimum Fit Function Value                          0.355
Population Discrepancy Function Value (F0)          0.190
90 Percent Confidence Interval for F0               (0.0732 ; 0.360)
Root Mean Square Error of Approximation (RMSEA)     0.0890
90 Percent Confidence Interval for RMSEA            (0.0552 ; 0.122)
P-Value for Test of Close Fit (RMSEA < 0.05)        0.0311


Expected Cross-Validation Index (ECVI)              0.645
90 Percent Confidence Interval for ECVI             (0.528 ; 0.815)
ECVI for Saturated Model                            0.621
ECVI for Independence Model                         3.612


Chi-Square for Independence Model (36 df)           505.767


Normed Fit Index (NFI)                              0.898
Non-Normed Fit Index (NNFI)                         0.912
Parsimony Normed Fit Index (PNFI)                   0.599
Comparative Fit Index (CFI)                         0.941
Incremental Fit Index (IFI)                         0.943
Relative Fit Index (RFI)                            0.847


Critical N (CN)                                     121.079



Root Mean Square Residual (RMR)                     16.195
Standardized RMR                                    0.0719
Goodness of Fit Index (GFI)                         0.930
Adjusted Goodness of Fit Index (AGFI)               0.869
Parsimony Goodness of Fit Index (PGFI)              0.496
```

The definitions of these fit measures are given in the appendix (see Chapter 15). Since the ML method was used to estimate the model in this case, LISREL uses the chi-square C_1 and the chi-square value for the independence model CI to compute these fit measures. The chi-square for the independence model CI is defined in the appendix (see Chapter 16).

What would the Normed Fit Index (NFI), or any of the other fit measures, be if, for example, C_{2NT}, or any of the other chi-squares, had been used instead of C_1? The answer to these type of questions is given in the **FTB** file which is obtained by putting FT on the OU line in **npv2b.lis**. The **FTB** file will then be **npv2b.ftb**. This looks like this

	C1	C2_NT	C2_NNT	C3	C4
Chi-Square for Model (CM)	51.542	48.952	64.648	50.061	35.134
Degrees of Freedom for Model (DFM)	24.000	24.000	24.000	24.000	16.844
Chi-Square for Independence Model (CI)	505.767	675.226	204.066	403.555	63.303
Degrees of Freedom for Independence Model (DFI)	36.000	36.000	36.000	36.000	5.647
CM/DFM	2.148	2.040	2.694	2.086	2.086
CI/DFI	14.049	18.756	5.668	11.210	11.210
Estimated Non-Centrality Parameter (NCP)	27.542	24.952	40.648	26.061	18.290
Population Discrepancy Function Value (F0)	0.190	0.172	0.280	0.180	0.126
Root Mean Square Error of Approximation (RMSEA)	0.089	0.085	0.108	0.087	0.087
Expected Cross-Validation Index (ECVI)	0.645	0.627	0.736	0.635	0.631
Model AIC (AIC)	93.542	90.952	106.648	92.061	91.446
Model BIC (BIC)	156.198	153.608	169.304	154.716	175.454
Normed Fit Index (NFI)	0.898	0.928	0.683	0.876	0.445
Non-Normed Fit Index (NNFI)	0.912	0.941	0.637	0.894	0.894
Parsimony Normed Fit Index (PNFI)	0.599	0.618	0.455	0.584	1.327
Comparative Fit Index (CFI)	0.941	0.961	0.758	0.929	0.683
Incremental Fit Index (IFI)	0.943	0.962	0.774	0.931	0.606
Relative Fit Index (RFI)	0.847	0.891	0.525	0.814	0.814
Goodness of Fit Index (GFI)	0.930	0.930	0.930	0.930	0.930
Adjusted Goodness of Fit Index (AGFI)	0.869	0.869	0.869	0.869	0.814
Parsimony Goodness of Fit Index (PGFI)	0.496	0.496	0.496	0.496	0.348

Estimating the Model by Other Methods

So far we have used the ML method to estimate the model. This is well justified in most examples with continuous variables in particular if used with robust estimates of standard errors and chi-squares. But since the variables in this data are slightly non-normal, one can also use any of the other methods available in LISREL. All of these, except WLS can also be used with robust estimates of standard errors.

To use any of the other methods, for example DWLS, put ME=DWLS on the OU line in LISREL syntax. In SIMPLIS syntax, write a line

```
Method of Estimation: Diagonally Weighted Least Squares
```

or include a line

```
Options: DWLS
```

See files **npv3a.spl** and **npv3b.lis**.

Table 7.6 summarizes the various chi-squares obtained by all methods.

7.3.4 Analyzing Correlations

Factor analysis was mainly developed by psychologists for the purpose of identifying mental abilities by means of psychological testing. Various theories of mental abilities and various procedures for analyzing the correlations among psychological tests emerged.

Table 7.6: Chi-Square Values for Different Methods Data: Grant-White School - CFA Model for Nine Psychological Variables The CI values are chi-squares for the independence model

	C1	C2_NT	C2_NNT	C3	C4	CI
ML	51.542	48.952	64.648	50.061	35.134	505.767
ULS	92.500	59.260	58.563	75.004	44.407	505.767
GLS	81.108	85.631	60.867	56.444	38.042	139.289
DWLS	62.148	48.449	60.560	57.906	38.769	505.767
WLS	175.205	-.—	-.—	-.—	-.—	204.066

Following this old tradition, users of **LISREL** might be tempted to analyze the correlation matrix of the nine psychological variables instead of the covariance matrix as we have done in the previous examples. However, analyzing the correlation matrix by maximum likelihood (ML) is problematic in several ways as pointed out by Cudeck (1989). In particular, the standard errors are wrong. One can obtain correct standard errors by formulating the model as a correlation structure, but this requires another extension of **LISREL** which is covered in Chapter 8. The analysis of correlation structures is considered in Section 8.7.

Here we consider some other ways to resolve this problem:

Approach 1 Use the same input as before and request the completely standardized solution (SC)[2] and add a line

```
Options SC
```

see file **npv4a.spl**. In **LISREL** just add SC on the OU line, see file **npv4b.lis**. In addition to the unstandardized solution as before, this gives the completely standardized solution in matrix form as

```
Completely Standardized Solution


        LAMBDA-X


              Visual    Verbal     Speed

              --------  --------  --------

  VISPERC      0.677      - -       - -
    Cubes      0.517      - -       - -
 LOZENGES      0.694      - -       - -
  PARCOMP       - -      0.866      - -
  SENCOMP       - -      0.829      - -
 WORDMEAN       - -      0.826      - -
 ADDITION       - -       - -      0.659
 COUNTDOT       - -       - -      0.796
   SCCAPS       - -       - -      0.701
```

---

[2]**LISREL** has two kinds of standardized solutions: the standardized solution (SS) in which only the latent variables are standardized and the completely standardized solution (SC) in which both the observed and the latent variables are standardized.

```
PHI


              Visual      Verbal      Speed
            --------    --------    --------
   Visual      1.000
   Verbal      0.541       1.000
    Speed      0.523       0.336       1.000



THETA-DELTA


      VISPERC       Cubes    LOZENGES     PARCOMP     SENCOMP    WORDMEAN
     --------    --------    --------    --------    --------    --------
        0.542       0.733       0.519       0.251       0.312       0.317
     ADDITION    COUNTDOT      SCCAPS
     --------    --------    --------
        0.566       0.367       0.509
```

The disadvantage with this alternative is that one does not get standard errors for the completely standardized solution.

Approach 2 Use the following **PRELIS** syntax file to standardize the original variables (file **npv2.prl**):

```
SY=NPV.LSF
SV ALL
OU RA=NPVstd.LSF
```

SV is a new **PRELIS** command to standardize the variables. One can standardize some of the variables by listing them on the SV line. **npv2.prl** produces a new .lsf file **NPVstd.lsf** in which all variables has sample means 0 and sample standard deviations 1.

Then use NPVstd.LSF instead of NPV.LSF in **npv2a.spl** or **npv2b.lis** to obtain a completely standardized solution with robust standard errors.

Approach 3 Use the sample correlation matrix with robust unweighted least squares (RULS) or with robust diagonally weighted least squares (RDWLS) This will use an estimate of the asymptotic covariance matrix of the sample correlations to obtain correct asymptotic standard errors and chi-squares under non-normality.

The following **SIMPLIS** command file demonstrates the Approach 3, see file **npv5a.spl**:

Estimation of the NPV Model by Robust Diagonally Weighted Least Squares
Using Correlations

Raw Data from File NPV.LSF
Analyze Correlations
Latent Variables: Visual Verbal Speed
Relationships:
    VISPERC - LOZENGES = Visual
    PARCOMP - WORDMEAN = Verbal
    ADDITION - SCCAPS = Speed
Robust Estimation
Options: DWLS
Path Diagram
End of Problem

Note the added line

Analyze Correlations

This gives the standardized solution as

LISREL Estimates (Robust Diagonally Weighted Least Squares)

```
 VISPERC = 0.726*Visual, Errorvar.= 0.472 , R² = 0.528
Standerr  (0.0712)                 (0.196)
Z-values   10.201                   2.408
P-values   0.000                    0.016


   CUBES = 0.481*Visual, Errorvar.= 0.769 , R² = 0.231
Standerr  (0.0815)                 (0.184)
Z-values   5.897                    4.175
P-values   0.000                    0.000


LOZENGES = 0.677*Visual, Errorvar.= 0.541 , R² = 0.459
Standerr  (0.0691)                 (0.191)
Z-values   9.794                    2.832
P-values   0.000                    0.005


 PARCOMP = 0.863*Verbal, Errorvar.= 0.255 , R² = 0.745
Standerr  (0.0329)                 (0.176)
Z-values   26.257                   1.448
P-values   0.000                    0.147


 SENCOMP = 0.836*Verbal, Errorvar.= 0.302 , R² = 0.698
Standerr  (0.0352)                 (0.177)
Z-values   23.743                   1.707
P-values   0.000                    0.088


WORDMEAN = 0.823*Verbal, Errorvar.= 0.323 , R² = 0.677
Standerr  (0.0364)                 (0.177)
Z-values   22.620                   1.826
P-values   0.000                    0.068
```

```
ADDITION = 0.611*Speed, Errorvar.= 0.627 , R² = 0.373
Standerr  (0.0662)                     (0.185)
Z-values   9.226                        3.384
P-values   0.000                        0.001


COUNTDOT = 0.711*Speed, Errorvar.= 0.494 , R² = 0.506
Standerr  (0.0588)                     (0.186)
Z-values  12.094                        2.650
P-values   0.000                        0.008


  SCCAPS = 0.842*Speed, Errorvar.= 0.290 , R² = 0.710
Standerr  (0.0588)                     (0.194)
Z-values  14.324                        1.498
P-values   0.000                        0.134
```

```
            Correlation Matrix of Independent Variables


                Visual      Verbal      Speed
               --------    --------    --------
    Visual       1.000

    Verbal       0.535       1.000
                (0.085)
                 6.292

     Speed       0.571       0.379       1.000
                (0.087)     (0.087)
                 6.591       4.354
```

Modifying the Model

The output file **npv2a.out** gives the following information about modification indices

```
        The Modification Indices Suggest to Add the
    Path to  from    Decrease in Chi-Square    New Estimate
    ADDITION  Visual          8.5                -6.90
    COUNTDOT  Verbal          8.3                -4.94
    SCCAPS    Visual         28.4                24.44
    SCCAPS    Verbal         10.8                11.11
```

This suggests that the fit can be improved by adding a path from Visual to SCCAPS. If this makes sense, one can add this path, see file **npv2aa.spl** and rerun the model. This gives a solution where

```
   SCCAPS = 16.559*Visual + 16.274*Speed, Errorvar.= 620.929, R² = 0.547
Standerr   (3.700)              (3.359)                      (98.281)
Z-values   4.475                4.845                        6.318
P-values   0.000                0.000                        0.000
```

and the chi-squares are now

```
Degrees of Freedom for (C1)-(C3)                           23
Maximum Likelihood Ratio Chi-Square (C1)                   28.293 (P = 0.2049)
Browne's (1984) ADF Chi-Square (C2_NT)                     27.898 (P = 0.2197)
Browne's (1984) ADF Chi-Square (C2_NNT)                    31.701 (P = 0.1065)
Satorra-Bentler (1988) Scaled Chi-square (C3)              28.221 (P = 0.2075)
Satorra-Bentler (1988) Adjusted Chi-square (C4)            20.437 (P = 0.2342)
Degrees of Freedom for C4                                  16.656
```

indicating a good fit.

To do this with **LISREL** syntax use **npv2bb.lis**. Note the added **LX(9,1) on the FR-** line.

The interpretation of this model is that the variable SCCAPS is not only a measure of Speed but a more complex variable involving also some degree of visual ability.

Interpreting Chi-Square Differences

The scaled chi-square statistic of Satorra and Bentler (1988) is often used in SEM to test model fit. This statistic is often called the SB statistic and is denoted C_3 in **LISREL**. However, as pointed out by Satorra (2000), the difference between two SB statistics is not suitable for testing the hypothesis of no difference between one restricted model M_0 and a more general model M_1. Unlike the ML chi-square (C_1 in **LISREL**) this SB difference does not have an approximate chi-square distribution. To work properly the SB difference for models M_0 and M_1 must be scaled and Satorra and Bentler (2001) develop two alternative approximate chi-square difference tests called the **original** scaled chi-square difference test and the **new** scaled chi-square difference test. Bryant and Satorra (2012) show how the **original** scaled chi-square difference test can be computed from the maximum likelihood chi-square C_1 and the C_3 in for both models. They also show that one can use a specific model M_{10} and use the C_1 and C_3 obtained for this model to compute also the new scaled chi-square difference test. The model M_{10} is the same as model M_1 but the chi-squares are evaluated at the solution for model M_0 so no iterations have to be used.

The calculation scheme described by Bryant and Satorra (2012) is unnecessarily complicated. **LISREL** 9 has a unique way of obtaining these chi-square differences **automatically** by running only Model M_0 and modifying this model in the path diagram. To do this, include the line

```
Path Diagram
```

in **npv2a.spl** or the line

```
pd
```

in the **npv2b.lis** for Model M_0 only. This gives a path diagram which looks like this

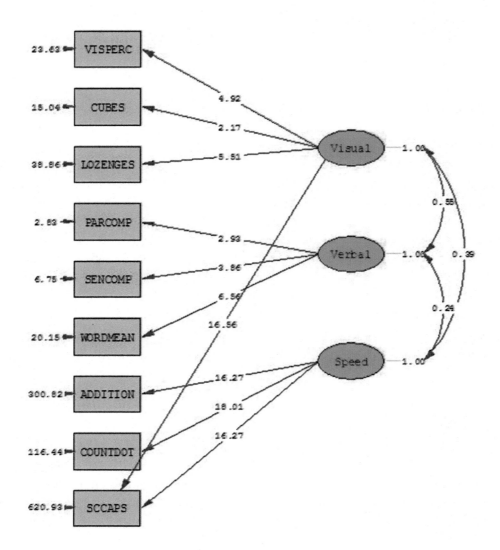

While the path diagram is still visible, select the one-way arrow in the drawing tools and use the mouse to add a path by dragging the mouse from the latent variable Visual to the observed variable SCCAPS. This will free the loading `lx(9,1)` (so it is no longer constrained to be equal to 0) and give temporary estimates of of all the loadings. Now click the **Run LISREL** button **L**. LISREL will then estimate the modified model M_1 and show a new path diagram which looks like this

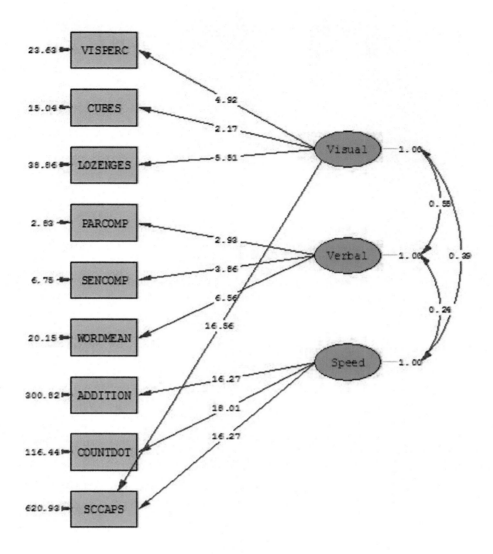

Close this path diagram and look at the output file. This contains the results for both Model M_0 and Model M_1. After the chi-square values for Model M_1, the following lines appear

```
Degrees of Freedom for Difference                       1
Chi-square Difference (C1)                              23.249 (P = 0.0000)
Chi-square Difference (C3)                              21.840 (P = 0.0000)
Original Scaled Chi-square Difference (C3)              14.078 (P = 0.0002)
Approxim New Scaled Chi-square Difference (C3)          22.581 (P = 0.0002)
```

7.3.5 Continuous Variables with Missing Values

7.3.6 Example: Longitudinal Data on Math and English Scores

The data for this example is the **jsp** data used in Chapter 4 to illustrate a multivariate multilevel analysis with the objective to estimate the within schools and between schools covariance matrix. For this example we will use the data aggregated over schools and students and omitting the variables gender and ravens. The data file is **jsp_aggregated.lsf** where the missing value code is -9. The first few rows of the data file are shown here

	math1	math2	math3	eng1	eng2	eng3
1	23.000	24.000	23.000	72.000	80.000	39.000
2	14.000	11.000	-9.000	7.000	17.000	-9.000
3	36.000	32.000	39.000	88.000	89.000	83.000
4	24.000	26.000	32.000	12.000	25.000	12.000
5	22.000	23.000	-9.000	67.000	78.000	-9.000
6	19.000	23.000	11.000	52.000	76.000	19.000
7	22.000	22.000	26.000	37.000	68.000	31.000
8	18.000	29.000	28.000	57.000	86.000	40.000
9	30.000	31.000	-9.000	42.000	59.000	-9.000
10	29.000	29.000	-9.000	46.000	79.000	-9.000
11	31.000	28.000	32.000	69.000	84.000	50.000
12	18.000	26.000	-9.000	54.000	74.000	-9.000
13	23.000	-9.000	27.000	63.000	-9.000	39.000

In this kind of data it is inevitable that there are missing values. For example, a student may be absent at one test occasion even if he was present at a previous test occasion. It is seen in that

- Students 2, 5, 9, 10, and 12 are missing at the third test occasion

- Student 13 is missing at the second test occasions

We discussed some general issues relating to missing values in Chapter 1. In the following analysis it is assumed that data are missing at random (MAR), although there may be a small probability that a student will be missing because his/her ability is low.

Data Screening

Whenever one starts an analysis of a new data set, it is recommended to begin with a data screening. To do so click on **Statistics** at the top of the screen and select **Data Screening** from the **Statistics** menu. This will reveal the following information about the data.

```
Number of Missing Values per Variable

    math1     math2     math3     eng1      eng2      eng3
  --------  --------  --------  --------  --------  --------
       38        63       239        38        63       239
```

This table says that there are 38 student missing at the first test, 63 are missing at the second test, and 239 are missing at the third test.

```
Distribution of Missing Values

Total Sample Size(N) =    1192

Number of Missing Values       0      1      2      3      4
          Number of Cases    887      0    270      0     35
```

This shows that there are 1192 students altogether but only 887 students took all six tests. There are $2 \times 270 + 4 \times 35 = 680$ missing values out of $6 \times 1192 = 7152$ elements in the data matrix. Thus, there are 9.5% missing values in the data but this percentage varies considerably over the test occasions.

We will use the full information maximum likelihood method (FIML) for dealing with missing values. This method does not delete any cases, nor does it impute any missing values. Instead, for the given model and assuming multivariate normality, it maximizes the likelihood of the sample as it is. If all the observed variables are multivariate normal also every subset of these variables will be multivariate normal and its likelihood under the model can therefore be computed. So the log-likelihood can be computed for each case in the data. The total log-likelihood is the sum of the log-likelihood over all cases, For mathematical details, see Section 16.1.10. FIML is the default method of estimation when there is incomplete data, just as ML is the default method with complete data. Both methods are maximum likelihood methods but it is convenient to use the terms ML and FIML to distinguish between the cases of complete versus incomplete data.

The model to be estimated is shown in Figure 7.4

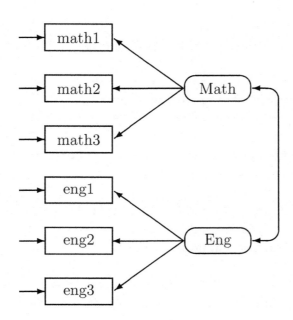

Figure 7.4: Path Diagram for CFA Model for Math and English Scores

As it is likely that students who are good in English are also good in Math, it is interesting to estimate the correlation between these two latent variables.

SIMPLIS Syntax

The **SIMPLIS** input file **jsp_aggregated1a.spl** are straight forward.

```
CFA Model for Math and English Scores
Raw Data from File jsp_aggregated.lsf
Latent Variables: Math Eng
Relationships
math1 - math3 = Math
eng1 - eng3 = Eng
Path Diagram
End of Problem
```

LISREL Syntax

The LISREL input file **jsp_aggregated1b.lis** is also straight forward

```
CFA Model for Math and English Scores
da ni=6
ra=jsp_aggregated.lsf
mo nx=6 nk=2
lk
Math Eng
fr lx(1,1) lx(2,1) lx(3,1) lx(4,2) lx(5,2) lx(6,2)
pd
ou
```

Both input files give the following information in the beginning of the output file:

```
        ---------------------------------
          EM Algorithm for missing Data:
        ---------------------------------


        Number of different missing-value patterns=        7
        Effective sample size:       1192

        Convergence of EM-algorithm in     4 iterations
        -2 Ln(L) =     46426.46767
        Percentage missing values=    9.51

    Note:
        The Covariances and/or Means to be analyzed are estimated
        by the EM procedure and are only used to obtain starting
        values for the FIML procedure
```

This clarifies that FIML has been used and can be explained as follows. To start iterations LISREL needs good starting values. If there is complete data, LISREL obtains these starting values automatically (except for very special models) for all methods of estimation. If there is incomplete data, LISREL begins by estimating the saturated model where μ and Σ are unconstrained, *i.e.*, the model $\mathbf{x} \sim N(\boldsymbol{\mu}, \boldsymbol{\Sigma})$ and LISREL then uses the estimated $\hat{\boldsymbol{\mu}}$ and $\hat{\boldsymbol{\Sigma}}$ to obtain the required starting values for the parameters. It took 4 EM iterations to obtain the estimated saturated model. This also gives the value of $2 \ln L$ for the saturated model, which is needed to obtain a FIML chi-square.

```
        -2 Ln(L) =     46426.46767
```

After this the iterations over the incomplete data can begin. We give the parameter estimates as they appear in the output file **jsp_aggregated1b.out**. The estimated $\boldsymbol{\Lambda}_x$ is:

LISREL Estimates (Maximum Likelihood)

LAMBDA-X

	Math	Eng
math1	6.096	- -
	(0.173)	
	35.321	
math2	6.740	- -
	(0.189)	
	35.751	
math3	5.664	- -
	(0.177)	
	31.928	
eng1	- -	22.094
		(0.573)
		38.580
eng2	- -	19.226
		(0.504)
		38.138
eng3	- -	19.554
		(0.558)
		35.061

The factor loadings for the Eng test are much higher than for the Math test, but this is just a reflection of the fact that these tests have different number of items. The estimated Φ matrix is

PHI

	Math	Eng
Math	1.000	
Eng	0.840	1.000
	(0.012)	
	67.281	

which shows that the estimated correlation between Math and Eng is 0.840 with a standard error of 0.012. This gives an approximate 95% confidence interval of $0.82 < \phi_{21} < 0.86$. Thus, it is not likely that the correlation is above 0.86.

The output file also gives the following fit measures:

```
                       Global Goodness of Fit Statistics, FIML case


                    -2ln(L) for the saturated model =        46426.468
                    -2ln(L) for the fitted model    =        46513.628

Degrees of Freedom = 8
Full Information ML Chi-Square                        87.160 (P = 0.0000)
Root Mean Square Error of Approximation (RMSEA)       0.0911
90 Percent Confidence Interval for RMSEA             (0.0744 ; 0.109)
P-Value for Test of Close Fit (RMSEA < 0.05)          0.000
```

indicating that the fit of the model is not good. We will therefore suggest an alternative model.

Modifying the Model

The modification indices for the model we have estimated suggests that there are many correlated measurement errors in $\boldsymbol{\delta}$. This is a common situation in longitudinal studies. But one should not just add error covariances for the purpose of getting a better fit. In the end one has to justify these error covariances and interpret them.

Many students in this study are in the same class and school. They are therefore exposed to the same environment, for example, same teachers and peers. This suggests that the errors of Math and Eng are correlated at least at the first test occasion. Furthermore, autocorrelation may be present for the same test over time. We therefore include correlated errors for the two adjacent occasions.

The model we propose has the following covariance pattern for $\boldsymbol{\Theta}_\delta$:

$$\boldsymbol{\Theta}_\delta = \begin{pmatrix} x & & & & & \\ x & x & & & & \\ 0 & x & x & & & \\ x & 0 & x & x & & \\ 0 & 0 & 0 & x & x & \\ 0 & 0 & 0 & 0 & x & x \end{pmatrix}, \tag{7.22}$$

where x is a free parameter to be estimated and 0 is a fixed xero.

In **LISREL** syntax one can specify this by a pattern matrix:

```
pa td
 1
 1 1
 0 1 1
 1 0 0 1
 0 0 0 1 1
 0 0 0 0 1 1
```

Alternatively, one can specify this by, see file **jsp_aggregated2b.lis**

```
fr td(4,1) td(2,1) td(3,2) td(5,4) td(6,5)
```

In **SIMPLIS** syntax one specifies these correlated errors as, see file **jsp_aggregated2a.spl**

```
Let the errors of math1 and eng1 correlate
Let the errors of math2 and math1 correlate
Let the errors of math3 and math2 correlate
Let the errors of eng2 and eng1 correlate
Let the errors of eng3 and eng2 correlate
Let the errors of math1 and eng1 correlate
```

If there are many correlated errors **LISREL** syntax has an advantage over **SIMPLIS** syntax for the model can be specified more compactly.

The output files are **jsp_aggregated2a.out** and **jsp_aggregated2b.out**. One can verify that all estimated error covariances are statistically significant. The fit measures for the modified models are

```
        -2ln(L) for the saturated model =        46426.468
        -2ln(L) for the fitted model    =        46444.658


Degrees of Freedom = 3
Full Information ML Chi-Square                 18.190 (P = 0.0004)
Root Mean Square Error of Approximation (RMSEA)    0.0652
90 Percent Confidence Interval for RMSEA      (0.0385 ; 0.0954)
P-Value for Test of Close Fit (RMSEA < 0.05)       0.160
```

showing much better fit. The *P*-value for the test of close fit may be considered acceptable.

FIML versus RML

A weakness of the FIML method is that the estimated standard errors and the FIML chi-square cannot be robustified as in the case of complete data. The FIML method depends on the assumption of multivariate normality and, at the time of writing, we are not aware of any study investigating the robustness of FIML against departure from normality.

The result of the data screening showed that the variables are highly non-normal. To get some indication of the severity of this problem we can get some indication from this example. Robust estimates can only be obtained from a complete dataset. Here we have the choice of working with the listwise sample of 887 cases or an imputed sample with 1192 cases. Since the imputation procedure also depends on the assumption of multivariate normality which is the assumption we want to avoid, we choose to use the listwise sample. The .lsf was created in 1 and will be called **jsp_aggregated_listwise.lsf**. To estimate the model for this dataset just change the file name accordingly, replace **FIML** in the second title line by **RML**, and add the line **RO** in the files **jsp_aggregated2a.spl** and **jsp_aggregated2b.lis**, see files **jsp_aggregated_listwise3a.spl** and **jsp_aggregated_listwise3b.lis**. In the **SIMPLIS** syntax the command for robust estimation is

```
Robust Estimation
```

The resulting chi-squares are now

```
Degrees of Freedom for (C1)-(C3)                        3
Maximum Likelihood Ratio Chi-Square (C1)                15.532 (P = 0.0014)
Browne's (1984) ADF Chi-Square (C2_NT)                  15.401 (P = 0.0015)
Browne's (1984) ADF Chi-Square (C2_NNT)                 14.937 (P = 0.0019)
Satorra-Bentler (1988) Scaled Chi-Square (C3)           14.300 (P = 0.0025)
Satorra-Bentler (1988) Adjusted Chi-Square (C4)         14.017 (P = 0.0027)
Degrees of Freedom for C4                               2.941
```

The C1 chi-square is somewhat smaller than the FIML chi-square but this may be a reflection of the smaller sample size. The small difference between C2_NT and C2_NNT indicates that the effect of non-normality is very minor.

The effect of non-normality can be seen in Table 7.7. It seems that the effect of non-normality on parameter estimates and standard errors are very small, but note that five out of seven standard errors are smaller for FIML than for RML.

Table 7.7: Parameter Estimates and Standard Errors for FIML and RML

Parameter		FIML	RML
math1 on Math	$\lambda_{11}^{(x)}$	5.993(0.183)	5.941(0.193)
math2 on Math	$\lambda_{21}^{(x)}$	6.387(0.215)	6.186(0.232)
math3 on Math	$\lambda_{31}^{(x)}$	5.640(0.185)	5.469(0.224)
eng1 on Eng	$\lambda_{42}^{(x)}$	20.956(0.619)	20.610(0.531)
eng2 on Eng	$\lambda_{52}^{(x)}$	19.088(0.578)	17.713(0.687)
eng3 on Eng	$\lambda_{62}^{(x)}$	20.497(0.575)	19.844(0.531)
Eng vs Math	ϕ_{21}	0.861(0.015)	0.857(0.017)

7.4   CFA with Ordinal Variables

There are many methods available for estimating structural equation models with ordinal variables, see Yang-Wallentin, Jöreskog, and Luo (2010) or Forero, Maydeu-Olivares, and Gallardo-Pujol (2009) and references given there. There are essentially two types of methods. One is a full information maximum likelihood method (FIML) using a probit, logit, or other link function. The other method fits the model to a matrix of polychoric correlation or covariance matrix using some fit function like ULS or DWLS. The focus is here on the two methods: FIML and DWLS. These methods are illustrated using confirmatory factor analysis models, but any LISREL model can be used, even those involving a mean structure.

7.4.1   Ordinal Variables without Missing Values

Estimation Using Adaptive Quadrature

Here we continue the analysis of the scitech data used in Chapter 6. The ML approach used there is only intended for exploratory factor analysis with one factor or two uncorrelated factors. For other cases one needs a more efficient way to evaluate the integrals and perform the iterations involved. For this we need to use adaptive quadrature.

For literature on applications of adaptive quadrature to CFA and other models, see Schilling and Bock (2005), Bryant and Jöreskog (2016) and Vasdekis, Cagnone, S., and Moustaki (2012).

Example: Attitudes towards Science and Technology

Following the results obtained by the exploratory factor analysis of the scitech data, we will use the confirmatory factor analysis shown in Figure 7.5

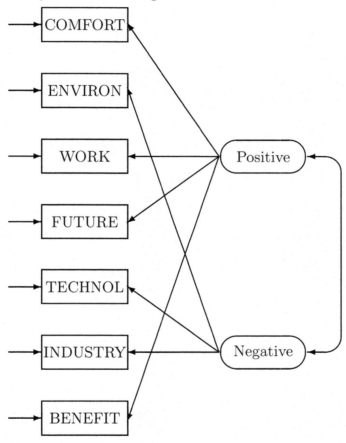

Figure 7.5: Path Diagram for CFA Model for Science and Technology Variables

To estimate this model by maximum likelihood use the following **SIMPLIS** syntax file **scitech3a.spl**:

```
Raw Data from file scitech.lsf
$ADAPQ(8) LOGIT
Latent Variables Positive Negative
Relationships
COMFORT WORK FUTURE BENEFIT = Positive
ENVIRON TECHNOL INDUSTRY   = Negative
End of Problem
```

For **LISREL** syntax use the file **scitech3b.lis**:

```
Scitech Run 3
da ni=7
ra=SCITECH.lsf
$ADAPQ(8) LOGIT GR(5)
mo nx=7 nk=2
lk
Positive Negative
fr lx(1,1) lx(2,2) lx(3,1) lx(4,1) lx(5,2) lx(6,2) lx(7,1)
pd
ou
```

The line

```
$ADAPQ(8) LOGIT
```

specifies that the adaptive quadrature procedure is to be used. One can specify the number of quadrature points to be used in the adaptive quadrature procedure and the link function. Follow the following guidelines to specify the number of quadrature points:

- For models with one latent variable, use 8 quadrature points

- For models with two or three latent variables, use 5-10 quadrature points

- For models with four or five quadrature points, use 5-6 quadrature points

- For models with six to ten latent variables use 3-4 quadrature points

The following link functions are available

- LOGIT

- PROBIT

- LOGLOG

- CLL (Complimentary log-log)

Both output files give

```
Number of quadrature points =               8
Number of free parameters =                29
Number of iterations used =                 8

-2lnL (deviance statistic) =       5841.79608
Akaike Information Criterion       5899.79608
Schwarz Criterion                  6014.96268
```

A weakness of the adapted quadrature approach is that it does not give a chi-square goodness-of-fit value. This is because there is no reasonable saturated model to compare its $-2\ln L$ value with. All one can do is to compare the values of $-2\ln L$ for different models.

The output file **scitech3b.out** gives the following parameter estimates

LISREL Estimates (Maximum Likelihood)

```
        LAMBDA-X

                Positive    Negative
                --------    --------
    COMFORT       1.046        - -
                 (0.190)
                  5.503


    ENVIRON         - -        1.622
                              (0.250)
                               6.490


       WORK       1.221        - -
                 (0.182)
                  6.715


     FUTURE       2.289        - -
                 (0.484)
                  4.725


    TECHNOL         - -        1.744
                              (0.279)
                               6.247


   INDUSTRY         - -        1.530
                              (0.242)
                               6.316


    BENEFIT       1.095        - -
                 (0.183)
                  5.970



        PHI

                Positive    Negative
                --------    --------
   Positive       1.000

   Negative       0.020       1.000
                 (0.081)
                  0.250
```

which shows that all factor loadings are statistically significant but the factor correlation is not. Hence, we can assume that the factors Positive and Negative are uncorrelated. The diagonal matrix Θ_δ appear in the output as

THETA-DELTA

COMFORT	ENVIRON	WORK	FUTURE	TECHNOL	INDUSTRY	BENEFIT
1.000	1.000	1.000	1.000	1.000	1.000	1.000

Since the ordinal variables have no scale there are no error terms in the model. The model is defined by equation (16.111) in the appendix (see Chapter 16). This model has unstandardized factor loadings β_{ij} and threshold parameters $\alpha_s^{(i)}$. So instead of Θ_δ we have estimated threshold parameters. They appear in the output as

Threshold estimates and standard deviations

Threshold	Estimates	S.E.	Est./S.E.
TH1_COMFORT	-4.86789	0.49163	-9.90162
TH2_COMFORT	-2.64287	0.22365	-11.81713
TH3_COMFORT	1.46795	0.15931	9.21422
TH1_ENVIRON	-3.43055	0.31632	-10.84525
TH2_ENVIRON	-1.25484	0.17941	-6.99409
TH3_ENVIRON	1.00321	0.16769	5.98253
TH1_WORK	-2.92080	0.23907	-12.21714
TH2_WORK	-0.90018	0.14270	-6.30835
TH3_WORK	2.26337	0.20288	11.15644
TH1_FUTURE	-5.23237	0.73195	-7.14854
TH2_FUTURE	-2.21286	0.35796	-6.18195
TH3_FUTURE	1.96282	0.32332	6.07088
TH1_TECHNOL	-4.16898	0.40088	-10.39964
TH2_TECHNOL	-1.49502	0.20337	-7.35119
TH3_TECHNOL	1.09062	0.18085	6.03045
TH1_INDUSTRY	-4.58387	0.42731	-10.72728
TH2_INDUSTRY	-2.42542	0.24091	-10.06765
TH3_INDUSTRY	0.45819	0.14732	3.11025
TH1_BENEFIT	-3.34883	0.27681	-12.09782
TH2_BENEFIT	-0.99225	0.14060	-7.05739
TH3_BENEFIT	1.68846	0.16873	10.00674

Since each observed ordinal variable has four categories, there are three thresholds for each variable.

Suppose we relax the loading of COMFORT on Negative, see file **scitech4b.lis**. As shown in the output file **scitech4b.out**

	Positive	Negative
COMFORT	1.094	0.443
	(0.199)	(0.159)
	5.506	2.775

This gives a solution where the estimate of $\lambda_{12}^{(x)}$ is statistically significant while at the same time all other loading are statistically significant. The deviance for this model is $-2\ln L = 5833.412$, compared with the previous $-2\ln L = 5841.800$. The difference 8,388 can be used as a chi-square for testing the hypothesis $\lambda_{12}^{(x)} = 0$. Clearly this hypothesis is rejected. This result can be interpreted as follows. Some fraction of people who generally have a positive attitude to science and technology will respond in the Strongly Disagree or Disagree to Some Extent categories to the COMFORT item.

Estimation using Polychoric Correlations

The FIML approach using adaptive quadrature has certain disadvantages:

- It does not provide an overall measure of fit.

- It does not give a completely standardized solution with standard errors

- Most importantly it is very time-consuming if the number of variables is large, particularly if the number of factors is large. In such cases it becomes impractical. Speed also depends on the number of categories, the fewer the better. For example, if all variables are dichotomous, it works much better.

An alternative approach which does not suffer from these problems is to use polychoric correlations and their asymptotic covariance matrix. This works well even for large number of variables and factors. As will be seen, this approaches give standardized solutions which are very close in the example. Although these is not a full maximum likelihood method, it seems to work well in practice.

As shown by Yang-Wallentin, Jöreskog, and Luo (2010) this approach can be used with ML or DWLS. We recommend using DWLS with robust estimation.

Given that the variables are defined as ordinal in the .lsf file, the estimation of the last model using SIMPLIS or LISREL syntax is straight forward. The SIMPLIS syntax file is **scitech5a.spl**:

```
Estimating the model using polychoric correlations
Raw Data from file scitech.lsf
Latent Variables Positive Negative
Relationships
COMFORT WORK FUTURE BENEFIT = Positive
COMFORT ENVIRON TECHNOL INDUSTRY   = Negative
Analyze Correlations
Robust Estimation
Method of Estimation: Diagonally Weighted Least Squares
Path Diagram
End of Problem
```

The corresponding LISREL syntax file is **scitech5b.lis**:

```
Estimating the model using polychoric correlations
da ni=7 ma=km
ra=SCITECH.lsf
mo nx=7 nk=2
lk
Positive Negative
fr lx(1,1) lx(1,2) lx(2,2) lx(3,1) lx(4,1) lx(5,2) lx(6,2) lx(7,1)
ro
pd
ou me=dwls
```

It should be understood that the two approaches estimate the same model but with methods based on different assumptions. The number of estimated parameters is the same in both approaches. But the parameters are not directly comparable. To compare them one has to obtain the completely standardized solution obtained as bi-product in the FIML approach and the DWLS solution which is already standardized. Here the FIML method assumes a logistic link function with intercept parameters allowed to vary across categories and variables but with slope parameters varying across variables but assumed to be the same for all categories. The polychoric approach is based on the assumption of underlying bivariate normality, This assumption can be tested using the information provided in the output.

Here we present some pieces of the output file **scitech5b.out**.

```
Total Sample Size(N) =      392
```

```
Univariate Marginal Parameters
```

Variable	Mean	St. Dev.	Thresholds		
COMFORT	0.000	1.000	-2.234	-1.314	0.749
ENVIRON	0.000	1.000	-1.447	-0.514	0.450
WORK	0.000	1.000	-1.377	-0.428	1.079
FUTURE	0.000	1.000	-1.803	-0.774	0.691
TECHNOL	0.000	1.000	-1.686	-0.589	0.464
INDUSTRY	0.000	1.000	-1.951	-1.056	0.219
BENEFIT	0.000	1.000	-1.611	-0.500	0.845

There are three estimated thresholds for each variable. These are estimated from the univariate marginal distributions

```
Univariate Distributions for Ordinal Variables
```

COMFORT	Frequency	Percentage
1	5	1.3
2	32	8.2
3	266	67.9
4	89	22.7

```
ENVIRON Frequency Percentage
   1        29        7.4
   2        90       23.0
   3       145       37.0
   4       128       32.7


  WORK  Frequency Percentage
   1        33        8.4
   2        98       25.0
   3       206       52.6
   4        55       14.0


 FUTURE Frequency Percentage
   1        14        3.6
   2        72       18.4
   3       210       53.6
   4        96       24.5


TECHNOL Frequency Percentage
   1        18        4.6
   2        91       23.2
   3       157       40.1
   4       126       32.1


INDUSTRY Frequency Percentage
   1        10        2.6
   2        47       12.0
   3       173       44.1
   4       162       41.3


BENEFIT Frequency Percentage
   1        21        5.4
   2       100       25.5
   3       193       49.2
   4        78       19.9
```

The tests of underlying bivariate normality is given as

Correlations and Test Statistics

(PE=Pearson Product Moment, PC=Polychoric, PS=Polyserial)

Variable vs. Variable	Correlation	Test of Model			Test of Close Fit	
		Chi-Squ.	D.F.	P-Value	RMSEA	P-Value
ENVIRON vs. COMFORT	0.099 (PC)	11.473	8	0.176	0.033	0.999
WORK vs. COMFORT	0.201 (PC)	21.505	8	0.006	0.066	0.951
WORK vs. ENVIRON	-0.083 (PC)	19.293	8	0.013	0.060	0.972
FUTURE vs. COMFORT	0.346 (PC)	9.316	8	0.316	0.020	1.000
FUTURE vs. ENVIRON	-0.028 (PC)	23.206	8	0.003	0.070	0.928

```
   FUTURE vs.      WORK  0.479 (PC)   9.497    8   0.302    0.022  1.000
  TECHNOL vs.   COMFORT  0.090 (PC)  18.633    8   0.017    0.058  0.977
  TECHNOL vs.   ENVIRON  0.464 (PC)  25.012    8   0.002    0.074  0.897
  TECHNOL vs.      WORK -0.104 (PC)  14.140    8   0.078    0.044  0.995
  TECHNOL vs.    FUTURE -0.036 (PC)  20.224    8   0.010    0.062  0.964
 INDUSTRY vs.   COMFORT  0.182 (PC)   7.637    8   0.470    0.000  1.000
 INDUSTRY vs.   ENVIRON  0.411 (PC)  35.824    8   0.000    0.094  0.588
 INDUSTRY vs.      WORK -0.008 (PC)  30.851    8   0.000    0.085  0.751
 INDUSTRY vs.    FUTURE  0.103 (PC)  27.730    8   0.001    0.079  0.837
 INDUSTRY vs.   TECHNOL  0.435 (PC)  44.382    8   0.000    0.108  0.313
  BENEFIT vs.   COMFORT  0.408 (PC)  13.815    8   0.087    0.043  0.996
  BENEFIT vs.   ENVIRON -0.037 (PC)  21.499    8   0.006    0.066  0.951
  BENEFIT vs.      WORK  0.209 (PC)  18.356    8   0.019    0.057  0.979
  BENEFIT vs.    FUTURE  0.377 (PC)  20.179    8   0.010    0.062  0.965
  BENEFIT vs.   TECHNOL -0.014 (PC)  14.915    8   0.061    0.047  0.994
  BENEFIT vs.  INDUSTRY  0.118 (PC)  20.728    8   0.008    0.064  0.959
```

For practical purposes use the P-values in the last column. None of these P-values reject underlying bivariate normality for any pair of variables.

The estimated polychoric correlation matrix is

```
        Correlation Matrix

          COMFORT  ENVIRON    WORK   FUTURE  TECHNOL INDUSTRY  BENEFIT
         -------- -------- -------- -------- -------- -------- --------
 COMFORT   1.000
 ENVIRON   0.099    1.000
    WORK   0.201   -0.083    1.000
  FUTURE   0.346   -0.028    0.479    1.000
 TECHNOL   0.090    0.464   -0.104   -0.036    1.000
INDUSTRY   0.182    0.411   -0.008    0.103    0.435    1.000
 BENEFIT   0.408   -0.037    0.209    0.377   -0.014    0.118    1.000
```

It took six iterations to minimize the fit function for DWLS to obtain the estimates that follow

```
Number of Iterations = 6

LISREL Estimates (Robust Diagonally Weighted Least Squares)

     LAMBDA-X

          Positive  Negative
         --------  --------
 COMFORT    0.528     0.197
          (0.074)   (0.076)
            7.167     2.595
```

ENVIRON	– –	0.661
		(0.069)
		9.609
WORK	0.536	– –
	(0.064)	
	8.310	
FUTURE	0.768	– –
	(0.061)	
	12.588	
TECHNOL	– –	0.692
		(0.061)
		11.365
INDUSTRY	– –	0.631
		(0.067)
		9.417
BENEFIT	0.535	– –
	(0.064)	
	8.322	

As judged by the Z-values, all factor loadings are statistically significant. The estimated factor correlation is very small and not statistically significant.

PHI

	Positive	Negative
	--------	--------
Positive	1.000	
Negative	-0.025	1.000
	(0.093)	
	-0.264	

The estimated variances of the δ_i are estimated as

THETA-DELTA

COMFORT	ENVIRON	WORK	FUTURE	TECHNOL	INDUSTRY	BENEFIT
--------	--------	--------	--------	--------	--------	--------
0.687	0.564	0.713	0.410	0.521	0.602	0.713
(0.127)	(0.136)	(0.122)	(0.138)	(0.132)	(0.132)	(0.122)
5.409	4.148	5.825	2.974	3.957	4.566	5.834

The robust DWLS gives the following goodness-of-fit chi-squares

Goodness-of-Fit Statistics

```
Degrees of Freedom for (C1)-(C3)                    12
Maximum Likelihood Ratio Chi-Square (C1)            47.235 (P = 0.0000)
Browne's (1984) ADF Chi-Square (C2_NT)              48.780 (P = 0.0000)
Browne's (1984) ADF Chi-Square (C2_NNT)             20.009 (P = 0.0669)
Satorra-Bentler (1988) Scaled Chi-Square (C3)       24.655 (P = 0.0165)
Satorra-Bentler (1988) Adjusted Chi-Square (C4)     21.719 (P = 0.0219)
Degrees of Freedom for C4                           10.571
```

As judged by these the model does not fit well but we shall not investigate this here.

It is of interest to compare the estimated factor loadings with this obtained in the standardized FIML estimates. These are shown in Table 7.8 It is seen that all FIML estimates are larger than

Table 7.8: Estimated Factor Loadings for FIML and DWLS

	FIML	DWLS
$\lambda_{11}^{(x)}$	0.712	0.528
$\lambda_{12}^{(x)}$	0.288	0.197
$\lambda_{22}^{(x)}$	0.854	0.661
$\lambda_{31}^{(x)}$	0.778	0.536
$\lambda_{41}^{(x)}$	0.913	0.768
$\lambda_{52}^{(x)}$	0.866	0.962
$\lambda_{62}^{(x)}$	0.837	0.631
$\lambda_{71}^{(x)}$	0.740	0.535
ϕ_{21}	-0.031	-0.025

the corresponding DWLS estimates with one single exception. This suggest that there might be some bias or inconsistency in the DWLS estimates. A conjecture is that this bias might be smaller if the probit link function had been used in the FIML estimation. A comprehensive simulation study is needed to settle these issues.

7.4.2   Ordinal Variables with Missing Values

To illustrate the analysis of ordinal variables in this section some data from the Political Action Survey will be used. This was a cross-national survey designed and carried out to obtain information on conventional and unconventional forms of political participation in industrial societies (Barnes & Kaase, 1979).

The first Political Action Survey was conducted between 1973 and 1975 in eight countries: Britain, West Germany, The Netherlands, Austria, the USA, Italy, Switzerland, and Finland. New cross-sections including a panel were obtained during 1980–81 in three of the original countries: West Germany, The Netherlands, and the USA. All data was collected through personal interviews on representative samples of the population 16 years and older[3].

The Political Action Survey contains several hundred variables. For the present purpose of illustration the six variables representing the operational definition of *political efficacy* will be used. These items have been previously analyzed by Aish and Jöreskog (1990), Jöreskog (1990), and Jöreskog and Moustaki (2001, 2006), among others Here we use the data from the first cross-section of the USA sample.

7.4.3   Example: Measurement of Political Efficacy

The conceptual definition of political efficacy *is the feeling that individual political action does have, or can have, an impact upon the political process* (Campbell, Gurin, & Miller, 1954). The operational definition of political efficacy is based on the responses to the following six items:[4]

NOSAY People like me have no say in what the government does

VOTING Voting is the only way that people like me can have any say about how the government runs things

COMPLEX Sometimes politics and government seem so complicated that a person like me cannot really understand what is going on

NOCARE we don't think that public officials care much about what people like me think

TOUCH Generally speaking, those we elect to Congress in Washington lose touch with the people pretty quickly

INTEREST Parties are only interested in people's votes but not in their opinions

Permitted responses to these statements were

AS agree strongly

A agree

D disagree

DS disagree strongly

DK don't know

NA no answer

---

[3]The data was made available by the Zentralarchiv für Empirische Sozialforschung, University of Cologne. The data was originally collected by independent institutions in different countries. Neither the original collectors nor the Zentralarchiv bear any responsibility for the analysis reported here.

[4]These are the questions that were used in the USA. In Britain, the same questions were used with *Congress in Washington* replaced by *Parliament*. In the other countries the corresponding questions were used in other languages.

These responses were coded 1, 2, 3, 4, 8, 9, respectively. The data used here is the USA sample from the 1973 the Political Action Survey which was a cross-national survey designed and carried out to obtain information on conventional and unconventional forms of political participation in industrial societies (Barnes & Kaase, 1979). The data file is **efficacy.dat**, a text file with spaces as delimiters.

Data Screening

Most raw data from surveys are downloaded from large files at data archives and stored on media like diskettes or tapes for analysis. The data file may contain many variables on many cases. Before doing more elaborate analysis of the data, it is important to do a careful data screening to check for coding errors and other mistakes in the data. Such a data screening will also reveal outliers and other anomalies, and detect if there are specific patterns of missing values in the data. The data screening gives a general idea of the character and quality of the data. To get a complete data screening of all values in the data file, use the following PRELIS command file, see **efficacy1.prl**:

```
Screening of Efficacy Data
DA NI=6
LA
NOSAY VOTING COMPLEX NOCARE TOUCH INTEREST
RA=EFFICACY.DAT
CL NOSAY-INTEREST 1=AS 2=A 3=D 4=DS 8=DK 9=NA
OU
```

PRELIS determines the sample size, all distinct data values for each variable and the absolute and relative frequency of occurrence of each value. The output file shows that there are 1719 cases in the data, that there are six distinct values on each variable, labeled AS, A, D, DS, DK, and NA, and the distribution of the data values over these categories.

The results are presented in compact form in Table 7.9.

Table 7.9: Univariate Marginal Distributions

	Frequency						Percentage					
	AS	**A**	**D**	**DS**	**DK**	**NA**	**AS**	**A**	**D**	**DS**	**DK**	**NA**
NOSAY	175	518	857	130	29	10	10.2	30.1	49.9	7.6	1.7	0.6
VOTING	283	710	609	80	26	11	16.5	41.3	35.4	4.7	1.5	0.6
COMPLEX	343	969	323	63	9	12	20.0	56.4	18.8	3.7	0.5	0.7
NOCARE	250	701	674	57	20	17	14.5	40.8	39.2	3.3	1.2	1.0
TOUCH	273	881	462	26	60	17	15.9	51.3	26.9	1.5	3.5	1.0
INTEREST	264	762	581	31	62	19	15.4	44.3	33.8	1.8	3.6	1.1

Note that there are more people responding *Don't Know* on Touch and Interest.

Obviously, the responses *Don't Know* and *No Answer* cannot be used as categories for the ordinal scale that goes from *Agree Strongly* to *Disagree Strongly*. To proceed with the analysis, one must first define the *Don't Know* and *No Answer* responses as missing values. This can be done by adding MI=8,9 on the DA line. In addition by adding RA=EFFICAY.LSF on the OU line, one will obtain a LISREL data system file **EFFICACY.LSF** which will serves as a basis for further analysis. The PRELIS command file now looks like this, see file **efficay2.prl**:

```
Creation of a lsf file for Efficacy Data
DA NI=6 MI=8,9
LA
NOSAY VOTING COMPLEX NOCARE TOUCH INTEREST
RA=EFFICACY.DAT
CL NOSAY-INTEREST 1=AS 2=A 3=D 4=DS
OU RA=EFFICACY.LSF
```

The first 15 lines of **efficacy.lsf** looks like this

efficacy.lsf	NOSAY	VOTING	COMPLEX	NOCARE	TOUCH	INTEREST
1	2.000	2.000	1.000	1.000	1.000	1.000
2	2.000	3.000	3.000	3.000	2.000	3.000
3	3.000	2.000	2.000	3.000	3.000	3.000
4	3.000	3.000	2.000	3.000	2.000	3.000
5	2.000	2.000	1.000	2.000	2.000	2.000
6	2.000	2.000	1.000	1.000	2.000	1.000
7	3.000	2.000	2.000	3.000	3.000	3.000
8	2.000	2.000	2.000	2.000	1.000	2.000
9	3.000	1.000	2.000	2.000	2.000	2.000
10	2.000	1.000	2.000	2.000	1.000	1.000
11	2.000	2.000	1.000	2.000	2.000	2.000
12	1.000	1.000	1.000	1.000	2.000	-999999.000
13	1.000	-999999.000	1.000	1.000	1.000	1.000
14	3.000	3.000	2.000	3.000	3.000	3.000
15	3.000	3.000	2.000	2.000	3.000	3.000

Note that missing values now appear as -999999.000 which is the global missing value code in **LISREL**.

To perform a data screening of **efficacy.lsf**, select **Data Screening** in the **Statistics** menu. This gives the following results.

The distribution of missing values over variables are given first.

```
Number of Missing Values per Variable

     NOSAY    VOTING    COMPLEX    NOCARE    TOUCH    INTEREST
        39        37         21        37       77          81
```

It is seen that there are only 21 missing values on COMPLEX whereas there are 77 and 81 on TOUCH and INTEREST, respectively. As we already know that most of the missing values on TOUCH and INTEREST are *Don't Know* rather than *No Answer* responses, it seems that these items are considered by the respondents to be more difficult to answer.

Next in the output is the distribution of missing values over cases.

```
Distribution of Missing Values

Total Sample Size =    1719

Number of Missing Values    0      1      2     3     4     5     6
         Number of Cases  1554    106     26    18     4     2     9
```

It is seen that there are only 1554 out of 1719 cases without any missing values. The other 165 cases have one or more missing values. With list-wise deletion this is the loss of sample size that will occur. Most, or 106, of the 165 cases with missing values have only one missing value. But note that there are 9 cases with 6 missing values, *i.e.*, these cases have either not responded or have responded *Don't Know* to all of the six items. These 9 cases are of course useless for any purpose considered here.

The next two tables of output give information about sample sizes for all variables and all pairs of variables. These sample sizes are given in absolute numbers as well as in percentages.

```
Effective Sample Sizes
Univariate (in Diagonal) and Pairwise Bivariate (off Diagonal)
```

	NOSAY	VOTING	COMPLEX	NOCARE	TOUCH	INTEREST
NOSAY	1680					
VOTING	1658	1682				
COMPLEX	1670	1674	1698			
NOCARE	1655	1656	1675	1682		
TOUCH	1620	1627	1635	1622	1642	
INTEREST	1619	1621	1632	1622	1598	1638

This table gives the univariate and bivariate sample sizes. Thus, there are 1680 cases with complete data on NOSAY but only 1638 cases with complete data on INTEREST. There are 1658 cases with complete data on *both* NOSAY and VOTING but only 1598 cases with complete data on *both* TOUCH and INTEREST.

The same kind of information, but in terms of percentage of missing data instead, is given in the following table.

```
Percentage of Missing Values
Univariate (in Diagonal) and Pairwise Bivariate (off Diagonal)
```

	NOSAY	VOTING	COMPLEX	NOCARE	TOUCH	INTEREST
NOSAY	2.27					
VOTING	3.55	2.15				
COMPLEX	2.85	2.62	1.22			
NOCARE	3.72	3.66	2.56	2.15		
TOUCH	5.76	5.35	4.89	5.64	4.48	
INTEREST	5.82	5.70	5.06	5.64	7.04	4.71

The next lines give all possible patterns of missing data and their sample frequencies. Each column under Pattern corresponds to a variable. A 0 means a complete data and a 1 means a missing data.

Missing Data Map

Frequency	PerCent	Pattern
1554	90.4	0 0 0 0 0 0
16	0.9	1 0 0 0 0 0
12	0.7	0 1 0 0 0 0
1	0.1	1 1 0 0 0 0
4	0.2	0 0 1 0 0 0
11	0.6	0 0 0 1 0 0
31	1.8	0 0 0 0 1 0
1	0.1	0 1 0 0 1 0
2	0.1	1 1 0 0 1 0
1	0.1	0 1 1 0 1 0
4	0.2	0 0 0 1 1 0
1	0.1	0 0 1 1 1 0
32	1.9	0 0 0 0 0 1
1	0.1	0 1 0 0 0 1
1	0.1	1 1 0 0 0 1
1	0.1	1 0 1 0 0 1
5	0.3	0 0 0 1 0 1
2	0.1	1 0 0 1 0 1
1	0.1	0 0 1 1 0 1
1	0.1	1 0 1 1 0 1
14	0.8	0 0 0 0 1 1
4	0.2	1 0 0 0 1 1
4	0.2	0 1 0 0 1 1
2	0.1	1 1 0 0 1 1
1	0.1	0 1 1 0 1 1
1	0.1	0 0 0 1 1 1
2	0.1	0 1 1 1 1 1
9	0.5	1 1 1 1 1 1

Thus, there are 1554 cases or 90.4% with no missing data, there are 16 cases or 0.9% with missing values on **NOSAY** only, and 1 case with missing values on both NOSAY and VOTING, etc. Note again that there are 9 cases with missing values on all 6 variables.

This kind of information is very effective in detecting specific patterns of missingness in the data. In this example there are no particular patterns of missingness. The only striking feature is that there are more missing values on **TOUCH** and **INTEREST**. We know from the first run that these are mainly *Don't know* responses.

The rest of the output (not shown here) gives the distribution of the 1554 cases of the listwise sample over the four ordinal categories for each variable. This shows that most people answer either *agree* or *disagree*. Fewer people answer with the *stronger* alternatives.

Estimating Models by FIML Using Adaptive Quadrature

First we investigate whether the six items measure one unidimensional latent variable, see Figure 7.6.

We begin with FIML, see **efficacy2a.spl**:

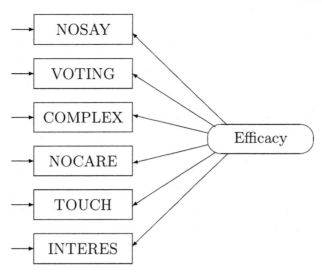

Figure 7.6: Measurement Model 1 for Political Efficacy

```
Efficacy: Model 1 Estimated by FIML
Raw Data from file EFFICACY.LSF
$ADAPQ(8) PROBIT
Latent Variable Efficacy
Relationships
NOSAY - INTEREST = Efficacy
End of Problem
```

The output gives the following factor loadings

```
    NOSAY = 0.739*Efficacy, Errorvar.= 1.000, R² = 0.353
Standerr  (0.0407)
Z-values   18.154
P-values   0.000


   VOTING = 0.377*Efficacy, Errorvar.= 1.000, R² = 0.124
Standerr  (0.0324)
Z-values   11.643
P-values   0.000


 COMPLEX = 0.601*Efficacy, Errorvar.= 1.000, R² = 0.265
Standerr  (0.0375)
Z-values   16.042
P-values   0.000


   NOCARE = 1.656*Efficacy, Errorvar.= 1.000, R² = 0.733
Standerr  (0.103)
Z-values   16.007
P-values   0.000


    TOUCH = 1.185*Efficacy, Errorvar.= 1.000, R² = 0.584
Standerr  (0.0632)
Z-values   18.754
P-values   0.000
```

```
INTEREST = 1.361*Efficacy, Errorvar.= 1.000, R² = 0.649
Standerr  (0.0744)
Z-values   18.290
P-values    0.000
```

Note the small loading on **VOTING**. This indicates very low validity and reliability of the **VOTING** item which might be explained as follows. If the six items really measure one unidimensional trait **Efficacy**, then people who are high on **Efficacy** are supposed to disagree or disagree strongly and people who are low on **Efficacy** should agree or agree strongly to all items. If this is the case, there would be a positive association between the latent variable **Efficacy** and each ordinal variable. But isn't something wrong with **VOTING**? If one is high on **Efficacy** and one believes that voting is the only way one can influence politics, then one would agree or agree strongly to the **VOTING** statement. This fact in itself is sufficient to suggest that the **VOTING** item should be excluded from further consideration.

The output also gives the following information:

```
Number of quadrature points =           8
Number of free parameters =            24
Number of iterations used =             7


-2lnL (deviance statistic) =      19934.56514
Akaike Information Criterion       19982.56514
Schwarz Criterion                  20113.22711
```

For the moment we note the value of the deviance statistic $-2 \ln L = 19934.465$. Since there is no value of $-2 \ln L$ for a saturated model, it is impossible to say whether this is large or small in some absolute sense. The deviance statistic can therefore only be used to compare different models for the same data.

The output also gives estimates of the thresholds, their standard errors and z-values. The thresholds are parameters of the model but are seldom useful in analysis of a single sample.

```
Threshold estimates and standard deviations
-----------------------------------------------
```

Threshold	Estimates	S.E.	Est./S.E.
TH1_NOSAY	-1.57282	0.05484	-28.67862
TH2_NOSAY	-0.26243	0.03847	-6.82165
TH3_NOSAY	1.74605	0.05834	29.92866
TH1_VOTING	-1.02350	0.03901	-26.23749
TH2_VOTING	0.24600	0.03301	7.45288
TH3_VOTING	1.78347	0.05658	31.52036
TH1_COMPLEX	-0.96727	0.04174	-23.17335
TH2_COMPLEX	0.87494	0.04015	21.79122
TH3_COMPLEX	2.04308	0.06656	30.69345
TH1_NOCARE	-2.01404	0.10803	-18.64379
TH2_NOCARE	0.35551	0.06034	5.89170
TH3_NOCARE	3.39121	0.16957	19.99895
TH1_TOUCH	-1.49811	0.06756	-22.17305

TH2_TOUCH	0.84076	0.05430	15.48431
TH3_TOUCH	3.21173	0.13297	24.15415
TH1_INTEREST	-1.65353	0.07801	-21.19712
TH2_INTEREST	0.56685	0.05494	10.31724
TH3_INTEREST	3.42422	0.15211	22.51147

It has been suggested in the political science literature that there are two components of Political Efficacy: Internal Efficacy (here called *Efficacy*) indicating *individuals self-perceptions that they are capable of understanding politics and competent enough to participate in political acts such as voting*, and External Efficacy (here called *Responsiveness* and abbreviated *Respons*) indicating *the belief that the public cannot influence political outcomes because government leaders and institutions are unresponsive* (Miller, W.E., Miller, A.H., & Schneider, 1980; Craig & Maggiotto, 1981). With this view, NOSAY and COMPLEX are indicators of Efficacy and TOUCH and INTEREST are indicators of Respons. The statement NOCARE contains two referents: *public officials* and *people like me.* This statement might elicit perceptions of the responsiveness of *government officials* to public opinion generally, in which case the emphasis is on the political actors, or it might express the opinions of *people like me* in which case the emphasis is on the respondent. In the first case, NOCARE measures Respons; in the second case, it measures Efficacy. I will therefore consider NOCARE as a complex variable, *i.e.*, as a variable measuring both Efficacy and Respons or a mixture of them. This is Model 2 shown in Figure 7.7.

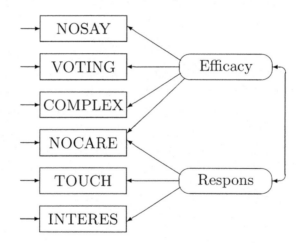

Figure 7.7: Measurement Model 2 for Political Efficacy

A SIMPLIS command file for Model 2 is, see file **efficacy3a.spl**:

```
Efficacy: Model 2 Estimated by FIML
Raw Data from file EFFICACY.LSF
$ADAPQ(8) PROBIT GR(5)
Latent Variables Efficacy Respons
Relationships
NOSAY - NOCARE = Efficacy
NOCARE - INTEREST = Respons
End of Problem
```

This gives the following estimated factor loadings

```
    NOSAY = 0.916*Efficacy, Errorvar.= 1.000, R² = 0.456
Standerr  (0.0601)
Z-values  15.253
P-values  0.000


   VOTING = 0.461*Efficacy, Errorvar.= 1.000, R² = 0.175
Standerr  (0.0385)
Z-values  11.981
P-values  0.000


 COMPLEX = 0.686*Efficacy, Errorvar.= 1.000, R² = 0.320
Standerr  (0.0455)
Z-values  15.091
P-values  0.000


   NOCARE = 0.821*Efficacy + 0.903*Respons, Errorvar.= 1.000, R² = 0.723
Standerr  (0.131)          (0.108)
Z-values  6.275            8.339
P-values  0.000            0.000


    TOUCH = 1.333*Respons, Errorvar.= 1.000, R² = 0.640
Standerr  (0.0778)
Z-values  17.138
P-values  0.000


 INTEREST = 1.607*Respons, Errorvar.= 1.000, R² = 0.721
Standerr  (0.107)
Z-values  14.996
P-values  0.000
```

Note that the loading of **NOCARE** on **Efficacy** is almost as large as that on **Respons**.

The correlation between the two components **Efficacy** and **Respons** is estimated as 0.75 as shown in the next part of the output:

```
        Correlation Matrix of Independent Variables

              Efficacy    Respons
              --------    --------

Efficacy       1.000

Respons        0.752      1.000
              (0.030)
              25.051
```

The correlation 0.75 is highly significant meaning that it is significant from 0. But more interestingly it is also significant from 1. An approximate 95% confidence interval for the correlation is from 0.69 to 0.81.

This model has two more parameters than the previous model. The deviance statistic for this model is 19858.058 as shown in the next part of the output:

```
Number of quadrature points =              8
Number of free parameters =                26
Number of iterations used =                8


-2lnL (deviance statistic) =        19858.05790
Akaike Information Criterion        19910.05790
Schwarz Criterion                   20051.60837
```

The difference between the deviance statistic for this model and the deviance statistic for the unidimensional model is 19934.565-19858.058=76.507, which suggests that Model 2 fits the data much better than Model 1.

Estimation by Robust Diagonally Weighted Least Squares (RDWLS)

An alternative approach to estimate models for ordinal variables from data with missing values is to impute the missing values first and then estimate the model by *Robust Unweighted Least Squares* (RULS) or *Robust Diagonally Weighted Least Squares* (RDWLS), see Yang-Wallentin, Jöreskog, and Luo (2010) or Forero, Maydeu-Olivares, and Gallardo-Pujol (2011).

Following the argument about the sf VOTING item on page 335 we now estimate the modified Model 2 in Figure 7.8

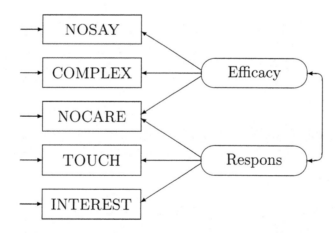

Figure 7.8: Modified Measurement Model 2 for Political Efficacy

This can be estimated using the following **SIMPLIS** syntax file, see file **efficacy4a.spl**:

```
Efficacy: Model 2 Estimated by Robust Diagonally Weighted Least Squares
Raw Data from file EFFICACY.LSF
Multiple Imputation with MC
Latent Variables Efficacy Respons
Relationships
NOSAY COMPLEX NOCARE = Efficacy
NOCARE - INTEREST = Respons
Robust Estimation
Method of Estimation: Diagonally Weighted Least Squares
Path Diagram
End of Problem
```

This approach is much faster than the FIML approach especially for large number of variables.

Chapter 8

Structural Equation Models (SEM) with Latent Variables

Factor analysis is used to investigate latent variables that are presumed to underlie a set of manifest variables. Understanding the structure and meaning of the latent variables in the context of their manifest variables is the main goal of traditional factor analysis. After a set of factors has been identified, it is natural to go on and use the factors themselves as predictors or outcome variables in further analyses. Broadly speaking, this is the goal of *structural equation modeling*.

A further extension of the classical factor analysis model is to allow the factors not only to be correlated, as in confirmatory factor analysis, but also to allow some latent variables to depend on other latent variables. Models of this kind are usually called structural equation models and there are many examples of this in the literature. For a recent bibliography see Wolfle (2003) and for the growth of structural equation modeling in 1994–2001, see Hershberger (2003).

The **LISREL** model presented in this chapter combines features of both econometrics and psychometrics into a single model. The first **LISREL** model was a linear structural equation model for latent variables, each with a single observed, possibly fallible, indicator, see Jöreskog (1973). This model was generalized to models with multiple indicators of latent variables, to simultaneous structural equation models in several groups, and to more general covariance structures, see Jöreskog (1974, 1981).

8.1   Example: Hypothetical Model

To illustrate the main features of a full **LISREL** model, we have created a hypothetical model. This is shown in **SIMPLIS** format in the path diagram in Figure 8.1.

In such a model the latent variables are classified into dependent and independent latent variables. The dependent latent variables are those which depend on other latent variables. In the path diagram these have one or more one-way (uni-directed) arrows pointing towards them. The independent latent variables are those which do not depend on other latent variables. In the path diagram they have no one-way arrows pointing to them. In Figure 8.1, **Eta1** and **Eta2** are the dependent latent variables and **Ksi1**, **Ksi2**, and **Ksi3** are the independent latent variables. The latter are freely correlated whereas the variances and covariances of **Eta1** and **Eta2** depend on their relationships with **Ksi1**, **Ksi2** and **Ksi3**.

© Springer International Publishing Switzerland 2016

K.G. Jöreskog et al., *Multivariate Analysis with LISREL*, Springer Series in Statistics,

DOI 10.1007/978-3-319-33153-9_8

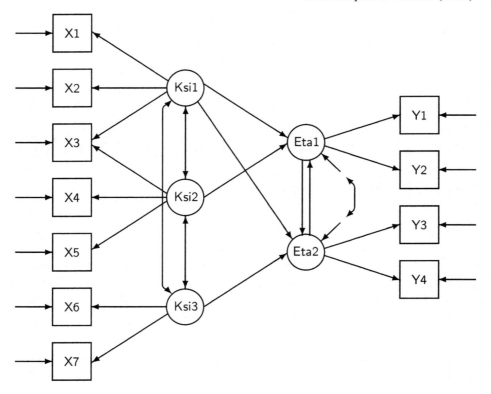

Figure 8.1: Path Diagram for a Hypothetical LISREL Model

The observed variables are also classified into two categories: y-variables and x-variables. The y-variables are those observed variables that depend on the dependent latent variables, and the x-variables are those that depend on the independent latent variables. In the path diagram in Figure 8.1, there are four y-variables Y1, Y2, Y3, Y4 and seven x-variables X1, X2, X3, X4, X5, X6, X7. There is a confirmatory factor model on the left side for the x-variables and a confirmatory factor model on the right side for the y-variables.

The relations in the middle of the path diagram represent a structural equation model with two jointly dependent variables **Eta1** and **Eta2** and independent variables **Ksi1, Ksi2, Ksi3**. This is of the same form as covered in Section 2.3 but with latent variables instead of observed variables.

There is an error term on each relationship (equation). For the observed variables, these are the uni-directed arrows on the left and the right side. The two structural relationships also have error terms on them. The hypothetical model illustrates the idea of reciprocal causation between **Eta1** and **Eta2** with two correlated error terms indicated by the two-way arrow in the middle.

8.1.1   Hypothetical Model with SIMPLIS Syntax

The model is hypothetical so there is no data for it. Instead of giving a full **SIMPLIS** syntax file we present only the **SIMPLIS** command lines that specify the model.

```
Latent Variables: Eta1 Eta2 Ksi1-Ksi3
Relationships
    Eta1 = Eta2 Ksi1 Ksi2
    Eta2 = Eta1 Ksi1 Ksi3
Let the Errors of Eta1 and Eta2 Correlate
    Y1 Y2 = Eta1
    Y3 Y4 = Eta2
    X1 - X3 = Ksi1
    X3 - X5 = Ksi2
    X6 - X7 = Ksi3
```

Note the following:

- The names of the variables, observed as well as latent, can be anything. **LISREL** figures out which latent variables are dependent and which are independent. Similarly, **LISREL** figures out which are y-variables and which are x-variables. This follows from the relationships.

- All error terms are automatically included and they are all uncorrelated by default. The correlated structural error terms are specified by the line

```
Let the Errors of Eta1 and Eta2 Correlate
```

  All error variances and this specified error covariance will be estimated.

- All latent variables are standardized by default. Since nothing is known about scale of the latent variables this seems most reasonable. See the next few sections for specifying other unit of measurement for the latent variables.

8.2 The General LISREL Model in LISREL Format

The hypothetical model in **LISREL** format is shown in the path diagram in Figure 8.2.

As before, the latent variables are classified into dependent and independent latent variables. The dependent latent variables are those which depend on other latent variables. In the path diagram these have one or more one-way arrows pointing towards them. The independent latent variables are those which do not depend on other latent variables. In the path diagram they have no one-way arrows pointing to them. In Figure 8.2, η_1 and η_2 are the dependent latent variables and ξ_1, ξ_2, and ξ_3 are the independent latent variables. The latter are freely correlated whereas the variances and covariances of η_1 and η_2 depend on their relationships with ξ_1, ξ_2, and ξ_3.

The observed variables are also classified into two categories: y-variables and x-variables. The y-variables are those observed variables that depend on the dependent latent variables, and the x-variables are those that depend on the independent latent variables. In Figure 8.2, there is a confirmatory factor model on the left side for the x-variables and a confirmatory factor model on the right side for the y-variables.

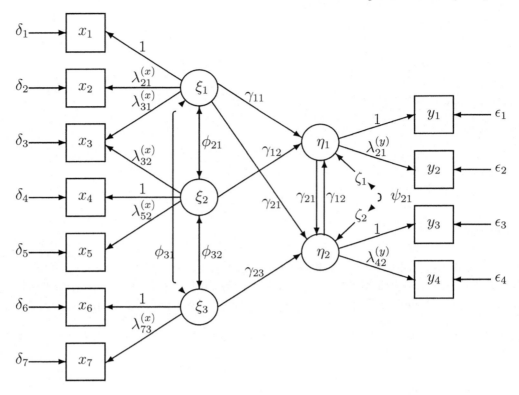

Figure 8.2: Path Diagram of a General LISREL Model

8.3   General Framework

In its most general form the LISREL model is defined as follows. Consider random vectors $\boldsymbol{\eta} = (\eta_1, \eta_2, \ldots, \eta_m)'$ and $\boldsymbol{\xi} = (\xi_1, \xi_2, \ldots, \xi_n)'$ of latent dependent and independent variables, respectively, and the following system of linear structural relations

$$\boldsymbol{\eta} = \boldsymbol{\alpha} + \mathbf{B}\boldsymbol{\eta} + \boldsymbol{\Gamma}\boldsymbol{\xi} + \boldsymbol{\zeta} \,, \tag{8.1}$$

where $\boldsymbol{\alpha}$ is a vector of intercept terms, $\mathbf{B}$ and $\boldsymbol{\Gamma}$ are coefficient matrices and $\boldsymbol{\zeta} = (\zeta_1, \zeta_2, \ldots, \zeta_m)'$ is a random vector of residuals (errors in equations, random disturbance terms). The elements of $\mathbf{B}$ represent direct effects of η-variables on other η-variables and the elements of $\boldsymbol{\Gamma}$ represent direct effects of ξ-variables on η-variables. It is assumed that $\boldsymbol{\zeta}$ is uncorrelated with $\boldsymbol{\xi}$ and that $\mathbf{I} - \mathbf{B}$ is non-singular.

Vectors $\boldsymbol{\eta}$ and $\boldsymbol{\xi}$ are not observed, but instead vectors $\mathbf{y} = (y_1, y_2, \ldots, y_p)'$ and $\mathbf{x} = (x_1, x_2, \ldots, x_q)'$ are observed, such that

$$\mathbf{y} = \boldsymbol{\tau}_y + \boldsymbol{\Lambda}_y\boldsymbol{\eta} + \boldsymbol{\epsilon} \,, \tag{8.2}$$

and

$$\mathbf{x} = \boldsymbol{\tau}_x + \boldsymbol{\Lambda}_x\boldsymbol{\xi} + \boldsymbol{\delta} \,, \tag{8.3}$$

where $\boldsymbol{\epsilon}$ and $\boldsymbol{\delta}$ are vectors of error terms (errors of measurement or measure-specific components) assumed to be uncorrelated with $\boldsymbol{\eta}$ and $\boldsymbol{\xi}$, respectively. These equations represent the multivariate regressions of $\mathbf{y}$ on $\boldsymbol{\eta}$ and of $\mathbf{x}$ on $\boldsymbol{\xi}$, respectively, with $\boldsymbol{\Lambda}_y$ and $\boldsymbol{\Lambda}_x$ as regression matrices and $\boldsymbol{\tau}_y$ and $\boldsymbol{\tau}_x$ as vectors of intercept terms. It is convenient to refer to $\mathbf{y}$ and $\mathbf{x}$ as the observed variables and $\boldsymbol{\eta}$ and $\boldsymbol{\xi}$ as the latent variables.

Let $\boldsymbol{\kappa}$ be the mean vector of $\boldsymbol{\xi}$, $\boldsymbol{\Phi}$ and $\boldsymbol{\Psi}$ the covariance matrices of $\boldsymbol{\xi}$ and $\boldsymbol{\zeta}$, $\boldsymbol{\Theta}_\epsilon$ and $\boldsymbol{\Theta}_\delta$ the covariance matrices of $\boldsymbol{\epsilon}$ and $\boldsymbol{\delta}$, and $\boldsymbol{\Theta}_{\delta\epsilon}$ the covariance matrix between $\boldsymbol{\delta}$ and $\boldsymbol{\epsilon}$. Then it follows that the mean vector $\boldsymbol{\mu}$ and covariance matrix $\boldsymbol{\Sigma}$ of $\mathbf{z} = (\mathbf{y}', \mathbf{x}')'$ are

$$\boldsymbol{\mu} = \left(\begin{array}{c} \boldsymbol{\tau}_y + \boldsymbol{\Lambda}_y(\mathbf{I} - \mathbf{B})^{-1}(\boldsymbol{\alpha} + \boldsymbol{\Gamma}\boldsymbol{\kappa}) \\ \boldsymbol{\tau}_x + \boldsymbol{\Lambda}_x\boldsymbol{\kappa} \end{array} \right) , \tag{8.4}$$

$$\boldsymbol{\Sigma} = \left(\begin{array}{cc} \boldsymbol{\Lambda}_y\mathbf{B}^\star(\boldsymbol{\Gamma}\boldsymbol{\Phi}\boldsymbol{\Gamma}' + \boldsymbol{\Psi})\mathbf{B}^{\star'}\boldsymbol{\Lambda}_y' + \boldsymbol{\Theta}_\epsilon & \boldsymbol{\Lambda}_y\mathbf{B}^\star\boldsymbol{\Gamma}\boldsymbol{\Phi}\boldsymbol{\Lambda}_x' + \boldsymbol{\Theta}_{\delta\epsilon}' \\ \boldsymbol{\Lambda}_x\boldsymbol{\Phi}\boldsymbol{\Gamma}'\mathbf{B}^{\star'}\boldsymbol{\Lambda}_y' + \boldsymbol{\Theta}_{\delta\epsilon} & \boldsymbol{\Lambda}_x\boldsymbol{\Phi}\boldsymbol{\Lambda}_x' + \boldsymbol{\Theta}_\delta \end{array} \right) , \tag{8.5}$$

where $\mathbf{B}^\star = (\mathbf{I} - \mathbf{B})^{-1}$.

The elements of $\boldsymbol{\mu}$ and $\boldsymbol{\Sigma}$ are functions of the elements of $\boldsymbol{\kappa}$, $\boldsymbol{\alpha}$, $\boldsymbol{\tau}_y$, $\boldsymbol{\tau}_x$, $\boldsymbol{\Lambda}_y$, $\boldsymbol{\Lambda}_x$, $\mathbf{B}$, $\boldsymbol{\Gamma}$, $\boldsymbol{\Phi}$, $\boldsymbol{\Psi}$, $\boldsymbol{\Theta}_\epsilon$, $\boldsymbol{\Theta}_\delta$, and $\boldsymbol{\Theta}_{\delta\epsilon}$ which are of three kinds:

- *fixed parameters* that have been assigned specified values,

- *constrained parameters* that are unknown but linear or non-linear functions of one or more other parameters, and

- *free parameters* that are unknown and not constrained.

8.3.1 Scaling of Latent Variables

Latent variables are unobservable and have no definite scales. Both the origin and the unit of measurement in each latent variable are arbitrary. To define the model properly, the origin and the unit of measurement of each latent variable must be defined.

Typically, there are two ways in which this is done. The most useful and convenient way of assigning the units of measurement of the latent variables is to assume that they are standardized so that they have zero means and unit variances in the population. **LISREL** does that by default.

Another way to assign a unit of measurement for a latent variable, is to fix a non-zero coefficient (usually one) in the relationship for one of its observed indicators. This defines the unit for each latent variable in relation to one of the observed variables, a so-called **reference variable**. In practice, one chooses as reference variable the observed variable, which, in some sense, best represents the latent variable.

LISREL defines the units of latent variables as follows

- If a reference variable is assigned to a latent variable by the specification of a fixed non-zero coefficient in the relationship between the reference variable and the latent variable, then this defines the scale for that latent variable.

- If no reference variable is specified for a latent variable, by assigning a fixed non-zero coefficient for an observed variable, the program will standardize this latent variable.

Further considerations are necessary when the observed variables in the model are ordinal, because then, even the observed variables do not have any units of measurement. Other considerations are necessary in longitudinal and multiple group studies in order to put the variables on the same scale at different occasions and in different groups. In such studies one can also relax the assumption of zero means for the latent variables.

8.3.2 Notation for LISREL Syntax

For **LISREL** syntax, the notation for the parameter matrices are summarized in Tables 8.1 and 8.2.

Table 8.1: Parameter Matrices in LISREL: Their Possible Forms and Default Values

Name	Math Symbol	LISREL Name	Order	Possible Forms	Default Form	Default Mode
LAMBDA-Y	$\mathbf{\Lambda}_y$	LY	NY × NE	ID,IZ,ZI,DI,FU	FU	FI
LAMBDA-X	$\mathbf{\Lambda}_x$	LX	NX × NK	ID,IZ,ZI,DI,FU	FU	FI
BETA	$\mathbf{B}$	BE	NE × NE	ZE,SD,FU	ZE	FI
GAMMA	$\mathbf{\Gamma}$	GA	NE × NK	ID,IZ,ZI,DI,FU	FU	FR
PHI	$\mathbf{\Phi}$	PH	NK × NK	ID,DI,SY,ST	SY	FR
PSI	$\mathbf{\Psi}$	PS	NE × NE	ZE,DI,SY	SY	FR
THETA-EPSILON	$\mathbf{\Theta}_\epsilon$	TE	NY × NY	ZE,DI,SY	DI	FR
THETA-DELTA	$\mathbf{\Theta}_\delta$	TD	NX × NX	ZE,DI,SY	DI	FR
THETA-DELTA-EPSILON	$\mathbf{\Theta}_{\delta\epsilon}$	TH	NX × NY	ZE	ZE	FI

The default value for the intercept and mean parameters in Table 8.2 is **0** (**ZE**). Technically it works like this, if **TY** and **TX** are default, **LISREL** sets $\boldsymbol{\tau}_y = \bar{y}$ and $\boldsymbol{\tau}_x = \bar{x}$ the sample mean vectors of the y- and x-variables, respectively. This implies that only the sample covariance matrix **S** is needed to estimate the model. For examples where intercepts and means are useful, see Chapter 9.

One can make any number of specifications of the form

```
... MN = AA,BB
```

Table 8.2: Intercept and Mean Parameter Matrices in LISREL

Name	Math Symbol	Order	LISREL Name	Possible Modes†
TAU-Y	$\boldsymbol{\tau}_y$	NY × 1	TY	FI,FR
TAU-X	$\boldsymbol{\tau}_x$	NX × 1	TX	FI,FR
ALPHA	$\boldsymbol{\alpha}$	NE × 1	AL	FI,FR
KAPPA	$\boldsymbol{\kappa}$	NK × 1	KA	FI,FR

† Since these are vectors, Form is not relevant, only Mode is

on the MO line, where MN is a matrix name (column 3), AA is a matrix form (column 5) and BB is FR (free) or FI (fixed) (column 7). Either AA or BB may be omitted in which case the defaults of Table 8.1 are used. Any element in the parameter matrices can be specified to be fixed or free by listing the element on a FI or a FR line.

8.4   Special Cases of the General LISREL Model

There are a number of useful special cases of the general LISREL model. Some of these have been covered in earlier chapters. Some others are presented in this chapter.

1. The multivariate general linear model. This is the model

$$\mathbf{y} = \boldsymbol{\alpha} + \boldsymbol{\Gamma}\mathbf{x} + \mathbf{z} \tag{8.6}$$

   defined in Chapter 2. This can be used for univariate and multivariate regression, ANOVA, ANCOVA, MANOVA, and MANCOVA. One can also include the term $\mathbf{By}$ on the right side to do recursive and nonrecursive models. In LISREL syntax, these models are recognized by omitting NE and NK on the MO line.

2. The factor model. This is the model

$$\mathbf{x} = \boldsymbol{\tau}_x + \boldsymbol{\Lambda}_x\boldsymbol{\xi} + \boldsymbol{\delta} \tag{8.7}$$

   defined in Chapter 7. This is used for confirmatory factor analysis and latent growth curves, see Chapter 9. In LISREL syntax, this model is recognized by omitting NY and NE on the MO line.

3. The model with $\boldsymbol{\eta}$, $\boldsymbol{\xi}$, and $\mathbf{y}$ but no $\mathbf{x}$. The second-order factor analysis model considered in Section 8.6 is an example of this. In LISREL syntax, this model is recognized by omitting NX on the MO line.

4. The model with $\boldsymbol{\eta}$, $\mathbf{y}$ and $\mathbf{x}$ but no $\boldsymbol{\xi}$. The MIMIC model in Section 8.8 is an example of this. This model is recognized by omitting NK on the MO line.

In SIMPLIS syntax it is even more simple. One will just specify the latent variables and write the equations one wants to estimate. LISREL will automatically figure out the model specification.

8.4.1   Matrix Specification of the Hypothetical Model

The structural equations are

$$\begin{aligned}
\eta_1 &= \beta_{12}\eta_2 + \gamma_{11}\xi_1 + \gamma_{12}\xi_2 + \zeta_1 \\
\eta_2 &= \beta_{21}\eta_1 + \gamma_{21}\xi_1 + \gamma_{23}\xi_3 + \zeta_2
\end{aligned}$$

or in matrix form

$$\begin{pmatrix} \eta_1 \\ \eta_2 \end{pmatrix} = \begin{pmatrix} 0 & \beta_{12} \\ \beta_{21} & 0 \end{pmatrix} \begin{pmatrix} \eta_1 \\ \eta_2 \end{pmatrix} + \begin{pmatrix} \gamma_{11} & \gamma_{12} & 0 \\ \gamma_{21} & 0 & \gamma_{23} \end{pmatrix} \begin{pmatrix} \xi_1 \\ \xi_2 \\ \xi_3 \end{pmatrix} + \begin{pmatrix} \zeta_1 \\ \zeta_2 \end{pmatrix}$$

This corresponds to the structural equation model (8.1) but with intercepts omitted.

The measurement model equations for y-variables are

$$
\begin{aligned}
y_1 &= \eta_1 + \epsilon_1 \\
y_2 &= \lambda_{21}^{(y)} \eta_1 + \epsilon_2 \\
y_3 &= \eta_2 + \epsilon_3 \\
y_4 &= \lambda_{42}^{(y)} \eta_2 + \epsilon_4
\end{aligned}
$$

or in matrix form

$$\begin{pmatrix} y_1 \\ y_2 \\ y_3 \\ y_4 \end{pmatrix} = \begin{pmatrix} 1 & 0 \\ \lambda_{21}^{(y)} & 0 \\ 0 & 1 \\ 0 & \lambda_{42}^{(y)} \end{pmatrix} \begin{pmatrix} \eta_1 \\ \eta_2 \end{pmatrix} + \begin{pmatrix} \epsilon_1 \\ \epsilon_2 \\ \epsilon_3 \\ \epsilon_4 \end{pmatrix}$$

and the measurement model equations for x-variables are

$$
\begin{aligned}
x_1 &= \xi_1 + \delta_1 \\
x_2 &= \lambda_{21}^{(x)} \xi_1 + \delta_2 \\
x_3 &= \lambda_{31}^{(x)} \xi_1 + \lambda_{32}^{(x)} \xi_2 + \delta_3 \\
x_4 &= \xi_2 + \delta_4 \\
x_5 &= \lambda_{52}^{(x)} \xi_2 + \delta_5 \\
x_6 &= \xi_3 + \delta_6 \\
x_7 &= \lambda_{73}^{(x)} \xi_3 + \delta_7
\end{aligned}
$$

or in matrix form

$$\begin{pmatrix} x_1 \\ x_2 \\ x_3 \\ x_4 \\ x_5 \\ x_6 \\ x_7 \end{pmatrix} = \begin{pmatrix} 1 & 0 & 0 \\ \lambda_{21}^{(x)} & 0 & 0 \\ \lambda_{31}^{(x)} & \lambda_{32}^{(x)} & 0 \\ 0 & 1 & 0 \\ 0 & \lambda_{52}^{(x)} & 0 \\ 0 & 0 & 1 \\ 0 & 0 & \lambda_{73}^{(x)} \end{pmatrix} \begin{pmatrix} \xi_1 \\ \xi_2 \\ \xi_3 \end{pmatrix} + \begin{pmatrix} \delta_1 \\ \delta_2 \\ \delta_3 \\ \delta_4 \\ \delta_5 \\ \delta_6 \\ \delta_7 \end{pmatrix}$$

These equations correspond to (8.2) and (8.3), respectively, but with intercept terms omitted.

One λ in each column of $\mathbf{\Lambda}_y$ and $\mathbf{\Lambda}_x$ has been set equal to 1 to fix the scales of measurement in the latent variables.

In these equations, note that the second subscript on each coefficient is always equal to the subscript of the variable that follows the coefficient. This can serve as a check that everything is correct. Furthermore, in the matrices $\mathbf{B}$, $\mathbf{\Gamma}$, $\mathbf{\Lambda}_y$, and $\mathbf{\Lambda}_x$, subscripts on each coefficient, which were originally defined in the path diagram, now correspond to the row and column of the matrix in which they appear. Also note that arrows which are not included in the path diagram correspond to zeros in these matrices.

Each of the parameter matrices contains fixed elements (the zeros and ones) and free parameters (the coefficients with two subscripts).

The four remaining parameter matrices are the symmetric matrices

$$\mathbf{\Phi} = \begin{pmatrix} \phi_{11} & & \\ \phi_{21} & \phi_{22} & \\ \phi_{31} & \phi_{32} & \phi_{33} \end{pmatrix},$$

the covariance matrix of $\mathbf{\xi}$,

$$\mathbf{\Psi} = \begin{pmatrix} \psi_{11} & \\ \psi_{21} & \psi_{22} \end{pmatrix},$$

the covariance matrix of $\mathbf{\zeta}$, and the diagonal matrices

$$\mathbf{\Theta}_\epsilon = diag(\theta_{11}^{(\epsilon)}, \theta_{22}^{(\epsilon)}, \ldots, \theta_{44}^{(\epsilon)}),$$

the covariance matrix of $\mathbf{\epsilon}$ and

$$\mathbf{\Theta}_\delta = diag(\theta_{11}^{(\delta)}, \theta_{22}^{(\delta)}, \ldots, \theta_{77}^{(\delta)}),$$

the covariance matrix of $\mathbf{\delta}$.

8.4.2 LISREL syntax for the Hypothetical Model

The hypothetical model can be specified by the following **LISREL** syntax lines

```
LE
Eta1 Eta2
LK
Ksi1 Ksi2 Ksi3
MO NY=4 NX=7 NE=2 NK=3 BE=FU PS=SY,FR
FR LY(2,1) LY(4,2)
FR LX(2,1) LX(3,1) LX(3,2) LX(5,2) LX(7,3)
FR BE(2,1) BE(1,2)
FI GA(1,3) GA(2,2)
VA 1 LY(1,1) LY(3,2) LX(1,1) LX(4,2) LX(6,3)
```

To explain this syntax, $\mathbf{\Lambda}_x$ and $\mathbf{\Lambda}_y$ are default, so they are fixed. The elements to be estimated in these matrices are specified on the two first **FR** lines. Since $\mathbf{B}$ is not $\mathbf{0}$, we set **BE=FU** on the **MO** line and the elements to be estimated β_{21} and β_{12} are specified on the next **FR** line. In the **SIMPLIS** syntax considered previously we standardized the latent variables. Here we use the other scaling of the latent variables mentioned in Section 8.3.1 and fix an element in each column of $\mathbf{\Lambda}_x$ and $\mathbf{\Lambda}_y$ to 1. These fixed values are specified on the **VA** line.

8.5   Measurement Errors in Regression

Consider the regression of y on x:

$$y = \gamma_{y.x} x + z \ , \tag{8.8}$$

and suppose x is measured with error, so that

$$x = \xi + \delta \ , \tag{8.9}$$

where δ is the measurement error and ξ is the true value of x. Suppose we are interested in the relationship between y and ξ:

$$y = \gamma \xi + \zeta \ . \tag{8.10}$$

The γ in (8.10) is not the same as $\gamma_{y.x}$ in (8.8). We say that $\gamma_{y.x}$ in (8.8) is a *regression parameter* and γ in (8.10) is a *structural parameter*. Assuming that ξ, ζ, and δ are mutually uncorrelated, the covariance matrix $\boldsymbol{\Sigma}$ of (y, x) is

$$\boldsymbol{\Sigma} = \begin{pmatrix} \gamma^2 \phi + \psi & \\ \gamma\phi & \phi + \theta \end{pmatrix}$$

where $\phi = \mathrm{Var}(\xi)$, $\psi = \mathrm{Var}(\zeta)$, and $\theta = \mathrm{Var}(\delta)$. The regression coefficient $\gamma_{y.x}$ in (8.8) is

$$\gamma_{y.x} = \frac{Cov(y, x)}{Var(x)} = \frac{\gamma\phi}{\phi + \theta} = \gamma\frac{\phi}{\phi + \theta} = \gamma\rho_{xx} \tag{8.11}$$

where $\rho_{xx} = \phi/(\phi + \theta)$ is the reliability of x. It is seen in (8.11) that the regression parameter $\gamma_{y.x}$ is not equal to the structural parameter γ but is smaller than γ if $\theta > 0$. If one estimates γ by the sample regression coefficient of y on x, one obtains an estimate which is biased downwards. This bias does not decrease as the sample size increases, i.e., $\hat{\gamma}_{y.x}$ is *not a consistent estimate* of γ.

Equation (8.11) suggests one way to *dis-attenuated* $\hat{\gamma}_{y.x}$ if the reliability ρ_{xx} of x is known, namely to use the estimate

$$\hat{\gamma} = \hat{\gamma}_{y.x}/\rho_{xx} \ . \tag{8.12}$$

In this context we refer to $\hat{\gamma}_{y.x}$ as the *attenuated* coefficient and $\hat{\gamma}$ as the *disattenuated* coefficient.

In practice, ρ_{xx} is not known but has to be estimated from data. If r_{xx} is a consistent estimate of ρ_{xx},

$$\hat{\gamma} = \hat{\gamma}_{y.x}/r_{xx} \tag{8.13}$$

can be used to estimate γ. This reduces the bias in $\hat{\gamma}_{y.x}$ at the expense of an increased sampling variance.

8.5.1   Example: Verbal Ability in Grades 4 and 5

Härnqvist (1962) provided data on a 40-item similarities test for 262 boys who were tested first in grade 4 (x) and then in grade 5 (y). The covariance matrix is

$$\mathbf{S} = \begin{array}{cc} y & x \\ \begin{pmatrix} 46.886 & \\ 45.889 & 59.890 \end{pmatrix} \end{array} .$$

The reliability of the test was estimated at $r_{xx} = 0.896$. The regression estimate is

$$\hat{\gamma}_{y.x} = \frac{45.889}{59.890} = 0.766 .$$

Corrected for attenuation, this becomes

$$\hat{\gamma} = \frac{0.766}{0.896} = 0.855 .$$

The issue considered above can be extended to the case of multiple regression with several explanatory variables x_1, x_2, ..., x_q. Consider the relationship

$$y = \gamma_1 \xi_1 + \gamma_2 \xi_2 + \ldots + \gamma_q \xi_q + \zeta ,$$

where $\xi_i = x_i - \delta_i$. The estimator of $\boldsymbol{\gamma}$ corresponding to (8.13) is

$$\hat{\boldsymbol{\gamma}} = (\mathbf{S}_{xx} - \hat{\boldsymbol{\Theta}}_\delta)^{-1} \mathbf{s}_{y.x} \tag{8.14}$$

where $\mathbf{S}_{xx}$ is the sample covariance matrix of the x's, $\hat{\boldsymbol{\Theta}}_\delta$ is a diagonal matrix of estimated error variances in the x's, and $\mathbf{s}_{y.x}$ is a vector of sample covariances between y and the x's. Formula (8.14) may not work well in practice because $\mathbf{S}_{xx}$ - $\hat{\boldsymbol{\Theta}}_\delta$ will not always be positive definite.

8.5.2 Example: Role Behavior of Farm Managers

Warren, White, and Fuller (1974) report on a study wherein a random sample of 98 managers of farmer cooperatives operating in Iowa was selected with the objective of studying managerial behavior.

The role behavior of a manager in farmer cooperatives, as measured by his Role Performance, was assumed to be linearly related to the four variables:

- x_1: Knowledge of economic phases of management directed toward profit-making in a business and product knowledge

- x_2: Value Orientation: tendency to rationally evaluate means to an economic end

- x_3: Role Satisfaction: gratification obtained by the manager from performing the managerial role

- x_4: Past Training: amount of formal education

To measure Knowledge, sets of questions were formulated by specialists in the relevant fields of economics and fertilizers and chemicals. The measure of rational Value Orientation to economic ends was a set of 30 items administrated to respondents by questionnaire on which the respondents were asked to indicate the strength of their agreement or disagreement. The respondents indicated the strength of satisfaction or dissatisfaction for each of 11 statements covering four areas of satisfaction:

Table 8.3: Covariance Matrix for Example: Role Behavior of Farm Managers

	y	x_1	x_2	x_3	x_4
ROLBEHAV	0.0209				
KNOWLEDG	0.0177	0.0520			
VALORIEN	0.0245	0.0280	0.1212		
ROLSATIS	0.0046	0.0044	-0.0063	0.0901	
TRAINING	0.0187	0.0192	0.0353	-0.0066	0.0946

(1) managerial role itself, (2) the position, (3) rewards and (4) performance of complementary role players. The amount of past training was the total number of years of formal schooling divided by six.

Role Performance was measured with a set of 24 questions covering the five functions of planning, organizing, controlling, coordinating and directing. The recorded verbal responses of managers on how they performed given tasks were scored by judges on a scale of 1 to 99 on the basis of performance leading to successful management. Responses to each question were randomly presented to judges and the raw scores were transformed by obtaining the "Z" value for areas of 0.01 to 0.99 from a cumulative normal distribution (a raw score of 40 received transformed score of -0.253). For each question, the mean of transformed scores of judges was calculated. The covariance matrix of the five variables is given in Table 8.3

Since the raw data is no longer available we use covariance matrices as input in this example. Ignoring measurement errors in the x-variables, the ordinary least squares regression estimates with standard errors can be obtained using the following syntax file, see file **rolebehavior1a.spl**:

```
Observed Variables: ROLBEHAV  KNOWLEDG  VALORIEN  ROLSATIS  TRAINING
Covariance Matrix:
0.0209
0.0177  0.0520
0.0245  0.0280  0.1212
0.0046  0.0044 -0.0063  0.0901
0.0187  0.0192  0.0353 -0.0066  0.0946
Sample Size: 98
Regress ROLBEHAV on KNOWLEDG  VALORIEN  ROLSATIS  TRAINING
```

This gives the following estimated regression

```
ROLBEHAV = 0.23*KNOWLEDG + 0.12*VALORIEN + 0.056*ROLSATIS + 0.11*TRAINING
Standerr   (0.053)         (0.036)         (0.037)          (0.039)
Z-values   4.32            3.37            1.50             2.81
P-values   0.000           0.001           0.133            0.005
```

Now suppose the reliabilities of the x-variables are

$$0.60, 0.64, 0.81, 1.00$$

and suppose we take these numbers to be known values. The error variance is 1 minus the reliability times the observed variance of x. This gives the error variances 0.0208, 0.0436, 0.0171,

0.0000. Subtracting these values from the observed variances give the following syntax file, see file **rolebehavior2a.spl**.

```
Observed Variables: ROLBEHAV  KNOWLEDG  VALORIEN  ROLSATIS  TRAINING
Covariance Matrix:
0.0209
0.0177  0.0312
0.0245  0.0280  0.0776
0.0046  0.0044 -0.0063  0.0730
0.0187  0.0192  0.0353 -0.0066  0.0946
Sample Size: 98
Regress ROLBEHAV on KNOWLEDG  VALORIEN  ROLSATIS  TRAINING
```

This is the same as before except for the diagonal elements of the covariance matrix for the last four variables. We can then re-estimate the regression to reduce or eliminate the effects of measurement errors. The re-estimated regression equation is

```
ROLBEHAV = 0.38*KNOWLEDG + 0.15*VALORIEN + 0.059*ROLSATIS + 0.068*TRAINING
Standerr  (0.069)        (0.045)        (0.037)        (0.035)
Z-values  5.49           3.40           1.61           1.91
P-values  0.000          0.001          0.108          0.056
```

Some of these estimates and standard errors differ considerably between the two solutions. Note that the regression coefficients for the fallible x-variables are all larger than the corresponding regression coefficients in the first solution, whereas the regression coefficient of x_4, which was assumed to be without error, is smaller than in the first solution. Thus measurement error in one variable can have an effect on a regression coefficient of another variable.

The problem with this analysis is that the error variances must be known *a priori*. First, these error variances cannot be estimated from the information provided in the covariance matrix if there is only one measure x for each explanatory.

To estimate the effect of measurement error in the observed variables, Rock et al. (1977) split each of the measures y, x_1, x_2 and x_3 randomly into two parallel halves. The full covariance matrix of all the split-halves is given in Table 8.4 (the number of items in each split-half is given in parentheses). This covariance matrix is available in the file **rock.cm**. This can be used to estimate the true regression equation

$$\eta = \gamma_1\xi_1 + \gamma_2\xi_2 + \gamma_3\xi_3 + \gamma_4\xi_4 + \zeta \tag{8.15}$$

using the following measurement models

$$\begin{pmatrix} y_1 \\ y_2 \end{pmatrix} = \begin{pmatrix} 1 \\ 1 \end{pmatrix} \eta + \begin{pmatrix} \epsilon_1 \\ \epsilon_2 \end{pmatrix} \tag{8.16}$$

$$\begin{pmatrix} x_{11} \\ x_{12} \\ x_{21} \\ x_{22} \\ x_{31} \\ x_{32} \\ x_4 \end{pmatrix} = \begin{pmatrix} 1 & 0 & 0 & 0 \\ 1 & 0 & 0 & 0 \\ 0 & 1 & 0 & 0 \\ 0 & 1 & 0 & 0 \\ 0 & 0 & 1 & 0 \\ 0 & 0 & 1.2 & 0 \\ 0 & 0 & 0 & 1 \end{pmatrix} \begin{pmatrix} \xi_1 \\ \xi_2 \\ \xi_3 \\ \xi_4 \end{pmatrix} + \begin{pmatrix} \delta_{11} \\ \delta_{12} \\ \delta_{21} \\ \delta_{22} \\ \delta_{31} \\ \delta_{32} \\ 0 \end{pmatrix} \tag{8.17}$$

Table 8.4: Covariance Matrix for Split-halves

		y_1	y_2	x_{11}	x_{12}	x_{21}	x_{22}	x_{31}	x_{32}	x_4
y_1	(12)	.0271								
y_2	(12)	.0172	.0222							
x_{11}	(13)	.0219	.0193	.0876						
x_{12}	(13)	.0164	.0130	.0317	.0568					
x_{21}	(15)	.0284	.0294	.0383	.0151	.1826				
x_{22}	(15)	.0217	.0185	.0356	.0230	.0774	.1473			
x_{31}	(5)	.0083	.0011	-.0001	.0055	-.0087	-.0069	.1137		
x_{32}	(6)	.0074	.0015	.0035	.0089	-.0007	-.0088	.0722	.1024	
x_4		.0180	.0194	.0203	.0182	.0563	.0142	-.0056	-.0077	.0946

The value 1.2 in the last equation reflects the fact that x_{32} has six items whereas x_{31} has only five.

The latent variables are

- η = role behavior

- ξ_1 = knowledge

- ξ_2 = value orientation

- ξ_3 = role satisfaction

- ξ_4 = past training

The observed variables are

- y_1 = a split-half measure of role behavior

- y_2 = a split-half measure of role behavior

- x_{11}= a split-half measure of knowledge

- x_{12}= a split-half measure of knowledge

- x_{21}= a split-half measure of value orientation

- x_{22}= a split-half measure of value orientation

- x_{31}= a split-half measure of role satisfaction

- x_{32}= a split-half measure of role satisfaction

- x_4 = ξ_4= a measure of past training

The input file **rolehavior3b.lis** is

```
Role Behavior of Farm Managers, Part B
DA NI=9 NO=98
CM=rock.cm
LA
Y1 Y2 X11 X12 X21 X22 X31 X32 X4
MO NY=2 NE=1 NX=7 NK=4
FI TD 7
VA 1 LY 1 LY 2 LX 1 1 LX 2 1 LX 3 2 LX 4 2 LX 5 3 LX 7 4
VA 1.2 LX 6 3
OU SE
```

The fit of the model is $\chi^2 = 26.97$ with 22 degrees of freedom, which represents a rather good fit. The ML estimates of the γ's and their standard errors (below) are

$$\hat{\gamma}' = \begin{matrix} (0.350 & 0.168 & 0.045 & 0.071). \\ 0.133 & 0.079 & 0.054 & 0.045 \end{matrix}$$

Comparing these with the ordinary least squares (OLS) estimates, previously given, for the regression of y on x_1, x_2, x_3 and x_4, it is seen that there is considerable bias in the OLS estimates although their standard errors are smaller.

Estimates of the true and error score variances for each observed measure are also obtained. These can be used to compute the reliabilities of the composite measures. The reliability estimates are

$$\begin{matrix} y & x_1 & x_2 & x_3 \\ 0.82 & 0.60 & 0.64 & 0.81 \end{matrix}$$

The model defined by (5.11) - (5.13) can be generalized directly to the case when there are several jointly dependent variables $\boldsymbol{\eta}$. The only differences will be that λ and $\boldsymbol{\gamma}$ will be replaced by matrices $\boldsymbol{\Lambda}_y$ and $\boldsymbol{\Gamma}$, respectively, and ψ by a full symmetric positive definite matrix $\boldsymbol{\Psi}$.

8.6   Second-Order Factor Analysis

The factors in an oblique factor solution may depend other factors, so called second-order factors. These in turn may depend on still higher-order factors. Among the first to discuss these ideas are Schmidt and Leiman (1957). An example of a second-order factor analysis model is shown in Figure 8.3.

In this model there are four latent variables η_1, η_2, η_3, and ξ and there is a dependence structure among them such that η_1, η_2, and η_3 depend on ξ. In equation form this model is given by two sets of equations;

$$\mathbf{y} = \boldsymbol{\Lambda}_y\boldsymbol{\eta} + \boldsymbol{\epsilon}\,, \tag{8.18}$$

$$\boldsymbol{\eta} = \boldsymbol{\Gamma}\boldsymbol{\xi} + \boldsymbol{\zeta}\,, \tag{8.19}$$

where $\mathbf{y} = (y_1, y_2, \ldots, y_9)'$, $\boldsymbol{\eta} = (\eta_1, \eta_2, \eta_3)'$ and $\boldsymbol{\xi} = \xi$. The horizontal arrows on the left side represent the measurement errors $\boldsymbol{\epsilon} = (\epsilon_1, \epsilon_2, \ldots, \epsilon_9)'$ and the vertical arrows in the middle represent the structural errors $\boldsymbol{\zeta} = (\zeta_1, \zeta_2, \zeta_3)$. The latent variables η_1, η_2, and η_3 are first-order factors and the latent variable ξ is a second-order factor. The factor loadings of the first order factors are in

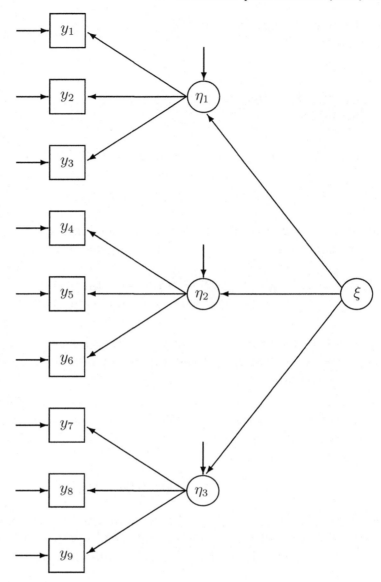

Figure 8.3: Path Diagram of a Second-order Factor Analysis Model

the 9×3 matrix $\boldsymbol{\Lambda}_y$ and the factor loadings of the second-order factor are in the 3×1 matrix $\boldsymbol{\Gamma}$. In Figure 8.3, $\mathbf{y}$ is supposed to satisfy a confirmatory factor model with three indicators of each of the first-order factors and $\boldsymbol{\eta}$ is supposed to satisfy a confirmatory factor model with one factor.

8.6.1   Example: Second-Order Factor of Nine Psychological Variables

To illustrate second-order factor analysis we use again the same data on nine psychological variables used previously in Chapter 7. The data file is **npv.lsf**.

The **SIMPLIS** syntax is intuitively very simple, see file **npvsecorder1a.spl**:

```
Estimation of the Second-Order NPV Model by Maximum Likelihood
Raw Data from File npv.lsf
Latent Variables: Visual Verbal Speed General
Relationships:
    VISPERC - LOZENGES = Visual
    PARCOMP - WORDMEAN = Verbal
    ADDITION - SCCAPS = Speed
    Visual - Speed = General
Path Diagram
End of Problem
```

This is a CFA model for the nine test scores VISPERC - SCCAPS with first order factor Visual - Speed and a factor model for Visual - Speed with second-order factor General. This is Spearman's g appearing as a second-order factor.

The corresponding input file in **LISREL** syntax is **npvsecorder1b.lis**:

```
Estimation of the NPV Model by Maximum Likelihood
DA NI=9
RA=NPV.LSF
MO NY=9 NE=3 NK=1 LX=FR
LE
Visual Verbal Speed
LK
General
FR LY(1,1) LY(2,1) LY(3,1) LY(4,2) LY(5,2) LY(6,2)LY(7,3) LY(8,3) LY(9,3)
PD
OU
```

Note the following

- The variables in this model are $\mathbf{x}$ $\boldsymbol{\eta}$ and $\boldsymbol{\xi}$, *but there are no x-variables.* Consequently NX is not specified on the MO line.

- Many **LISREL** users believe that one has to fix a non-zero value, usually 1, in each column of $\boldsymbol{\Lambda}_y$ and $\boldsymbol{\Lambda}_x$, but this is not necessary. **LISREL** will automatically scale all latent variables so that they have variance 1. Other types of scaling the latent variables is only necessary if one wants to define the unit of measurement in the latent variables in relation to observed variables.

- Although the solution is standardized in the sense that the latent variables are standardized, it is not a completely standardized solution for the observed variables are analyzed in their original units of measurements. To obtain a completely standardized solution put SC on the OU line, or alternatively put MA=KM on the DA line. In SIMPLIS syntax one can request a completely standardized solution by adding a line

```
Options SC
```

or alternatively add the line

```
Analyze Correlations
```

In both the SIMPLIS output file **npvsecorder1a.out** and the LISREL output file **npvsecorder1b.out** the covariance matrix of the latent variables appear as

```
Covariance Matrix of ETA and KSI

              Visual      Verbal       Speed     General
            --------    --------    --------    --------
   Visual     1.000
   Verbal     0.541       1.000
    Speed     0.523       0.336       1.000
  General     0.917       0.589       0.570       1.000
```

In this case this is a correlation matrix.

In **npvsecorder1a.out** the second order factor ladings appear as

```
   Visual = 0.917*General, Errorvar.= 0.158 , R² = 0.842
Standerr   (0.176)                     (0.246)
Z-values   5.225                        0.644
P-values   0.000                        0.519

   Verbal = 0.589*General, Errorvar.= 0.653 , R² = 0.347
Standerr   (0.117)                     (0.145)
Z-values   5.037                        4.488
P-values   0.000                        0.000

    Speed = 0.570*General, Errorvar.= 0.675 , R² = 0.325
Standerr   (0.132)                     (0.199)
Z-values   4.327                        3.385
P-values   0.000                        0.001
```

and in **npvsecorder1b.out** they appear as

GAMMA

```
                General
                --------
    Visual       0.917
                (0.176)
                 5.225


    Verbal       0.589
                (0.117)
                 5.037


     Speed       0.570
                (0.132)
                 4.327
```

Both output files give the chi-squares C1 and C2_NT as

```
Degrees of Freedom for (C1)-(C2)                    24
Maximum Likelihood Ratio Chi-Square (C1)            51.542 (P = 0.0009)
Browne's (1984) ADF Chi-Square (C2_NT)              48.952 (P = 0.0019)
```

Note that these are the same chi-squares as in **npv1a.out** and **npv1b.out** in Chapter 7. Interestingly the first-order factor model and the second-order factor model has the same fit. The reason for this is that one factor will always fit the covariance matrix of three variables perfectly. So the second-order factor model is a tautology.

8.7   Analysis of Correlation Structures

The model for a correlation structure is defined by the correlation matrix $\mathbf{P}(\boldsymbol{\theta})$, with diagonal elements equal to 1, *i.e.*, the diagonal elements are not functions of parameters. It may be formulated as a covariance structure as follows

$$\boldsymbol{\Sigma} = \mathbf{D}_\sigma \mathbf{P}(\boldsymbol{\theta}) \mathbf{D}_\sigma , \tag{8.20}$$

where $\mathbf{D}_\sigma$ is a diagonal matrix of population standard deviations $\sigma_1, \sigma_2, \ldots, \sigma_k$ of the observed variables, which are regarded as free parameters. The covariance structure (8.20) has parameters $\sigma_1, \sigma_2, \ldots, \sigma_k, \theta_1, \theta_2, \ldots, \theta_t$. Such a model may be estimated correctly using the sample covariance matrix $\mathbf{S}$. However, the standard deviations $\sigma_1, \sigma_2, \ldots, \sigma_k$, as well as $\boldsymbol{\theta}$ must be estimated from the data and the estimate of σ_i does not necessarily equal the corresponding standard deviation s_i in the sample.

Illustrating the case of $k = 4$, here are some examples of correlation structures
Example 1

$$\mathbf{P} = \begin{pmatrix} 1 & & & \\ \rho_{21} & 1 & & \\ \rho_{31} & \rho_{32} & 1 & \\ \rho_{41} & \rho_{42} & \rho_{43} & 1 \end{pmatrix} \tag{8.21}$$

with $\boldsymbol{\theta} = (\rho_{21}, \rho_{31}, \rho_{32}, \rho_{41}, \rho_{42}, \rho_{43})$.

Example 2

$$\mathbf{P} = \begin{pmatrix} 1 \\ \rho_{21} & 1 \\ \rho_{31} & \rho_{21} & 1 \\ \rho_{41} & \rho_{31} & \rho_{21} & 1 \end{pmatrix} \tag{8.22}$$

with $\boldsymbol{\theta} = (\rho_{21}, \rho_{31}, \rho_{41})$.

Example 3

$$\mathbf{P} = \begin{pmatrix} 1 \\ \rho & 1 \\ \rho^2 & \rho & 1 \\ \rho^3 & \rho^2 & \rho & 1 \end{pmatrix} \tag{8.23}$$

with $\boldsymbol{\theta} = (\rho)$. Obviously, example 3 is a special case of 2 and 2 is a special case of 1.

Each of these models can be estimated using LISREL syntax but example 2 requires equality constraints and example 3 requires more general constraints.

8.7.1   Example: CFA Model for NPV Estimated from Correlations

We now return to the problem of estimating a CFA model to obtain correct standard errors when the sample correlation matrix **R** is used.

Recall the LISREL syntax file **npv5b.lis** used in Chapter 7. For the purpose here we list this file as **npv8b.lis**

```
Estimation of the NPV Model by Maximum Likelihood
Using Correlations
DA NI=9 MA=KM
RA=NPV.LSF
MO NX=9 NK=3
LK
Visual Verbal Speed
FR LX(1,1) LX(2,1) LX(3,1) LX(4,2) LX(5,2) LX(6,2)LX(7,3) LX(8,3) LX(9,1) LX(9,3)
PD
OU
```

The only difference between **npv5b.lis** and **npv8b.lis** is that the line RO is omitted in **npv8b.lis**. In **npv8b.lis** we estimate the modified CFA model for the NPV data by ML fitted to the sample correlation matrix. Unfortunately, this gives incorrect standard errors

To obtain correct standard errors for the standardized factor loadings, one must use the following **npv8bb.lis**:

```
Estimation of the NPV Model by Maximum Likelihood
Using Correlations
DA NI=9 MA=KM
RA=NPV.LSF
MO NY=9 NE=9 NK=3 LY=DI,FR GA=FI PH=ST PS=DI TE=ZE
LK
Visual Verbal Speed
FR GA(1,1) GA(2,1) GA(3,1) GA(4,2) GA(5,2) GA(6,2)GA(7,3) GA(8,3) GA(9,1) GA(9,3)
CO PS(1 1) = 1 - GA(1 1) ** 2
CO PS(2 2) = 1 - GA(2 1) ** 2
CO PS(3 3) = 1 - GA(3 1) ** 2
CO PS(4 4) = 1 - GA(4 2) ** 2
CO PS(5 5) = 1 - GA(5 2) ** 2
CO PS(6 6) = 1 - GA(6 2) ** 2
CO PS(7 7) = 1 - GA(7 3) ** 2
CO PS(8 8) = 1 - GA(8 3) ** 2
CO PS(9 9) = 1 - GA(9 1)**2 - 2*GA(9 1)*GA(9 3)*PH(3 1) - GA(9 3)**2
OU AD=OFF SO
```

The factor loadings with correct standard errors can now be found in the GAMMA matrix. It may verified that these two different formulations of the same confirmatory factor analysis model give exactly the same parameter estimates and fit statistics; only the standard errors and t-values differ. It can also be verified that if MA=KM is omitted on the DA line, so that the sample covariance matrix $\mathbf{S}$ is analyzed instead of the correlation matrix $\mathbf{R}$, then the factor loadings and factor correlations will be the same; only the estimates of the population standard deviations will differ.

To explain the input file **npv8bb.lis**, the model (8.20) is written in LISREL notation as

$$\boldsymbol{\Sigma} = \boldsymbol{\Lambda}_y(\boldsymbol{\Gamma}\boldsymbol{\Phi}\boldsymbol{\Gamma}' + \boldsymbol{\Psi})\boldsymbol{\Lambda}_y' , \tag{8.24}$$

where $\boldsymbol{\Lambda}_y$ is the diagonal matrix $\mathbf{D}_\sigma$, and

$$\mathbf{P}(\boldsymbol{\theta}) = \boldsymbol{\Gamma}\boldsymbol{\Phi}\boldsymbol{\Gamma}' + \boldsymbol{\Psi} , \tag{8.25}$$

where

$$\boldsymbol{\Gamma} = \begin{pmatrix} \gamma_{11} & 0 & 0 \\ \gamma_{21} & 0 & 0 \\ \gamma_{31} & 0 & 0 \\ 0 & \gamma_{42} & 0 \\ 0 & \gamma_{52} & 0 \\ 0 & \gamma_{62} & 0 \\ 0 & 0 & \gamma_{73} \\ 0 & 0 & \gamma_{83} \\ \gamma_{91} & 0 & \gamma_{93} \end{pmatrix} , \tag{8.26}$$

$$\boldsymbol{\Phi} = \begin{pmatrix} 1 & & \\ \phi_{21} & 1 & \\ \phi_{31} & \phi_{32} & 1 \end{pmatrix} , \tag{8.27}$$

and the diagonal matrix $\boldsymbol{\Psi}$ is chosen such that $\boldsymbol{\Psi}$ has 1's in the diagonal, *i.e.*,

$$\psi_{11} = 1 - \gamma_{11}^2$$

$$\psi_{22} = 1 - \gamma_{21}^2$$

$$\psi_{33} = 1 - \gamma_{31}^2$$

$$\psi_{44} = 1 - \gamma_{42}^2$$

$$\psi_{55} = 1 - \gamma_{52}^2$$

$$\psi_{66} = 1 - \gamma_{62}^2$$

$$\psi_{77} = 1 - \gamma_{73}^2$$

$$\psi_{88} = 1 - \gamma_{83}^2$$

$$\psi_{99} = 1 - \gamma_{91}^2 - 2\gamma_{91}\gamma_{93}\phi_{31} - \gamma_{93}^2$$

The parameter vector $\boldsymbol{\theta}$ is

$$\boldsymbol{\theta}' = \left(\gamma_{11}, \gamma_{21}, \gamma_{31}, \gamma_{42}, \gamma_{52}, \gamma_{62}, \gamma_{73}, \gamma_{83}, \gamma_{91}, \gamma_{93}, \phi_{21}, \phi_{31}, \phi_{32} \right) . \tag{8.28}$$

Unless there is a Heywood case, no matter what values the parameters in $\boldsymbol{\theta}$ has

$$\boldsymbol{\Gamma}\boldsymbol{\Phi}\boldsymbol{\Gamma}' + \boldsymbol{\Psi} \tag{8.29}$$

will always be a correlation matrix with 1's in the diagonal.

Table 8.5 summarizes the results from **npv8b.out** and **npv8bb.out** regarding standard errors. It can be seen that **npv8b.out** uniformly overestimates the correct standard errors in **npv8bb.out**.

Table 8.5: Parameter Estimates and Standard Errors based on Correlations

Parameter	Estimate	Standard Errors	
Factor Loadings		**npv8b.out**	**npv8bb.out**
$\lambda_{11}^{(x)}$	0.711	0.086	0.063
$\lambda_{21}^{(x)}$	0.489	0.090	0.078
$\lambda_{31}^{(x)}$	0.662	0.087	0.066
$\lambda_{42}^{(x)}$	0.867	0.070	0.032
$\lambda_{52}^{(x)}$	0.829	0.071	0.035
$\lambda_{62}^{(x)}$	0.825	0.071	0.035
$\lambda_{73}^{(x)}$	0.684	0.087	0.065
$\lambda_{83}^{(x)}$	0.858	0.089	0.064
$\lambda_{91}^{(x)}$	0.447	0.087	0.081
$\lambda_{93}^{(x)}$	0.439	0.087	0.081

8.8 MIMIC Models

The mimic model is a model with multiple indicators and multiple causes of a single latent variable. This type of model was introduced by Hauser and Goldberger (1971) and discussed by Goldberger (1972). Since these papers were published there has been many applications of MIMIC models in the literature, for example, Van de Ven and Van der Gaag (1982), Smith and Patterson (1984), and Gallo, Rabins, and Anthony (1999).

Jöreskog and Goldberger (1975) pointed out that the model can be viewed as a multivariate regression model with two specific constraints:

- the regression matrix must have rank 1

- the residual covariance matrix must satisfy the congeneric measurement model

The MIMIC model is an example with x-variables but no ξ-variables.

A prototype MIMIC model with three causes and three indicators is shown in Figure 8.4

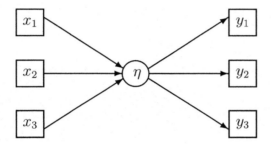

Figure 8.4: A MIMIC Model with Three Causes and Three Indicators

There are numerous applications of MIMIC models in the literature but none that provide any raw data that we can use here. We shall therefore give only the relationships in the model corresponding to Figure 8.4.

The **SIMPLIS** syntax is particularly simple:

```
Y1 - Y3 = Eta
Eta = X1 - X3
```

8.8.1 Example: Peer Influences and Ambition

Sociologists have often called attention to the way in which one's peers for example, best friends, influence one's decisions, for example, choice of occupation. They have recognized that the relation must be reciprocal, *i.e.*, if my best friend influences my choice, I must influence his. Duncan, Haller, and Portes (1968) present a simultaneous equation model of peer influences on occupational choice, using a sample of Michigan high-school students paired with their best friends. The authors interpret educational and occupational choice as two indicators of a single latent variable "ambition," and specify the choices. This model with reciprocal causation between two latent variables is displayed in Figure 8.5. Note that the variables in this figure are symmetrical with respect to a horizontal line in the middle.

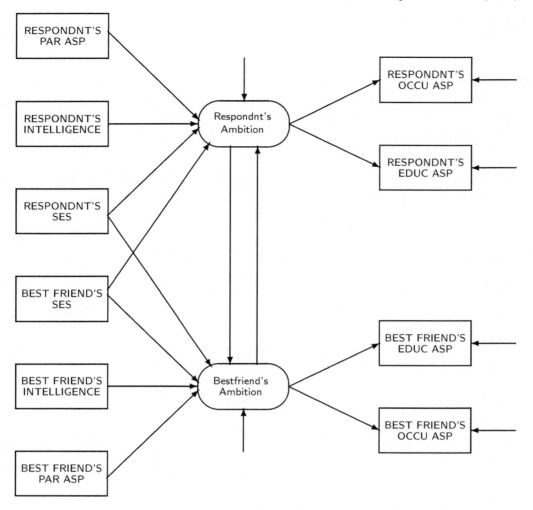

Figure 8.5: Path Diagram for Peer Influences on Ambition

In Figure 8.5 we have omitted all the two-way arrows among the x-variables, i.e., the variances and covariances of six variables to the left in the figure. In terms of the general notation in Section 8.2, the observed variables are

$x_1 = $ respondent's parental aspiration (REPARASP)

$x_2 = $ respondent's intelligence (REINTGCE)

$x_3 = $ respondent's socioeconomic status (RESOCIEC)

$x_4 = $ best friend's socioeconomic status (BFSOCIEC)

$x_5 = $ best friend's intelligence (BFINTGCE)

$x_6 = $ best friend's parental aspiration (BFPARASP)

$y_1 = $ respondent's occupational aspiration (REOCCASP)

$y_2 = $ respondent's educational aspiration (REEDASP)

$y_3 = $ best friend's educational aspiration (BFEDASP)

$y_4 = $ best friend's occupational aspiration (BFOCCASP)

The two latent variables η_1 and η_2 are respondent's ambition (Reambitn) and best friend's ambition (Bfambitn), respectively. Note there are no ξ-variables in this model. Strictly speaking LISREL interprets it as the ξ-variables are identical to the x-variables.

Note that, although long names are used in the path diagram, only eight-character names are allowed in the input files as the program can only use labels of at most eight characters. Note also that the order of the variables in the data do not correspond with the order of the variables in the path diagram. To make the path diagram produced by LISREL look like the path diagram in Figure 8.5 we reorder the variables in the following input files. In SIMPLIS syntax this is done by a Reorder command, and in LISREL syntax this is done by a SE line.

A SIMPLIS syntax file for this model with equal reciprocal causation is **peerinfluences1a.spl**:

```
Peer Influences on Ambition
---------------------------
Raw Data from file peerinfluences.lsf
Reorder Variables: REPARASP REINTGCE RESOCIEC BFSOCIEC BFINTGCE
                   BFPARASP REOCCASP REEDASP BFEDASP BFOCCASP
Latent Variables: Reambitn  Bfambitn
Relationships
   REOCCASP  = 1*Reambitn
   REEDASP = Reambitn
   BFEDASP = Bfambitn
   BFOCCASP  = 1*Bfambitn
   Reambitn  = Bfambitn REPARASP - BFSOCIEC
   Bfambitn  = Reambitn RESOCIEC - BFPARASP
Set Reambitn -> Bfambitn = Bfambitn -> Reambitn
Set the Error Covariance between Reambitn and Bfambitn to 0
Path Diagram
End of Problem
```

The corresponding input file in LISREL syntax is **peerinfluences1b.lis**:

```
Peer Influencies on Ambition: Model with BE(2,1)=BE(1,2) and PS(2,1)=0
DA NI=10
RA=peerinfluences.lsf
SE
4 5 10 9 2 1 3 8 6 7
MO NY=4 NE=2 NX=6 FI PS=DI BE=FU
LE
Reambitn Bfambitn
FR LY(2,1) LY(3,2) BE(1,2)
FI GA(5)-GA(8)
VA 1 LY(1) LY(8)
EQ BE(1,2) BE(2,1)
PD
OU AD=OFF
```

This gives a chi-square C1 of 27.043 with 17 degrees of freedom. This has a P- value of 0.0574.

One may also be interested in the hypothesis of complete symmetry between best friend and respondent, i.e., that the model is completely symmetric above and below a horizontal line in the middle of the path diagram. To test this hypothesis, one must include all the following equality constraints in the input file (see file **peerinfluences2a.spl**).

```
Set Reambitn -> Bfambitn = Bfambitn -> Reambitn
Set REPARASP -> Reambitn = BFPARASP -> Bfambitn
Set REINTGCE -> Reambitn = BFINTGCE -> Bfambitn
Set RESOCIEC -> Reambitn = BFSOCIEC -> Bfambitn
Set BFSOCIEC -> Reambitn = RESOCIEC -> Bfambitn
Set Reambitn -> 'RE EDASP' = Bfambitn -> 'BF EDASP'
```

In **LISREL** syntax this corresponds to

```
EQ BE(1,2) BE(2,1)
EQ LY(2,1) LY(3,2)
EQ GA(1,1) GA(2,6)
EQ GA(1,2) GA(2,5)
EQ GA(1,3) GA(2,4)
EQ GA(1,4) GA(2,3)
EQ PS(1) PS(2)
EQ TE(1) TE(4)
EQ TE(2) TE(3)
```

This gives a chi-square C1 of 33.540 with 25 degrees of freedom. This has a P- value of 0.1181. This shows that a model with constraints can have a larger P-value than a model without these constraints. In this case this means that there is strong evidence that the model of complete symmetry holds.

```
Root Mean Square Error of Approximation (RMSEA)         0.0322
90 Percent Confidence Interval for RMSEA               (0.0 ; 0.0581)
P-Value for Test of Close Fit (RMSEA < 0.05)           0.856
```

The effects of the explanatory variables on the latent variables are given in the output file **peerinfluences2b.out** as

```
        GAMMA

          REPARASP   REINTGCE   RESOCIEC   BFSOCIEC   BFINTGCE   BFPARASP
          --------   --------   --------   --------   --------   --------
Reambitn     0.153      0.278      0.222      0.074      - -        - -
           (0.026)    (0.028)    (0.029)    (0.030)
             5.977     10.027      7.613      2.492


Bfambitn      - -        - -       0.074      0.222      0.278      0.153
                                 (0.030)    (0.029)    (0.028)    (0.026)
                                   2.492      7.613     10.027      5.977
```

An interesting observation is that the effect of intelligence is stronger than the effect of socio-economic status. It seems that these students go more for intelligence than socio-economic status when they choose their best friends.

The new feature in the next example is that the indicators of the latent variables are ordinal rather than continuous.

8.8.2 Example: Continuous Causes and Ordinal Indicators

We now return to the analysis of the Efficacy variables in the Political Action Survey described in Section 7.4.2. The data file for this illustration is a text file **USA.RAW**. This contains 10 variables in free format. The first six are the six Efficacy variables; the other four variables are (the original variable names are given in parenthesis):

YOB Year of birth with *Don't Know* coded as 1998 and *No Answer* coded as 1999 (V0146). Recall that the interviews were done in 1974.

GENDER Gender coded as 1 for *Male*, 2 for *Female*, and 9 for *No Answer* (V0283).

LEFTRIGH A left-right scale from 1 to 10 with *Don't Know* coded as 98 and *No Answer* coded as 99 (V0020).

EDUCAT Education coded as 1 for *Compulsory level only*, 2 for *Middle level*, 3 for *Higher or Academic level*, and 9 for *No Answer* (V0214)

As always, it is a good idea to begin with a data screening. This can be done by running the following **PRELIS** command file **usa1.prl**:

```
Screening the Data in USA.RAW
DA NI=10
LA
NOSAY VOTING COMPLEX NOCARE TOUCH INTEREST YOB GENDER LEFTRIGH EDUCAT
RA = USA.RAW
CL NOSAY - INTEREST 1=AS 2=A 3=D 4=DS 8=DK 9=NA
CL GENDER 1=MALE 2=FEMA 9=NA
CL LEFTRIGH 98=DK 99=NA
CL EDUCAT 1=COMP 2=MIDD 3=HIGH 9=NA
OU
```

The output reveals that

- There are 1719 cases in the USA sample, 736 males and 983 females.

- The marginal distributions of the six efficacy variables are those reported in Section 7.4.2.

- There are more than 15 different birthyears in the sample. The oldest person was born in 1882. Only one person did not report his/her birthyear and nobody reported not knowing his/her year of birth.

- As many as 547 persons or 31.8% did not place themselves on the left-right scale.

- Only 8 persons did not answer the education question.

Before one can proceed one must decide how to treat the *Don't Know* and *No Answer* responses. In Chapter 1, we discussed various alternative ways of dealing with missing values. We do not repeat that discussion here. The major difficulty is to decide how to treat the 547 people who did not answer the **LEFTRIGH** variable. Does this mean that these people are in the middle of the scale, or that the concept of left-right has no meaning for them, or what? Given the data, it is impossible to know. For **LEFTRIGH**, there are various alternatives:

- One could replace the missing values 98 and 99 on **LEFTRIGH** by 5 or 6 [1].

- One could delete the variable **LEFTRIGH**.

- One could use imputation by matching or multiple imputation by EM or MCMC.

- One can use FIML to deal with the missing value problem. This will use all the data available on each case.

- One could investigate whether the probability of missing on **LEFTRIGH** depends on any of the other variables. This turns out not to be the case. We will therefore use listwise deletion.

---

[1] Had there been 9 categories we would use 5. Had there been 11 categories we would use 6.

The file **usa2.prl**:

```
DA NI=10
LA
NOSAY VOTING COMPLEX NOCARE TOUCH INTEREST YOB GENDER LEFTRIGH EDUCAT
RA=USA.RAW
MI 8,9 NOSAY - INTEREST
MI 9998,9999 YOB
CL NOSAY - INTEREST 1=AS 2=A 3=D 4=DS
MI 9 GENDER EDUCAT
CL GENDER 1=MALE 2=FEMA
MI 98,99 LEFTRIGH
CL EDUCAT 1=COMP 2=MIDD 3=HIGH
NEW AGE = 1974 - YOB
SD YOB
OU RA=USA.LSF
```

eliminates all cases with *Don't Know* and *No Answer* responses (listwise deletion) and saves the data on all complete cases in a **LISREL** system file called **USA.LSF**. In addition, AGE is computed as 1974 - YOB. This is a proxy for age. The output file shows that the resulting listwise sample size is 1076. Thus, 643 cases were lost.

The idea of a MIMIC model is that a set of possibly explanatory variables (covariates) affects latent variables which are indicated by other observed variables, in this case ordinal variables. Thus there are multiple indicators and multiple causes of latent variables, see Jöreskog and Goldberger (1975). Since this model has two latent variables, we call it a Double MIMIC model. The MIMIC model considered here is shown in Figure 8.6.

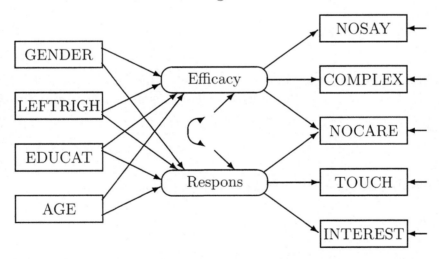

Figure 8.6: Double MIMIC Model for Efficacy and Respons

For reasons stated in Section 7.4.2, we do not include the variable **VOTING** in the model although it is in the data. With **SIMPLIS** syntax, selection of variables is automatic in the sense that only the variables included in the relationships will be used in the estimation of the model. With **LISREL** syntax, one has to use a select line **SE** to select the variables to be used in the model.

A **SIMPLIS** syntax file for estimating the model in Figure 8.6 is **usa2a.spl**:

```
Estimating a Double MIMIC Model
Raw Data from File usa.lsf
Latent Variables: Efficacy Respons
Relationships:
   NOSAY COMPLEX NOCARE =  Efficacy
   NOCARE TOUCH INTEREST = Respons
   NOSAY = 1*Efficacy
   INTEREST = 1*Respons
   Efficacy Respons = GENDER LEFTRIGH EDUCAT AGE
Let the errors of Efficacy and Respons correlate
Robust Estimation
Path Diagram
End of Problem
```

The corresponding input file in **LISREL** syntax is **usa2b.lis**:

```
Estimating a Double MIMIC Model
da ni=10
ra=usa.lsf
se
1 3 4 5 6 7 8 9 10 /
mo ny=5 ne=2 nx=4 nk=4 lx=id td=ze ps=sy,fr
le
Efficacy Respons
fr ly(2,1) ly(3,1) ly(3,2) ly(4,2)
va 1 ly(1,1) ly(5,2)
ro
pd
ou
```

These input files use RML to estimate the model. One can also use RULS or RDWLS. One could also use FIML with adaptive quadrature. For this alternative, put the line

`$ADAPQ(8) POBIT`

after the **ra** line.

The output **usa2b.out** gives the structural equations as

GAMMA

	GENDER	LEFTRIGH	EDUCAT	AGE
Efficacy	-0.074	0.026	0.394	0.002
	(0.051)	(0.014)	(0.042)	(0.002)
	-1.447	1.873	9.378	1.454
Respons	0.028	0.002	0.202	0.004
	(0.045)	(0.012)	(0.034)	(0.001)
	0.624	0.178	5.954	2.728

which shows that only **EDUCAT** has statistically significant effects on both **Efficacy** and **Respons**. Higher education increases the feelings of efficacy and responsiveness. AGE has a statistically significant effect on **Respons**. AGE increases the feeling of responsiveness. None of the other explanatory variables has any statistically significant effects.

8.9 A Model for the Theory of Planned Behavior

The theory of planned behavior of Ajzen (1991) is a theory for how latent variables such as attitudes, norms, control, and intentions affect behavior. For an evaluation of different methods for estimating this model, see for example Jöreskog and Yang (1996), Jöreskog (1998), and Reinecke (2002).

8.9.1 Example: Attitudes to Drinking and Driving

In a study designed to determine the predictors of drinking and driving behavior among 18- to 24-year-old males, the model shown in the path diagram in Figure 8.9.1 was proposed.

The latent variables shown in the figure are as follows:

Attitude attitude toward drinking and driving

Norms social norms pertaining to drinking and driving

Control perceived control over drinking and driving

Intention intention to drink and drive

Behavior drinking and driving behavior

Attitude is measured by five indicators X_1-X_5, Norms is measured by three indicators X_6-X_8, Control is measured by four indicators X_9-X_{12}, Intention is measured by two indicators Y_1-Y_2, and behavior is measured by two indicators Y_3-Y_4. Fictitious data based on the theory of planned behavior is given in file **drinkdata.lsf**.

This example illustrates a possible strategy of analysis. Begin by testing the measurement model for Attitude, see file **drink11a.spl**.

```
Drinking and Driving
Testing the Measurement Model for Attitude
Raw Data from File drinkdata.lsf
Latent Variables Attitude
Relationships
X1-X5 = Attitude
Path Diagram
End of Problem
```

Note that although the data file **drinkdata.lsf** contains many variables, LISREL automatically selects the subset of variables used in the model.

To test the measurement model for Attitude and Norms simultaneously, add Norms in the list of Latent Variables and add the line, see file **drink12a.spl**.

```
X6-X8 = Norms
```

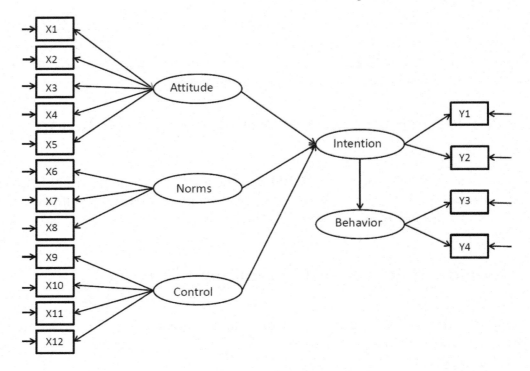

Figure 8.7: Conceptual Path Diagram for Attitudes to Drinking and Driving

To test the measurement model for Attitude, Norms, and Control simultaneously, add Control in the list of Latent Variables and add the line, see file **drink13a.spl**.

```
X9-X12= Control
```

Finally, to test the measurement model for all latent variables simultaneously, add Intention and Behavior in the list of Latent Variables and add the two lines, see file **drink14a.spl**.

```
Y1-Y2 = Intention
Y3-Y4 = Behavior
```

If any of these analysis shows a large modification index (see Sörbom, 1989) for an indicator, the measurement model must be reconsidered and modified. For example, suppose there is a large modification index for the path from Attitude to X8. This might mean that X8 is not entirely an indicator of Norms but to some extent also a measure of Attitude. If this idea makes sense then the model should be modified by letting X8 be a composite measure of both Norms and Attitude.

One can now test the full model in Figure 8.9.1 by adding the two lines, see file **drink15a.spl**.

```
Intention = Attitude - Control
Behavior = Intention
```

defining the structural relationships among the latent variables The full **SIMPLIS** command file is now

```
Drinking and Driving
Raw Data from File drinkdata.lsf
Latent Variables Attitude Norms Control Intention Behavior
Relationships
Y1-Y2 = Intention
Y3-Y4 = Behavior
X1-X5 X8 = Attitude
X6-X8 = Norms
X9-X12= Control
Intention = Attitude - Control
Behavior = Intention
Path Diagram
End of Problem
```

According to Browne and Cudeck (1993) one can use the following fit measures, see file **drink15a.out**:

```
Root Mean Square Error of Approximation (RMSEA)        0.0282
90 Percent Confidence Interval for RMSEA              (0.0196 ; 0.0363)
P-Value for Test of Close Fit (RMSEA < 0.05)           1.00
```

to conclude that the model fits at least approximately.

The LISREL syntax file corresponding to **drink15a.spl** is **drink15b.lis**:

```
Drinking and Driving
da ni=16
ra=drinkdata.lsf
mo ny=4 ne=2 nx=12 nk=3 be=sd ga=fi
le
Intention Behavior
lk
Attitude Norms Control
fr ly(1,1) ly(2,1) ly(3,2) ly(4,2)
fr lx(1,1) lx(2,1) lx(3,1) lx(4,1) lx(5,1)
fr lx(6,2) lx(7,2) lx(8,1) lx(8,2)
fr lx(9,3) lx(10,3) lx(11,3) lx(12,3)
fr ga(1,1) ga(1,2) ga(1,3)
pd
ou
```

To explain this syntax, note that `ly` and `lx` are default on the `mo` line which means that they are fixed. So we need to specify the elements that should be estimated which is done on the first four `fr` lines. `be=sd` means that `be` is subdiagonal. Since $\mathbf{B}$ is a 2×2 matrix this means that β_{21} will be estimated. `ga=fi` which means that $\mathbf{\Gamma}$ is fixed. So we need to specify which elements of $\mathbf{\Gamma}$ that should be estimated. This is done on the last `fr` line. Note that all elements in the second row of $\mathbf{\Gamma}$ are all fixed at 0 because according to the model Behavior does not depend on Attitude, Norms, and Control.

8.10   Latent Variable Scores

There are several methods available for estimating latent variable scores[2], *e.g.*, Lawley and Maxwell (1971, Chapter 8) or Bartholomew and Knott (1999, pp. 65–68). The two most commonly used methods for estimating latent variable scores are the regression method of Thomson (1939) and the Bartlett method (Bartlett, 1937). In **LISREL** we use the procedure of Anderson and Rubin (1956). This procedure has the advantage of producing latent variable scores that have the same covariance matrix as the latent variables themselves.

One can also estimate individual scores for all the error terms, measurement errors as well as structural errors, in any single-group **LISREL** model. Following Bollen and Arminger (1991) we use the general term *observational residuals* for this. The observational residuals depend on the method used for estimating the latent variable scores. Bollen and Arminger (1991) gave formulas which are valid for any method of estimating latent variable scores as linear combinations of observed variables. The results reported here are based on latent variable scores estimated by the Anderson & Rubin method.

This section describes how latent variable scores and observational residuals can be obtained with **LISREL** and illustrates their use with one example.

8.10.1   Example: Panel Model for Political Democracy

Bollen (1989a, p. 17) presents a panel model of political democracy and industrialization in 75 countries. Bollen and Arminger (1991) used the same model in their discussion of observational residuals. The model is shown in the path diagram in Figure 8.10.1.

The variables in the model are

- y_1 Freedom of press 1960

- y_2 Freedom of political opposition 1960

- y_3 Fairness of elections 1960

- y_4 Effectiveness of legislature 1960

- y_5 Freedom of press 1965

- y_6 Freedom of political opposition 1965

- y_7 Fairness of elections 1965

- y_8 Effectiveness of legislature 1965

- x_1 GNP per capita 1960

- x_2 Energy consumption per capita 1960

---

[2]In classical exploratory factor analysis these are usually called factor scores

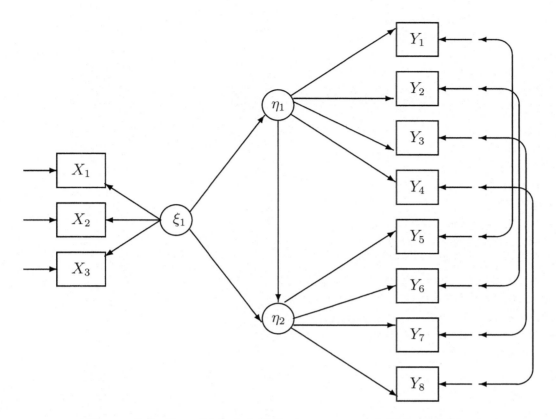

Figure 8.8: Panel Model of Democracy and Industrialization

- x_3 Percentage of labor force in industry 1960

- η_1 Democracy in 1960 (Latent variable Dem60)

- η_2 Democracy in 1965 (Latent variable Dem65)

- ξ Level of industrialization in 1960 (Latent variable Indus)

$y_1 - y_4$ are taken as indicators of the latent variable Dem60 (Democracy 1960) and $y_5 - y_8$ are taken as indicators of the latent variable Dem65 (Democracy 1965). $x_1 - x_3$ are taken as indicators of the latent variable Indus (Industrialization 1960). Data on $y_1 - x_3$ are available for 75 developing countries. These data are in the file **POLIDEM.LSF**[3].

The following **SIMPLIS** syntax file (**BA1a.SPL**) will estimate scores on Demo60, Demo65, and Indus for each country in file **POLIDEMnew.LSF**.

---

[3]Ken Bollen kindly provided the data but unfortunately he was unable to give the names of the 75 countries.

```
Industrialization-Democracy Example
Raw Data from file POLIDEM.LSF
Latent Variables: Dem60 Dem65 Indus
Relationships:
Y1= 1*Dem60
Y2-Y4 = Dem60
Y5 = 1*Dem65
Y6-Y8 = Dem65
X9 = 1*Indus
X10-X11 = Indus
Dem60  = Indus
Dem65 = Dem60 Indus
Set Dem60 -> Y2 = Dem65 -> Y6
Set Dem60 -> Y3 = Dem65 -> Y7
Set Dem60 -> Y4 = Dem65 -> Y8
Let the errors of Y5 and Y1 be correlated
let the errors of Y6 and Y2 be correlated
Let the errors of Y7 and Y3 be correlated
Let the errors of Y8 and Y4 be correlated
LSFfile POLIDEMnew.LSF
Path Diagram
End of Problem
```

The variables y_1–y_4 are the same variables as y_5–y_8 measured at two points in time. So Bollen and Arminger (1991) assume that their loadings on η_1 and η_2 are the same. This is specified by the lines

```
Set Dem60 -> Y2 = Dem65 -> Y6
Set Dem60 -> Y3 = Dem65 -> Y7
Set Dem60 -> Y4 = Dem65 -> Y8
```

The loadings of y_1 and y_5 are set to 1. Furthermore, they assume that the measurement errors of corresponding y-variables are correlated. This is specified by the lines

```
Let the errors of Y5 and Y1 be correlated
let the errors of Y6 and Y2 be correlated
Let the errors of Y7 and Y3 be correlated
Let the errors of Y8 and Y4 be correlated
```

After this run is completed the file **POLIDEMnew.LSF** contains the following variables

```
Y1 Y2 Y3 Y4 Y5 Y6 Y7 Y8 X1 X2 X3 Dem60 Dem65 Indus
R_Y1 R_Y2 R_Y3 R_Y4 R_Y5 R_Y6 R_Y7 R_Y8 R_X1 R_X2 R_X3 R_Dem60 R_Dem65
```

For example, R_Y5 is the estimate of the measurement error ϵ_5, R_X3 is the estimate of the measurement error δ_3, and R_Dem64 is the estimate of the structural error ζ_2 in the LISREL model.

The LISREL syntax file corresponding to **ba1a.spl** is **ba1b.lis**:

```
Industrialization-Democracy Example
da ni=11
ra=polidem.lsf
mo ny=8 nx=3 ne=2 nk=1 be=sd te=sy
le
Dem60 Dem65
lk
Indus
fr lx(2) lx(3)
fr ly(2,1) ly(3,1) ly(4,1)
fr ly(6,2) ly(7,2) ly(8,2)
eq ly(2,1) ly(6,2)
eq ly(3,1) ly(7,2)
eq ly(4,1) ly(8,2)
va 1 ly(1,1) ly (5,2) lx(1,1)
fr te(5,1) te(6,2) te(7,3) te(8,4)
lsffile polidemnew.lsf
pd
ou
```

It would be useful to know which country each row belongs to. Unfortunately, this information is not available. However, one can construct a variable COUNTRY which runs from 1 to 75[4] This is done with the following **PRELIS** syntax file (**BA2.PRL**) which at the same time constructs scores on another latent variable Diff = Dem65 - Dem60.

```
SY=POLIDEMnew.LSF
New COUNTRY=TIME
New Diff=Dem65-Dem60
CO ALL
Select COUNTRY Dem60 Dem65 Diff
OU RA=DEMDIFF.LSF
```

The file **DEMDIFF.LSF** contains the following variables

```
COUNTRY Dem60 Dem65 Diff
```

With this file one can do various things to find out which countries have most democracy or which countries increased or decreased their democracy between 1960 and 1965. For example, do a bivariate line plot of Diff against COUNTRY. This shows that country 2 increased democracy most and country 30 had the largest decrease. This can also be seen by sorting Diff in descending order. This shows that country 2 has a Diff value of 1.69 and country 30 has a Diff value of -1.51. These are the best and worst countries. The second best and second worst countries are the countries 22 and 34. These have Diff values of 1.39 and -1.38, respectively.

---

[4]The variable TIME is always available in **PRELIS**. It assigns values $1, 2, \ldots, N$ to the cases. It is intended mainly for time series. Hence its name TIME.

One can add the country code and plot any of the observational residuals against COUNTRY or sort any of these residuals in ascending or descending values. This can be done for the purpose of finding errors in the data or other outliers or to determine countries with largest or smallest residuals.

One can also obtain estimates of the standardized latent variables. This is easy to do. Just delete the 1* in three places in **BA1a.SPL** and rerun that file. This gives a solution in which the three latent variables Dem60, Dem65, and Indus have unit variances. The correlation matrix of the latent variables is given in the output file as

```
Correlation Matrix

                Dem60       Dem65       Indus
              --------    --------    --------
   Dem60        1.000
   Dem65        0.945       1.000
   Indus        0.449       0.560       1.000
```

Furthermore, the resulting file **POLIDEMnew.LSF** now contains estimates of the standardized latent variables. Next run the following PRELIS syntax file (**BA3.PRL**):

```
SY=POLIDEMnew.LSF
Select Dem60 Dem65 Indus
OU MA=KM XU
```

This will compute the correlation matrix of the latent variable scores. The output file **BA3.OUT** verifies that this correlation matrix is indeed equal to the correlation matrix of the latent variables in the model as stated in the introduction of this section.

Chapter 9

Analysis of Longitudinal Data

The characteristic feature of a longitudinal research design is that the same measurement instruments are used on the same people at two or more occasions. The purpose of a longitudinal is to assess the changes that occur between the occasions, and to attribute these changes to certain background characteristics and events existing or occurring before the first occasion and/or to various treatments and developments that occur after the first occasion. Often, when the same variables are used repeatedly, there is a tendency for the measurement errors in these variables to correlate over time because of specific factors, memory or other retest effects. Hence there is a need to consider models with correlated measurement errors.

There are several types of longitudinal data. Here we concentrate on two types:

- two-wave data, sometimes called panel data, where data are collected at two points in time, which maybe near or far apart. Models for two-wave data are considered in Section 9.1.

- Repeated measurements data which are collected over a number of time points usually at equidistant time points such as daily, weekly, monthly or yearly. Two types of models for repeated measurements data will be considered: One is the SIMPLEX model in Section 9.2. The other is so called latent growth curves which are covered in Section 9.3

9.1 Two-wave Models

9.1.1 Example: Stability of Alienation

Wheaton et al. (1977) report on a study concerned with the stability over time of attitudes such as alienation, and the relation to background variables such as education and occupation. Data on attitude scales were collected from 932 persons in two rural regions in Illinois at three points in time: 1966, 1967, and 1971. The variables used for the present example are the Anomia sub-scale and the Powerlessness sub-scale, taken to be indicators of Alienation. This example uses data from 1967 and 1971 only. The background variables are the respondent's education (years of schooling completed) and Duncan's Socioeconomic Index (SEI). These are taken to be indicators of the respondent's socioeconomic status (Ses). The data file is **stability.lsf**.

© Springer International Publishing Switzerland 2016
K.G. Jöreskog et al., *Multivariate Analysis with LISREL*, Springer Series in Statistics,
DOI 10.1007/978-3-319-33153-9_9

The research questions are: Has the feeling of alienation increased or decreased between 1967 and 1971? Is the feeling of alienation stable over time? Stability is indicated by the variance of alienation. If the variance has changed much between 1967 and 1971 it means that for some people the feeling of alienation has increased while it has decreased for other people.

The model to be considered here is shown in Figure 9.1. For Models A and B we estimate this model as a covariance structure. For Model A we assume uncorrelated measurement errors. For Model B we assume that the measurement errors on ANOMIA67 and ANOMIA71 and the measurement errors on POWERL67 and POWERL71 are correlated as in Figure 9.1. For Model C we add a mean structure to estimate the latent mean difference between Alien71 and Alien67.

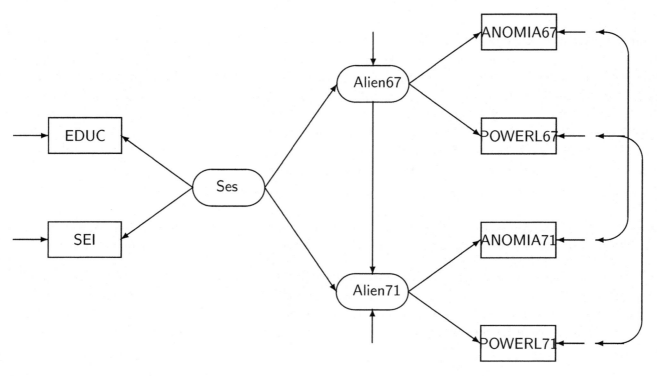

Figure 9.1: Model for Stability of Alienation

We specify the error terms of ANOMIA and POWERL to be correlated over time to take specific factors into account. The four one-way arrows on the right side represent the measurement errors in ANOMIA67, POWERL67, ANOMIA71, and POWERL71, respectively. The two-way arrows on the right side indicate that some of these measurement errors are correlated. The covariance between the two error terms for each variable can be interpreted as a specific error variance. To set up Model A for SIMPLIS is straightforward as shown in the following input file **stability1a.spl**:

```
Stability of Alienation
Raw Data from File stability.lsf
Latent Variables  Alien67 Alien71 Ses
Relationships
    ANOMIA67 POWERL67 = Alien67
    ANOMIA71 POWERL71 = Alien71
    EDUC SEI = Ses
    Alien67 = Ses
    Alien71 = Alien67 Ses
End of Problem
```

The model is specified in terms of relationships. The first three lines specify the relationships between the observed and the latent variables. The last two lines specify the structural relationships. For example,

```
ANOMIA71 POWERL71 = Alien71
```

means that the observed variables ANOMIA71 and POWERL71 depend on the latent variable Alien71, i.e., that ANOMIA71 and POWERL71 are indicators of Alien71. The line

```
Alien71 = Alien67 Ses
```

means that the latent variable Alien71 depends on the two latent variables Alien67 and Ses. This is one of the two structural relationships.

One can specify the model in terms of its paths instead of its relationships:

```
Paths
    Alien67 -> ANOMIA67 POWERL67
    Alien71 -> ANOMIA71 POWERL71
    Ses -> EDUC SEI
    Alien67 -> Alien71
    Ses -> Alien67 Alien71
```

The output file **stability1a.out** reveals that the model does not fit. Chi-square is 71.546 with 6 degrees of freedom.

For Model B we add the two lines, see file **stability2a.spl**.

```
Let the Errors of ANOMIA67 and ANOMIA71 Correlate
Let the Errors of POWERL67 and POWERL71 Correlate
```

The output file **stability2a.out** reveals that the model fits very well. Chi-square is 4.738 with 4 degrees of freedom. Thus, by adding two parameters chi-square decreased from 71.546 to 4.738. The estimated error covariances are

```
Error Covariance for ANOMIA71 and ANOMIA67 = 1.625
                                            (0.314)
                                             5.174


Error Covariance for POWERL71 and POWERL67 = 0.339
                                            (0.261)
                                             1.297
```

which shows that only one of the two error covariances is statistically significant. Thus, the decrease in chi-square is largely due to one single parameter. The two structural equations are estimated as:

```
        Structural Equations

  Alien67 =  - 0.563*Ses, Errorvar.= 0.683  , R² = 0.317
  Standerr      (0.0466)                (0.0659)
  Z-values      -12.092                  10.359
  P-values       0.000                    0.000


  Alien71 = 0.567*Alien67 - 0.208*Ses, Errorvar.= 0.503  , R² = 0.497
  Standerr  (0.0477)        (0.0459)              (0.0498)
  Z-values  11.897          -4.518                10.104
  P-values   0.000           0.000                 0.000
```

The effect of socio-economic status on alienation is negative but when mediated by alienation 1967 it is smaller. the total effects of socio-economic status on alienation is shown in the reduced form:

```
        Reduced Form Equations

  Alien67 =  - 0.563*Ses, Errorvar.= 0.683, R² = 0.317
  Standerr      (0.0466)
  Z-values      -12.086
  P-values       0.000


  Alien71 =  - 0.527*Ses, Errorvar.= 0.722, R² = 0.278
  Standerr      (0.0451)
  Z-values      -11.693
  P-values       0.000
```

The estimated covariance matrix of all the latent variables is also given in the output file as:

```
        Covariance Matrix of Latent Variables

                  Alien67      Alien71        Ses
                  --------     --------     --------
  Alien67          1.000
  Alien71          0.684        1.000
      Ses         -0.563       -0.527       1.000
```

Since the latent variables are standardized in this solution, this is a correlation matrix.

The solution just presented is in terms of standardized latent variables. LISREL automatically standardizes all latent variables unless some other units of measurement are specified, see Section 8.3.1). In this example, when a covariance matrix is analyzed and the units of measurement are the same at the two occasions, it would be more meaningful to assign units of measurement to the latent variables in relation to the observed variables. This will make the variance of Alien67 and Alien71 directly comparable. Also, in order to estimate latent means, one must include an intercept term in each relationship. The means of Alien67 and Alien71 are still zero, but by specifying equality of intercepts for the same variable across time, one can estimate the difference in the latent means of Alien67 and Alien71. Technically, the mean of Alien67 is zero and the mean of Alien71 is estimated.

The SIMPLIS syntax file for Model C is **stability3a.spl**:

```
Stability of Alienation Model C
Estimating Latent Difference Alien71-Alien67
Raw Data from File stability.lsf
Latent Variables  Alien67 Alien71 Ses
Relationships
   ANOMIA67 = CONST 1*Alien67
   POWERL67 = CONST Alien67
   ANOMIA71 = CONST 1*Alien71
   POWERL71 = CONST Alien71
   EDUC = CONST 1*Ses
   SEI = CONST Ses

Set CONST -> ANOMIA67 = CONST -> ANOMIA71
Set CONST -> POWERL67 = CONST -> POWERL71

Set Alien67 -> POWERL67 = Alien71 -> POWERL71

   Alien67 = Ses
   Alien71 = CONST Alien67 Ses

Let the Errors of ANOMIA67 and ANOMIA71 Correlate
Let the Errors of POWERL67 and POWERL71 Correlate

Path Diagram

End of Problem
```

The 1* in the first measurement relation specifies a fixed coefficient of 1 in the relationship between ANOMIA67 and Alien67. The effect of this is to fix the unit of measurement in Alien67 in relation to the unit in the observed variable ANOMIA67. Similarly, in the third relationship, the unit of measurement in Alien71 is fixed in relation to the unit in the observed variable ANOMIA71. Since ANOMIA67 and ANOMIA71 are measured in the same units, this puts Alien67 and Alien71 on the same scale. The fifth relationship specifies Ses to be on the same scale as EDUC.

The variable **CONST** is included in all relationships to estimate the intercepts. The equality of intercepts mentioned earlier is specified by the two lines

```
Set CONST -> ANOMIA67 = CONST -> ANOMIA71
Set CONST -> POWERL67 = CONST -> POWERL71
```

The line

```
Set Alien67 -> POWERL67 = Alien71 -> POWERL71
```

sets the two loadings **POWERL** equal. This is not necessary for identification but adds more power to the analysis.

The output file **stability3a.out** gives the following information about the chi-squares

```
Degrees of Freedom for (C1)-(C2)                    6
Maximum Likelihood Ratio Chi-Square (C1)            16.625 (P = 0.0108)
Due to Covariance Structure                         5.974
Due to Mean Structure                               10.651
```

which shows that chi-square is 16.625 to which the covariance structure contributes 5.974 and the mean structure contributes 10.651. However, as judged by

```
Root Mean Square Error of Approximation (RMSEA)     0.0436
90 Percent Confidence Interval for RMSEA            (0.0193 ; 0.0692)
P-Value for Test of Close Fit (RMSEA < 0.05)        0.621
```

the fit can be judged as acceptable.

The two structural equations are now estimated as

```
Alien67 =  - 0.597*Ses, Errorvar.= 5.105 , R² = 0.321
Standerr    (0.0566)               (0.471)
Z-values    -10.549                10.834
P-values    0.000                  0.000
```

```
Alien71 = 0.339   + 0.593*Alien67 - 0.226*Ses, Errorvar.= 4.062 , R² = 0.501
Standerr  (0.0924)  (0.0476)        (0.0525)              (0.377)
Z-values  3.666     12.459          -4.306                10.772
P-values  0.000     0.000           0.000                 0.000
```

The estimated mean difference between **Alien71** and **Alien67** is 0.339 which is highly significant. Thus, we can be fairly confident that the feeling of alienation has increased between 1967 and 1971.

From

```
           Covariance Matrix of Latent Variables

              Alien67     Alien71       Ses
             --------    --------    --------
Alien67         7.517
Alien71         5.371       8.135
    Ses        -4.039      -3.925       6.763
```

we see that the variance of alienation has increased slightly but we cannot judge whether this is statistically significant.

9.1.2   Example: Panel Model for Political Efficacy

In this section we develop a panel model for the efficacy items introduced in Section 7.4.2. The measurement model involves the two components of efficacy called **Efficacy** and **Respons**. The model is the same as considered in Section 7.4.2 where it was estimated from cross-sectional data.

Here this measurement model is applied at two time points. Recall from Section 7.4.2 that the Political Action Survey was conducted in the first wave between 1973 and 1975 and then repeated 1980–81 in three of the original countries: West Germany, The Netherlands, and the USA. Here we use only the data from USA. Thus, the same people were interviewed at both occasions with about a five years interval.

The objective of the panel model is to answer such questions as: Has the level of efficacy increased or decreased over time? Has the variance of efficacy increased or decreased over time?

A conceptual path diagram of the panel model for efficacy is shown in Figure 9.2.

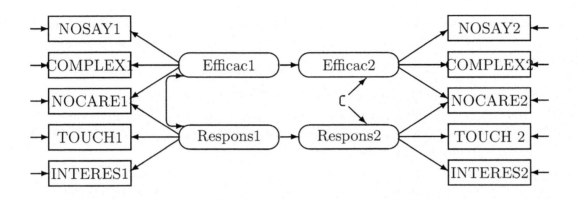

Figure 9.2: Panel Model for Efficacy and Respons

The model also involves a structural model in the middle of the path diagram in which Efficacy at time 2 is predicted by Efficacy at time 1 without the use of Respons at time 1, and Respons at time 2 is predicted by Respons at time 1 without the use of Efficacy at time 1. In addition to these features, the model includes several features not visible in Figure 9.2.

- The measurement error in each variable at time 1 correlates with the measurement error in the corresponding variable at time 2 due to a specific factor in each item. To explain this further, we take COMPLEX as an example. Let x be COMPLEX1 and y be COMPLEX2. Then the measurement equations for COMPLEX1 and COMPLEX2 can be written

$$\text{Time 1}: \ x = \lambda_1 \xi + \delta = \lambda_1 \xi + s + d$$

$$\text{Time 2}: \ y = \lambda_2 \eta + \epsilon = \lambda_2 \eta + s + e \ ,$$

where ξ is Efficac1 and η is Efficac2, δ and ϵ are the so called measurement errors in the LISREL model. Each of these error terms are the sum of two components, one specific factor s unique to the item COMPLEX, and one pure random error component, d and e, respectively, where d and e are uncorrelated. It follows that δ and ϵ are correlated and that

$$Cov(\delta, \epsilon) = Var(s) \ .$$

Thus, the specific error variance can be estimated as the covariance between the measurement errors for the same variables.

- The loading of NOSAY1 on Efficac1 and of NOSAY2 on Efficac2 are fixed to 1 to fix the unit of measurement for Efficac1 and Efficac2. Since NOSAY1 and NOSAY2 have the same unit of measurement, `Efficac1` and `Efficac2` will also have the same unit of measurement. Similarly, the loadings of `INTERES1` on `Respons1` and of `INTERES2` on `Respons2` are fixed to 1 to fix the unit of measurement for `Respons1` and `Respons2`. Since `INTERES1` and `INTERES2` have the same unit of measurement, `Respons1` and `Respons2` will also have the same unit of measurement.

- The other four loadings on the latent variables are constrained to be the same across time.

- There is also an intercept term (not visible in the path diagram) in each measurement equation. These intercept terms are also constrained to be equal across time.

- The equality of intercepts and factor loadings across time is necessary in order to compare the latent variables over time on the same scale, *i.e.*, with the same origin and unit of measurement.

The data file is **panelusa.lsf**, where the item **VOTING** is included. However, for reason stated on page 335 this item is not included in the model. Missing values originally coded as 8, for **Don't Know** and 9 for **No Answer** have been recoded to the global missing value code -999999. The first 20 rows of the data file looks like this

	NOSAY1	VOTING1	COMPLEX1	NOCARE1	TOUCH1	INTERES1	NOSAY2	VOTING2	COMPLEX2	NOCARE2	TOUCH2	INTERES2
1	2.000	2.000	1.000	1.000	1.000	1.000	-999999.000	2.000	2.000	-999999.000	2.000	2.000
2	2.000	3.000	3.000	3.000	2.000	3.000	2.000	3.000	2.000	2.000	2.000	2.000
3	3.000	2.000	2.000	3.000	3.000	3.000	3.000	2.000	2.000	2.000	2.000	2.000
4	2.000	2.000	1.000	1.000	2.000	1.000	2.000	2.000	2.000	2.000	1.000	2.000
5	3.000	2.000	2.000	3.000	3.000	3.000	3.000	2.000	2.000	3.000	2.000	2.000
6	2.000	2.000	2.000	2.000	1.000	2.000	3.000	2.000	1.000	3.000	2.000	2.000
7	3.000	1.000	2.000	2.000	2.000	2.000	2.000	2.000	3.000	3.000	2.000	2.000
8	2.000	1.000	2.000	2.000	1.000	1.000	3.000	3.000	1.000	2.000	-999999.000	3.000
9	3.000	3.000	2.000	3.000	3.000	3.000	3.000	3.000	1.000	3.000	2.000	2.000
10	2.000	2.000	3.000	1.000	1.000	1.000	2.000	2.000	2.000	2.000	1.000	2.000
11	3.000	2.000	1.000	1.000	2.000	2.000	3.000	2.000	2.000	2.000	2.000	2.000
12	1.000	1.000	1.000	1.000	1.000	1.000	3.000	3.000	2.000	3.000	2.000	2.000
13	2.000	2.000	2.000	1.000	2.000	2.000	1.000	1.000	1.000	2.000	2.000	2.000
14	3.000	3.000	2.000	3.000	2.000	2.000	3.000	2.000	2.000	3.000	3.000	3.000
15	3.000	3.000	3.000	3.000	3.000	3.000	2.000	3.000	3.000	3.000	2.000	2.000
16	3.000	3.000	2.000	2.000	3.000	2.000	2.000	2.000	1.000	2.000	2.000	2.000
17	3.000	3.000	4.000	2.000	1.000	1.000	2.000	2.000	4.000	1.000	1.000	1.000
18	4.000	2.000	3.000	4.000	-999999.000	-999999.000	3.000	2.000	1.000	3.000	3.000	3.000
19	3.000	3.000	3.000	3.000	3.000	3.000	3.000	3.000	3.000	3.000	2.000	4.000
20	3.000	2.000	1.000	2.000	2.000	2.000	3.000	2.000	3.000	2.000	2.000	2.000

With all the conditions involved in the panel model, **LISREL** syntax is to be preferred over **SIMPLIS** syntax. With correlated error terms and equality constraints **SIMPLIS** syntax is rather inconvenient. Nevertheless, it can be used, see file **panelusa2a.spl** given below, and it has a certain advantage, namely that the variables to be included in the model are automatically selected in the sense that only the variables mentioned in the input file will be used. By contrast, with **LISREL** syntax, the variables to be included in the model must be selected in such a way that the y-variables come first.

Here is the **SIMPLIS** syntax for the panel model in Figure 9.2:

```
Estimating the Panel Model by FIML
Raw Data from file panelusa.lsf
Latent Variables: Efficac1 Respons1 Efficac2 Respons2
Relationships

  NOSAY1 = CONST 1*Efficac1
  COMPLEX1 NOCARE1 = CONST Efficac1
```

```
NOCARE1 TOUCH1 = CONST Respons1
INTERES1 = CONST 1*Respons1

NOSAY2 = CONST 1*Efficac2
COMPLEX2 NOCARE2 = CONST Efficac2
NOCARE2 TOUCH2 = CONST Respons2
INTERES2 = CONST 1*Respons2

Let the errors of NOSAY1 and NOSAY2 correlate
Let the errors of COMPLEX1 and COMPLEX2 correlate
Let the errors of NOCARE1 and NOCARE2 correlate
Let the errors of TOUCH1 and TOUCH2 correlate
Let the errors of INTERES1 and INTERES2 correlate

Set Efficac1 -> COMPLEX1 = Efficac2 -> COMPLEX2
Set Efficac1 -> NOCARE1 = Efficac2 -> NOCARE2
Set Respons1 -> NOCARE1 = Respons2 -> NOCARE2
Set Respons1 -> TOUCH1 = Respons2 -> TOUCH2

Set CONST -> NOSAY1 = CONST -> NOSAY2
Set CONST -> COMPLEX1 = CONST -> COMPLEX2
Set CONST -> NOCARE1 = CONST -> NOCARE2
Set CONST -> TOUCH1 = CONST -> TOUCH2
Set CONST -> INTERES1 = CONST -> INTERES2

 Efficac2 = CONST Efficac1
 Respons2 = CONST Respons1

Let the errors of Efficac2 and Respons2 correlate
Path Diagram
End of Problem
```

The -> is used to symbolize a path (a uni-directed arrow) in the path diagram.

While a SIMPLIS syntax can often be setup directly from the path diagram, for LISREL syntax it is best to write the matrix formulation corresponding to the equations (8.1), (8.2) and (8.3. For the panel model considered here, these equations are

$$
\begin{pmatrix} \text{Efficacy2} \\ \text{Respons2} \end{pmatrix} = \begin{pmatrix} \alpha_1 \\ \alpha_2 \end{pmatrix} + \begin{pmatrix} \gamma_{11} & 0 \\ 0 & \gamma_{22} \end{pmatrix} \begin{pmatrix} \text{Efficacy1} \\ \text{Respons1} \end{pmatrix} + \begin{pmatrix} \zeta_1 \\ \zeta_2 \end{pmatrix} \tag{9.1}
$$

$$
\begin{pmatrix} \text{NOSAY2} \\ \text{COMPLEX2} \\ \text{NOCARE2} \\ \text{TOUCH2} \\ \text{INTERES2} \end{pmatrix} = \begin{pmatrix} \tau_1^{(y)} \\ \tau_2^{(y)} \\ \tau_3^{(y)} \\ \tau_4^{(y)} \\ \tau_5^{(y)} \end{pmatrix} + \begin{pmatrix} 1 & 0 \\ \lambda_{21}^{(y)} & 0 \\ \lambda_{31}^{(y)} & \lambda_{32}^{(y)} \\ 0 & \lambda_{42}^{(y)} \\ 0 & 1 \end{pmatrix} \begin{pmatrix} \text{Efficacy2} \\ \text{Respons2} \end{pmatrix} + \begin{pmatrix} \epsilon_1 \\ \epsilon_2 \\ \epsilon_3 \\ \epsilon_4 \\ \epsilon_5 \end{pmatrix} \tag{9.2}
$$

$$
\begin{pmatrix} \text{NOSAY1} \\ \text{COMPLEX1} \\ \text{NOCARE1} \\ \text{TOUCH1} \\ \text{INTERES1} \end{pmatrix} = \begin{pmatrix} \tau_1^{(x)} \\ \tau_2^{(x)} \\ \tau_3^{(x)} \\ \tau_4^{(x)} \\ \tau_5^{(x)} \end{pmatrix} + \begin{pmatrix} 1 & 0 \\ \lambda_{21}^{(x)} & 0 \\ \lambda_{31}^{(x)} & \lambda_{32}^{(x)} \\ 0 & \lambda_{42}^{(x)} \\ 0 & 1 \end{pmatrix} \begin{pmatrix} \text{Efficacy1} \\ \text{Respons1} \end{pmatrix} + \begin{pmatrix} \delta_1 \\ \delta_2 \\ \delta_3 \\ \delta_4 \\ \delta_5 \end{pmatrix} \tag{9.3}
$$

The matrix $\boldsymbol{\Phi}$ is

$$
\boldsymbol{\Phi} = \begin{pmatrix} \phi_{11} & \\ \phi_{21} & \phi_{22} \end{pmatrix}
$$

This is the covariance matrix of

$$
\begin{pmatrix} \text{Efficacy1} \\ \text{Respons1} \end{pmatrix}
$$

The matrix $\boldsymbol{\Psi}$ is

$$
\boldsymbol{\Phi} = \begin{pmatrix} \psi_{11} & \\ \psi_{21} & \psi_{22} \end{pmatrix}
$$

This is the covariance matrix of

$$
\begin{pmatrix} \zeta_1 \\ \zeta_2 \end{pmatrix}
$$

The matrix $\boldsymbol{\Theta}_\epsilon$ is

$$
\boldsymbol{\Theta}_\epsilon = diag(\theta_{11}^{(\epsilon)}, \theta_{22}^{(\epsilon)}, \theta_{33}^{(\epsilon)}, \theta_{44}^{(\epsilon)}, \theta_{55}^{(\epsilon)})
$$

These are the error variances of

$$
\begin{pmatrix} \text{NOSAY2} \\ \text{COMPLEX2} \\ \text{NOCARE2} \\ \text{TOUCH2} \\ \text{INTERES2} \end{pmatrix}
$$

The matrix $\boldsymbol{\Theta}_\delta$ is

$$
\boldsymbol{\Theta}_\delta = diag(\theta_{11}^{(\delta)}, \theta_{22}^{(\delta)}, \theta_{33}^{(\delta)}, \theta_{44}^{(\delta)}, \theta_{55}^{(\delta)})
$$

These are the error variances of

$$
\begin{pmatrix} \text{NOSAY1} \\ \text{COMPLEX1} \\ \text{NOCARE1} \\ \text{TOUCH1} \\ \text{INTERES1} \end{pmatrix}
$$

The new feature in this model is the matrix $\boldsymbol{\Theta}_{\delta\epsilon}$. This is a diagonal matrix containing the specific error variances $Cov(\epsilon_i, \delta_i)$, $i = 1, 2, 3, 4, 5$.

A **LISREL** command file for estimating this panel model is **panelusa2b.lis**:

```
Estimating the Panel Model by FIML
da ni=12
ra=panelusa.lsf
se
7 9 10 11 12 1 3 4 5 6 /
mo ny=5 nx=5 ne=2 nk=2 ga=di ps=sy,fr th=fi ty=fr tx=fr al=fr
le
Efficac2 Respons2
lk
Efficac1 Respons1
fr lx(2,1) lx(3,1) lx(3,2) lx(4,2)
va 1 lx(1,1) lx(5,2)
fr ly(2,1) ly(3,1) ly(3,2) ly(4,2)
va 1 ly(1,1) ly(5,2)
fr th(1,1) th(2,2) th(3,3) th(4,4) th (5,5)
eq lx(2,1) ly(2,1)
eq lx(3,1) ly(3,1)
eq lx(3,2) ly(3,2)
eq lx(4,2) ly(4,2)
eq tx(1) ty(1)
eq tx(2) ty(2)
eq tx(3) ty(3)
eq tx(4) ty(4)
eq tx(5) ty(5)
pd
ou
```

The output from file **panelusa2b.out** gives the following information.

```
          --------------------------------
              EM Algorithm for missing Data:
          --------------------------------


       Number of different missing-value patterns=        42
       Effective sample size:        933

       Convergence of EM-algorithm in      4 iterations
       -2 Ln(L) =     19604.71957
       Percentage missing values=    2.15
```

Since there are missing values in the data, FIML is used to estimate the model. This uses all the information available in the data without imputing any values.

The covariance matrix is given as

Covariance Matrix

	NOSAY2	COMPLEX2	NOCARE2	TOUCH2	INTERES2	NOSAY1
NOSAY2	0.436					
COMPLEX2	0.126	0.435				
NOCARE2	0.204	0.130	0.418			
TOUCH2	0.127	0.065	0.201	0.380		
INTERES2	0.159	0.100	0.248	0.223	0.413	
NOSAY1	0.176	0.101	0.165	0.105	0.141	0.565
COMPLEX1	0.086	0.175	0.099	0.067	0.094	0.174
NOCARE1	0.140	0.083	0.184	0.109	0.150	0.282
TOUCH1	0.087	0.055	0.133	0.120	0.126	0.204
INTERES1	0.121	0.082	0.168	0.141	0.178	0.228

Covariance Matrix

	COMPLEX1	NOCARE1	TOUCH1	INTERES1
COMPLEX1	0.521			
NOCARE1	0.190	0.538		
TOUCH1	0.139	0.271	0.488	
INTERES1	0.159	0.308	0.300	0.516

This is a polychoric covariance matrix estimated by fixing the mean and the variance of the underlying ordinal variables to 0 and 1.

The parameter specifications are given as follows

Parameter Specifications

LAMBDA-Y

	Efficac2	Respons2
NOSAY2	0	0
COMPLEX2	1	0
NOCARE2	2	3
TOUCH2	0	4
INTERES2	0	0

LAMBDA-X

	Efficac1	Respons1
NOSAY1	0	0
COMPLEX1	1	0
NOCARE1	2	3
TOUCH1	0	4
INTERES1	0	0

GAMMA

	Efficac1	Respons1
Efficac2	5	0
Respons2	0	6

PHI

	Efficac1	Respons1
Efficac1	7	
Respons1	8	9

PSI

	Efficac2	Respons2
Efficac2	10	
Respons2	11	12

THETA-EPS

NOSAY2	COMPLEX2	NOCARE2	TOUCH2	INTERES2
13	14	15	16	17

THETA-DELTA-EPS

	NOSAY2	COMPLEX2	NOCARE2	TOUCH2	INTERES2
NOSAY1	18	0	0	0	0
COMPLEX1	0	20	0	0	0
NOCARE1	0	0	22	0	0
TOUCH1	0	0	0	24	0
INTERES1	0	0	0	0	26

THETA-DELTA

NOSAY1	COMPLEX1	NOCARE1	TOUCH1	INTERES1
19	21	23	25	27

TAU-Y

NOSAY2	COMPLEX2	NOCARE2	TOUCH2	INTERES2
28	29	30	31	32

```
          TAU-X

                 NOSAY1    COMPLEX1    NOCARE1    TOUCH1    INTERES1
                --------   --------   --------   --------   --------
                     28         29         30         31         32

          ALPHA

                 Efficac2    Respons2
                 --------    --------
                      33          34
```

This lists the independent parameters in sequence. This specification should be inspected to make sure that all equality constraints and other requirements are correct.

The measurement equations are estimated as

```
LISREL Estimates (Maximum Likelihood)

          LAMBDA-Y

                 Efficac2    Respons2
                 --------    --------
     NOSAY2         1.000       - -

   COMPLEX2         0.651       - -
                  (0.047)
                   13.882

    NOCARE2         0.609       0.521
                  (0.092)     (0.066)
                    6.642       7.909

     TOUCH2          - -        0.838
                              (0.033)
                               25.643

   INTERES2          - -        1.000
```

```
          LAMBDA-X

                 Efficac1    Respons1
                 --------    --------

     NOSAY1        1.000       - -

   COMPLEX1        0.651       - -
                 (0.047)
                  13.882

    NOCARE1        0.609       0.521
                 (0.092)     (0.066)
                  6.642       7.909

     TOUCH1        - -         0.838
                              (0.033)
                               25.643

   INTERES1        - -         1.000
```

The structural parameters are in the GAMMA matrix

```
          GAMMA

                 Efficac1    Respons1
                 --------    --------
   Efficac2        0.515       - -
                 (0.041)
                  12.466

   Respons2        - -         0.483
                              (0.035)
                               13.710
```

The matrices Φ and Ψ are estimated as

```
          PHI

                 Efficac1    Respons1
                 --------    --------
   Efficac1        0.266
                 (0.027)
                  9.750

   Respons1        0.233       0.345
                 (0.017)     (0.024)
                  13.350      14.119
```

PSI

	Efficac2	Respons2
	--------	--------
Efficac2	0.125	
	(0.017)	
	7.483	
Respons2	0.097	0.195
	(0.010)	(0.016)
	9.398	11.995

The error variances of δ and ξ are estimated as

THETA-EPS

NOSAY2	COMPLEX2	NOCARE2	TOUCH2	INTERES2
--------	--------	--------	--------	--------
0.239	0.355	0.159	0.196	0.138
(0.020)	(0.018)	(0.011)	(0.012)	(0.013)
12.083	19.537	13.868	16.150	10.771

THETA-DELTA

NOSAY1	COMPLEX1	NOCARE1	TOUCH1	INTERES1
--------	--------	--------	--------	--------
0.295	0.404	0.212	0.235	0.171
(0.023)	(0.021)	(0.015)	(0.014)	(0.015)
12.749	19.498	14.183	16.450	11.524

and the estimated specific error variances appear in the matrix $\Theta_{\delta\epsilon}$.

THETA-DELTA-EPS

	NOSAY2	COMPLEX2	NOCARE2	TOUCH2	INTERES2
	--------	--------	--------	--------	--------
NOSAY1	0.030	- -	- -	- -	- -
	(0.007)				
	4.358				
COMPLEX1	- -	0.116	- -	- -	- -
		(0.007)			
		16.359			
NOCARE1	- -	- -	0.016	- -	- -
			(0.005)		
			3.635		
TOUCH1	- -	- -	- -	0.013	- -
				(0.005)	
				2.859	
INTERES1	- -	- -	- -	- -	0.007
					(0.005)
					1.532

It is seen the measurement model is the same at both time points and that almost all parameters are statistically significant. All the estimates of error covariances are positive which is in line with the interpretation of them as variances of the specific factors. The only non-significant specific error variance is that of INTERES. This does not mean that it does not exist, only that the sample is not large enough to make it significant.

The goodness of fit statistics are given as

```
Global Goodness of Fit Statistics, FIML case

    -2ln(L) for the saturated model =          16045.447
    -2ln(L) for the fitted model    =          16077.746


Degrees of Freedom = 31
Full Information ML Chi-Square                  32.299 (P = 0.4024)
Root Mean Square Error of Approximation (RMSEA)  0.00670
90 Percent Confidence Interval for RMSEA        (0.0 ; 0.0257)
P-Value for Test of Close Fit (RMSEA < 0.05)    1.00
```

indicating that the fit of the model is very good despite all constraints imposed.

Further information about the four latent variables in the output is summarized in Tables 9.1 and 9.2.

Table 9.1: Estimated Means and Covariance Matrix for Efficacy

	Efficacy1	Efficacy2	*Means*
Efficacy1	0.266		0.000
Efficacy2	0.137	0.196	0.036

Table 9.2: Estimated Means and Covariance Matrix for Respons

	Respons1	Respons2	*Means*
Respons1	0.345		0.000
Respons2	0.166	0.275	0.003

The estimated mean differences appear in the vector α

```
        ALPHA

    Efficac2    Respons2
    --------    --------

       0.036       0.003
     (0.022)     (0.023)
       1.624       0.154
```

Since these differences are not significant it appears as there is no significant increase or decrease in Efficacy or Respons between the two occasions.

9.2   Simplex Models

The research question and experimental design that leads to a simplex covariance structure is very different from that of the theoretical rationale of linear factor analysis that leads to its covariance structure in Chapters 6 and 7. However, the covariance structures for some versions of both models are formally identical, so a simplex model can be viewed as a special kind of factor analysis structure. Consequently, issues of model identification and estimation are the same. One of the useful features of factor analysis is that it provides a general framework for investigating models that initially seem to be unrelated.

A simplex model is a type of covariance structure which often occurs in longitudinal studies when the same variable is measured repeatedly on the same individuals over several occasions. The simplex model is equivalent to the covariance structure generated by a first-order non-stationary autoregressive process. Guttman (1954b) used the term simplex also for variables which are not ordered through time but by other criteria. One of his examples concerns tests of verbal ability ordered according to increasing complexity. The typical feature of a simplex correlation structure is that the entries in the correlation matrix decrease as one moves away from the main diagonal. Such a correlation pattern is inconsistent with the factor analysis model with one factor.

Following Anderson (1960), Jöreskog (1970) formulated various simplex models in terms of the well-known Wiener and Markov stochastic processes. A distinction was made between a perfect simplex and a quasi-simplex. A *perfect simplex* is reasonable only if the measurement errors in the observed variables are negligible. A *quasi-simplex*, on the other hand, allows for sizable errors of measurement. Jöreskog (1970) developed procedures for estimation and testing of various simplex models.

Consider p fallible variables $y_1, y_2, \ldots, y_p$. The unit of measurement in the true variables η_i may be chosen to be the same as in the observed variables y_i. The equations defining the model are then

$$y_i = \eta_i + \epsilon_i, \qquad i = 1, 2, \ldots, p , \tag{9.4}$$

$$\eta_i = \gamma_i \eta_{i-1} + \zeta_i, \qquad i = 2, 3, \ldots, p , \tag{9.5}$$

where the ϵ_i are uncorrelated among themselves and uncorrelated with all the η_i and where ζ_i is uncorrelated with η_{i-1} for $i = 2, 3, \ldots, p$. A path diagram of the simplex model with $p = 4$ is given in Figure 9.2.

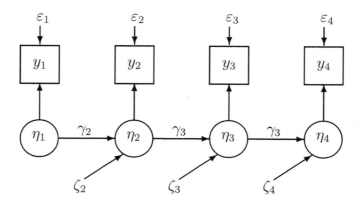

The parameters of the model are $\omega_1 = \mathsf{Var}(\eta_1)$, $\psi_i = \mathsf{Var}(\zeta_i)$ $(i = 2, 3, \ldots, p)$, $\theta_i = \mathsf{Var}(\epsilon_i)$ $(i = 1, 2, \ldots, p)$ and γ_2, γ_3, $\ldots$, γ_p. To see the implied covariance structure it is convenient to reparameterize. Let $\omega_i = \mathsf{Var}(\eta_i) = \gamma_i^2 \omega_{i-1} + \psi_i$ $(i = 2, 3, \ldots, p)$. Then there is a one-to-one correspondence between the parameters $\gamma_2, \gamma_3, \ldots, \gamma_p, \omega_1, \psi_2, \psi_3, \ldots, \psi_p$ and the parameters $\gamma_2,$ $\gamma_3, \ldots, \gamma_p, \omega_1, \omega_2, \ldots, \omega_p$. The covariance matrix of $y_1, y_2, \ldots, y_p$ has the form

$$\Sigma = \begin{pmatrix} \omega_1 + \theta_1 & & & \\ \gamma_2\omega_1 & \omega_2 + \theta_2 & & \\ \gamma_2\gamma_3\omega_1 & \gamma_3\omega_2 & \omega_3 + \theta_3 & \\ \gamma_2\gamma_3\gamma_4\omega_1 & \gamma_3\gamma_4\omega_2 & \gamma_4\omega_3 & \omega_4 + \theta_4 \end{pmatrix}. \tag{9.6}$$

Identification

We consider the identification problem here for the case of $p = 4$. From (9.6), it is seen that, although the product $\beta_2\omega_1 = \sigma_{21}$ is identified, β_2 and ω_1 are not separately identified. The product $\beta_2\omega_1$ is involved in the off-diagonal elements in the first column (and row) only. One can multiply β_2 by a non-zero constant and divide ω_1 by the same constant without changing the product. The change induced by ω_1 in σ_{11} can be absorbed in θ_1 in such a way that σ_{11} remains unchanged. Hence $\theta_1 = \mathsf{Var}(\epsilon_1)$ is not identified. For η_2 and η_3 we have

$$\omega_2 = \frac{\sigma_{32}\sigma_{21}}{\sigma_{31}}, \qquad \omega_3 = \frac{\sigma_{43}\sigma_{32}}{\sigma_{42}},$$

so that ω_2 and ω_3, and hence also θ_2 and θ_3, are identified. With ω_2 and ω_3 identified, β_3 and β_4 are identified by σ_{32} and σ_{43}. The middle coefficient β_3 is overidentified since

$$\beta_3\omega_2 = \frac{\sigma_{31}\sigma_{42}}{\sigma_{41}} = \sigma_{32}.$$

Since both ω_4 and θ_4 are involved in σ_{44} only, these are not identified. Only their sum, σ_{44}, is identified.

This analysis of the identification problem shows that for the "inner" variables y_2 and y_3, the parameters $\omega_2, \omega_3, \theta_2, \theta_3$, and β_3 are identified, whereas there is an indeterminacy associated with each of the "outer" variables y_1 and y_4. To eliminate these indeterminacies one condition must be imposed on the parameters ω_1, θ_1, and β_2, and another on the parameters ω_4 and θ_4. In terms of the original LISREL parameters, $\beta_2, \psi_1 = \omega_1, \psi_2, \psi_4, \theta_1$, and θ_4 are not identified whereas $\beta_3,$ $\beta_4, \psi_3, \theta_2$, and θ_3 are identified. One indeterminacy is associated with β_2, ψ_1, ψ_2 and θ_1 and another indeterminacy is associated with ψ_4 and θ_4. The parameters β_2, ψ_1, ψ_2, and θ_1 are only determined by the three equations

$$\sigma_{11} = \psi_1 + \theta_1, \qquad \sigma_{21} = \beta_2\psi_1, \qquad \omega_2 = \beta_2^2\psi_1 + \psi_2,$$

where ω_2 is identified. The parameters ψ_4 and θ_4 are only determined by the single equation

$$\sigma_{44} = \beta_4^2\omega_3 + \psi_4 + \theta_4,$$

where ω_3 is identified. The most natural way of eliminating the indeterminacies is to set $\theta_1 = \theta_2$ and $\theta_4 = \theta_3$, which makes sense if the y-variables are on the same scale. It is not necessary to assume that *all* error variances are equal, only that the error variances for the first two and last two variables are each equal. The assumption of equal error variances across all variables is in fact testable with $p - 3$ degrees of freedom.

In the general simplex model with p variables, there are $3p - 3$ independent parameters and the degrees of freedom are $\frac{1}{2}p(p+1) - 3p + 3$. If $p = 3$, this is zero and the model is a tautology. For testing a simplex model, p must be at least 4.

Table 9.3: Correlations between Grade Point Averages, High School Rank, and an Aptitude Test

	x_1	x_2	y_1	y_2	y_3	y_4	y_5	y_6	y_7	y_8
HSR	1.000									
ACT	.393	1.000								
GPA1	.387	.375	1.000							
GPA2	.341	.298	.556	1.000						
GPA3	.278	.237	.456	.490	1.000					
GPA4	.270	.255	.439	.445	.562	1.000				
GPA5	.240	.238	.415	.418	.496	.512	1.000			
GPA6	.256	.252	.399	.383	.456	.469	.551	1.000		
GPA7	.240	.219	.387	.364	.445	.442	.500	.544	1.000	
GPA8	.222	.173	.342	.339	.354	.416	.453	.482	.541	1.000

9.2.1   Example: A Simplex Model for Academic Performance

Humphreys (1968) presented the correlation matrix shown in Table 9.3 based on 1600 undergraduate students at the University of Illinois. The variables include eight semesters of grade-point averages, $y_1, y_2, \ldots, y_8$, high school rank x_1 and a composite score on the American College Testing test x_2.

The data we use here has been generated to have the same correlation matrix as in Table 9.3 and is available in the file **humphreys.lsf**. We consider two models. In Model 1 we use only the eight GPA measures. In Model 2 we consider HSR and ACT as indicators of ξ_1 which directly affects η_1.

File **humphreys1b.lis** can be used to estimate and test Model 1.

```
SIMPLEX Model for GPA   Model 1
DA NI=10
RA=humphreys.lsf
SE
3 4 5 6 7 8 9 10 /
MO NY=8 NE=8 LY=ID BE=FU
FR BE 2 1 BE 3 2 BE 4 3 BE 5 4 BE 6 5 BE 7 6 BE 8 7
EQ TE 1 TE 2
EQ TE 7 TE 8
OU
```

This is a **LISREL** model with only y and η-variables. The parameter matrices are $\mathbf{B}$, $\boldsymbol{\Psi}$, and $\boldsymbol{\Theta}_\epsilon$, where both $\boldsymbol{\Psi}$ and $\boldsymbol{\Theta}_\epsilon$ are diagonal and

$$\mathbf{B} = \begin{pmatrix} 0 & 0 & 0 & 0 & 0 & 0 & 0 & 0 \\ \beta_{21} & 0 & 0 & 0 & 0 & 0 & 0 & 0 \\ 0 & \beta_{32} & 0 & 0 & 0 & 0 & 0 & 0 \\ 0 & 0 & \beta_{43} & 0 & 0 & 0 & 0 & 0 \\ 0 & 0 & 0 & \beta_{54} & 0 & 0 & 0 & 0 \\ 0 & 0 & 0 & 0 & \beta_{65} & 0 & 0 & 0 \\ 0 & 0 & 0 & 0 & 0 & \beta_{76} & 0 & 0 \\ 0 & 0 & 0 & 0 & 0 & 0 & \beta_{87} & 0 \end{pmatrix}.$$

This specification automatically defines ζ_1 as η_1 so that $\psi_1 = \omega_1 = Var(\eta_1)$. To achieve identification we set $\theta_1 = \theta_2$ and $\theta_7 = \theta_8$. This model has a chi-square C1 of 23.924 with 15 degrees of freedom.

For Model 2 we use the following input file **humphreys2b.lis**:

```
Simplex Model for Academic Performance   Model 2
DA NI=10
RA=humphreys.lsf
SE
3 4 5 6 7 8 9 10 1 2
MO NY=8 NX=2 NE=8 NK=1 LY=ID BE=FU GA=FI
FR LX(2,1) GA 1
FR BE 2 1 BE 3 2 BE 4 3 BE 5 4 BE 6 5 BE 7 6 BE 8 7
VA 1 LX 1 1
EQ TE 7 - TE 8
OU
```

The only identification condition for Model 2 is $\theta_7 = \theta_8$.

This has a chi-square C1 of 36.947 with 28 degrees of freedom and a P-value of 0.1200. Hence the fit is acceptable. One can test the hypothesis that the process is stationary (Model 3) by adding the line

```
EQ BE 2 1 BE 3 2 BE 4 3 BE 5 4 BE 6 5 BE 7 6 BE 8 7
```

see file **humphreys3b.lis**. This gives a chi-square C1 = 46.633 with 34 degrees of freedom. The chi-square difference for Model 3 vs Model 2 is 9.686 with 6 degrees of freedom which is not significant. So the hypothesis of stationarity cannot be rejected. The auto-covariance coefficient is estimated as 0.913 with a standard error of 0.007.

9.3 Latent Curve Models

A common application of SEM and the **LISREL** model is in the analysis of repeated measurements data where each unit of observation is followed over a number of successive time points (occasions). For example, in psychology one studies the development of cognitive abilities in children, in economics one studies the growth of nations or sales of companies, and in sociology one studies the change in crime rates across communities. Such data is often called longitudinal but we shall use this term in a more general sense than repeated measurements where the occasions occur at regular intervals.

The time intervals may be daily, weekly, monthly, quarterly or annually. The unit of observation may be individuals, businesses, regions, or countries. "Regardless of the subject area or the time interval, social and behavioral scientists have a keen interest in describing and explaining the time trajectories of their variables." Bollen and Curran (2006, p. 1). Bollen and Curran (2006) describe these types of models and give many examples of how such models can be handled with structural equation modeling.

A general form for a linear growth curve model is as follows. If there are k occasions (time points), the observed value y_{it} is

$$y_{it} = a_i + b_i T_t + e_{it} , \tag{9.7}$$

where

$$i = 1, 2, \ldots, N \ \text{ individuals} , \tag{9.8}$$

and

$$T_t = \text{Time at occasion } t = 1, 2, \ldots, k \tag{9.9}$$

In (9.7) a_i is an intercept, b_i is a slope and e_{ij} is an error term. These are random variables in a population of individuals.

If there is a covariate z_i observed for each individual i, assumed to be constant over time, the model may also include

$$a_i = \alpha + \gamma_a z_i + u_i , \tag{9.10}$$

$$b_i = \beta + \gamma_b z_i + v_i , \tag{9.11}$$

where γ_a and γ_b are regression coefficients and u_i and v_i are error terms with

$$\begin{pmatrix} u_i \\ v_i \end{pmatrix} \sim N(\mathbf{0}, \mathbf{\Phi}) . \tag{9.12}$$

If the measurements y_{it} are on the same scale for all t it is reasonable to assume that

$$e_{it} \sim N(0, \sigma_e^2) , \tag{9.13}$$

σ_e^2 is the error variance, but this assumption can easily be relaxed. The usual assumptions are that u_i and v_i are uncorrelared with z_i and that all error terms are uncorrelated except u_i and v_i.

The form of the growth curve need not be linear but can be quadratic, cubic or other forms of functions of t and the covariate z_i can be multivariate.

In this chapter we give several examples with a variation of ingredients.

The principal idea of these kinds of models can be illustrated as follows. Suppose there are four time points. For individual i the observed values are shown as

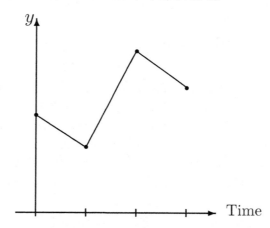

Fitting a linear regression to these four values gives a straight line as shown in the figure.

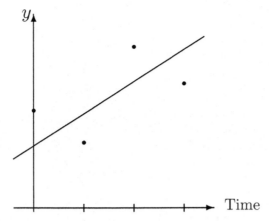

The fitted line is represented as

$$y_{it} = a_i + b_i T_t \, , \tag{9.14}$$

where, for example, $T_t = 0, 1, 2, 3$. This is the growth curve for individual i.

For four individuals their growth curves are shown in the figure

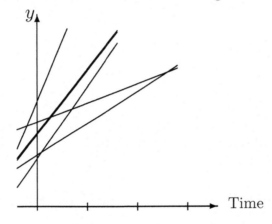

where the thick line is the average regression line. If the population consists of these four persons, the thick line represents the population regression line

$$E(y_{it}) = \alpha + \beta T_t \, , \tag{9.15}$$

where $\alpha = E(a_i)$ and $\beta = E(b_i)$.

Path diagrams for the models without and with a covariate are illustrated in Figures 9.3 and 9.4, respectively, with $T_t = t - 1$ for four occasions. It is clear from these that a and b are latent variables, hence the term Latent Growth Curves.

An interpretation of this is as follows. Each individual has his/her own linear growth curve represented by (9.7) which is the regression of y_{it} on time with intercept a_i and slope b_i varying across individuals. In principle, the intercepts a_i and slopes b_i could all be different across individuals. It is of interest to know if the intercepts and/or the slopes are equal across individuals. The four cases are illustrated in Figure 9.5. If there is variation in intercepts and/or the slopes across individuals, one is interested in whether a covariate z_i can predict the intercept and/or the slope.

Models of the type in Figures 9.3 and 9.4 are particularly easy to do with **SIMPLIS** syntax but they are also straight forward to do with **LISREL** syntax as the following examples will demonstrate.

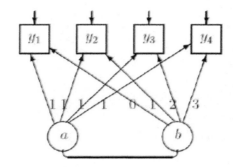

Figure 9.3: The Linear Curve Model

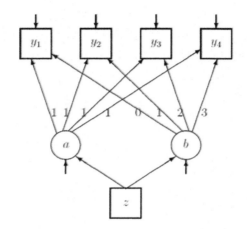

Figure 9.4: The Linear Curve Model with Covariate

9.3.1   Example: Treatment of Prostate Cancer

A medical doctor offered all his patients diagnosed with prostate cancer a treatment aimed at reducing the cancer activity in the prostate. The severity of prostate cancer is often assessed by a plasma component known as prostate specific antigen (PSA), an enzyme that is elevated in the presence of prostate cancer. The PSA level was measured regularly every three months. The data contains five repeated measurements of PSA. The age of the patient is also included in the data. Not every patient accepted the offer directly and several patients chose to enter the program after the first occasion. Some patients, who accepted the initial offer, are absent at some later occasions for various reasons. Thus, there are missing values in the data.

The aim of this study is to answer the following questions: What is the average initial PSA value? Do all patients have the same initial PSA value? Is there an overall effect of treatment. Is there a decline of PSA values over time, and, if so, what is the average rate of decline? Do all patients have the same rate of decline? Does the individual initial PSA value and/or the rate of decline depend on the patient's age?

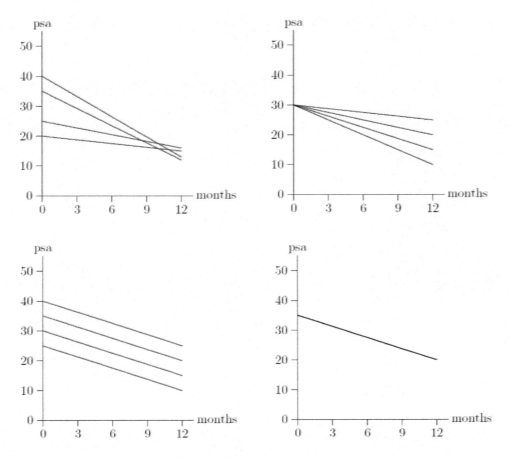

Figure 9.5: Four Cases of Intercepts and Slopes

This is a typical example of repeated measurements data, the analysis of which is sometimes done within the framework of multilevel analysis. It represents the simplest type of two-level model but it can also be analyzed as a structural equation model, see Bollen and Curran (2006). In this context it illustrates a mean and covariance structure model estimated from longitudinal data with missing values.

The data file for this example is **psavar.lsf**, where missing values are shown as -9.000[1].

---

[1]If the data is imported from an external source which already have a missing value code, the missing values will show up in the **.lsf** file as -999999.000, which is the global missing data code in LISREL.

	PSA0	PSA3	PSA6	PSA9	PSA12	Age
1	30.400	28.000	26.900	25.200	19.600	69.000
2	27.800	26.700	20.500	18.700	18.800	58.000
3	26.600	21.800	17.800	17.900	14.500	53.000
4	24.800	24.500	20.200	19.800	18.800	61.000
5	33.700	30.300	25.400	27.300	20.100	63.000
6	26.500	24.600	20.900	-9.000	18.900	49.000
7	26.200	24.400	21.800	22.200	18.400	63.000
8	24.800	19.500	18.000	16.100	12.500	49.000
9	28.400	-9.000	22.500	19.400	22.900	63.000
10	26.100	-9.000	23.300	22.000	14.600	56.000
11	28.800	31.300	-9.000	23.100	22.800	68.000
12	29.800	-9.000	25.600	24.500	21.000	67.000
13	22.900	23.900	-9.000	19.400	15.600	47.000
14	30.100	27.700	25.700	20.400	20.800	56.000
15	26.500	-9.000	-9.000	20.000	17.400	57.000
16	-9.000	-9.000	17.100	12.900	-9.000	43.000

We analyzed this data in Chapter 4, see Section 4.6.1, using multilevel analysis methods. Although the data is the same, the data is organized in different forms for multilevel analysis and for repeated measurement analysis. It is convenient to refer to the different forms as multilevel form and longitudinal form, respectively. To explain the difference we write the data for the first 16 patients in ASCII (text) form.

PSA Data in Multilevel Form

```
Patient Months PSA     Age
    1      0   30.4     69
    1      3   28.0     69
    1      6   26.9     69
    1      9   25.2     69
    1     12   19.6     69
    .      .     .       .
    9      0   28.4     63   Patients 9 and 10 are missing at 3 months
    9      6   22.5     63
    9      9   19.4     63
    9     12   22.9     63
   10      0   26.1     56
   10      6   23.3     56
   10      9   22.0     56
   10     12   14.6     56
    .      .     .       .
   15      0   26.5     57   Patient 15 is missing at 3 and 6 months
   15      9   20.0     57
   15     12   17.4     57
   16      6   17.1     43   Patient 16 is missing at 0, 3, and 12 months
   16      9   12.9     43
```

PSA Data in Longitudinal Form

Patient	PSA0	PSA3	PSA6	PSA9	PSA12	Age
1	30.4	28.0	26.9	25.2	19.6	69
2	27.8	26.7	20.5	18.7	18.8	58
3	26.6	21.8	17.8	17.9	14.5	53
4	24.8	24.5	20.2	19.8	18.8	61
5	33.7	30.3	25.4	27.3	20.1	63
6	26.5	24.6	20.9	-9	18.9	49
7	26.2	24.4	21.8	22.2	18.4	63
8	24.8	19.5	18.0	16.1	12.5	49
9	28.4	-9	22.5	19.4	22.9	63
10	26.1	-9	23.3	22.0	14.6	56
11	28.8	31.3	-9	23.1	22.8	68
12	29.8	-9	25.6	24.5	21.0	67
13	22.9	23.9	-9	19.4	15.6	47
14	30.1	27.7	25.7	20.4	20.8	56
15	26.5	-9	-9	20.0	17.4	57
16	-9	-9	17.1	12.9	-9	43

In this kind of data it is inevitable that there are missing values. For example, a patient may be on vacation or ill or unable to come to the doctor for any reason at some occasion or a patient may die and therefore will not come to the doctor after a certain occasion. It is seen in that

- Patients 9 and 10 are missing at 3 months

- Patient 15 is missing at 3 and 6 months

- Patient 16 is missing at 0, 3, and 12 months

In the following analysis it is assumed that data are missing at random (MAR), although there may be a small probability that a patient will be missing because his PSA value is high. The data file in longitudinal form is **psavar.lsf**.

Data Screening

Whenever one starts an analysis of a new data set, it is recommended to begin with a data screening. To do so click on **Statistics** at the top of the screen and select **Data Screening** from the **Statistics** menu. This will reveal the following information about the data.

```
Number of Missing Values per Variable
```

PSA0	PSA3	PSA6	PSA9	PSA12	Age
17	14	13	12	11	0

This table says that there are 17 patients missing initially, 14 missing at 3 months, 13 at 6 months, etc.

```
Distribution of Missing Values

Total Sample Size =    100

Number of Missing Values    0     1     2     3
          Number of Cases   46    43    9     2
```

This table says that there are only 46 patients with complete data on all six occasions. Thus, if one uses listwise deletion 54% of the sample will be lost. 43 patients are missing on one occasion, 9 patients are missing at two occasions, 2 patients are missing on three occasions. This table does not tell on which occasions the patients are missing. The next table gives more complete information about the missing data patterns.

```
Missing Data Map

Frequency PerCent    Pattern
       46    46.0    0 0 0 0 0 0
        9     9.0    1 0 0 0 0 0
        8     8.0    0 1 0 0 0 0
        2     2.0    1 1 0 0 0 0
        8     8.0    0 0 1 0 0 0
        2     2.0    1 0 1 0 0 0
        2     2.0    0 1 1 0 0 0
        9     9.0    0 0 0 1 0 0
        1     1.0    1 0 0 1 0 0
        1     1.0    1 1 0 1 0 0
        1     1.0    0 0 1 1 0 0
        9     9.0    0 0 0 0 1 0
        1     1.0    1 0 0 0 1 0
        1     1.0    1 1 0 0 1 0
```

The columns under **Pattern** correspond to the variables in the order they are in **psavar.lsf**. A 0 means a non-missing value and a 1 means a missing value. Recall that the last variable is the patient's age. This has no missing values. Here one can see for example that two patients are missing at both 0 and 3 months and another patient is missing at 6 and 9 months.

The following information about the univariate distributions of the variables have been obtained using all available data for each variable,*i.e.*, 83 patients for **PSA0**, 86 patients for **PSA3**, etc.

```
Univariate Summary Statistics for Continuous Variables
```

Variable	Mean	St. Dev.	Skewness	Kurtosis	Minimum	Freq.	Maximum	Freq.
PSA0	31.164	5.684	0.068	-0.852	19.900	1	44.100	1
PSA3	30.036	6.025	-0.248	-0.732	14.500	1	42.100	1
PSA6	27.443	6.084	-0.335	-0.961	13.700	1	37.600	1
PSA9	25.333	6.391	-0.331	-1.066	10.600	1	36.200	1
PSA12	23.406	6.306	-0.309	-1.069	9.600	1	35.800	1
Age	55.450	7.896	-0.329	-0.234	32.000	1	70.000	1

It is seen that the mean age is 55.45 years and that average initial PSA value is 31.164 with a minimum at 19.9 and maximum at 44.1. At 12 months the corresponding values are 23.406, 9.6, and 35.8, respectively. Thus there is some evidence that the PSA values are decreasing over time.

The first model to be estimated is

$$
\begin{pmatrix} PSA0_i \\ PSA3_i \\ PSA6_i \\ PSA9_i \\ PSA12_i \end{pmatrix} = \begin{pmatrix} 1 & 0 \\ 1 & 3 \\ 1 & 6 \\ 1 & 9 \\ 1 & 12 \end{pmatrix} \begin{pmatrix} a_i \\ b_i \end{pmatrix} + \begin{pmatrix} e_{i1} \\ e_{i2} \\ e_{i3} \\ e_{i4} \\ e_{i5} \end{pmatrix}
\tag{9.16}
$$

SIMPLIS Syntax

The model in Figure 9.3 can be estimated with FIML using the following **SIMPLIS** syntax file (**psavar1a.spl**):

```
Linear Growth Curve for psavar Data
Raw Data from File psavar.LSF
Latent Variables: a b
Relationships
PSA0  = 1*a 0*b
PSA3  = 1*a 3*b
PSA6  = 1*a 6*b
PSA9  = 1*a 9*b
PSA12 = 1*a 12*b
a b = CONST
Equal Error Variances: PSA0 - PSA12
Path Diagram
End of Problem
```

There are two latent variables a and b in the model. They represent the intercept and slope of the patients linear growth curves. The objective is to estimate the mean vector and covariance matrix of a and b and the error variance of the psa measures. The error variance is assumed to be the same at all occasions.

In the current example, a and b are latent variables, and the line in the input file **psavar1a.spl**

```
a b = CONST
```

specifies that the means of a and b should be estimated.

The output gives the following information

```
        --------------------------------
            EM Algorithm for missing Data:
        --------------------------------

    Number of different missing-value patterns=      14
    Effective sample size:       100
```

```
Convergence of EM-algorithm in      9 iterations
-2 Ln(L) =      1997.49237
Percentage missing values=  13.40
```

The EM algorithm is first used to estimate a saturated model where both the mean vector and covariance matrix are unconstrained. These are used to obtain starting values for the FIML method. This also gives the value $-2\ln(L) = 1997.4924$ for the saturated which is used to obtain a chi-square. After convergence the FIML method gives the following information about the fit of the model.

```
        Global Goodness of Fit Statistics, FIML case


        -2ln(L) for the saturated model =        1997.492
        -2ln(L) for the fitted model    =        2008.601
```

```
Degrees of Freedom = 14
Full Information ML Chi-Square                    11.108 (P = 0.6775)
Root Mean Square Error of Approximation (RMSEA)  0.0
90 Percent Confidence Interval for RMSEA         (0.0 ; 0.0775)
P-Value for Test of Close Fit (RMSEA < 0.05)     0.844
```

Note that the value of `-2ln(L) for the fitted model = 2008.601` is the same as obtained in Section 4.6.1 using multilevel modeling. Also the parameter estimates and their standard errors are approximately the same. This shows that one can use either methodology to estimate the model.

The FIML estimates of the model parameters are given as

```
        Covariance Matrix of Independent Variables
                a           b

                --------    --------

        a       30.899
                (4.612)
                6.700
        b       0.302       0.004
                (0.108)     (0.005)
                2.811       0.728



        Mean Vector of Independent Variables
                a           b

                --------    --------

                31.934      -0.742
                (0.571)     (0.019)
                55.927      -39.871
```

The conclusions from this analysis are

- The average initial PSA value is 31.9 with a variance of 30.9.

- Thus, the initial PSA value varies considerably from patient to patient

- The effect of treatment is highly significant.

- The PSA value decreases by 0.7 per quarter (0.23 per year) and this rate of decrease is the same for all patients.

To estimate the model in Figure 9.4 one can just add Age on the lines for a and b. The SIMPLIS syntax file is **psavar2a.spl**:

```
Linear Model with Covariate for psavar Data
Raw Data from File psavar.LSF
Latent Variables: a b
Relationships
PSA0 = 1*a 0*b
PSA3 = 1*a 3*b
PSA6 = 1*a 6*b
PSA9 = 1*a 9*b
PSA12 = 1*a 12*b
a b = CONST Age
Let the Errors on a and b correlate
Equal Error Variances: PSA0 - PSA12
Path Diagram
End of Problem
```

However, since we already know that all patients have the same slope b, it is not meaningful to predict b from **Age**. Thus instead of the line

```
a b = CONST Age
```

one should use, see file **psavar2aa.spl**

```
a = CONST Age
b = CONST 0*Age
```

The prediction equation for the intercept **a** is estimated as

```
        a = 15.288 + 0.300*Age, Errorvar.= 25.818, R² = 0.177
Standerr  (3.709)  (0.0662)              (3.911)
Z-values   4.121    4.533                 6.601
P-values   0.000    0.000                 0.000
```

Thus, the intercept a depends on age. The intercept increases by 0.30 per year of age, on average.

There is an alternative method of estimation, based on the same assumptions. One can use multiple imputation to obtain a complete data set and then analyze this by maximum likelihood or robust maximum likelihood method. For more information on multiple imputation, see Section 1.6. Since the sample size $N = 100$ is small it is not recommended to use robust estimation. It is best to use maximum likelihood.

For the model in Figure 9.4 the SIMPLIS syntax will be, see file **psavar3a.spl**:

```
Linear Model with Covariate for psavar Data
Estimated by ML using Multiple Imputation
Raw Data from File psavar.lsf
Multiple Imputation with MC
Latent Variables: a b
Relationships
PSA0  = 1*a 0*b
PSA3  = 1*a 3*b
PSA6  = 1*a 6*b
PSA9  = 1*a 9*b
PSA12 = 1*a 12*b
a b = CONST Age
Let the Errors of a and b correlate
Equal Error Variances: PSA0 - PSA12
Path Diagram
End of Problem
```

The only difference between this input file and **psavar2aa.spl** is the line

```
Multiple Imputation with MC
```

which has been added. The output gives the following estimated equation for a:

```
          a = 15.000 + 0.306*Age, Errorvar.= 26.020, R² = 0.183
Standerr  (3.570)  (0.0637)              (3.849)
Z-values   4.202    4.798                 6.761
P-values   0.000    0.000                 0.000
```

which is very similar to previous results.

An advantage of this approach is the one can get more measures of goodness of fit:

Log-likelihood Values

	Estimated Model	Saturated Model
Number of free parameters(t)	10	27
-2ln(L)	1786.231	1764.336
AIC (Akaike, 1974)*	1806.231	1818.336
BIC (Schwarz, 1978)*	1832.283	1888.675

*LISREL uses AIC= 2t - 2ln(L) and BIC = tln(N)- 2ln(L)

Goodness-of-Fit Statistics

Degrees of Freedom for (C1)-(C2)	17
Maximum Likelihood Ratio Chi-Square (C1)	21.895 (P = 0.1888)
Due to Covariance Structure	17.959
Due to Mean Structure	3.936
Browne's (1984) ADF Chi-Square (C2_NT)	22.020 (P = 0.1839)
Estimated Non-centrality Parameter (NCP)	4.895
90 Percent Confidence Interval for NCP	(0.0 ; 21.126)
Minimum Fit Function Value	0.219
Population Discrepancy Function Value (F0)	0.0490
90 Percent Confidence Interval for F0	(0.0 ; 0.211)
Root Mean Square Error of Approximation (RMSEA)	0.0537
90 Percent Confidence Interval for RMSEA	(0.0 ; 0.111)
P-Value for Test of Close Fit (RMSEA < 0.05)	0.421

LISREL Syntax

For **LISREL** syntax of the model in Figure 9.3 one should regard equation (9.7) as

$$\mathbf{x} = \mathbf{\Lambda}_x \xi + \boldsymbol{\delta} \, , \tag{9.17}$$

where

$$\mathbf{\Lambda}_x = \begin{pmatrix} 1 & 0 \\ 1 & 3 \\ 1 & 6 \\ 1 & 9 \\ 1 & 12 \end{pmatrix} \, , \tag{9.18}$$

and

$$\xi = \begin{pmatrix} a \\ b \end{pmatrix} \, . \tag{9.19}$$

The **LISREL** syntax file is **psavar1b.lis**:

```
Linear Growth Curve for psavar Data
da ni=6
ra=psavar.lsf
mo nx=5 nk=2 ka=fr
lk
a b
ma lx
1  0
1  3
1  6
1  9
1 12
eq td(1)-td(5)
pd
ou
```

The results are the same as for **psavar1a.spl**. The mean vector of a and b is in $\boldsymbol{\kappa}$:

```
KAPPA
            a            b

       --------     --------

         31.934       -0.742
        (0.571)      (0.019)
         55.933      -39.870
```

and the covariance matrix of a and b is in $\boldsymbol{\Phi}$:

```
PHI
            a            b

       --------     --------

   a     30.899
         (4.611)
          6.701


   b      0.302        0.004
         (0.108)      (0.005)
          2.812        0.728
```

The LISREL syntax for the model with Age as a covariate is **psavar2b.lis**, not shown here, and the LISREL syntax corresponding to **psavar3a.spl** is **psavar3b.lis**:

```
Linear Model with Covariate for psavar Data
Estimated from Imputed Data
da ni=6
ra=psavar.lsf
mu mc
mo ny=5 ne=2 nx=1 nk=1 lx=id td=ze ps=sy,fr al=fr ka=fr
le
a b
ma ly
1  0
1  3
1  6
1  9
1 12
eq te(1)-te(5)
pd
ou
```

The `mo` line may need some explanation. In Figure 9.4 a and b are η-variables. All one-way paths pointing to η must come from ξ-variables. So we have to define ξ to be identical to the x-variable Age. This is done by the specification `lx=id td=ze` on the `mo` line. Thus, the variance of ξ which is the variance of Age is in the 1×1 matrix $\boldsymbol{\Phi}$.

9.3.2 Example: Learning Curves for of Traffic Controllers

The data used in this example are described by Kanfer and Ackerman (1989)[2]. The data consists of information for 141 U.S. Air Force enlisted personnel who carried out a computerized air traffic controller task developed by Kanfer and Ackerman.

The subjects were instructed to accept planes into their hold pattern and land them safely and efficiently on one of four runways, varying in length and compass directions, according to rules governing plane movements and landing requirements. For each subject, the success of a series of between three and six 10-minute trials was recorded. The measurement employed was the number of correct landings per trial.

The Armed Services Vocational Battery (ASVB) was also administered to each subject. A global measure of cognitive ability, obtained from the sum of scores on the 10 sub-scales of ASVB, is also included in the data.

We analyzed this data in Section 4.6.2 using multilevel modeling methods. The data file there was **kanfer.lsf** but for now we need the data in longitudinal form as in **trial.lsf**. We show the data for the first 10 controllers and the last 10 controllers here:

	TRIAL1	TRIAL2	TRIAL3	TRIAL4	TRIAL5	TRIAL6	ABILITY
1	24.000	27.000	30.000	32.000	38.000	41.000	1.422
2	2.000	3.000	9.000	13.000	13.000	14.000	-0.076
3	12.000	18.000	24.000	24.000	27.000	28.000	-0.674
4	7.000	19.000	23.000	30.000	30.000	32.000	0.115
5	17.000	24.000	28.000	29.000	33.000	34.000	-0.025
6	14.000	23.000	27.000	26.000	29.000	28.000	2.320
7	18.000	24.000	29.000	31.000	34.000	37.000	0.922
8	9.000	14.000	18.000	22.000	25.000	24.000	-1.221
9	3.000	4.000	27.000	22.000	27.000	24.000	-1.111
10	6.000	27.000	25.000	34.000	39.000	44.000	0.139

	TRIAL1	TRIAL2	TRIAL3	TRIAL4	TRIAL5	TRIAL6	ABILITY
132	14.000	14.000	23.000	27.000	26.000	26.000	-0.005
133	5.000	11.000	16.000	29.000	24.000	25.000	-0.871
134	13.000	25.000	27.000	27.000	37.000	45.000	0.353
135	25.000	20.000	22.000	29.000	29.000	30.000	-0.402
136	13.000	22.000	34.000	44.000	46.000	51.000	0.260
137	20.000	27.000	29.000	29.000	31.000	33.000	1.311
138	5.000	12.000	19.000	18.000	17.000	-9.000	-0.249
139	13.000	19.000	30.000	32.000	29.000	-9.000	-0.586
140	7.000	15.000	21.000	24.000	25.000	-9.000	-0.847
141	2.000	7.000	12.000	-9.000	-9.000	-9.000	-1.245

Missing values appear as -9.000. It is seen that controllers 138, 139, 140 only completed the first 5 trials, and controller 141 only completed the first 3 trials. Thus there are missing values also in this data. Of course, the ordering of the controllers in the data file is irrelevant. So the controllers with missing data could have been anywhere in the data.

One can deal with the missing values using FIML in the same way as in the previous example. But since there are only 7 missing values out of $141 \times 7 = 987$ values in the data file we will use a different approach here. We will replace the missing values by the means of the variables. This data file is called **trial_filled.lsf**. The last 11 controllers are now

---

[2]Permission for SSI to use the copyrighted raw data was provided by R. Kanfer and P.L. Ackerman. The data are from experiments reported in Kanfer and Ackerman (1989). The data remain the copyrighted property of Ruth Kanfer and Phillip L. Ackerman. Further publication or further dissemination of these data is not permitted without the expressed consent of the copyright owners.

131	19.000	25.000	30.000	31.000	38.000	38.000	0.643
132	14.000	14.000	23.000	27.000	26.000	26.000	-0.005
133	5.000	11.000	16.000	29.000	24.000	25.000	-0.871
134	13.000	25.000	27.000	27.000	37.000	45.000	0.353
135	25.000	20.000	22.000	29.000	29.000	30.000	-0.402
136	13.000	22.000	34.000	44.000	46.000	51.000	0.260
137	20.000	27.000	29.000	29.000	31.000	33.000	1.311
138	5.000	12.000	19.000	18.000	17.000	33.919	-0.249
139	13.000	19.000	30.000	32.000	29.000	33.919	-0.586
140	7.000	15.000	21.000	24.000	25.000	33.919	-0.847
141	2.000	7.000	12.000	30.777	32.299	33.919	-1.245

In Section 4.6.2 we used a box plot to show that the growth curves could be non-linear. Since the time T_t is not included in the data in longitudinal form, it is not possible to produce such a box plot as was done in Section 4.6.2. But one can plot the values of selected individual controllers to show that the growth curve might be non-linear. Figure 9.6 shows cubic growth curves fitted to three controllers. It is seen that there may be some variation of the growth curves across

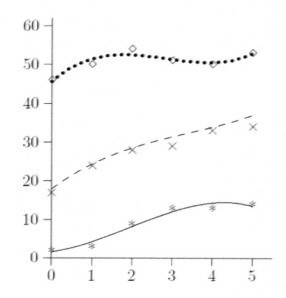

Figure 9.6: Cubic Growth Curves Fitted to Three Controllers Data

controllers. In particular, it is clear that there will be a large variation in the intercept. The following analysis will confirm these initial observations.

In Section 4.6.2 we considered three different models:

Linear Model: $y_{it} = a_i + b_{i1}T_{it} + e_{it}$

Quadratic Model: $y_{it} = a_i + b_{i1}T_{it} + b_{i2}T_{it}^2 + e_{it}$

Cubic Model: $y_{it} = a_i + b_{i1}T_{it} + b_{i2}T_{it}^2 + b_{i3}T_{it}^3 + e_{it}$

where $T_t = t - 1$, $t = 1, 2, \ldots, 6$.

Since the linear and quadratic growth curves are special cases of the cubic growth curve we consider only the cubic here. The objective of the analysis is to estimate the mean vector and covariance matrix of $(a_i, b_{i1}, b_{i2}, b_{i3})$ and the error variance σ_e^2.

Let $\mathbf{x} = (y_{i1}, y_{i2}, \ldots, y_{i6})'$ and $\boldsymbol{\xi} = (a_i, b_{i1}, b_{i2}, b_{i3})'$. With LISREL notation the model is

$$\mathbf{x} = \boldsymbol{\Lambda}_x \boldsymbol{\xi} + \boldsymbol{\delta} ,$$ (9.20)

where

$$\Lambda_x = \begin{pmatrix} 1 & 0 & 0 & 0 \\ 1 & 1 & 1 & 1 \\ 1 & 2 & 4 & 8 \\ 1 & 3 & 9 & 27 \\ 1 & 4 & 16 & 64 \\ 1 & 5 & 25 & 125 \end{pmatrix}, \qquad (9.21)$$

$$\Phi = Cov(\boldsymbol{\xi}), \qquad (9.22)$$

and $Cov(\boldsymbol{\delta}) = \sigma_e^2 \mathbf{I} = Var(e_{it})\mathbf{I}$. Then the mean vector and covariance matrix of $\mathbf{x}$ is

$$\boldsymbol{\mu} = \Lambda_x \boldsymbol{\kappa}, \qquad (9.23)$$

$$\boldsymbol{\Sigma} = \Lambda_x \Phi \Lambda_x' + \sigma_e^2 \mathbf{I}. \qquad (9.24)$$

The parameter matrices in the model are $\boldsymbol{\kappa}$, Φ and σ_e^2.

Following the previous example, the **SIMPLIS** syntax is straight forward, see file **trial1a.spl**:

```
Cubic Growth Curve for TRIAL Data
Raw Data from File TRIAL_filled.LSF
Latent Variables: a b1 b2 b3
Relationships
TRIAL1 = 1*a 0*b1    0*b2     0*b3
TRIAL2 = 1*a 1*b1    1*b2     1*b3
TRIAL3 = 1*a 2*b1    4*b2     8*b3
TRIAL4 = 1*a 3*b1    9*b2    27*b3
TRIAL5 = 1*a 4*b1   16*b2    64*b3
TRIAL6 = 1*a 5*b1   25*b2   125*b3
a b1 b2 b3= CONST
Equal Error Variances: TRIAL1 - TRIAL6
Path Diagram
End of Problem
```

The sample means, variances and covariances are given in the output file **trial1a.out**:

Means

	TRIAL1	TRIAL2	TRIAL3	TRIAL4	TRIAL5	TRIAL6
	11.631	21.156	27.305	30.878	32.392	34.196

Covariance Matrix

	TRIAL1	TRIAL2	TRIAL3	TRIAL4	TRIAL5	TRIAL6
TRIAL1	53.249					
TRIAL2	49.079	76.090				
TRIAL3	39.356	63.388	79.913			
TRIAL4	33.649	58.070	72.854	85.335		
TRIAL5	31.581	54.008	66.862	76.019	81.839	
TRIAL6	32.392	56.316	68.276	78.350	80.401	94.133

Note that both the mean and variance of TRIAL1 are somewhat smaller than the other means and variances. Note also that both the means and the variances increases over the learning process. Thus, although the controllers get better and better on average there is also a greater variation in the learning process as it proceeds.

The estimated mean vector of $(a_i, b_{i1}, b_{i2}, b_{i3})$ is

```
Mean Vector of Independent Variables
```

a	b1	b2	b3
11.584	11.913	-2.401	0.184
(0.621)	(0.648)	(0.284)	(0.036)
18.643	18.392	-8.448	5.162

Since all the estimated means are highly significant the estimated average growth curve is a cubic. The estimated growth curve is shown in Figure 9.7 together with the six observed means marked as ×. It is seen that the fit is almost perfect.

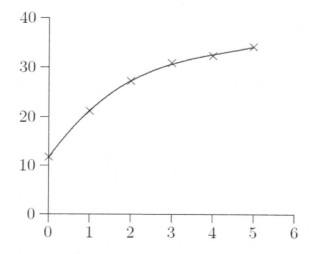

Figure 9.7: Average Cubic Growth Curve and Observed Means

The estimated covariance matrix of $(a_i, b_{i1}, b_{i2}, b_{i3})$ is

Covariance Matrix of Independent Variables

	a	b1	b2	b3
a	46.766			
	(6.516)			
	7.177			
b1	5.349	30.196		
	(4.864)	(7.455)		
	1.100	4.050		
b2	-5.607	-9.650	4.238	
	(2.125)	(3.162)	(1.484)	
	-2.639	-3.052	2.856	
b3	0.798	0.913	-0.471	0.056
	(0.267)	(0.379)	(0.185)	(0.024)
	2.985	2.411	-2.539	2.358

All the variances are highly significant which suggests that there is considerable variation of the shape of the growth curves across controllers.

The error variance is estimated as 7.992. Also this is highly significant.

Some goodness-of-fit measures are given in the output file as

```
Degrees of Freedom for (C1)-(C2)              12
Maximum Likelihood Ratio Chi-Square (C1)      26.971 (P = 0.0078)
Due to Covariance Structure                   25.147
Due to Mean Structure                         1.823
Browne's (1984) ADF Chi-Square (C2_NT)        27.875 (P = 0.0058)
```

This indicates that most of the lack of fit is due to the covariance structure (9.24). The most reasonable way to improve the fit is to allow the error variances of to be different. This can be done by deleting the line

```
Equal Error Variances: TRIAL1 - TRIAL6
```

see file **trial1aa.spl**. This gives the following fit measures

```
Degrees of Freedom for (C1)-(C2)              7
Maximum Likelihood Ratio Chi-Square (C1)      7.540 (P = 0.3749)
Due to Covariance Structure                   5.415
Due to Mean Structure                         2.125
Browne's (1984) ADF Chi-Square (C2_NT)        7.430 (P = 0.3856)
```

indicating an acceptable fit. However, the error variance of **TRIAL1** is negative, although not significant. To resolve this problem one can set this error variance to some small value, such as 1 as is done in file **trial1aaa.spl**. This gives the following fit measures

```
Degrees of Freedom for (C1)-(C2)                        8
Maximum Likelihood Ratio Chi-Square (C1)                10.883 (P = 0.2084)
Due to Covariance Structure                             8.769
Due to Mean Structure                                   2.113
Browne's (1984) ADF Chi-Square (C2_NT)                  11.072 (P = 0.1976)
```

still indicating an acceptable fit.

Next we investigate whether the shape of the growth curve can be predicted by the ability measure. To do this we regress each of $(a_i, b_{i1}, b_{i2}, b_{i3})$ on ABILITY. The SIMPLIS file for this is **trial2a.spl**:

```
Cubic Growth Curve with Covariate for TRIAL Data
Raw Data from File TRIAL_filled.LSF
Latent Variables: a b1 b2 b3
Relationships
TRIAL1 = 1*a 0*b1    0*b2      0*b3
TRIAL2 = 1*a 1*b1    1*b2      1*b3
TRIAL3 = 1*a 2*b1    4*b2      8*b3
TRIAL4 = 1*a 3*b1    9*b2      27*b3
TRIAL5 = 1*a 4*b1    16*b2     64*b3
TRIAL6 = 1*a 5*b1    25*b2     125*b3
a b1 b2 b3= CONST ABILITY
Let the errors of a-b3 correlate
Set the Error Variance of TRIAL1 to 1
Path Diagram
End of Problem
```

The four regression equations are estimated as

```
        a = 11.625 + 3.686*ABILITY, Errorvar.= 38.764, R² = 0.259
Standerr  (0.531)  (0.508)                       (7.754)
Z-values  21.892   7.256                          4.999
P-values   0.000   0.000                          0.000


        b1 = 11.982 + 0.624*ABILITY, Errorvar.= 39.047, R² = 0.00988
Standerr  (0.649)  (0.644)                        (6.770)
Z-values  18.470   0.969                           5.768
P-values   0.000   0.333                           0.000


       b2 =  - 2.453  - 0.554*ABILITY, Errorvar.= 4.925 , R² = 0.0587
Standerr      (0.278)  (0.278)                      (1.294)
Z-values      -8.827   -1.992                         3.805
P-values       0.000    0.046                         0.000


       b3 = 0.192    + 0.0784*ABILITY, Errorvar.= 0.0632 , R² = 0.0887
Standerr   (0.0344)   (0.0335)                       (0.0217)
Z-values    5.577      2.340                           2.916
P-values    0.000      0.019                           0.004
```

Since the R^2s are very small it is clear that ABILITY explains very little of the variation in $(a_i, b_{i1}, b_{i2}, b_{i3})$. For the intercept a 26% of the variance is explained, but for the other coefficients less than 10% of the variance is explained.

To use LISREL syntax, let $\mathbf{y} = (y_{i1}, y_{i2}, \ldots, y_{i6})'$ and $\boldsymbol{\eta} = (a_i, b_{i1}, b_{i2}, b_{i3})'$ and write the model as

$$\mathbf{y} = \boldsymbol{\Lambda}_y \boldsymbol{\eta} + \boldsymbol{\delta} , \tag{9.25}$$

where

$$\boldsymbol{\eta} = \boldsymbol{\alpha} + \boldsymbol{\Gamma}\xi + \boldsymbol{\zeta}, , \tag{9.26}$$

and set $\xi \equiv x$ which is the covariate ABILITY. The identity $\xi \equiv x$ is obtained by setting $\boldsymbol{\Lambda}_x = \mathbf{I}$ and $\boldsymbol{\Theta}_\delta = \mathbf{0}$. The matrix $\boldsymbol{\Gamma}$ is a 4×1 vector.

The LISREL syntax file corresponding to **trial2a.spl** is **trial2b.lis**:

```
Cubic Growth Curve with Covariate for TRIAL Data
da ni=7
ra=TRIAL_filled.lsf
mo ny=6 ne=4 nx=1 nk=1 lx=id td=ze ps=sy,fr al=fr ka=fr
le
a b1 b2 b3
ma ly
1  0   0    0
1  1   1    1
1  2   4    8
1  3   9   27
1  4  16   64
1  5  25  125
fix te(1)
va 1 te(1)
pd
ou
```

The result is the same as for **trial2a.spl**. The structural regression coefficients appear in GAMMA:

```
        GAMMA

             ABILITY
             --------
     a        3.686
             (0.531)
              6.941


     b1       0.624
             (0.649)
              0.962
```

```
b2       -0.554
        (0.278)
        -1.994


b3        0.078
        (0.034)
         2.282
```

9.4   Latent Growth Curves and Dyadic Data

Dyadic data is when the unit of observation is a pair, where the same data is available for each individual in the pair. Often there is an interest in symmetry between the individuals in the pair. We have already seen an example of this in Section 8.8.1, where the pair was the respondent and his/her best friend. Another example is given in Section 10.2.8, where the pair is a twin pair. For a thorough analysis of dyadic data analysis, see the book by Kenny, Kashy, and Cook (2006).

Here we consider an example of the combination of latent growth curves and dyadic data.

9.4.1   Example: Quality of Marriages

Husbands and wives rated the quality of their marriage over five annual assessment using Spanier's Dyadic Adjustment Scale which ranges from 0 - 151.[3] The data file in longitudinal form is **dyad.lsf**, where missing values are coded as -9.000. The first few lines of the data file looks like this

HQUAL1	HQUAL2	HQUAL3	HQUAL4	HQUAL5	WQUAL1	WQUAL2	WQUAL3	WQUAL4	WQUAL5
110.000	108.000	109.000	118.000	117.000	111.000	106.000	101.000	109.000	104.000
117.000	-9.000	-9.000	-9.000	-9.000	141.000	-9.000	-9.000	-9.000	-9.000
131.000	-9.000	-9.000	-9.000	-9.000	131.000	-9.000	-9.000	-9.000	-9.000
125.000	121.000	118.000	118.000	117.000	116.000	109.000	96.000	112.000	107.000
135.000	126.000	126.000	126.000	120.000	136.000	126.000	132.000	120.000	130.000
120.000	109.000	101.000	109.000	105.000	116.000	107.000	106.000	121.000	106.000
115.000	-9.000	-9.000	-9.000	-9.000	117.000	-9.000	-9.000	-9.000	-9.000
106.000	-9.000	-9.000	-9.000	-9.000	90.000	-9.000	-9.000	-9.000	-9.000
111.000	109.000	114.000	114.000	99.000	108.000	113.000	112.000	100.000	106.000
117.000	-9.000	-9.000	-9.000	-9.000	129.000	-9.000	-9.000	-9.000	-9.000

A data screening gives the following information about missing values.

```
Number of Missing Values per Variable
```

HQUAL1	HQUAL2	HQUAL3	HQUAL4	HQUAL5	WQUAL1	WQUAL2	WQUAL3
0	132	219	266	299	0	132	219

```
Number of Missing Values per Variable
```

WQUAL4	WQUAL5
266	299

```
Total Sample Size(N) =    538
```

---

[3]The data was collected and made available by the late Larry Kurdek.

There are 538 married couples and all of them rated their marriage quality the first year but after that the number of missing values increases each year. In the fifth year as many as 299 couples did not rate their marriage. We will not speculate in the reasons for this increase in missing values. We assume MAR although the individual probability of missingness may depend on the his/her quality measure.

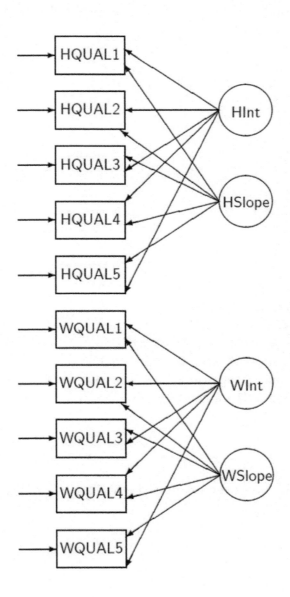

Figure 9.8: Growth Curve Model for Ratings of Quality of Marriage

We assume linear growth curves and the model we want to estimate is shown in Figure 9.8, where the intercepts are labeled HInt and Wint and the slopes are labeled HSlope and WSlope. Although not shown in the path diagram, we assume that HInt and HSlope are correlated and also that HSlope and WSlope are correlated. Except for these correlations all other correlations between the latent variables are zero. All path from the latent variables to the observed variables are fixed; those emanating from **HInt** and **WInt** are all fixed equal to 1 and those emanating from **HSlope** and **WSlope** are 0, 1, 2, 3, and 4 and similarly for **HSlope** and **WSlope**.

The objective is to estimate the means and variances of the intercepts and slopes and to test whether they are the same for husbands and wives. We begin by estimating the model without imposing any equality constraints. This is done by the **SIMPLIS** syntax file **dyad1a.spl**:

```
Quality of Marriages
Raw Data from File dyad.lsf
Latent Variables: HInt HSlope WInt WSlope
Relationships
HQUAL1 = 1*HInt 0*HSlope
HQUAL2 = 1*HInt 1*HSlope
HQUAL3 = 1*HInt 2*HSlope
HQUAL4 = 1*HInt 3*HSlope
HQUAL5 = 1*HInt 4*HSlope
WQUAL1 = 1*WInt 0*WSlope
WQUAL2 = 1*WInt 1*WSlope
WQUAL3 = 1*WInt 2*WSlope
WQUAL4 = 1*WInt 3*WSlope
WQUAL5 = 1*WInt 4*WSlope
Equal Error Variances HQUAL1 - HQUAL5
Equal Error Variances WQUAL1 - WQUAL5
Set the covariance between HInt and WInt to 0
Set the covariance between HInt and WSlope to 0
Set the covariance between HSlope and WInt to 0
Set the covariance between HSlope and WSlope to 0
HInt HSlope WInt WSlope = CONST
Path Diagram
End of Problem
```

The output file **dyad1a.out** gives the estimated covariance matrix for the latent variables as

	HInt	HSlope	WInt	WSlope
HInt	136.227			
	(10.597)			
	12.855			
HSlope	-4.605	5.238		
	(2.547)	(0.929)		
	-1.808	5.638		
WInt	- -	- -	119.153	
			(9.846)	
			12.102	
WSlope	- -	- -	2.992	6.928
			(2.653)	(1.130)
			1.128	6.133

We conclude that there is significant variation of the intercepts and slopes for both husbands and wives. The mean vector of the latent variables is given as

HInt	HSlope	WInt	WSlope
118.903	-2.435	120.687	-2.858
(0.565)	(0.185)	(0.544)	(0.207)
210.320	-13.164	222.003	-13.837

So the estimated average growth curve is

$$\text{HQUAL} = 118.903 - 2.435T_t$$

for husbands and

$$\text{WQUAL} = 120.687 - 2.858T_t \, ,$$

for wives, where $T_t = t - 1$, $t = 1, 2, 3, 4, 5$.

For **LISREL** syntax instead of **SIMPLIS** syntax one can use the following **dyad1b.lis**:

```
Quality of Marriages
da ni=10
ra=dyad.lsf
mo nx=10 nk=4 ph=sy,fr ka=fr
lk
HInt HSlope WInt WSlope
ma lx
1 0 0 0
1 1 0 0
1 2 0 0
1 3 0 0
1 4 0 0
0 0 1 0
0 0 1 1
0 0 1 2
0 0 1 3
0 0 1 4
pa ph
1
1 1
0 0 1
0 0 1 1
eq td(1)-td(5)
eq td(6)-td(10)
pd
ou
```

This gives the same results. The covariance matrix of the latent variables is given in the PHI matrix:

PHI

	HInt	HSlope	WInt	WSlope
	--------	-------	--------	-------
HInt	136.227			
	(10.597)			
	12.855			
HSlope	-4.605	5.238		
	(2.547)	(0.929)		
	-1.808	5.638		
WInt	- -	- -	119.153	
			(9.846)	
			12.102	
WSlope	- -	- -	2.992	6.928
			(2.653)	(1.130)
			1.128	6.133

and the mean vector of the latent variables is given in the vector KAPPA:

```
KAPPA

        HInt      HSlope      WInt      WSlope
      --------    --------   --------   --------
      118.903     -2.435     120.687    -2.858
      (0.565)     (0.185)    (0.544)    (0.207)
      210.320    -13.164     222.003   -13.837
```

As judged by the FIML chi-square the model does not fit the data with chi-square $= 778.11$ with 53 degrees of freedom. The fit can be considerably improved by using quadratic growth curves instead of linear and by allowing some correlated measurement errors, see file **dyad1bb.lis**. However, we shall not consider this model here but rather demonstrate how one can test the hypothesis of equal average growth curves for husbands and wives.

The hypothesis of equal average intercept can be tested by adding the line

```
eq ka(1) ka(3)
```

in **dyad1b.lis**, see file **dyad2b.lis**.

The hypothesis of equal average slopes can be tested by adding the line

```
eq ka(2) ka(4)
```

in **dyad1b.lis**, see file **dyad3b.lis**.

One can also test the hypothesis of equal variances of the intercepts by adding the line

```
eq ph(1,1) ph(3,3)
```

in **dyad1b.lis**, see file **dyad4b.lis**.

Similarly one can test the hypothesis of equal variances of the slopes by adding the line

```
eq ph(2,2) ph(4,4)
```

in **dyad1b.lis**, see file **dyad5b.lis**.

By comparing the overall FIML chi-square for each of these hypotheses with 53 degrees of freedom with the chi-square 778.11 with 54 degrees of freedom and computing the chi-square differences, we obtain the chi-squares with 1 degree of freedom, with the results shown in Table 9.4.

Table 9.4: Various Chi-Squares for Testing Equality of Intercepts, Slopes, and their Variances

Hypothesis	Chi-Square
$\kappa_1 = \kappa_3$	5.16
$\kappa_2 = \kappa_4$	2.27
$\phi_{11} = \phi_{33}$	1.37
$\phi_{22} = \phi_{44}$	0.97

From these chi-squares we conclude that the average growth curve is the same for both husbands and wives. But each individual has its own growth curve. Although the variation around the average growth curve is significant for both husbands and wives, this variation is the same for both husbands and wives.

To specify $\kappa_1 = \kappa_3$ in **SIMPLIS** syntax use the line

```
Set CONST -> HInt = CONST -> WInt
```

see file **dyad2a.spl**.

To specify $\kappa_2 = \kappa_4$ in **SIMPLIS** syntax use the line

```
Set CONST -> HSlope = CONST -> WSlope
```

see file **dyad3a.spl**.

To specify $\phi_{11} = \phi_{33}$ in **SIMPLIS** syntax use the line

```
Set the Variance of HInt = the Variance of WInt
```

see file **dyad4a.spl**.

To specify $\phi_{22} = \phi_{44}$ in **SIMPLIS** syntax use the line

```
Set the Variance of HSlope = the Variance of WSlope
```

see file **dyad5a.spl**.

The resulting chi-squares are the same whether **SIMPLIS** or **LISREL** syntax is used.

Chapter 10

Multiple Groups

Consider a set of G populations. These may be different nations, states or regions, culturally or socioeconomically different groups, groups of individuals selected on the basis of some known selection variables, groups receiving different treatments, and control groups, etc. In fact, they may be any set of mutually exclusive groups of individuals that are clearly defined. It is assumed that a number of variables have been measured on a number of individuals from each population. This approach is particularly useful in comparing a number of treatment and control groups regardless of whether individuals have been assigned to the groups randomly or not.

Any LISREL model may be specified and fitted for each group of data. However, LISREL assumes by default that the models are identically the same over groups, i.e., all relationships and all parameters are the same in each group. Thus, only differences between groups need to be specified. Our first example of a multi-group analysis clarifies how this is done.

The most common application of multi-group analysis is factorial invariance. The hypothesis of factorial invariance states that the factor loadings are the same in all groups. Group differences in variances and covariances of the observed variables are due only to differences in variances and covariances of the factors and different error variances. The idea of factorial invariance is that the factor loadings are attributes of the tests and they should therefore be independent of the population sampled, whereas the distribution of the factors themselves could differ across populations. A stronger assumption is to assume that the error variances are also equal across groups:

Meredith (1964) discussed the problem of factorial invariance from the point of view of multivariate selection. Jöreskog (1971) considered several alternative models for studying differences in covariance matrices across groups. He also showed how the model of factorial invariance can be estimated by the maximum likelihood method. Further problems and issues in factorial invariance are discussed by Millsap and Meredith (2007). Sörbom (1974) extended the model of factorial invariance to include intercepts and means of latent variables.

10.1 Factorial Invariance

Consider the situation where the same tests have been administered in G different groups and the factor analysis model is applied in each group:

$$\mathbf{x}_g = \boldsymbol{\Lambda}_g \boldsymbol{\xi}_g + \boldsymbol{\delta}_g, \; g = 1, 2, \ldots, G \,, \tag{10.1}$$

© Springer International Publishing Switzerland 2016

K.G. Jöreskog et al., *Multivariate Analysis with LISREL*, Springer Series in Statistics,
DOI 10.1007/978-3-319-33153-9_10

where, as before, $\boldsymbol{\xi}_g$ and $\boldsymbol{\delta}_g$ are uncorrelated. The covariance matrix of $\mathbf{x}_g$ in group g is[1]

$$\boldsymbol{\Sigma}_g = \boldsymbol{\Lambda}_g \boldsymbol{\Phi}_g \boldsymbol{\Lambda}'_g + \boldsymbol{\Theta}_g^2 \ . \tag{10.2}$$

The hypothesis of factorial invariance is:

$$\boldsymbol{\Lambda}_1 = \boldsymbol{\Lambda}_2 = \cdots = \boldsymbol{\Lambda}_G \ . \tag{10.3}$$

This states that the factor loadings are the same in all groups. Group differences in variances and covariances of the observed variables are due only to differences in variances and covariances of the factors and different error variances. The idea of factorial invariance is that the factor loadings are attributes of the tests and they should therefore be independent of the population sampled, whereas the distribution of the factors themselves could differ across populations. A stronger assumption is to assume that the error variances are also equal across groups:

$$\boldsymbol{\Theta}_1 = \boldsymbol{\Theta}_2 = \cdots = \boldsymbol{\Theta}_G \ . \tag{10.4}$$

Sörbom (1974) extended the model of factorial invariance to include intercepts $\boldsymbol{\tau}$ in (10.1):

$$\mathbf{x}_g = \boldsymbol{\tau}_g + \boldsymbol{\Lambda}_g \boldsymbol{\xi}_g + \boldsymbol{\delta}_g, \ g = 1, 2, \ldots, G \ . \tag{10.5}$$

The mean vector $\boldsymbol{\mu}_g$ of $\mathbf{x}_g$ is

$$\boldsymbol{\mu}_g = \boldsymbol{\tau}_g + \boldsymbol{\Lambda} \boldsymbol{\kappa}_g \ , \tag{10.6}$$

where $\boldsymbol{\kappa}_g$ is the mean vector of $\boldsymbol{\xi}_g$.

The model of complete factorial invariance is:

$$\boldsymbol{\tau}_1 = \boldsymbol{\tau}_2 = \cdots = \boldsymbol{\tau}_G \ , \tag{10.7}$$

$$\boldsymbol{\Lambda}_1 = \boldsymbol{\Lambda}_2 = \cdots = \boldsymbol{\Lambda}_G \ . \tag{10.8}$$

Sörbom (1974) showed that one can estimate the mean vector and covariance matrix of $\boldsymbol{\xi}$ in each group on a scale common to all groups. In particular, this makes it possible to estimate group differences in means of the latent variables.

Let $\bar{\mathbf{x}}_g$ and $\mathbf{S}_g$ be the sample mean vector and covariance matrix of $\mathbf{x}_g$ in group g, and let $\boldsymbol{\mu}_g(\boldsymbol{\theta})$ and $\boldsymbol{\Sigma}_g(\boldsymbol{\theta})$ be the corresponding population mean vector and covariance matrix, where $\boldsymbol{\theta}$ is a vector of all independent parameters in $\boldsymbol{\tau}$, $\boldsymbol{\Lambda}$ and $\boldsymbol{\kappa}_g$, $\boldsymbol{\Phi}_g$, $\boldsymbol{\Theta}_g$, $g = 1, 2, \ldots, G$. The fit function for the multigroup case is defined as

$$F(\boldsymbol{\theta}) = \sum_{g=1}^{G} \frac{N_g}{N} F_g(\boldsymbol{\theta}) \ , \tag{10.9}$$

where $F_g(\boldsymbol{\theta}) = F(\bar{\mathbf{z}}_g, \mathbf{S}_g, \boldsymbol{\mu}_g(\boldsymbol{\theta}), \boldsymbol{\Sigma}_g(\boldsymbol{\theta}))$ is any of the fit functions defined for a single group. Here N_g is the sample size in group g and $N = N_1 + N_2 + \ldots + N_G$ is the total sample size. To estimate the model, one usually fixes a one in each column of $\boldsymbol{\Lambda}$ and the mean vector of $\boldsymbol{\xi}$ to 0 in one group and leaves $\boldsymbol{\Phi}_g$ and $\boldsymbol{\Theta}_g$ free in each group, see e.g., Sörbom (1974). If the ML fit function is used, one can use N times the minimum of F as a χ^2 with degrees of freedom $d = Gp(p+1)/2 - t$ for testing the model. Here p is the number of observed variables and t is the number of independent parameters estimated.

---

[1]For simplicity we write $\boldsymbol{\Theta}$ instead of $\boldsymbol{\Theta}_\delta$ in this presentation.

10.2 Multiple Groups with Continuous Variables

In this section we consider the estimation of models for multiple groups where the variables are continuous. The case of multiple groups and ordinal variables are considered in Section 10.3.

10.2.1 Equal Regressions

In this section, we consider the problem of testing whether a regression equation is the same in several populations. Suppose that a dependent variable y and a number of explanatory variables $x_1, x_2, \ldots, x_q$ are observed in two or more groups. We are interested in determining the extent to which the regression equation

$$y = \alpha + \gamma_1 x_1 + \gamma_2 x_2 + \cdots + \gamma_q x_q + z \qquad (10.10)$$

is the same in different groups. We say that the regressions are:

- equal if $\alpha, \gamma_1, \gamma_2, \ldots, \gamma_q$, are the same in all groups

- parallel if $\gamma_1, \gamma_2, \ldots, \gamma_q$, are the same in all groups

Normally the covariance matrix of the x-variables is not expected to be the same across groups and often one finds that the intercept terms differ between groups. It may also be the case that *only some* of the regression coefficients are the same across groups.

10.2.2 Example: STEP Reading and Writing Tests in Grades 5 and 7

The data for this example are based on scores on the ETS Sequential Test of Educational Progress (STEP) for two groups of boys who took the test in both Grade 5 and Grade 7. The two groups were defined according to whether or not they were in the academic curriculum in Grade 12. We call the groups Boys Academic (BA) and Boys Non-Academic (BNA). We will use this data to demonstrate how one can test the equality of various regressions.

The data files are **stepba.lsf** and **stepbna.lis** where all the variables are continuous.

As a first example, we consider the testing of equal regressions of STEP Reading at Grade 7 on STEP Reading and Writing at Grade 5. STEP Writing at Grade 7 is not involved in this example. For example, we may want to predict STEP Reading at Grade 7 from STEP Reading and Writing at Grade 5 and to see if the prediction equation is the same for both groups. A path diagram is shown in Figure 10.1, where CONST is a variable which is constant equal to 1 for every case. The intercept α in the regression equation is the coefficient of CONST.

The input file for this example is **step1a.spl**:

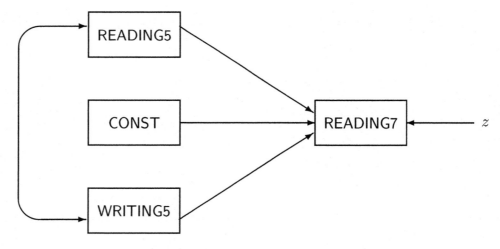

Figure 10.1: Path Diagram for Regression of READING7

```
Group BA: STEP Reading and Writing, Grades 5 and 7
Raw Data from File stepba.lsf
Equation: READING7 = CONST READING5 WRITING5

Group BNA: STEP Reading and Writing, Grades 5 and 7
Raw Data from File stepbna.lsf
Set the Error Variance of READING7 free
End of Problem
```

Note the following:

- Four variables are read but only three variables are used in the model. The program automatically selects the variables for analysis on the basis of the variables included on the **Equation** line.

- The word **Equation** is synonymous with **Relationship** but allows the equation to be written on the same line as the **Equation** line.

- The intercept α in the regression equation is the regression coefficient of the variable CONST. The line

  ```
  Equation: READING7 = CONST READING5 WRITING5
  ```

  is interpreted in the usual way, i.e., three coefficients α, γ_1, and γ_2 will be estimated.

- The line

  ```
  Set the Error Variance of READING7 Free
  ```

  is needed in the second group; otherwise the regression error variance will be assumed to be the same in both groups.

The test of equal regressions gives a chi-square of 34.859 with three degrees of freedom. To test for parallel regressions rather than equal regressions, one must allow α to be different in the two groups. This is done by adding the line

```
Set the Path from CONST to READING7 Free
```

in the second group (see file **step2a.spl**). Another way of specifying the same thing is to include the relationship

```
READING7 = CONST
```

in the second group. The meaning of this is that α, the coefficient of CONST, will be re-estimated in the second group, while the remaining part of the regression equation remains the same as in group 1, i.e., the coefficients of READING5 and WRITING5 will be the same in both groups.

The test of parallel regressions gives a chi-square of 3.240 with two degrees of freedom. Thus, it is evident that the regressions are parallel but not equal.

As a second example, consider estimating two regression equations simultaneously in two groups and testing whether both regressions are parallel in the two groups. We estimate the regressions of READING7 and WRITING7 on READING5 and WRITING5 (**step3a.spl**):

```
Group BA: STEP Reading and Writing, Grades 5 and 7
Raw data from file stepba.lsf
Equations: READING7 - WRITING7 = CONST READING5 WRITING5
Set the Error Covariance between READING7 and WRITING7 free

Group BNA: STEP Reading and Writing, Grades 5 and 7
Raw data from file stepbna.lsf
Equations: READING7 - WRITING7 = CONST
Set the Error Variances of READING7 - WRITING7 Free
Set the Error Covariance Matrix of READING7 and WRITING7 Free
Path Diagram
End of Problem
```

- The line

  ```
  Set the Error Covariance between READING7 and WRITING7 Free
  ```

  in the first group, specifies that the two error terms z_1 and z_2 may be correlated, i.e., we don't believe that READING5 and WRITING5 will account for the whole correlation between READING7 and WRITING7.

- The line

  ```
  Equations: READING7 - WRITING7 = CONST
  ```

  in the second group, specifies that the intercept terms for the second group do not have to equal those of the first group.

- The line

  ```
  Set the Error variances of READING7 - WRITING7 Free
  ```

in the second group specifies that the variances of z_1 and z_2 in the second group are not constrained to be equal to those of the first group.

- The line

  ```
  Set the Error Covariance between READING7 and WRITING7 Free
  ```

  in the second group specifies that the covariance between z_1 and z_2 in the second group is not constrained to be equal to that of the first group.

Chi-square for this model is 8.789 with 4 degrees of freedom and a P-value of 0.067.

10.2.3   Estimating Means of Latent Variables

In this section we continue the analysis of the data on STEP Reading and Writing tests used in the previous analysis.

Another way of modeling the data is to take measurement error in the observed variables into account and to regard Reading and Writing as two indicators of a latent variable Verbal Ability, say, and to estimate the regression of Verbal Ability at Grade 7 on Verbal Ability at Grade 5. However, first we must learn how to estimate means of latent variables.

Since a latent variable is unobservable, it does not have an intrinsic scale. Neither the origin nor the unit of measurement are defined. In a single population the origin is fixed by assuming that all observed variables are measured in deviations from their means and that the means of all latent variables are zero. The unit of measurement of each latent variable is usually fixed either by assuming that it is a standardized variable with variance 1 or by fixing a non-zero loading for a reference variable.

In multi-group studies, these restrictions can be relaxed somewhat by assuming that the latent variables are on the same scale in all groups. The common scale may be defined by assuming that the means of the latent variables are zero in one group and that the loadings of the observed variables on the latent variables are invariant over groups, with one loading for each latent variable fixed for a reference variable. Under these assumptions it is possible to estimate the means and covariance matrices of the latent variables relative to this common scale. To illustrate this we will first consider a small example based on the same data used in the previous section, and then we consider a larger example.

In the first part of this example we take READING5 and WRITING5 to be indicators of a latent variable Verbal5 (Verbal Ability at Grade 5) and estimate the mean difference in Verbal5 between groups.

The measurement model for READING5 and WRITING5 is:

$$\text{READING5} = \alpha_1 + \lambda_1 \text{Verbal5} + D1$$

$$\text{WRITING5} = \alpha_2 + \lambda_2 \text{Verbal5} + D2$$

Note that these relationships now have intercept terms α_1 and α_2. The scale for Verbal5 is fixed by assuming that the mean is zero in Group BA and that $\lambda_1 = 1$. It is further assumed that $\alpha_1, \alpha_2, \lambda_1, \lambda_2$ are invariant over groups. We can then estimate the mean of Verbal5 in Group BNA as well as all the other parameters. The mean of Verbal5 is interpreted as the mean difference in verbal ability between the groups.

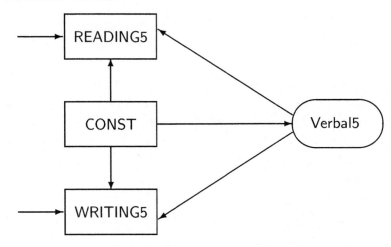

Figure 10.2: Path Diagram for Estimating the Mean of Verbal5

A path diagram is shown in Figure 10.2. Note that the path from CONST to READING5 corresponds to α_1, the path from CONST to WRITING5 corresponds to α_2, and the path from CONST to Verbal5 corresponds to the mean of Verbal5.

The input file for this model is (**step4a.spl**):

```
Group BA: STEP Reading and Writing, Grades 5 and 7
Raw Data from File stepba.lsf
Latent Variable: Verbal5
Relationships:
   READING5 = CONST + 1*Verbal5
   WRITING5 = CONST + Verbal5

Group BNA: STEP Reading and Writing, Grades 5 and 7
Raw Data from File stepbna.lsf
Relationship: Verbal5 = CONST
Set the Error Variances of READING5 - WRITING5 free
Set the Variance of Verbal5 free
Path Diagram
End of Problem
```

The two measurement equations correspond to:

```
READING5 = CONST + 1*Verbal5
WRITING5 = CONST + Verbal5
```

The intercept terms α_1 and α_2 are the coefficients of CONST in these relationships. In the first relationship, λ_1 is given as a fixed coefficient of 1.

The mean of Verbal5 is the coefficient of CONST in the expression

```
Relationship: Verbal5 = CONST
```

in the second group. Since this is not present in the first group, the mean of Verbal5 will be zero in the first group but estimated in the second group. The variance of Verbal5 is estimated in each group.

The output file shows that the mean of Verbal5 is -13.56 with a standard error of 1.22, from which we can conclude that the non-academic group is below the academic group in verbal ability as measured by these two tests. The model is saturated. So it fits the data perfectly.

As the second part of the example we extend the previous model to include also READING7 and WRITING7 as indicators of Verbal7. A path diagram is shown in Figure 2.5, where, for simplicity, we have omitted the variable CONST. It is tacitly assumed that there is a path from CONST to all the variables in the path diagram (as in Figure 10.3). We now estimate the mean difference between groups in the two latent variables, Verbal5 and Verbal7.

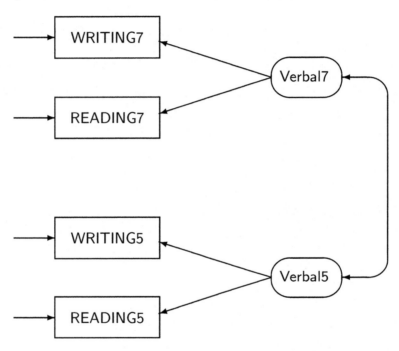

Figure 10.3: Path Diagram for Estimating the Means of Verbal5 and Verbal7

The input file for this analysis is (**step5a.spl**):

```
Group BA: STEP Reading and Writing, Grades 5 and 7
Raw Data from File stepba.lsf
Latent Variable: Verbal5 Verbal7
Relationships:
    READING5 = CONST + 1*Verbal5
    WRITING5 = CONST + Verbal5
    READING7 = CONST + 1*Verbal7
    WRITING7 = CONST + Verbal7

Group BNA: STEP Reading and Writing, Grades 5 and 7
Raw Data from File stepbna.lsf
Relationships:
    Verbal5 = CONST
    Verbal7 = CONST
Set the Error Variances of READING5 - WRITING7 free
Set the Variances of Verbal5 - Verbal7 free
Set the Covariance between Verbal5 and Verbal7 free
Path Diagram
End of Problem
```

The estimated group means, variances and covariances of the two latent variables are shown in Table 10.1. It can be seen that the non-academic group mean is below the mean of the academic group both in Grade 5 and in Grade 7. Furthermore, the non-academic group has a less favorable development from Grade 5 to Grade 7; its mean has decreased compared to the academic group.

Table 10.1: Estimated Means and Covariance Matrices of Verbal5 and Verbal7

Boys Academic (N = 373)		
	Verbal5	Verbal7
Verbal5	220.33	
Verbal7	212.20	233.51
Means	0.00	0.00

Boys Non-Academic (N = 249)		
	Verbal5	Verbal7
Verbal5	156.45	
Verbal7	126.96	153.67
Means	−13.80	−17.31

One way to describe the group differences graphically is to draw elliptical regions as shown in Figure 2.6. Based on the results Table 10.1, these regions have been constructed such that approximately 95% of the student are located inside the ellipses. The ellipses can differ in origin, shape and orientation. For any score on Verbal5, one can see the most likely range of values of Verbal7 for each group.

It can be seen from Figure 10.4 that the slope of the regression lines of Verbal7 on Verbal5 is similar. This leads to the problem of estimating these regression lines and testing whether they are the same.

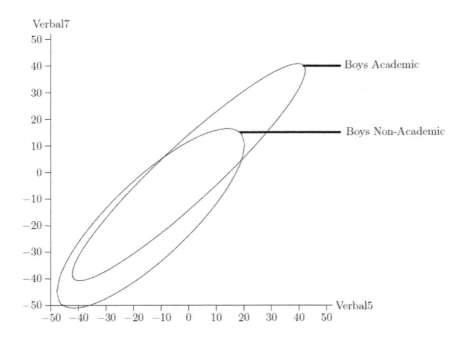

Figure 10.4: Concentration Ellipses for Verbal5 and Verbal6

10.2.4   Confirmatory Factor Analysis with Multiple Groups

10.2.5   Example: Chicago Schools Data

In this example we reconsider the analysis of the Holzinger and Swineford (1939) data. Recall from Chapter 1 that the data comes from two schools in Chigago: The Pasteur School and the Grant White school. As in Chapters 5, 6, and 7 we will use only the selected nine psychological variables. The data files are **pasteur.lsf** and **grantwhite.lsf**. We will also use the label 1 for the Pasteur school and the label 2 for the Grant-White school.

There are several approaches for studying factorial invariance in the literature. One approach is to test the following hypotheses in sequence:

$$H_{\boldsymbol{\Lambda}} : \boldsymbol{\Lambda}_1 = \boldsymbol{\Lambda}_2 \tag{10.11}$$

$$H_{\boldsymbol{\Lambda}\boldsymbol{\Theta}_\delta} : \boldsymbol{\Lambda}_1 = \boldsymbol{\Lambda}_2 \ \boldsymbol{\Theta}_{\delta 1} = \boldsymbol{\Theta}_{\delta 2} \tag{10.12}$$

$$H_{\boldsymbol{\Lambda}\boldsymbol{\Theta}_\delta\boldsymbol{\Phi}} : \boldsymbol{\Lambda}_1 = \boldsymbol{\Lambda}_2 \ \boldsymbol{\Theta}_{\delta 1} = \boldsymbol{\Theta}_{\delta 2} \ \boldsymbol{\Phi}_1 = \boldsymbol{\Phi}_2 \tag{10.13}$$

We will follow a different approach here.

The model we use here is the modified model in Section 7.3.4 where the variable SCCAPS is allowed to load on both factors Visual and Speed. We are interested to determine the extent to which this model holds for both schools. So we begin by fitting this model separately for each group. This gives a chi-square C1=53.253 for the Pasteur school and C1=28.293 for the Grant-White school, both with 23 degrees of freedom. This indicates that the model fits well in the Grant-White school but not so well in the Pasteur school. We will return to this issue later but for the moment we are primarily interested in demonstrating how to test the hypothesis $H_{\boldsymbol{\Lambda}}$ and to estimate the mean vector and covariance matrix of the three latent variables Visual, Verbal, and Speed.

Since the groups are independent we can add the two chi-squares C1=53.253+28.293=81.546 and use this as a chi-square C1 with 46 degrees. This is the baseline model for testing further restrictions on the two groups. This baseline model imposes no constraints across groups. It only states that the number of factors and the pattern of fixed and free loadings in Λ are the same. This model makes good sense and, as an inspection of the output files for each group verifies, all parameters are statistically significant.

We begin by testing factorial invariance in the sense of hypothesis H_Λ. The LISREL syntax file for this is **twoschools1b.lis**:

```
Group: Pasteur
DA NG=2 NI=9
RA=PASTEUR.LSF
MO NX=9 NK=3 PH=FR TD=DI
LK
Visual Verbal Speed
FR LX(2,1) LX(3,1) LX(5,2) LX(6,2) LX(8,3) LX(9,3),LX(9,1)
VA 1 LX(1,1) LX(4,2) LX(7,3)
OU

Group: Grant White
DA
RA=GRANTWHITE.LSF
MO LX=IN
PD
OU
```

In multigroup analysis one gives syntax lines for each group. Each group begins with the word **Group** followed by anything or nothing on the same line. The general rule is that everything is the same as in the previous group unless specified.

The line

```
DA NG=2 NI=9
```

specifies the number of groups and the number of variables. This is the same in the second group so the DA line is left empty in the second group. As in single group analysis, the MO line specifies the number of x-variables and the number of ξ-variables. Since LX is default it means that it is fixed. The PH=FR TD=DI specifies that Φ is a free covariance matrix and that Θ_δ is a diagonal matrix with free diagonal elements. This is just as in a single group analysis. The λ-parameters to be estimated are specified on the FR line. The new thing here is that the scales of the latent variables are defined by fixing one factor loading to 1 in each column of Λ. This is done by the VA line. This way of specifying the scales for the latent variables is necessary in multiple group analysis in order to put the latent variables on the same scale in each group. This makes it possible to compare the variances of the latent variables in the two groups. Each group must end with an OU line. The LX=IN on the MOline in the second group specifies that Λ_x should be invariant. The PH=FR TD=DI need not be included on this MO line since it is automatically the same as in the first group.

The output file **twoschools1b.out** gives

Global Goodness-of-Fit Statistics

```
Degrees of Freedom for (C1)-(C2)                    53
Maximum Likelihood Ratio Chi-Square (C1)            88.975 (P = 0.0014)
```

The difference between this chi-square of 88.975 and the chi-square of the baseline model 81.546 is 7.429 with 7 degrees of freedom. So it is clear that the hypothesis H_Λ cannot be rejected. Note that the degrees of freedom 7 equals the number of factor loadings in Λ that are equal across groups.

We proceed to test the hypothesis

$$H_{\Lambda\tau_x} : \Lambda_1 = \Lambda_2 \; \tau_{x1} = \tau_{x2} \tag{10.14}$$

This means that the intercepts in the measurement equations are the same in both groups. Since the means of the latent variables are 0 this is the same as saying that the means of the observed variables are equal across groups. From Chapter 1 we know that this is not likely to hold.

The LISREL input file is **twoschools2b.lis**:

```
Group: Pasteur
DA NG=2 NI=9
RA=PASTEUR.LSF
MO NX=9 NK=3 PH=FR TD=DI TX=FR
LK
Visual Verbal Speed
FR LX(2,1) LX(3,1) LX(5,2) LX(6,2) LX(8,3) LX(9,3),LX(9,1)
VA 1 LX(1,1) LX(4,2) LX(7,3)
OU

Group: Grant White
DA
RA=GRANTWHITE.LSF
MO LX=IN TX=IN
PD
OU
```

The only difference compared with the previous input file is that `TX=FR` is added on the first `MO` line and that `TX=IN` is added on the second `MO` line. This gives a chi-square C1=170.034 with 62 degrees of freedom. The difference from the previous chi-square of 88.975 is 81.059 with 9 degrees of freedom. So it is clear that the hypothesis that $\tau_{x1} = \tau_{x2}$ must be rejected.

We proceed by investigating whether the difference in means of the observed variables can be explained by differences in means of the latent variables. This is done by adding `KA=FR` on the `MO` line in the second group, see file **twoschools3b.lis**. One cannot estimate the means of the latent variables in an absolute sense but by setting the means equal to 0 in the first group (which they are by default) one can estimate them in the second group.

This model has a C1 chi-square of 130.137 with 59 degrees of freedom. Compared with the previous chi-square of 170.034 with 62 degrees of freedom this represents a very large improvement in fit with the addition of just three parameters. Thus, the chi-square for testing the hypothesis of no difference in factor means between the schools is 39.907 with 3 degrees of freedom. So the hypothesis is strongly rejected.

The estimates of the latent mean difference appear in the output file **twoschools3b.out** as

KAPPA

Visual	Verbal	Speed
-0.714	1.729	-4.585
(0.722)	(0.353)	(2.262)
-0.990	4.895	-2.027

The interpretation of this result is that the students in the Grant-White school are better than the students in the Pasteur school in Verbal ability, on average, whereas it is the other way around in Speed ability. In the Visual factor the difference in means are not statistically significant, although the sign suggests that the students of the Pasteur school might be better. Recall that the Pasteur school recruited children from immigrant families mostly from France and Germany whereas the children in the Grant-White school came from middle-income American native families.

The C1 chi-square of 130.137 with 59 degrees of freedom for the last model cannot be regarded as an acceptable fit. But the model of factorial invariance should be regarded as an approximation to the population model. Therefore it would be reasonable to measure fit in terms of RMSEA instead:

```
Root Mean Square Error of Approximation (RMSEA)      0.0897
90 Percent Confidence Interval for RMSEA             (0.0688 ; 0.111)
P-Value for Test of Close Fit (RMSEA < 0.05)         0.00149
```

But even so the fit is not acceptable. So, how can the fit be improved?

The identification conditions imposed on the last model are far more than what is needed to achieve identification of the factor means differences. It is possible to relax the equality constraints on all the seven non-fixed λ's and and to keep the equality constraints on the intercepts only for the three reference variables VISPERC, PARCOMP, and ADDITION. This gives a model with the minimum conditions for identification of the factor means differences. To estimate such a model use file **twoschools4b.lis**. This gives a chi-square C1 = 81.281 with 46 degrees of freedom which is almost identical to the chi-square for the baseline model we started out with. It can be shown that these two models are equivalent, *i.e.*, there is one-to-one correspondence between the parameters of these models.

The RMSEA values given in the output file **twoschools4b.out** are

```
Root Mean Square Error of Approximation (RMSEA)      0.0715
90 Percent Confidence Interval for RMSEA             (0.0451 ; 0.0966)
P-Value for Test of Close Fit (RMSEA < 0.05)         0.0851
```

which is just barely acceptable.

10.2.6   MIMIC Models for Multiple Groups

Subjective and Objective Status

Kluegel, Singleton, and Starnes (1977) analyzed data on subjective and objective social class based on samples of white and black people.

There are three variables measuring objective status and four measuring subjective status. The objective status measures are:

- education, measured by five categories ranging from less than ninth grade to college graduate.

- occupation, measured by the two-digit Duncan SEI score.

- income, measured as the total yearly family income before taxes in 1967, coded in units of $2,000 and ranging from under $2,000 to $16,000 or more.

All subjective class indicators were structured questions asking respondents to place themselves in one of four class categories: lower, working, middle, or upper. The questions asked the respondents to indicate which social class they felt their occupation, income, way of life, and influence were most like. The criteria, in terms of which class self-placements were made, correspond directly to the Weberian dimensions of economic class (occupation and income), status (lifestyle), and power (influence). The data files are **status_whites.lsf** and **status_blacks.lsf**

In this example we estimate the MIMIC model shown in Figure 10.5 for whites and blacks and test the invariance of various parts of the model.

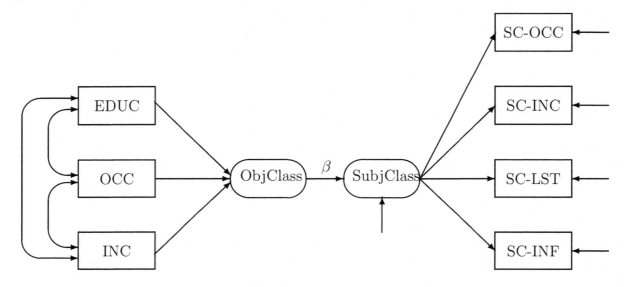

Figure 10.5: Mimic Model for Objective and Subjective Class

We begin by running each group separately as shown in files **status0w.spl** and **status0b.spl**. This gives C1 = 56.784 for whites and C1 = 57.942 for blacks with 14 degrees of freedom indicating that the model fits about equally bad in both groups. However, we will not be concerned about fit or the reason for lack of fit here. To disentangle these issues, see Jöreskog and Goldberger (1975). To put it simply: the problem is that not all effects of the objective measures on the subjective measures are mediated via the latent variables. Some of the objective measures have a direct effect on the subjective measures in addition to the indirect effect mediated via the latent variables.

For the moment we will disregard issues of fit and use the sum $56.784 + 57.942 = 114.726$ as a baseline chi-square C1 for testing the invariance of various part of the model. Writing

$$\mathbf{x}' = (\text{EDUC}, \text{OCC}, \text{INC}),$$

$$\mathbf{y}' = (\text{SC} - \text{OCC}, \text{SC} - \text{INC}, \text{SC} - \text{LST}, \text{SC} - \text{INF}),$$

and
$$\eta = \text{Status} ,$$
the model can be written
$$\begin{pmatrix} y_1 \\ y_2 \\ y_3 \\ y_4 \end{pmatrix} = \begin{pmatrix} \lambda_1 \\ \lambda_2 \\ \lambda_3 \\ \lambda_4 \end{pmatrix} \eta + \begin{pmatrix} \epsilon_1 \\ \epsilon_2 \\ \epsilon_3 \\ \epsilon_4 \end{pmatrix} , \qquad (10.15)$$
$$\eta = \alpha + \gamma_1 x_1 + \gamma_2 x_2 + \gamma_3 x_3 + \zeta , \qquad (10.16)$$

where the λ's are elements of $\boldsymbol{\Lambda}_y$. The element λ_1 is set equal to 1 to fix the scale of η. Since SC-OCC is on the same scale in both groups, this makes it possible to compare the variances of η for each group.

There is no unique way to go about testing the invariance of the various parts of the model. We begin by testing the measurement model for Status,*i.e.*, that λ_2, λ_3, and λ_4 are the same for whites and blacks. This can be done by using following SIMPLIS syntax file **status1a.spl**:

```
Group Whites
Raw Data from File status_whites.lsf
Latent Variable Status
Relationships
'SC-OCC' = 1*Status
'SC-INC' - 'SC-INF' = Status
Status = CONST EDUC - INC

Group Blacks
Raw Data from File status_blacks.lsf
Latent Variable Status
Relationships
Status = CONST EDUC - INC
Set the error variance of Status free
Set the error variances of 'SC-OCC' free
Set the error variances of 'SC-INC' free
Set the error variances of 'SC-LST' free
Set the error variances of 'SC-INF' free
LISREl Output
Path Diagram
End of Problem
```

Note the following

- The ' around the name SC-OCC and similar names are necessary; otherwise LISREL interprets the - as a range sign.

- The fact that the lines

```
'SC-OCC' = 1*Status
'SC-INC' - 'SC-INF' = Status
```

are omitted in the second group means that these relationships should be the same in the second group as in the first group.

- The line

```
Set the error variance of Status free
```

in the second group is included to allow $Var(\zeta)$ to be different in the two groups.

- The lines

```
Set the error variances of 'SC-OCC' free
Set the error variances of 'SC-INC' free
Set the error variances of 'SC-LST' free
Set the error variances of 'SC-INF' free
```

in the second group are included to allow the measurement error variances $Var(\delta_i)$ to be different in the two groups.

The output file **status1a.out** gives the chi-square C1 as

```
Degrees of Freedom for (C1)-(C2)                    31
Maximum Likelihood Ratio Chi-Square (C1)            125.314 (P = 0.0000)
```

The difference between 125.314 and the baseline chi-square 114.726 can be used as chi-square for testing the hypothesis that λ_2, λ_3, and λ_4 are the same for whites and blacks. This chi-square is 10.588 with 3 degrees of freedom. This has a P value less than 0.025. This is barely significant but we reject the hypothesis and allow λ_2, λ_3, and λ_4 to be different for whites and blacks. We will look at these differences later.

In the next run we investigate whether γ_1, γ_2, and γ_3 are the same for whites and blacks but allowing the intercept α to be different. This can be done by using the following SIMPLIS syntax file **status2a.spl**:

```
Group Whites
Raw Data from File status_whites.lsf
Latent Variable Status
Relationships
'SC-OCC' = 1*Status
'SC-INC' - 'SC-INF' = Status
Status = CONST EDUC - INC

Group Blacks
Raw Data from File status_blacks.lsf
Latent Variable Status
Relationships
'SC-OCC' = 1*Status
'SC-INC' - 'SC-INF' = Status
Status = CONST
Set the error variance of Status free
```

```
Set the error variace of 'SC-OCC' free
Set the error variace of 'SC-INC' free
Set the error variace of 'SC-LST' free
Set the error variace of 'SC-INF' free
Path Diagram
End of Problem
```

This gives

```
Degrees of Freedom for (C1)-(C2)                 31
Maximum Likelihood Ratio Chi-Square (C1)         128.347 (P = 0.0000)
```

The difference $128.347 - 114.726 = 13.587$ can be used as a chi-square with 3 degrees of freedom for testing the hypothesis that γ_1, γ_2, and γ_3 are the same for whites and blacks is also rejected. It is clear that this must be rejected. We will therefore keep γ_1, γ_2, and γ_3 as different between groups, and proceed to test whether α is the same in both groups. This can be done by replacing the line

```
Status =CONST
```

in the second group by the line

```
Status = EDUC - INC
```

This gives a chi-square of 150.965 with 32 degrees of freedom. Obviously, the hypothesis that α is invariant must also be rejected.

So we end up with the conclusion that no part of the MIMIC model is invariant over whites and blacks. Although the model differences between whites and blacks are statistically different, these difference may still be small from a practical point of view. So we will take a closer look at the differences. Here we compare the results obtained in **status0w.out** and **status0b.out** when the two groups are run separately.

For the measurement model (10.15) we obtain

```
              Whites                        Blacks
              Status                        Status
            --------                      --------

SC-OCC        1.000          SC-OCC         1.000


SC-INC        1.020          SC-INC         0.923
             (0.022)                       (0.028)
             45.991                        33.058


SC-LST        1.016          SC-LST         1.004
             (0.021)                       (0.027)
             48.520                        37.727


SC-INF        1.045          SC-INF         1.048
             (0.022)                       (0.031)
             47.501                        33.608
```

This suggests that the measurement model for whites and blacks are very similar.

For the explanatory model (10.16) we obtain

```
          Whites

   Status = 1.052     + 0.0859*EDUC + 0.0141*OCC + 0.0557*INC, R² = 0.230
 Standerr  (0.0597)   (0.0212)       (0.0115)      (0.0106)
 Z-values  17.619      4.048          1.230         5.268
 P-values   0.000      0.000          0.219         0.000


          Blacks

   Status = 0.592     + 0.0651*EDUC + 0.0653*OCC + 0.104*INC, R² = 0.235
 Standerr  (0.0762)   (0.0314)       (0.0212)      (0.0156)
 Z-values   7.770      2.075          3.089         6.644
 P-values   0.000      0.038          0.002         0.000
```

Although both models have about the same low explanatory power as judged by the R^2's it seems that OCC and INC has more weight for the blacks than for the whites in the formation of the Status variable. If we control for all the explanatory variables, it is clear that whites has a much higher level of subjective status. Note the large difference of the intercepts.

10.2.7   Twin Data Models

Models for twin data or other types of family relationships were common in the 1980's and 1990's and LISREL was often used for the analysis of these types of data, see *e.g.*, Neale & Cardon (1992). We consider only the simplest type of model here.

From biometrical genetic theory (see *e.g.*, Neale & Cardon (1992, Chapter 3)), we can formulate the simplest type of twin design as follows. The observed phenotypes P_1 and P_2 of twin 1 and twin 2 in a twin pair are postulated to depend on additive genes A_1 and A_2, common environments C_1 and C_2 (environmental influences shared by twins reared in the same family), dominance genetic deviations D_1 and D_2(dominance effects of alleles at multiple loci) and unique environments E_1 and E_2:

$$P_1 = h_1 A_1 + c_1 C_1 + d_1 D_1 + E_1 , \tag{10.17}$$

$$P_2 = h_2 A_2 + c_2 C_2 + d_2 D_2 + E_2 , \tag{10.18}$$

where D_1, A_1, C_1, C_2, A_2, and D_2 are latent random variables with the covariance matrix

$$\mathbf{\Phi} = \begin{pmatrix} \phi_{11} & & & & & \\ 0 & \phi_{22} & & & & \\ 0 & 0 & \phi_{33} & & & \\ 0 & 0 & \phi_{43} & \phi_{44} & & \\ 0 & \phi_{52} & 0 & 0 & \phi_{55} & \\ \phi_{61} & 0 & 0 & 0 & 0 & \phi_{66} \end{pmatrix} , \tag{10.19}$$

and E_1 and E_2 are uncorrelated random variables uncorrelated with all the other latent variables. It is assumed that the effects of the latent variables are the same for twin 1 and twin 2, so that $h_1 = h_2 = h$, $c_1 = c_2 = c$, and $d_1 = d_2 = d$. It is also assumed that the variances of E_1 and E_2 are equal. A path diagram is shown in Figure 10.6, where we combine information for monozygotic (MZ) and dizygotic (DZ) twins.

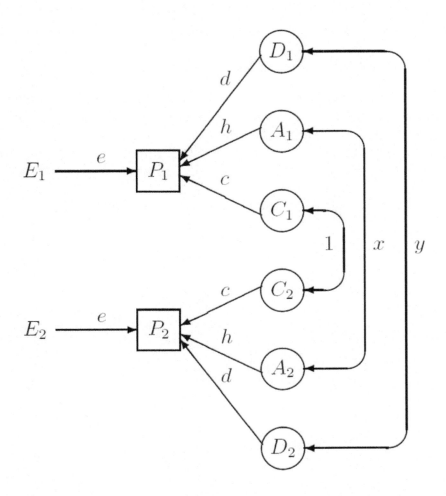

Figure 10.6: Simple Twin Model with Dominance Effects, Additive Genes, Common and Unique Environments ($x = 1$ for MZ, $x = \frac{1}{2}$ for DZ, $y = 1$ for MZ, $y = \frac{1}{4}$ for DZ)

This model may be represented in various ways in LISREL. We treat all six latent variables as ξ-variables and E_1 and E_2 as δ-variables. Denote the variances of E_1 and E_2 as e^2.

The three parameter matrices Λ_x, Φ, and Θ_δ are specified as

$$\Lambda_x = \begin{pmatrix} d & h & c & 0 & 0 & 0 \\ 0 & 0 & 0 & c & h & d \end{pmatrix},$$

Φ is a fixed matrix with all non-zero ϕ_{ij} in (10.19) equal to 1 for MZ but with $\phi_{52} = \frac{1}{2}$, and $\phi_{61} = \frac{1}{4}$ for DZ, and Θ_δ is a diagonal matrix with both diagonal elements equal to e^2 for both MZ and DZ.

It follows that the covariance matrices of P_1 and P_2 are

$$\Sigma_{\mathrm{MZ}} = \begin{pmatrix} h^2 + c^2 + d^2 + e^2 & \\ h^2 + c^2 + d^2 & h^2 + c^2 + d^2 + e^2 \end{pmatrix}$$

$$\Sigma_{\mathrm{DZ}} = \begin{pmatrix} h^2 + c^2 + d^2 + e^2 & \\ \frac{1}{2}h^2 + c^2 + \frac{1}{4}d^2 & h^2 + c^2 + d^2 + e^2 \end{pmatrix} .$$

There are four parameters h, c, d, and e. But note that

- Σ is only a function of the squares of these parameters.

- Although there are six manifest equations corresponding to the six elements of the two covariance matrices, only three of these equations are linearly independent. So one can only estimate at most three parameters. In practice, one resolves this problem by setting either c or d to 0. If $c = 0$ the model is called an ADE model. If $d = 0$ the model is called an ACE model.

Path diagrams of the ADE and ACE models are shown in Figure 10.7.

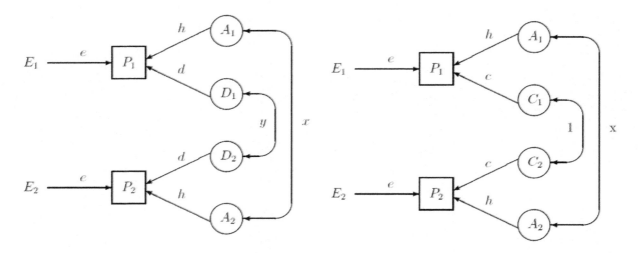

Figure 10.7: The ADE Model (left) and the ACE model (right) ($x = 1$ for MZ, $x = \frac{1}{2}$ for DZ, $y = 1$ for MZ, $y = \frac{1}{4}$ for DZ)

We consider the ADE model and the ACE model separately. For each model there are three different parameterizations.

ADE Model

The ADE model is specified as

$$\begin{pmatrix} P_1 \\ P_2 \end{pmatrix} = \begin{pmatrix} h & d & 0 & 0 \\ 0 & 0 & d & h \end{pmatrix} \begin{pmatrix} A_1 \\ D_1 \\ D_2 \\ A_2 \end{pmatrix} + \begin{pmatrix} E_1 \\ E_2 \end{pmatrix} = \Lambda_x \xi + \delta , \qquad (10.20)$$

In the first parameterization, often called the **path analysis** parameterization, we assume that all ξ-variables are standardized and estimate the parameters $\lambda_{1,1} = \lambda_{2,4} = h$, $\lambda_{1,2} = \lambda_{2,3} = d$, and $\theta_{11} = \theta_{22} = e^2$. Since Σ depends only on the squares of h and d, Σ is the same for all four combinations $(\pm h, \ \pm d)$. Typically this gives numerical problems in LISREL's choice of starting values and iterations. This parameterization is not recommended.

In the second parameterization, often called the **variance components** parameterization, we set $\lambda_{1,1} = \lambda_{1,2} = \lambda_{2,3} = \lambda_{2,4} = 1$ and estimate $\boldsymbol{\Phi}$ such that $\phi_{11} = \phi_{44} = h^2$, $\phi_{41} = xh^2$ and $\phi_{22} = \phi_{33} = d^2$, $\phi_{32} = yd^2$. The implied covariance matrix of the observed phenotypes is

$$\boldsymbol{\Sigma} = \begin{pmatrix} h^2 + d^2 + e^2 & \\ xh^2 + yd^2 & h^2 + d^2 + e^2 \end{pmatrix} .$$

ACE Model

The ACE model is specified as

$$\begin{pmatrix} P_1 \\ P_2 \end{pmatrix} = \begin{pmatrix} h & c & 0 & 0 \\ 0 & 0 & c & h \end{pmatrix} \begin{pmatrix} A_1 \\ C_1 \\ C_2 \\ A_2 \end{pmatrix} + \begin{pmatrix} E_1 \\ E_2 \end{pmatrix} = \boldsymbol{\Lambda}_x \xi + \boldsymbol{\delta} , \tag{10.21}$$

As for the ADE model, the path analysis parameterization of the ACE model is not recommended.

The variance components parameterization is to set $\lambda_{1,1} = \lambda_{1,2} = \lambda_{2,3} = \lambda_{2,4} = 1$ and estimate $\boldsymbol{\Phi}$ such that $\phi_{11} = \phi_{44} = h^2$, $\phi_{41} = xh^2$ and $\phi_{22} = \phi_{33} = \phi_{32} = c^2$. The implied covariance matrix of the observed phenotypes is

$$\boldsymbol{\Sigma} = \begin{pmatrix} h^2 + c^2 + e^2 & \\ xh^2 + c^2 & h^2 + c^2 + e^2 \end{pmatrix} .$$

For both the ACE and the ADE model we propose a third parameterization not mentioned in the literature. This is very easy to estimate with LISREL. In this parameterization we set $\boldsymbol{\Lambda}_x = \mathbf{I}$, $\boldsymbol{\Theta}_\delta = \mathbf{0}$ and use the parameters α, β, and γ defined by

$$\boldsymbol{\Phi}_{\text{MZ}} = \begin{pmatrix} \alpha & \\ \beta & \alpha \end{pmatrix} \quad \boldsymbol{\Phi}_{\text{DZ}} = \begin{pmatrix} \alpha & \\ \gamma & \alpha \end{pmatrix} ,$$

For the ADE model $\alpha = h^2 + d^2 + e^2$, $\beta = h^2 + d^2$, and $\gamma = \frac{1}{2}h^2 + \frac{1}{4}d^2$. Solving these equations for h^2, d^2, and e^2 in terms of α, β, and γ, gives $h^2 = 4\gamma - \beta$, $d^2 = 2\beta - 4\gamma$, $e^2 = \alpha - \beta$.

For the ACE model $\alpha = h^2 + c^2 + e^2$, $\beta = h^2 + c^2$, and $\gamma = \frac{1}{2}h^2 + c^2$. Solving these equations for h^2, c^2, and e^2 in terms of α, β, and γ, gives $h^2 = 2(\beta - \gamma)$, $c^2 = 2\gamma - \beta$, $e^2 = \alpha - \beta$.

Using the estimates of α, β, and γ, we can get estimates of h^2, d^2, and e^2 or h^2, c^2, and e^2, respectively. However, it is not necessary to do this calculation by hand since one can use the *additional parameters* feature of LISREL to obtain estimates of h^2, d^2, and e^2 or h^2, c^2, and e^2, respectivel, directly. This also gives the standard errors of the estimates.

10.2.8   Example: Heredity of BMI

The phenotypes P_1 and P_2 in the twin models can be anything observed or measured on each twin in a twin pair, such as the presence or absence of symptoms or illnesses or physiological or physical measures like blood pressure or weight. In the following we use BMI measured as the ratio of weight in kg and the square of length in meters (kg/m^2).

Neale & Cardon (1992, Table 8.1) report univariate characteristics (means, variances, skewness and kurtosis) of BMI values for ten subgroups of Australian twin pairs. The striking characteristics are the large skewness and kurtosis values, indicating non-normality. To achieve less non-normality these authors made a logarithmic transformation of the BMI values. We do not have access to their original data. So we will use data from UK in the UK Twin Data Archive which is maintained by the Department of Twin Research & Genetic Epidemiology, King's College London, UK. The data is available to researchers free of charge.

As part of an ongoing data collection project all twins in the UK twin register received a letter asking them to see their local doctor to have their length and weight measured. Alternatively, they could fill in these values on a form that could be sent directly by mail. Most twins chose to go to the doctor as they would then get an extra bonus by getting their diastolic and systolic blood preasure measured. The data file available to us in January 2016 is a Microsoft Excel file **UKdataonBMI.xlsx**. This contains data on 6747 twins but their are only 1552 twin pairs which can be used for our example[2]. The oldest twin pair was born 1924 and the youngest twin pair was born 1991. The sex of each twin was also recorded, but we will not use sex or age in our example. Among the 1552 twin pairs there were 794 MZ twins and 758 DZ twin pairs. We have created a LISREL system file for the MZ and DZ twins separately. These are called **ukbmimz.lsf** and **ukbmidz.lsf**, respectively.

A data screening of **ukbmimz.lsf** shows

```
Total Sample Size(N) =      794
```

Univariate Summary Statistics for Continuous Variables

Variable	Mean	St. Dev.	Skewness	Kurtosis	Minimum	Freq.	Maximum	Freq.
BMI1	25.766	4.936	1.197	2.278	15.122	1	52.707	1
BMI2	25.841	4.857	1.047	1.497	16.602	1	47.743	1

Test of Univariate Normality for Continuous Variables

	Skewness		Kurtosis		Skewness and Kurtosis	
Variable	Z-Score	P-Value	Z-Score	P-Value	Chi-Square	P-Value
BMI1	11.177	0.000	6.770	0.000	170.763	0.000
BMI2	10.143	0.000	5.321	0.000	131.197	0.000

```
Relative Multivariate Kurtosis = 1.744
```

---

[2]We can only speculate about the reason why the twin mate of so many as 3643 twins is not included in the data. They may have died or left the country, or they may be unable to register the data for any reason. If the reason is related to their BMI values the reults of the example may be biased.

```
Test of Multivariate Normality for Continuous Variables
```

	Skewness			Kurtosis			Skewness and Kurtosis	
Value	Z-Score	P-Value	Value	Z-Score	P-Value		Chi-Square	P-Value
2.632	14.798	0.000	13.951	10.156	0.000		322.116	0.000

It is seen that the BMI values are highly non-normal. All skewness and kurtosis values are highly significant. Similar results are obtained for the DZ twins.

LISREL can be used to effectively normalize any variable with 100 or more distinct values, no matter what the sample distribution is. This will be demonstrated here.

We now describe the procedure we have used to normalize the variables. The procedure is called *normal scores* and is described in the appendix (see Chapter 14). To normalize BMI1 and BMI2, use the following **PRELIS** syntax file **normalizemz.prl**:

```
sy=ukbmimz.lsf
new P1=BMI1
new P2=BMI2
ns P1 P2
ou ma=cm ra=ukbmimz_normalized.lsf
```

The lines

```
new P_1=BMI1 new P_2=BMI2
```

make copies of the variables BMI1 and BMI2 so that we can keep both the untransformed and the normalized variables. The line

```
ns P_1 P_2
```

normalizes the variables P_1 and P_2. The output file **normalizemz.out** shows that

```
Total Sample Size(N) =    794
```

```
Univariate Summary Statistics for Continuous Variables
```

Variable	Mean	St. Dev.	Skewness	Kurtosis	Minimum	Freq.	Maximum	Freq.
BMI1	25.766	4.936	1.197	2.278	15.122	1	52.707	1
BMI2	25.841	4.857	1.047	1.497	16.602	1	47.743	1
P1	25.766	4.936	0.000	-0.005	9.472	1	42.061	1
P2	25.841	4.857	0.000	-0.005	9.809	1	41.872	1

```
Test of Univariate Normality for Continuous Variables

             Skewness             Kurtosis        Skewness and Kurtosis

Variable Z-Score P-Value   Z-Score P-Value    Chi-Square P-Value

    BMI1  11.177  0.000      6.770  0.000        170.763  0.000
    BMI2  10.143  0.000      5.321  0.000        131.197  0.000
      P1   0.000  1.000      0.055  0.956          0.003  0.999
      P2   0.000  1.000      0.055  0.956          0.003  0.999
```

which verifies that P1 and P2 have been normalized. Similar results are obtained for DZ as well.

The output also gives the covariance matrix of all four variables. For MZ this is

```
             BMI1        BMI2        P1          P2

             --------    --------    --------    --------
BMI1         24.366
BMI2         18.797      23.587
P1           23.503      18.301      24.366
P2           18.270      22.888      18.697      23.587
```

In the same way we can normalize the variables in **ukbmidz.lsf**, see file **normalizedz.prl**. For DZ the covariance matrix is

```
             BMI1        BMI2        P1          P2

             --------    --------    --------    --------
BMI1         28.379
BMI2         12.657      25.751
P1           27.222      12.309      28.379
P2           12.493      25.060      12.604      25.751
```

For both groups, note that the variances of the untransformed and the normalized variables are the same but the covariance between them have changed from 18.797 to 18.697 for MZ and from 12.657 to 12.604 for DZ.

To see the transformation of BMI1 to P_1, for the MZ sample, say, make a line plot of P_1 on BMI1 as shown in Figure 10.8:

We are now ready to estimate the ACE model. Only LISREL syntax can be used for this example. In these syntax file we need only the transformed variables P1 and P2. They are selected by the lines

```
se
P1 P2 /
```

A LISREL syntax file for estimating the ACE model using the variance components parameterization is **bmi2_ace.lis**:

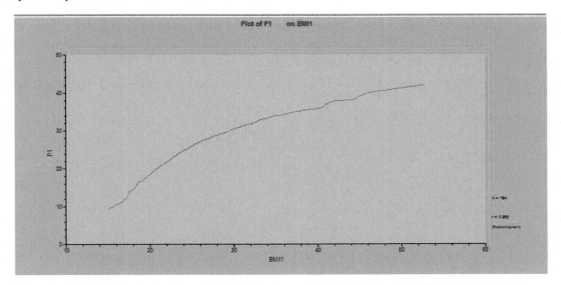

Figure 10.8: Line Plot of P_1 on PMI1

```
Group MZ: ACE Model
da ng=2 ni=4
ra=ukbmimz_normalized.lsf
se
P1 P2 /
mo nx=2 nk=4 ph=fi
va 1 lx(1,1) lx(1,2) lx(2,3) lx(2,4)
pa ph
2
0 3
0 3 3
2 0 0 2
eq td(1) td(2)
ou

Group DZ
da
ra=ukbmidz_normalized.lsf
se
P1 P2 /
mo nx=2 nk=4 lx=in td=in
pa ph
2
0 3
0 3 3
2 0 0 2
co ph(4,1)=.5*ph(1,4,1)
ou ad=off
```

This is a somewhat tricky syntax using constraints and labeling parameters by numbering them. In the output file **bmi2_ace.out** the estimate $\hat{h}^2$ appears as PH(1,1) and $\hat{c}^2$ appears as ph(2,2). These estimates are 17.595 and 2.797, respectively.

A LISREL syntax file for estimating the ACE model using the third parametrization is **bmi3_ace.lis**:

```
Group MZ: ACE Model
da ng=2 ni=4
ra=ukbmimz_normalized.lsf
se
P1 P2 /
mo nx=2 nk=2 lx=id td=ze
eq ph(1,1) ph(2,2)
ou

Group DZ
da
ra=ukbmidz_normalized.lsf
se
P1 P2 /
mo nx=2 nk=2 lx=id td=ze
eq ph(1,1) ph(2,2)
eq ph(1,1) ph(1,1,1)
ou
```

This is considerably easier The output file **bmi3_ace.out** gives the estimate as

$$\hat{\alpha} = 25.717 \quad \hat{\beta} = 20.390 \quad \hat{\gamma} = 11.594 \tag{10.22}$$

. From these estimates we get the following estimates by hand calculation

$$\hat{h}^2 = 17.594 \quad \hat{c}^2 = 2.797 \quad \hat{e}^2 = 5.326 \tag{10.23}$$

To obtain these estimates by LISREL, use the following input file **bmi3b_ace.lis**:

```
Group MZ: ACE Model
da ng=2 ni=4 AP=3
ra=ukbmimz_normalized.lsf
se
P1 P2 /
mo nx=2 nk=2 lx=id td=ze
co ph(1,1)=pa(1)+pa(2)+pa(3)
co ph(2,2)=pa(1)+pa(2)+pa(3)
co ph(2,1)=pa(1)+pa(2)
ou
```

```
Group DZ
da
ra=ukbmidz_normalized.lsf
se
P1 P2 /
mo nx=2 nk=2 lx=id td=ze
co ph(1,1)=pa(1)+pa(2)+pa(3)
co ph(2,2)=pa(1)+pa(2)+pa(3)
co ph(2,1)=.5*pa(1)+pa(2)
ou
```

Here ap=3 on the first da line, tells LISREL that there are 3 parameters called pa(1), pa(2), and pa(3). These correspond to h^2, c^2, and e^2. The elements of $\boldsymbol{\Phi}$ are expressed as functions of these using co commands.

The estimated pa(1), pa(2), and pa(3) with their standard errors and z-values appears in the output as

```
    ADDITIONAL PARAMETERS

         PA(1)       PA(2)       PA(3)
       --------    --------    --------

        17.594       2.797       5.326
       (1.498)     (1.441)     (0.266)
        11.745       1.941      19.993
```

According to these results, the herdidity of BMI can be estimated as

$$\frac{\hat{h}^2}{\hat{h}^2 + \hat{c}^2 + \hat{e}^2} = \frac{17.594}{25.719} = 0.684 ,$$

i.e., 68.4%.

This estimate is surprisingly large. However, Neale & Cardon (1992, Table 8.5) also report such large heredity coefficients of BMI for two groups of older twins in the Australian data. They also report much smaller values for younger twins. This suggest that the age of the twins can have an effect on the estimated heredity. The UK data we have analyzed has a wide range of age levels in the sample. Although the age of the twins is not included in the data (only the year of birth is), it would be possible to include a proxy for age in the analysis but we shall not pursue such an analysis here.

For the ADE model the estimates of α, β, and γ are the same as for the ACE model but when solving for the estimates of h^2, d^2, and e^2 we see that the estimate $\hat{d}^2 = 40.786 - 46.376 = -5.590$ is negative. Thus, the ADE model is non-admissible.

In cases where both the ACE and the ADE models are admissible, they will have the same fit.

10.3   Multiple Groups with Ordinal Variables

In the previous sections of this chapter we have assumed that all observed variables are continuous. In this section we consider multiple groups with observed ordinal variables.

Consider the situation where data on the same ordinal variables have been collected in several groups. These groups may be different nations, states or regions, culturally or socioeconomically different groups, groups of individuals selected on the basis of some known selection variables, groups receiving different treatments, etc. In fact, they may be any set of mutually exclusive groups of individuals which are clearly defined. It is assumed that the data is a random sample of individuals from each group. The objective is to compare different characteristics across groups. In particular, the procedure to be described can be used for testing factorial invariance and for estimating differences in factor means.

10.3.1   Example: The Political Action Survey

In this section we continue the analysis of the efficacy variables in the Political Action Survey which was carried out in eight countries. For information about this survey and the efficacy variables, see Section 7.4.2. In that section and in the example **Panel Model for Political Efficacy** in Section 9.1 we analyzed data from the USA only. Here we will analyze the data from all eight countries. The procedure to be described here makes it possible to answer questions like these:

- Do the efficacy items measure the same latent variables in all countries?

- If so, are the factor loadings invariant over countries?

- Are the intercepts invariant over countries?

If these conditions are satisfied one can estimate differences in means, variances, and covariances of the latent variables Efficacy and Respons between countries. Recall from Section 7.4.2 that Efficacy and Respons are two different components of Political Efficacy, where Efficacy indicates *individuals' self-perceptions that they are capable of understanding politics and competent enough to participate in political acts such as voting*, and Respons (short for *Responsiveness*) indicates *the belief that the public cannot influence political outcomes because government leaders and institutions are unresponsive*. People who are low on Efficacy or low on Respons are expected to agree or agree strongly with the items. Hence, the items measure these components from low to high.

Complete factorial invariance over all eight countries should not be expected to hold for the following reasons:

- The items are stated in different languages.

- Words may have different connotations in different languages.

- Other cultural differences between countries may lead to different response styles or response patterns in different countries.

These reasons may imply that the items are interpreted differently in different countries.

10.3.2 Data Screening

The data for all countries is in the datafile **EFFITOT.RAW** in free format. The first variable is **COUNTRY** coded as 1 = USA (USA), 2 = Germany (GER), 3 = The Netherlands (NET), 4 = Austria (AUS), 5 = Britain (BTN), 6 = Italy (ITY), 7 = Switzerland (SWI), and 8 = Finland (FIN). The other variables are the six efficacy variables described in Section 7.4.2. The item **VOTING** is included here but will not be included in the model. The response categories and their codings are those described in Section 7.4.2, but in Italy there was an additional response category *Don't Understand* (DU) coded as 6.

A data screening of all the data can be obtained by running the following **PRELIS** command file (file **effitot0.prl**):

```
!Data Screening of EFFITOT.RAW
da ni=7
la
COUNTRY NOSAY VOTING COMPLEX NOCARE TOUCH INTEREST
ra = EFFITOT.RAW
cl COUNTRY 1=USA 2=GER 3=NET 4=AUS 5=BTN 6=ITY 7=SWI 8=FIN
cl NOSAY - INTEREST 1=AS 2=A 3=D 4=DS 6=DU 8=DK 9=NA
ou
```

The output gives the numbers in the right and bottom margins of Tables 10.2 and 10.3. Just like in the USA sample, there are more people in the *Don't Know* than in the *No answer* categories.

To screen the data for one country, use the following **PRELIS** command file, here illustrated with the USA (file **effitot1.prl**):

```
!Creating lsf file for USA
da ni=7
la
COUNTRY NOSAY VOTING COMPLEX NOCARE TOUCH INTEREST
ra = EFFITOT.RAW
cl COUNTRY 1=USA 2=GER 3=NET 4=AUS 5=BTN 6=ITY 7=SWI 8=FIN
cl NOSAY - INTEREST 1=AS 2=A 3=D 4=DS 6=DU 8=DK 9=NA
sd COUNTRY = 1
ou ra=effiusa.lsf
```

To repeat this for another country, just change the line

Table 10.2: Observed Frequency Distributions

NOSAY

Response Category	USA	GER	NET	AUS	BTN	ITY	SWI	FIN	Total
Agree Strongly	175	721	171	804	211	306	347	295	3030
Agree	518	907	479	432	693	797	463	413	4702
Disagree	857	464	460	215	488	365	314	425	3588
Disagree Strongly	130	133	49	50	33	57	137	50	639
Don't Understand	-	-	-	-	-	74	-	-	74
Don't Know	29	27	40	82	42	173	26	35	454
No Answer	10	3	2	2	16	7	3	6	49
All Responses	1719	2255	1201	1585	1483	1779	1290	1224	12536

VOTING

Response Category	USA	GER	NET	AUS	BTN	ITY	SWI	FIN	Total
Agree Strongly	283	790	195	903	218	341	371	429	3530
Agree	710	861	634	431	884	902	403	512	5337
Disagree	609	448	285	135	289	271	310	234	2581
Disagree Strongly	80	103	38	28	19	45	150	18	481
Don't Understand	-	-	-	-	-	57	-	-	57
Don't Know	26	49	43	85	57	159	53	26	498
No Answer	11	4	6	3	16	4	3	5	52
All Responses	1719	2255	1201	1585	1483	1779	1290	1224	12536

COMPLEX

Response Category	USA	GER	NET	AUS	BTN	ITY	SWI	FIN	Total
Agree Strongly	343	688	262	531	312	484	495	386	3483
Agree	969	801	592	557	777	858	394	588	5536
Disagree	323	516	273	283	310	225	267	198	2395
Disagree Strongly	63	214	45	125	40	55	112	37	691
Don't Understand	-	-	-	-	-	39	-	-	39
Don't Know	9	30	25	86	28	114	18	28	338
No Answer	12	6	4	3	16	4	4	5	54
All Responses	1719	2255	1201	1585	1483	1779	1290	1224	12536

Table 10.3: Observed Frequency Distributions

NOCARE

Response Category	USA	GER	NET	AUS	BTN	ITY	SWI	FIN	Total
Agree Strongly	250	569	156	590	205	402	354	226	2752
Agree	701	880	487	474	756	853	418	552	5121
Disagree	674	638	421	334	404	286	364	354	3475
Disagree Strongly	57	103	36	86	33	24	82	26	467
Don't Understand	-	-	-	-	-	45	-	-	25
Don't Know	20	57	95	98	68	184	48	60	630
No Answer	17	8	6	3	17	5	8	6	66
All Responses	1719	2255	1201	1585	1483	1779	1290	1224	12536

TOUCH

Response Category	USA	GER	NET	AUS	BTN	ITY	SWI	FIN	Total
Agree Strongly	273	697	197	602	276	481	325	319	3170
Agree	881	978	575	533	737	869	523	578	5674
Disagree	462	425	267	257	347	167	262	230	2417
Disagree Strongly	26	49	25	49	18	15	56	21	259
Don't Understand	-	-	-	-	-	45	-	-	45
Don't Know	60	101	133	140	88	198	116	68	904
No Answer	17	5	4	4	17	4	8	8	67
All Responses	1719	2255	1201	1585	1483	1779	1290	1224	12536

INTEREST

Response Category	USA	GER	NET	AUS	BTN	ITY	SWI	FIN	Total
Agree Strongly	264	541	147	629	280	469	343	331	3004
Agree	762	792	443	458	709	834	432	519	4949
Disagree	581	698	443	309	377	247	318	280	3253
Disagree Strongly	31	149	44	76	24	26	93	28	471
Don't Understand	-	-	-	-	-	24	-	-	24
Don't Know	62	66	120	110	76	172	96	58	760
No Answer	19	9	4	3	17	7	8	8	75
All Responses	1719	2255	1201	1585	1483	1779	1290	1224	12536

`sd COUNTRY = 1`

The **sd** line (short for Select and Delete) first selects all cases with `COUNTRY = 1` and then deletes the variable `COUNTRY` (it is no use to keep the variable `COUNTRY` after selection of cases since all cases have the same value 1 on this variable). In addition to the data screening, file **effitot1.prl** creates an **.lsf** file for USA which will be used later.

One can do the same procedure for all countries to create **.lsf** files for all countries. Collecting the results from these output files gives the results shown in Tables 10.2 and 10.3.

It is seen in Tables 10.2 and 10.3 that there are considerable differences between countries in the univariate marginal distribution of these variables but these distributions are rather similar across variables. Countries like the USA and Britain which use the same language (English) are rather similar. Germany and Austria, where the German language was used, are also rather similar. But there is a considerable difference between these two pairs of countries. Most notably is the distribution in Austria where many people respond in the *Agree Strongly* category. We shall see that these manifest differences may be viewed as reflections of differences in the means of the latent variables between countries.

10.3.3   Multigroup Models

The model we use here is the one-factor as originally suggested by Campbell et al. (1954). This model is shown in Figure 7.6 in Chapter 7 but for reasons given on page 335, the variable `VOTING` is not included in the model. For simplicity of presentation here, this variable has been deleted in all eight **lsf** files. Furthermore, the codes 6, 8, and 9, which are regarded as missing values, have been recoded to the global missing value code -999999.000 in these **.lsf** files.

The models used in this section are defined by the cumulative response function

$$Pr\{x_i \le c_i \mid \boldsymbol{\xi}\} = F(\alpha_{c_i}^{(i)} + \sum_{j=1}^{k} \beta_{ij}\xi_j) , \qquad (10.24)$$

where F is a distribution function and

$$-\infty = \alpha_0^{(i)} < \alpha_1^{(i)} < \alpha_2^{(i)} \cdots < \alpha_{m_i-1}^{(i)} < +\infty .$$

The left hand side of (10.24) is the probability that an individual chooses a response in category c_i or less on variable i. In factor analysis terminology the $\alpha_s^{(i)}$ are thresholds and the β_{ij} are factor loadings.

This model is the same as was used in the single group examples in Sections 7.4.1 and 7.4.2 and is described in more detail in Section 16.2 in the appendix (see Chapter 16). Here we use the Probit link function, *i.e*,

$$F(u) = \Phi(u) = \int_{-\infty}^{u} \frac{1}{\sqrt{2\pi}} e^{-\frac{1}{2}t^2} dt , \qquad (10.25)$$

but it is possible to use any of the other link functions available, see Section 16.2. When applied to multiple groups the thresholds $\alpha_{c_i}^{(i)}$ as well as the factor loadings β_{ij} are allowed to vary across groups but we are particularly interested in testing whether they are the same in all groups.

To apply this model in LISREL we use the standard LISREL notation and specify the model as

$$\mathbf{x}^{(g)} = \mathbf{\Lambda}_x{}^{(g)}\boldsymbol{\xi}^{(g)} , \tag{10.26}$$

where $\mathbf{x}^{(g)}$ is a vector of the observed variables of NOSAY, COMPLEX, NOCARE, TOUCH, and INTEREST in group g, In this case the vector of latent variables $\boldsymbol{\xi}^{(g)}$ is a scalar representing the latent variable Efficacy in group g, and $\mathbf{\Lambda}_x$ is a vector corresponding to the matrix (β_{ij}) of order 5×1. The mean vector of $\boldsymbol{\xi}^{(g)}$ is $\boldsymbol{\kappa}^{(g)}$.

In the examples that follow, $\mathbf{\Lambda}_x$ is the vector

$$\begin{pmatrix} 1 \\ \lambda_2^{(x)} \\ \lambda_3^{(x)} \\ \lambda_4^{(x)} \\ \lambda_5^{(x)} \end{pmatrix} ,$$

Note that there are no intercept terms and no error terms in (10.26). Instead there are thresholds $\alpha_{c_i}^{(i)}$ in the model. These varies over variables, response categories and groups. For the examples considered here the thresholds are assumed to be the same in all groups which is necessary to be able to estimate the means of the latent variable on the same scale for all groups.

The parameter matrix $\mathbf{\Lambda}_x$ is regarded as an attribute of the observed items and is therefore assumed to be invariant over groups. The unit of measurement in the latent variable is defined by the element 1 in $\mathbf{\Lambda}_x$. Thus these units are the same across groups which makes it possible to compare the variances of the latent variable across groups.

USA vs Britain

We begin with the comparison of the two countries USA and Britain (BTN) where the same language, English, was used in the survey.[3]

Since the data contain missing values, we use FIML and since the variables are ordinal with four-point scales, we use Adaptive Quadrature to estimate the model. A LISREL command file for analysis of the two samples USA and BTN is (file **effi1b.lis**):

---

[3]The same wording was used in these countries, except the USA had *Congress in Washington* whereas Britain had *Parliament* in the TOUCH item, see Section 7.4.2.

```
Group: USA
!---------------------------------------------------------------------
! Comparing USA and BTN
! Two-group Confirmatory Factor Analysis Model of Political Efficacy Items
!---------------------------------------------------------------------
DA NG=2 NI=5
RA=effi_usa.lsf
$ADAPQ(10) PROBIT EQTH GR(5)
MO NX=5 NK=1 LX=FR KA=FR
LK
Efficacy
FI LX(1,1)
VA 1 LX(1,1)
OU

Group: BTN
DA
RA=effi_btn.lsf
MO LX=IN KA=FR
LK
Efficacy
OU
```

The **ADAPQ** command was explained in Section 7.4.1. The only new thing here is the **EQTH** keyword. This is used to specify that the thresholds should be equal across groups. Since there are 5 ordinal variables in the model and each has 4 categories there are 15 thresholds in each group. All these are constrained to be equal in both groups.

The rest of the input file is standard multigroup **LISREL** syntax as explained earlier in this chapter. The **LX=IN** on the **MO** line in the second group specifies factorial invariance *i.e.*, that Λ_x is the same in both groups.

The corresponding input file in **SIMPLIS** syntax is **effi1a.spl**:

```
Group USA
Raw Data from File effi_usa.lsf
$ADAPQ(10) PROBIT EQTH GR(5)
Latent Variables: Efficacy
Relationships:
   NOSAY = 1*Efficacy
   COMPLEX - INTEREST = Efficacy
Efficacy = CONST

Group BTN
Raw Data from File effibtn.lsf
Latent Variables: Efficacy Response
Efficacy = CONST
Set the variance of Efficacy free
End of Problem
```

Both the **SIMPLIS** and the **LISREL** syntax files give the same solution but in slightly different forms.

For USA the output file **effi1b.out** reports

```
Note: there are      9 cases with missing values
on all of the selected variables.
These cases are excluded from the analysis.
Effective sample size:     1710
```

The data file for USA has 1719 cases but 9 of these have not responded on the ordinal scale on any of the five variables. So the actual sample size used is 1710. For BTN there are 27 such cases out of 1483 cases giving an effective sample size of 1456:

```
Note: there are     27 cases with missing values
on all of the selected variables.
These cases are excluded from the analysis.
Effective sample size:     1456
```

The matrix of factor loadings Λ_x common to both groups is estimated

```
        LAMBDA-X

                Efficacy
      NOSAY      1.000

    COMPLEX      0.844
               (0.050)
                16.811

     NOCARE      2.217
               (0.121)
                18.394

      TOUCH      1.991
               (0.114)
                17.533


   INTEREST      2.397
               (0.145)
                16.489
```

The estimated variance and mean of the latent variable `Efficacy` are, for USA

```
        PHI

            Efficacy
            --------
               0.448
             (0.089)
               5.051

        KAPPA

            Efficacy
            --------
              -0.068
             (0.059)
              -1.154
```

and for BTN

```
        PHI

            Efficacy
            --------
               0.419
             (0.089)
               4.728

        KAPPA

            Efficacy
            --------
               0.087
             (0.060)
               1.451
```

Unfortunately there is no chi-square available for FIML with adaptive quadrature. The only thing we can go by to judge the fit of the model is the following information

```
    Number of quadrature points =              10
    Number of free parameters =                23
    Number of iterations used =                 8

    -2lnL (deviance statistic) =      29615.52401
    Akaike Information Criterion      29661.52401
    Schwarz Criterion                 29800.90917
```

The problems converged in 8 iterations. The number of free parameters is calculated as follows. There are 4 estimated parameters in the common $\mathbf{\Lambda}_x$. There is 1 parameter in $\mathbf{\Phi}$ in each group and 1 parameter in $\boldsymbol{\kappa}$ in each group. There are 15 thresholds common in both groups. Thus $4 + 2 + 2 + 15 = 23$

The estimated thresholds appear in the output as

```
Threshold estimates and standard deviations
-------------------------------------------
    Threshold          Estimates           S.E.        Est./S.E.

    TH1_NOSAY           -1.38974          0.06499       -21.38279
    TH2_NOSAY            0.05325          0.06183         0.86112
    TH3_NOSAY            1.91481          0.07332        26.11455
    TH1_COMPLEX         -0.92635          0.05473       -16.92484
    TH2_COMPLEX          0.83089          0.05572        14.91171
    TH3_COMPLEX          2.07541          0.06980        29.73301
    TH1_NOCARE          -1.86838          0.12670       -14.74684
    TH2_NOCARE           0.56501          0.13503         4.18427
    TH3_NOCARE           3.23096          0.17966        17.98417
    TH1_TOUCH           -1.50420          0.12339       -12.19055
    TH2_TOUCH            0.96193          0.11908         8.07783
    TH3_TOUCH            3.46062          0.15450        22.39939
    TH1_INTEREST        -1.71136          0.18047        -9.48272
    TH2_INTEREST         0.81913          0.12247         6.68836
    TH3_INTEREST         3.80682          0.00000         0.00000
```

One can test the hypothesis of factorial invariance, i.e., that $\mathbf{\Lambda}_x$ is the same in both groups by relaxing this constraint. To do so replace LX=IN on the MO line in the second group by LX=FR and add the two lines

```
FI LX(1,1)
VA 1 LX(1,1)
```

in the second group, see file **effi1b1.lis**. Running this gives

```
    Number of quadrature points =            10
    Number of free parameters =              27
    Number of iterations used =              10


    -2lnL (deviance statistic) =     29483.35782
    Akaike Information Criterion     29537.35782
    Schwarz Criterion               29700.98388
```

This model has four more parameters than the previous model and the difference in the values of $-2 \ln L$ between the two models can be used as a chi-square for testing the hypothesis of factorial invariance. This chi-square is 29615.524 - 29483.358 = 132.166 with 4 degrees of freedom. It is clear that the hypothesis must be rejected.

Germany vs. Austria

Next we consider comparing the two countries Germany (GER) and Austria (AUS) where the same language, German, was used in the survey.

One can follow the same procedure as for USA and BTN to test factorial invariance, see files **effi2b.lis** and **effi2b1.lis**. The output files for these runs give the following values of $-2 \ln L$, respectively:

```
Number of free parameters =                    23
-2lnL (deviance statistic) =        39837.98466

Number of free parameters =                    27
-2lnL (deviance statistic) =        39712.58930
```

The difference in $-2 \ln L$-value is 125.396. Regarding as a chi-square with 4 degrees of freedom, we must reject the hypothesis of factorial invariance for these two countries also.

All Countries

Since we have found that factorial invariance does not hold for USA and BTN and also not for GER and AUS it does not seem meaningful to test this for all eight countries. Nevertheless we will do this here as an illustration of how it can be done. The result will illustrate how one can estimate the latent mean of Efficacy for each country on the same scale.

The input file in LISREL syntax is **effi3b.lis**:

```
Group: USA
! Comparing All Eight Countries
! Eight-group Confirmatory Factor Analysis Model of Political efficacy Items
! Estimating Factor Means and Variances for Each Group
DA NG=8 NI=5
RA=effi_usa.lsf
$ADAPQ(10) PROBIT EQTH GR(5)
MO NX=5 NK=1 LX=FR KA=FR
LK
Efficacy
FI LX(1,1)
VA 1 LX(1,1)
OU

Group: GER
DA NI=5
RA=effi_ger.lsf
MO LX=IN KA=FR
LK
Efficacy
OU

Group: NET
DA NI=5
RA=effi_net.lsf
MO LX=IN KA=FR
LK
Efficacy
OU
```

```
Group: AUS
DA NI=5
RA=effi_aus.lsf
MO LX=IN KA=FR
LK
Efficacy
OU

Group: BTN
DA NI=5
RA=effi_btn.lsf
MO LX=IN KA=FR
LK
Efficacy
OU

Group: ITY
DA NI=5
RA=effi_ity.lsf
MO LX=IN KA=FR
LK
Efficacy
OU

Group: SWI
DA NI=5
RA=effi_swi.lsf
MO LX=IN KA=FR
LK
Efficacy
OU

Group: FIN
DA NI=5
RA=effi_fin.lsf
MO LX=IN KA=FR
LK
Efficacy
OU
```

It does not matter which country is chosen as the first country, and, in fact the ordering of the countries does not matter. However, the results will depend on which variable is chosen (here NOSAY) to define the scale of the latent variable Efficacy.

This model has 35 parameters and it converged in 8 iterations. From the output file **effi3b.out** we find the estimated factor loadings common to all countries:

```
           LAMBDA-X
            Efficacy
  NOSAY        1.000

COMPLEX        0.896
              (0.025)
              36.223

 NOCARE        1.927
              (0.052)
              37.246

  TOUCH        1.742
              (0.050)
              34.876

INTEREST       1.863
              (0.056)
              33.459
```

The estimated latent means for each country are:

```
            Efficacy
            --------
USA          -0.145
            (0.028)
            -5.187

GER           0.103
            (0.029)
             3.568

NET          -0.139
            (0.030)
            -4.686

AUS           0.402
            (0.034)
            11.852

BTN           0.000
            (0.029)
             0.017

ITA           0.266
            (0.029)
             9.136
```

```
SWI              0.068
                (0.035)
                 1.966


FIN              0.111
                (0.031)
                 3.537
```

To see more clearly the extent to which the means of Efficacy are different across countries, Figure 10.9 shows approximate 95% confidence intervals for each country, where the countries have been ordered in increasing means of Efficacy from left to right.

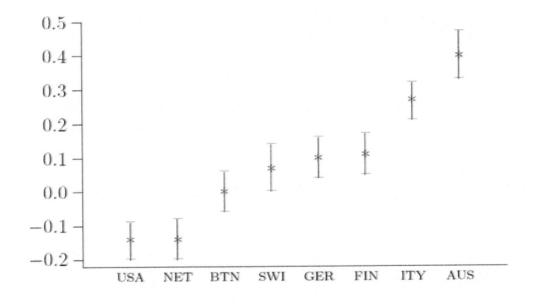

Figure 10.9: Approximate 95% Confidence Intervals for the Mean of Efficacy

The standard errors of the means are approximately the same over countries. So the length of the confidence intervals are approximately the same for all countries.

From Figure 10.9 it appears to be three groups of countries with approximately the same means within each group. The first group consists of USA and Netherlands which have the smallest means. The second group consists of Britain, Switzerland, Germany, and Finland. These countries have significantly larger means than USA and Netherlands. The third group consists of Italy and Austria, both of which have significantly larger means than the other countries. Since the confidence intervals for Italy and Austria do not overlap, it is clear that Austria have a significantly larger mean than Italy.

If one believes these results one could say that the people of USA and Netherlands are the least efficacious and the people of Austria are the most efficacious.

Chapter 11

Appendix A: Basic Matrix Algebra and Statistics

To follow the part of statistical theory included in each chapter and section of this book readers must have a basic understanding of matrix algebra and some statistical concepts. Such basic requirements are provided in this appendix.

11.1   Basic Matrix Algebra

Definition of a Matrix

Definition: A *matrix* $\mathbf{A}$ is a rectangular array of numbers. If $\mathbf{A}$ has p rows and q columns we say that $\mathbf{A}$ is of order $p \times q$.

Notation:

$$\mathbf{A} = \begin{bmatrix} a_{11} & a_{12} & \cdots & a_{1q} \\ a_{21} & a_{22} & \cdots & a_{2q} \\ \cdot & \cdot & & \cdot \\ \cdot & \cdot & & \cdot \\ \cdot & \cdot & & \cdot \\ a_{p1} & a_{p2} & \cdots & a_{pq} \end{bmatrix} . \tag{11.1}$$

or

$$\mathbf{A} = (a_{ij}) \tag{11.2}$$

The numbers a_{ij} are called the elements of $\mathbf{A}$. In this book we assume that all elements are real numbers. But matrices with complex numbers can also be considered. The element a_{ij} is the jth element in row i of $\mathbf{A}$. To specify that $\mathbf{A}$ is of order $p \times q$ we write $\mathbf{A}(p \times q)$. A matrix of order $p \times p$ is called a *square matrix*.

The Transpose of a Matrix

Definition: The *transpose* of $\mathbf{A}(p \times q)$, denoted $\mathbf{A}'$, is the matrix of order $q \times p$ with rows and columns $\mathbf{A}$ interchanged:

© Springer International Publishing Switzerland 2016
K.G. Jöreskog et al., *Multivariate Analysis with LISREL*, Springer Series in Statistics,
DOI 10.1007/978-3-319-33153-9_11

$$\mathbf{A}' = \begin{bmatrix} a_{11} & a_{21} & \cdots & a_{p1} \\ a_{12} & a_{22} & \cdots & a_{p2} \\ \cdot & \cdot & & \cdot \\ \cdot & \cdot & & \cdot \\ \cdot & \cdot & & \cdot \\ a_{1q} & a_{2q} & \cdots & a_{pq} \end{bmatrix}. \tag{11.3}$$

Properties

1. $(\mathbf{A}')' = \mathbf{A}$

2. $(\mathbf{A} + \mathbf{B})' = \mathbf{A}' + \mathbf{B}'$

3. $(\mathbf{AB})' = \mathbf{B}'\mathbf{A}'$

Definition of a Vector

Definition: A $p \times 1$ matrix is called a *column vector* and a $1 \times q$ matrix is called a *row vector*. In this book we follow the convention that a vector $\mathbf{x}$ is always a column vector and a row vector is always denoted $\mathbf{x}'$.

A matrix $\mathbf{A}$ of order $p \times q$ consists of q column vectors and p row vectors. Let $\mathbf{a}_{(1)}, \mathbf{a}_{(2)}, \ldots, \mathbf{a}_{(q)}$ be the columns of $\mathbf{A}$ and let $\mathbf{a}_1, \mathbf{a}_2, \ldots, \mathbf{a}_p$ the rows of $\mathbf{A}$ written as column vectors. Then

$$\mathbf{A} = (\mathbf{a}_{(1)}, \mathbf{a}_{(2)}, \ldots, \mathbf{a}_{(q)}) = \begin{bmatrix} \mathbf{a}_1' \\ \mathbf{a}_2' \\ \cdot \\ \cdot \\ \cdot \\ \mathbf{a}_p' \end{bmatrix}. \tag{11.4}$$

Addition and Multiplication of Matrices

If $\mathbf{A}$ and $\mathbf{B}$ are matrices of order $p \times q$ and c and d are constants, then

1. $c\mathbf{A} = (ca_{ij})$

2. $\mathbf{A} \pm \mathbf{B} = (a_{ij} \pm b_{ij})$

3. $c\mathbf{A} \pm d\mathbf{B} = (ca_{ij} \pm db_{ij})$

If $\mathbf{A}$ is $p \times q$ and $\mathbf{B}$ is $q \times r$, then the product of $\mathbf{A}$ and $\mathbf{B}$, denoted $\mathbf{AB}$, is defined as

$$\mathbf{AB} = (\mathbf{a}_i'\mathbf{b}_{(j)}) \tag{11.5}$$

Let $\mathbf{C} = \mathbf{AB}$, then

$$c_{ik} = \sum_{j=1}^{q} a_{ij} b_{jk} \tag{11.6}$$

Note that $\mathbf{AB}$ is not in general equal to $\mathbf{BA}$. Because of this fact, one must distinguish between *pre-multiplication* and *post-multiplication*. In $\mathbf{AB}$ we say that $\mathbf{B}$ is pre-multiplied by $\mathbf{A}$ and $\mathbf{A}$ is post-multiplied by $\mathbf{B}$.

Symmetric Matrix

Definition A square matrix $\mathbf{A}(p \times p)$ is *symmetric* if $\mathbf{A} = \mathbf{A}'$.

For any $p \times q$ matrix $\mathbf{A}$, $\mathbf{AA}'$ is a symmetric matrix of order $p \times p$ and $\mathbf{A}'\mathbf{A}$ is a symmetric matrix of order $q \times q$.

Some Types of Matrices

Name	Definition	Notation
Scalar	$p = q = 1$	a,b
Unit vector	$(1, 1, \ldots, 1)$	$\mathbf{1}$ or $\mathbf{1}_p$
Diagonal	$p = q, a_{ij} = 0, i \neq j$	$diag\mathbf{A}$
Identity	$diag(\mathbf{1})$	$\mathbf{I}$ or $\mathbf{I}_p$
Unit matrix	$p = q, a_{ij} = 1$	$\mathbf{J}_p = \mathbf{11}'$
Null matrix	$a_{ij} = 0$	$\mathbf{0}$
Triangular	$a_{ij} = 0, j > i$	$\mathbf{T}$

In particular, if $\mathbf{x}$ and $\mathbf{y}$ are $n \times 1$ vectors, *i.e.*, $\mathbf{x}' = (x_1, x_2, \ldots, x_n)$ and $\mathbf{y}' = (y_1, y_2, \ldots, y_n)$, then

$$\mathbf{x}'\mathbf{1} = \sum_{i=1}^{n} x_i$$

$$\mathbf{x}'\mathbf{x} = \sum_{i=1}^{n} x_i^2$$

$$\mathbf{x}'\mathbf{y} = \sum_{i=1}^{n} x_i y_i$$

but $\mathbf{xx}'$ and $\mathbf{xy}'$ are $n \times n$ matrices with elements $(x_i x_j)$ and $(x_i y_j)$, respectively.

Orthogonal and Orthonormal Vectors

Definition: Two vectors $\mathbf{x}_1$ and $\mathbf{x}_2$ are orthogonal if $\mathbf{x}_1'\mathbf{x}_2 = 0$.

Definition: A vector $\mathbf{x}$ is normalized if $\mathbf{x}'\mathbf{x} = 1$

Definition: Two vectors are orthonormal if they are normalized and orthogonal.

Definition: A set of vectors $\mathbf{x}_1, \mathbf{x}_2, \ldots, \mathbf{x}_k$ is said to be orthonormal if $\mathbf{x}_i'\mathbf{x}_j = \delta_{ij}$, where $\delta_{ij} = 0$, $i \neq j$ and $\delta_{ij} = 1$, $i = j$.

Linear Dependence

Definition: A set of vectors $\mathbf{x}_1, \mathbf{x}_2, \ldots, \mathbf{x}_k$ is said to be linearly dependent if there exists scalars $a_1, a_2, \ldots, a_k$, not all zero, such that

$$a_1\mathbf{x}_1 + a_2\mathbf{x}_2 + \cdots + a_k\mathbf{x}_k = \mathbf{0} , \tag{11.7}$$

otherwise it is said to be linearly independent.

1. Any set of vectors containing $\mathbf{0}$ is a linearly dependent set.

2. A set of vectors $\mathbf{x}_1, \mathbf{x}_2, \ldots, \mathbf{x}_k$ is linearly dependent if and only if one vector is a linear combination of the others.

3. If the nonzero vectors $\mathbf{x}_1, \mathbf{x}_2, \ldots, \mathbf{x}_k$ are pairwise orthogonal, *i.e.*, if $\mathbf{x}_i'\mathbf{x}_j = 0$ for all i and j with $i \neq j$, they are linearly independent.

4. If $\mathbf{x}_1, \mathbf{x}_2, \ldots, \mathbf{x}_p$ are p linearly independent vectors of order p and $\mathbf{y}$ is any vector of order p, there is a unique representation of $\mathbf{y}$ as

$$\mathbf{y} = a_1\mathbf{x}_1 + a_2\mathbf{x}_2 + \cdots + a_p\mathbf{x}_p \tag{11.8}$$

Rank of a Matrix

Definition: The rank $r(\mathbf{A})$ of a matrix $\mathbf{A}$ is the number of independent rows or columns it contains.

1. If $\mathbf{A}$ is of order $p \times q$, then $r(\mathbf{A}) \leq min\{p, q\}$.

2. $r(\mathbf{AB}) \leq min\{r(\mathbf{A}), r(\mathbf{B})\}$.

Determinant of a Matrix

Definition: The determinant $|\mathbf{A}|$ of a square matrix $\mathbf{A}$ of order $p \times p$ is a scalar function $D(\mathbf{a}_1, \mathbf{a}_2, \ldots, \mathbf{a}_p)$ of its columns $\mathbf{a}_1, \mathbf{a}_2, \ldots, \mathbf{a}_p$ such that

1. $D(\mathbf{a}_1, \mathbf{a}_2, \ldots, c\mathbf{a}_k \ldots, \mathbf{a}_p) = cD(\mathbf{a}_1, \mathbf{a}_2, \ldots, \mathbf{a}_p)$

2. $D(\mathbf{a}_1, \mathbf{a}_2, \ldots, \mathbf{a}_m + \mathbf{a}_k \ldots, \mathbf{a}_p) = D(\mathbf{a}_1, \mathbf{a}_2, \ldots, \mathbf{a}_p)$

3. $D(\mathbf{i}_1, \mathbf{i}_2, \ldots, \mathbf{i}_p) = 1$, where $\mathbf{i}_1, \mathbf{i}_2, \ldots, \mathbf{i}_p$ are the columns of the identity matrix $\mathbf{I}$.

It can be shown that such a function exists and defines a unique value.

Singular Matrix

Definition: A square matrix $\mathbf{A}$ is said to be singular if $|\mathbf{A}| = 0$, otherwise it is said to be non-singular.

1. If $\mathbf{A}$ is a square matrix of order $p \times p$, then $\mathbf{A}$ is singular if and only if $r(\mathbf{A}) < p$.

2. $|\mathbf{A}| = |\mathbf{A}'|$.

3. If $\mathbf{A}$ is diagonal or triangular, then $|\mathbf{A}| = a_{11}a_{22} \cdots a_{pp}$.

Inverse of a Matrix

Definition: If $\mathbf{A}$ is a non-singular matrix, there exists a unique matrix $\mathbf{B}$ such that $\mathbf{BA} = \mathbf{AB} = \mathbf{I}$. The matrix $\mathbf{B}$ is called the inverse of $\mathbf{A}$ and is denoted $\mathbf{A}^{-1} = (a^{ij})$.

1. $(\mathbf{A}')^{-1} = (\mathbf{A}^{-1})'$.

2. If both $\mathbf{A}$ and $\mathbf{B}$ are non-singular, then $(\mathbf{AB})^{-1} = \mathbf{B}^{-1}\mathbf{A}^{-1}$.

3. If $\mathbf{A}$ and $\mathbf{D}$ are non-singular matrices of orders $p \times p$ and $k \times k$, respectively and $\mathbf{B}$ and $\mathbf{C}'$ are matrices of order $p \times k$, then

$$(\mathbf{A} + \mathbf{BDC})^{-1} = \mathbf{A}^{-1} - \mathbf{A}^{-1}\mathbf{B}(\mathbf{CA}^{-1}\mathbf{B} + \mathbf{D}^{-1})^{-1}\mathbf{CA}^{-1} \tag{11.9}$$

4. The rank of a matrix is not changed if the matrix is pre- or post-multiplied by a non-singular matrix.

Homogeneous Linear Equations

Definition: A system of homogeneous equations is a set of p equations in q unknowns of the form

$$\mathbf{Ax} = \mathbf{0} , \tag{11.10}$$

where $\mathbf{A}$ is of order $p \times q$ and $\mathbf{x}$ is of order $q \times 1$.

1. If $\mathbf{x}$ is a solution to (11.10), then $c\mathbf{x}$ is also a solution, for any scalar c.

2. The system (11.10) always has the trivial solution $\mathbf{x} = \mathbf{0}$.

3. If $\mathbf{x}$ is a solution to (11.10), then $\mathbf{x}$ is orthogonal to each row of $\mathbf{A}$.

4. If $r(\mathbf{A}) = q$, then (11.10) has only the trivial solution $\mathbf{x} = \mathbf{0}$.

5. If $r(\mathbf{A}) = r < q$, there exists $q - r$ linearly independent solutions to (11.10).

6. If $p = q$, the system (11.10) has a non-trivial solution if and only if $\mathbf{A}$ is singular.

Eigenvalues and Eigenvectors

Definition: A non-zero vector $\mathbf{u}$ is said to be an eigenvector of a *symmetric* matrix $\mathbf{A}$ if for some scalar λ

$$\mathbf{A}\mathbf{u} = \lambda\mathbf{u} . \tag{11.11}$$

The number λ is said to be an eigenvalue of $\mathbf{A}$ corresponding to $\mathbf{u}$. Since the eigenvector is only determined up to a non-zero constant c, it is assumed that it is normalized such that $\mathbf{u}'\mathbf{u} = 1$.

1. The vector $\mathbf{u}$ is an eigenvector of $\mathbf{A}$ if and only if $|\mathbf{A} - \lambda\mathbf{I}| = 0$.

2. If $\mathbf{A}$ is of order $p \times p$, $|\mathbf{A} - \lambda\mathbf{I}|$ is a polynomial in λ of degree p. The equation $|\mathbf{A} - \lambda\mathbf{I}| = 0$ has p real roots $\lambda_1, \lambda_2, \ldots, \lambda_p$ which are the eigenvalues of $\mathbf{A}$.

1. If the eigenvalues are all distinct such that $\lambda_1 > \lambda_2 \ldots > \lambda_p$, there is a corresponding set of eigenvectors $\mathbf{u}_1, \mathbf{u}_2 \ldots, \mathbf{u}_p$, such that

$$\mathbf{A}\mathbf{u}_i = \lambda_i\mathbf{u}_i \quad , i = 1, 2, \ldots, p , \tag{11.12}$$

and the eigenvectors $\mathbf{u}_1, \mathbf{u}_2 \ldots, \mathbf{u}_p$ form an orthonormal set, *i.e.*, $\mathbf{u}_i'\mathbf{u}_j = \delta_{ij}$, where $\delta_{ij} = 0 , i \neq j$ and $\delta_{ij} = 1 , i = j$.

1. If $\mathbf{A}$ is a symmetric matrix of order $p \times p$, there exists a matrix $\mathbf{U}$ of order $p \times p$ such that

$$\mathbf{U}'\mathbf{A}\mathbf{U} = \mathbf{\Lambda} , \tag{11.13}$$

$$\mathbf{U}'\mathbf{U} = \mathbf{U}\mathbf{U}' = \mathbf{I} , \tag{11.14}$$

where $\mathbf{\Lambda}$ is a diagonal matrix of order $p \times p$. The diagonal elements of $\mathbf{\Lambda}$ are the eigenvalues of $\mathbf{A}$ and the columns of $\mathbf{U}$ are the corresponding eigenvectors.

2. Every symmetric matrix $\mathbf{A}$ can be written as

$$\mathbf{A} = \mathbf{U}\mathbf{\Lambda}\mathbf{U}' , \tag{11.15}$$

where $\mathbf{U}$ is orthonormal and $\mathbf{\Lambda}$ is diagonal.

3. The rank $r(\mathbf{A})$ of a symmetric matrix $\mathbf{A}$ equals the number of non-zero eigenvalues of $\mathbf{A}$.

4. The determinant $|\mathbf{A}|$ of a symmetric matrix $\mathbf{A}$ equals the product of its eigenvalues,*i.e.*, $|\mathbf{A}| = \prod \lambda_i$.

Positive Definite and Semidefinite Matrices

Definition: A symmetric matrix $\mathbf{A}$ is said to be positive definite (positive semi-definite) if all eigenvalues of $\mathbf{A}$ are positive (non-negative with at least one eigenvalue 0).

1. A symmetric matrix $\mathbf{A}$ is positive definite (positive semi-definite) if and only if $\mathbf{x}'\mathbf{A}\mathbf{x} > 0$ for all non-zero vectors $\mathbf{x}$ ($\mathbf{x}'\mathbf{A}\mathbf{x} \geq 0$ for all non-zero $\mathbf{x}$ and $= 0$ for at least one non-zero $\mathbf{x}$).

2. If $\mathbf{A}$ is positive definite, all its diagonal elements are positive.

3. If λ is an eigenvalue of a non-singular symmetric matrix $\mathbf{A}$, then $1/\lambda$ is an eigenvalue of $\mathbf{A}^{-1}$, and the corresponding eigenvector $\mathbf{u}$ of $\mathbf{A}$ and $\mathbf{A}^{-1}$ are the same. Moreover, if $\mathbf{A} = \mathbf{U}\boldsymbol{\Lambda}\mathbf{U}'$, then $\mathbf{A}^{-1} = \mathbf{U}\boldsymbol{\Lambda}^{-1}\mathbf{U}'$, where $\mathbf{U}$ is orthonormal and $\boldsymbol{\Lambda}$ is diagonal.

4. If $\mathbf{A}$ is positive definite, so is $\mathbf{A}^{-1}$.

Definition: If $\mathbf{A}$ is positive definite and $\mathbf{A} = \mathbf{U}\boldsymbol{\Lambda}\mathbf{U}'$, a symmetric square root $\mathbf{A}^{\frac{1}{2}}$ of $\mathbf{A}$ can be defined as $\mathbf{A} = \mathbf{U}\boldsymbol{\Lambda}^{\frac{1}{2}}\mathbf{U}'$ and a symmetric inverse square root of $\mathbf{A}$ can be defined as $\mathbf{A}^{-\frac{1}{2}} = \mathbf{U}\boldsymbol{\Lambda}^{-\frac{1}{2}}\mathbf{U}'$.

1. $\mathbf{A}^{\frac{1}{2}}\mathbf{A}^{\frac{1}{2}} = \mathbf{A}$

2. $\mathbf{A}^{-\frac{1}{2}} = (\mathbf{A}^{\frac{1}{2}})^{-1}$

Singular Value Decomposition

For any matrix $\mathbf{A}(p \times q)$ there exists three matrices $\mathbf{U}(p \times p)$, $\mathbf{V}(q \times q)$ and $\boldsymbol{\Gamma}(p \times q)$ such that

$$\mathbf{A} = \mathbf{U}\boldsymbol{\Gamma}\mathbf{V}' \tag{11.16}$$

$$\mathbf{U}'\mathbf{U} = \mathbf{U}\mathbf{U}' = \mathbf{I}_p \tag{11.17}$$

$$\mathbf{V}'\mathbf{V} = \mathbf{V}\mathbf{V}' = \mathbf{I}_q \tag{11.18}$$

$$\gamma_{ij} = 0 \, , i \neq j \ \gamma_{ii} \geq 0 \tag{11.19}$$

For any matrix $\mathbf{A}(p \times q)$ of rank r there exists three matrices $\mathbf{U}(p \times r)$, $\mathbf{V}(q \times r)$ and $\boldsymbol{\Gamma}(r \times r)$ such that

$$\mathbf{A} = \mathbf{U}\boldsymbol{\Gamma}\mathbf{V}' \tag{11.20}$$

$$\mathbf{U}'\mathbf{U} = \mathbf{I}_r \tag{11.21}$$

$$\mathbf{V}'\mathbf{V} = \mathbf{I}_r \tag{11.22}$$

$$\gamma_{ij} = 0 \, , i \neq j \ \gamma_{ii} > 0 \, , \tag{11.23}$$

The $\gamma_{ii} = \sqrt{\lambda_i} \, , i = 1, 2, \ldots, r$, where $\lambda_i \, , i = 1, 2, \ldots, r$ are the positive eigenvalues of $\mathbf{A}\mathbf{A}' = \mathbf{A}'\mathbf{A}$. The $\gamma_{ii} \, , i = 1, 2, \ldots, r$ are called the singular values of $\mathbf{A}$ and the columns of $\mathbf{U}$ and $\mathbf{V}$ are called singular vectors.

1. For any matrix $\mathbf{A}(p \times q)$, the positive eigenvalues of $\mathbf{A}\mathbf{A}'$ and $\mathbf{A}'\mathbf{A}$ are identical and if $\mathbf{u}$ is an eigenvector of $\mathbf{A}\mathbf{A}'$ corresponding to a positive eigenvalue, then $\mathbf{A}'\mathbf{u}$ is an eigenvector of $\mathbf{A}'\mathbf{A}$ corresponding to the same eigenvalue.

2. For any matrix $\mathbf{A}(p \times q)$, $|\mathbf{I}_p + \mathbf{A}\mathbf{A}'| = |\mathbf{I}_q + \mathbf{A}'\mathbf{A}|$

Least Squares Properties

Problem 1: Given a symmetric positive matrix $\mathbf{A}$ of order $p \times p$, find a symmetric positive semi-definite matrix $\mathbf{B}$ of rank $k < p$ which approximates $\mathbf{A}$ in the least squares sense.

Solution: $\mathbf{B} = \mathbf{U}_k \mathbf{\Lambda}_k \mathbf{U}'_k$, where $\mathbf{\Lambda}_k$ consists of the k largest eigenvalues of $\mathbf{A}$ and $\mathbf{U}_k$ consists of the corresponding eigenvectors.

Problem 2: Given a matrix $\mathbf{A}$ of order $p \times q$ of rank q, find a matrix matrix $\mathbf{B}$ of order $p \times k$ of rank $k < q$ which approximates $\mathbf{A}$ in the least squares sense.

Solution: $\mathbf{B} = \mathbf{U}_k \mathbf{\Gamma}_k \mathbf{V}'_k$, where $\mathbf{\Gamma}_k$ consists of the k largest singular values of $\mathbf{A}$ and $\mathbf{U}_k$ and $\mathbf{V}_k$ consist of the corresponding singular vectors.

Trace of a Matrix

Definition: The *trace* of a square matrix $\mathbf{A}(p \times p)$, $tr(\mathbf{A})$, is defined as the sum of its diagonal elements, *i.e.*,

$$tr(\mathbf{A}) = a_{11} + a_{22} + \cdots + a_{pp} \ . \tag{11.24}$$

1. $tr(\mathbf{AB}) = tr(\mathbf{BA})$

2. $tr(\mathbf{ABC}) = tr(\mathbf{CAB}) = tr(\mathbf{BCA})$

3. If $\mathbf{P}$ is non-singular,

$$tr(\mathbf{P}^{-1}\mathbf{AP}) = tr(\mathbf{A}) \tag{11.25}$$

4. If $\mathbf{U}$ is orthonormal,

$$tr(\mathbf{U}'\mathbf{AU}) = tr(\mathbf{A}) \tag{11.26}$$

5. If $\mathbf{A}$ has eigenvalues $\lambda_1, \lambda_2, \ldots, \lambda_p$, then

$$tr(\mathbf{A}) = \sum_{i=1}^{p} \lambda_i \ , \quad tr(\mathbf{A}^2) = \sum_{i=1}^{p} \lambda_i^2 \ , \tag{11.27}$$

6. If $\mathbf{A}$ is of order $p \times q$, then

$$tr(\mathbf{AA}') = tr(\mathbf{A}'\mathbf{A}) = \sum_{i=1}^{p} \sum_{j=1}^{q} a_{ij}^2 \ . \tag{11.28}$$

Maximum or Minimum of Analytical Functions

Unconstrained Maximum or Minimum

Let $y = f(\mathbf{x})$ be a continuous function of $\mathbf{x}' = (x_1, x_2, \ldots, x_p)$ with derivatives of first and second order. For y to have a maximum or minimum at $\mathbf{x} = \mathbf{x}_0$, it is necessary that $\partial y / \partial \mathbf{x} = \mathbf{0}$ at $\mathbf{x} = \mathbf{x}_0$.

$\mathbf{x}_0$ is a minimum if $\partial^2 y / \partial \mathbf{x} \partial \mathbf{x}'$ is positive definite at $\mathbf{x} = \mathbf{x}_0$ and a maximum if $-\partial^2 y / \partial \mathbf{x} \partial \mathbf{x}'$ is positive definite at $\mathbf{x} = \mathbf{x}_0$.

Constrained Maximum or Minimum

Let $y = f(\mathbf{x})$ be a continuous function of $\mathbf{x}' = (x_1, x_2, \ldots, x_p)$ with derivatives of first and second order and let $\mathbf{g}'(\mathbf{x}) = (g_1(\mathbf{x}), g_2(\mathbf{x}), \ldots, g_q(\mathbf{x}))$ be continuous functions of $\mathbf{x}$ with first derivatives. Let

$$z(\mathbf{x}, \boldsymbol{\gamma}) = f(\mathbf{x}) - \boldsymbol{\gamma}'\mathbf{g}(\mathbf{x}) \tag{11.29}$$

For y to have a maximum or minimum at $\mathbf{x} = \mathbf{x}_0$, subject to the constraint $\mathbf{g}(\mathbf{x}) = \mathbf{0}$ it is necessary that both $\partial z/\partial \mathbf{x} = \mathbf{0}$ and $\partial z/\partial \boldsymbol{\gamma} = \mathbf{0}$ at $\mathbf{x}_0$.

$\mathbf{x}_0$ is a constrained minimum if $\mathbf{v}'(\partial^2 y/\partial \mathbf{x}\partial \mathbf{x}')\mathbf{v} > 0$ at $\mathbf{x}_0$ for every non-zero vector $\mathbf{v}$ such that $(\partial \mathbf{g}\partial \mathbf{x}')\mathbf{v} = \mathbf{0}$ at $\mathbf{x}_0$ and a maximum if $\mathbf{v}'(\partial^2 y/\partial \mathbf{x}\partial \mathbf{x}')\mathbf{v} < 0$ at $\mathbf{x}_0$ for every non-zero vector $\mathbf{v}$ such that $(\partial \mathbf{g}\partial \mathbf{x}')\mathbf{v} = \mathbf{0}$ at $\mathbf{x}_0$.

11.2 Basic Statistical Concepts

Probability Distributions

Discrete Distribution

A discrete random variable is a random variable that takes only a finite number of variables. For each value x_i there is a probability p_i of x_i occuring.

x_1	x_2	$\cdots$	x_k
p_1	p_2	$\cdots$	p_k

where $p_i > 0$ and $\sum_i^k p_i = 1$

Examples of discrete random variables are the Bernoulli, Binomial, and the Poisson distributions.

Continuous Distribution

A continuous random variable is a random variable that can take any value in interval $a \leq x \leq b$. It is characterized by a density function $p(x)$ such that $p(x) > 0$ and $\int_a^b p(x)dx = 1$

Examples of continuous random variables are the Normal and Chi-Square Distributions.

Expectation of a Random Variable

Discrete Distribution
$E(x) = \sum_i^k x_i p_i$

Continuous Distribution

$E(x) = \int_{-\infty}^{+\infty} x p(x)dx$

Expectation of a Function of a Random Variable

Discrete Distribution
$E[f(x)] = \sum_i^k f(x_i)p_i$

Continuous Distribution

$E[f(x)] = \int_{-\infty}^{+\infty} (f(x)p(x)dx$

Mean and Variance of a Random Variable

$$\mu = E(x) \quad \sigma^2 = E[(x - \mu)^2] = E(x^2) - \mu^2$$

Examples

2	3	4	5	6	7	8
1/16	2/16	3/16	4/16	3/16	2/16	1/16

$\mu = 2 \times 1/16 + 3 \times 2/16 + 4 \times 3/16 + 5 \times 4/16 + 6 \times 3/16 + 7 \times 2/16 + 8 \times 1/16 = 80/16 = 5$

$\sigma^2 = 9 \times 1/16 + 4 \times 2/16 + 1 \times 3/16 + 0 \times 4/16 + 1 \times 3/16 + 4 \times 2/16 + 9 \times 1/16 = 40/16 = 2.5$

$$p(x) = \frac{1}{2}e^{-\frac{1}{2}(x-3)^2}$$

$$\mu = \int_{-\infty}^{+\infty} xp(x)dx = 3 \quad \sigma^2 = \int_{-\infty}^{+\infty} (x-3)^2 p(x)dx = 1$$

Linear Combinations of Random Variables

Let x_1 and x_2 be random variables with means μ_1, μ_2, variances σ_1^2, σ_2^2, and covariance σ_{12}, let a, b, c, and d be constants, and let

$$y_1 = ax_1 + bx_2$$
$$y_2 = cx_1 + dx_2$$

Then the means, variances and covariances of y_1 and y_2 are

$$\mu_{y_1} = a\mu_1 + b\mu_2$$

$$\mu_{y_2} = c\mu_1 + d\mu_2$$
$$\sigma_{y_1}^2 = a\sigma_1^2 + b\sigma_2^2 + 2ab\sigma_{12}$$
$$\sigma_{y_2}^2 = c\sigma_1^2 + d\sigma_2^2 + 2cd\sigma_{12}$$
$$\sigma_{y_1y_2} = (ac + bd)\sigma_{12}$$

11.3   Basic Multivariate Statistics

Definition: A *random matrix* is a matrix whose element are random variables.

Definition: The *expectation* of random matrix $\mathbf{X}$ is the matrix $E(\mathbf{X}) = (E[x_{ij}])$.

If $\mathbf{A}$ and $\mathbf{B}$ are fixed matrices and $\mathbf{X}$ is a random matrix, then $E(\mathbf{AXB}) = \mathbf{A}E(\mathbf{X})\mathbf{B}$.

Definition: If $\mathbf{x}$ is random vector, the *mean vector* $\boldsymbol{\mu}$ and the *covariance matrix* $\boldsymbol{\Sigma}$ of $\mathbf{x}$ are defined as

$$\boldsymbol{\mu} = E(\mathbf{x}) \quad \boldsymbol{\Sigma} = E[(\mathbf{x} - \boldsymbol{\mu})(\mathbf{x} - \boldsymbol{\mu})'] = E(\mathbf{x}\mathbf{x}') - \boldsymbol{\mu}\boldsymbol{\mu}'$$

Definition: If $\mathbf{x}$ is random vector, the *correlation matrix* $\mathbf{P}$ of $\mathbf{x}$ is defined as

$$\mathbf{P} = (diag\boldsymbol{\Sigma})^{-\frac{1}{2}}\boldsymbol{\Sigma}(diag\boldsymbol{\Sigma})^{-\frac{1}{2}}$$

If $\mathbf{x}$ is $p \times 1$, then $\boldsymbol{\mu}$ is $p \times 1$ and $\boldsymbol{\Sigma}$ is a $p \times p$ symmetric matrix whose diagonal element σ_{ii} is the variance of x_i and off-diagonal element σ_{ij} is the covariance between x_i and x_j. $\mathbf{P}$ is a $p \times p$ symmetric matrix with 1's in the diagonal and the off-diagonal element

$$\rho_{ij} = \frac{\sigma_{ij}}{\sqrt{\sigma_{ii}\sigma_{jj}}}$$

is the *correlation* between x_i and x_j.

Let $\mathbf{x}$ be a random vector with mean vector $\boldsymbol{\mu}$ and covariance matrix $\boldsymbol{\Sigma}$ and let

$$\mathbf{y} = \mathbf{a} + \mathbf{Bx}$$

then the mean vector $\boldsymbol{\mu}_y$ and covariance matrix $\boldsymbol{\Sigma}_y$ of $\mathbf{y}$ are

$$\boldsymbol{\mu}_y = \mathbf{a} + \mathbf{B}\boldsymbol{\mu} \quad \boldsymbol{\Sigma}_y = \mathbf{B}\boldsymbol{\Sigma}\mathbf{B}'$$

Sample Statistics

Sample: $\mathbf{z}_1, \mathbf{z}_2, \ldots, \mathbf{z}_n \quad iid$

Data Matrix:

$$\mathbf{Z} = \begin{bmatrix} z_{11} & z_{12} & \cdots & z_{1k} \\ z_{21} & z_{22} & \cdots & z_{2k} \\ . & . & \cdots & . \\ . & . & \cdots & . \\ z_{n1} & z_{n2} & \cdots & z_{nk} \end{bmatrix} = \begin{bmatrix} \mathbf{z}_1' \\ \mathbf{z}_2' \\ . \\ . \\ \mathbf{z}_n' \end{bmatrix}$$

Mean Vector: $\bar{\mathbf{z}} = (1/n)\sum_{i=1}^{n}\mathbf{z}_i = (1/n)\mathbf{Z}'\mathbf{1}$

Covariance Matrix: $\mathbf{S} = (1/n)\sum_{i=1}^{n}(\mathbf{z}_i - \bar{\mathbf{z}})(\mathbf{z}_i - \bar{\mathbf{z}})' = (1/n)\mathbf{Z}'\mathbf{Z} - \bar{\mathbf{z}}\bar{\mathbf{z}}'$

11.4   Measurement Scales

Measurement is a process by which numbers or symbols are attached to given characteristics or properties according to predetermined rules or procedures.

Nominal Gender, Nationality, Eye Color, Marital Status

Ordinal Ranks, Likert Scales. For example: Rank the different brands of colas in preference order or express your degree of agreement to a statement as agree strongly, agree, disagree, disagree strongly.

Interval A scale that has equal distances between scale points but no origin to measure distances from. For example: number of bus stops on a bus or metro line

Ratio Scale A scale with a unit of measurement and an origin of zero. For example: height, weight, length, and age.

Other classifications:

Discrete A scale that has only a countable number of distinct values. For example: number of children in a family or exam result on a course.

Continuous A scale that can have values between any two distinct values. For example: height, weight, length, and age.

Non-metric A scale without a unit of measurement and an origin. Nominal, ordinal and interval scales are non-metric scales.

Metric A scale with a unit of measurement and an origin. Ratio scales are metric scales.

Chapter 12

Appendix B: Testing Normality

Most methods for estimating structural equation models are based on the assumption of normality in one way or another. This assumption should be tested whenever it is possible and LISREL users should be aware of the issues of non-normality and ways to deal with it. This section is about diagnosis of non-normality. Ways to deal with non-normality are described in later sections.

12.1 Univariate Skewness and Kurtosis

To present the formulas used to calculate skewness and kurtosis, some population quantities are first defined.

Let Z be a continuous random variable with moments existing up through order four. Let $\mu = E(Z)$ be the mean of Z and denote

$$\mu_i = E(Z - \mu)^i, \qquad i = 2, 3, 4.$$ (12.1)

These are the population central moments of order 2, 3 , and 4. μ is a location parameter; it tells where the distribution is located. μ_2 is the variance; it tells something about the variation in the distribution. $\sqrt{\mu_2}$, the standard deviation, is a scale parameter; it can be used to define a unit of measurement for Z.

Following the notation of Cramer (1957, eqs. 15.8.1 and 15.8.2), skewness and kurtosis (kurtosis is sometimes called excess) are defined as follows.

$$\gamma_1 = \frac{\mu_3}{\mu_2^{3/2}} = \frac{\mu_3}{\mu_2\sqrt{\mu_2}}$$ (12.2)

$$\gamma_2 = \frac{\mu_4}{\mu_2^2} - 3$$ (12.3)

These are parameters describing the shape of the distribution independent of location and scale. In the literature, sometimes the notation $\beta_1 = \gamma_1^2$ and $\beta_2 = \gamma_2 + 3$ is used instead of γ_1 and γ_2, see, *e.g.*, Kendall and Stuart (1952, p. 85). A normal distribution has $\gamma_1 = 0$ and $\gamma_2 = 0$. All symmetric distributions have $\gamma_1 = 0$, but they vary in terms of γ_2 which can be positive or negative. Non-symmetric distributions have positive or negative γ_1.

γ_1 and γ_2 are population parameters. To decide whether the distribution is normal or non-normal, one can estimate γ_1 and γ_2 from the sample and decide whether the estimates differ

© Springer International Publishing Switzerland 2016
K.G. Jöreskog et al., *Multivariate Analysis with LISREL*, Springer Series in Statistics,
DOI 10.1007/978-3-319-33153-9_12

significantly from zero. If they do, this is an indication of non-normality. There are several ways of doing this. The one used in **LISREL** is as follows.

Let $z_1, z_2, \ldots, z_N$ be a random sample of size N from the distribution of Z. One can then use the sample quantities

$$m_i = (1/N) \sum_{a=1}^{N} (z_a - \bar{z})^i \,, \qquad i = 2, 3, 4 \,. \tag{12.4}$$

to estimate μ_i, where

$$\bar{z} = (1/N) \sum_{a=1}^{N} z_a \,, \tag{12.5}$$

is the sample mean.

Cramer (1957, p. 356) defined the sample skewness and kurtosis as

$$g_1 = \frac{m_3}{m_2{}^{3/2}} = \frac{m_3}{m_2 \sqrt{m_2}} = \sqrt{b_1} \tag{12.6}$$

$$g_2 = \frac{m_4}{m_2^2} - 3 = b_2 - 3 \tag{12.7}$$

In analogy with the use of β's instead of γ's, more recent literature commonly uses the notation[1] $\sqrt{b_1}$ for g_1 and b_2 for $g_2 + 3$, see, e.g., D'Agostino (1970, 1971, 1986), D'Agostino, et al. (1990), Mardia (1980), Bollen (1989, eqs. 9.74 and 9.75), and DeCarlo (1997)[2]. For those who are used to the notation $\sqrt{b_1}$ and b_2, the formulas that follow use the "old" notation on the left, the definition in the middle and the "new" notation on the right.

To test if skewness and kurtosis are zero in the population, one would like to know the mean and variance of these estimates and transform them to a z-statistic which can be used as a test statistic. In the general case, the exact mean and variance of g_1 and g_2 are not available. For a general distribution, Cramer (1957, eq. 27.7.8) gives expressions for the mean and variance accurate to the order of n^{-1}. However, to test normality, the normal distribution is the assumed distribution under the null hypothesis, and Cramer (1957, eq. 29.3.7) gives expressions for the mean and variance that are exact under normality. These expressions show that g_1 is unbiased and that g_2 and $\hat{\gamma}_2$ both have bias $-6/(n+1)$.

Cramer (1957, eq. 29.3.8) also gives the following two alternative estimates due to Fisher (1930), see also Fisher (1973, p. 75), both of which are unbiased under normality.

$$G_1 = \frac{\sqrt{n(n-1)}}{n-2} g_1 = \frac{\sqrt{n(n-1)}}{n-2} \sqrt{b_1} \tag{12.8}$$

$$G_2 = \frac{n-1}{(n-2)(n-3)} [(n+1)g_2 + 6] = \frac{n-1}{(n-2)(n-3)} [(n+1)b_2 - 3(n-1)] \tag{12.9}$$

In **LISREL**, as in SAS and SPSS, G_1 and G_2 are used to compute skewness and kurtosis.

---

[1] $\sqrt{b_1}$ is to be interpreted as a single entity. This notation is awkward. Mathematically $\sqrt{b_1}$ means the positive square root of b_1, with $b_1 = (m_3^2/m_2^3)$ positive. According to this definition, skewness cannot be negative. But skewness can be positive or negative as indicated by the sign of m_3. So the proper definition of skewness should $\sqrt{b_1} \times sign(m_3)$.

[2] The difference in notation seems to be one between the "old" and the "young" generation of statisticians.

Under normality, $E(G_1) = E(G_2) = 0$ and

$$\text{Var}(G_1) = \frac{6N(N-1)}{(N-2)(N+1)(N+3)} , \tag{12.10}$$

$$\text{Var}(G_2) = \frac{24N(N-1)^2}{(N-3)(N-2)(N+3)(N+5)} . \tag{12.11}$$

The standardized z-statistic for testing the hypothesis $\gamma_1 = 0$ is

$$z_s = \sqrt{\frac{(N-2)(N+1)(N+3)}{6N(N-1)}} \, G_1 . \tag{12.12}$$

This can be used for $N \geq 150$. By making a logarithmic transformation of skewness D'Agostino (1986) and D'Agostino, et al. (1990) developed another z-statistic that can be used with N as small as 8. Using different notation, Bollen (1989, Table 9.2)[3] summarized these formulas. For the convenience of **LISREL** users, the same formulas are given here in slightly different form. Compute

$$a_2 = \frac{3(N^2 + 27N - 70)(N+1)(N+3)}{(N-2)(N+5)(N+7)(N+9)} , \tag{12.13}$$

$$a_3 = \sqrt{2(a_2 - 1)} - 1 , \tag{12.14}$$

$$a_4 = \sqrt{\frac{2}{\ln a_3}} , \tag{12.15}$$

$$a_5 = \sqrt{\frac{2}{a_3 - 1}} , \tag{12.16}$$

Then the z-statistic for testing $\gamma_1 = 0$ is

$$z_s = a_4 \ln\{(a_1/a_5) + \sqrt{1 + (a_1/a_5)^2}\} , \tag{12.17}$$

where a_1 is the standardized z_s in (12.12).

To test the hypothesis $\gamma_2 = 0$, the standardized z-statistic is

$$z_k = \sqrt{\frac{(N-3)(N-2)(N+3)(N+5)}{24N(N-1)^2}} \, G_2 . \tag{12.18}$$

This is valid for $N \geq 1000$. For smaller N, compute

$$c_2 = \frac{6(N^2 - 5N + 2)}{(N+7)(N+9)} \sqrt{\frac{6(N+3)(N+5)}{N((N-2)(N-3)}} , \tag{12.19}$$

$$c_3 = 6 + (8/c_2) \left[(2/c_2) + \sqrt{(1 + (4/c_2^2)}\, \right] , \tag{12.20}$$

$$z_k = \frac{1 - (2/9c_3) - \left\{ [1 - (2/c_3)]/ \left[1 + c_1\sqrt{2/(c_4 - 4)}\, \right] \right\}^{1/3}}{\sqrt{(2/9c_3)}} , \tag{12.21}$$

---

[3]The formulas are correct in the first edition of the book but in subsequent editions there is an error in the formula for a_4; the numerator should be 2 not 1.

where c_1 is the standardized z_k in (12.18). The statistic z_k in (12.21) works well for $N \geq 20$.

There is also an omnibus test for skewness and kurtosis simultaneously. This is simply the sum of squares of the z-scores for skewness and kurtosis:

$$c_{sk} = z_s^2 + z_k^2 \ . \tag{12.22}$$

Under normality this has a chi-square distribution with 2 degrees of freedom.

12.2   Multivariate Skewness and Kurtosis

Mardia (1970) has defined multivariate skewness and kurtosis as follows. Let $\mathbf{z}' = (z_1, z_2, \ldots, z_k)$ be a random vector with mean vector $\boldsymbol{\mu}$ and a positive definite covariance matrix $\boldsymbol{\Sigma}$. Then the multivariate skewness of $\mathbf{z}$ is

$$\beta_{1k} = E\{(\mathbf{z} - \boldsymbol{\mu})'\boldsymbol{\Sigma}^{-1}(\mathbf{z} - \boldsymbol{\mu})\}^3 \ , \tag{12.23}$$

and the multivariate kurtosis of $\mathbf{z}$ is

$$\beta_{2k} = E\{(\mathbf{z} - \boldsymbol{\mu})'\boldsymbol{\Sigma}^{-1}(\mathbf{z} - \boldsymbol{\mu})\}^2 \ . \tag{12.24}$$

For a multivariate normal distribution, $\beta_{1k} = 0$ and $\beta_{2k} = k(k + 2)$.

Mardia (1970) also defined analogous sample measures of multivariate skewness and kurtosis. Let $\bar{\mathbf{z}}$ and $\mathbf{S}$ be the sample mean vector and covariance matrix, respectively, and let

$$g_{rs} = (\mathbf{z}_r - \bar{\mathbf{z}})'\mathbf{S}^{-1}(\mathbf{z}_s - \bar{\mathbf{z}}) \ . \tag{12.25}$$

Then the sample skewness and kurtosis are defined as

$$b_{1k} = (1/N^2) \sum_{r=1}^{N} \sum_{s=1}^{N} g_{rs}^3 \ , \tag{12.26}$$

$$b_{2k} = (1/N) \sum_{r=1}^{N} g_{rr}^2 \ . \tag{12.27}$$

Note that b_{1k} requires the evaluation of a quadratic form for each pair of observations. It can therefore be timeconsuming to compute if k and N is large.

A first indication of multivariate non-normality may be obtained by computing the relative multivariate kurtosis

$$\text{RMK} = \frac{b_{2k}}{k(k + 2)} \ . \tag{12.28}$$

If this is much smaller or larger than 1, this is an indication of excess multivariate kurtosis. For a more accurate assessment of multivariate non-normality, one can use the following test statistics.

Mardia (1985) showed that one can use the z-statistic

$$Z_s = \frac{\left\{[27Nk^2(k+1)^2(k+2)^2 b_{1,k}]^{1/3} - 3k(k+1)(k+2) + 4\right\}}{\sqrt{12k(k+1)(k+2)}} \ , \tag{12.29}$$

to test the hypothesis $\beta_{1k} = 0$.

Under multivariate normality, Mardia & Foster (1983) showed that

$$E(b_{2k}) = \frac{(N-1)k(k+2)}{N+1} , \tag{12.30}$$

$$\mathrm{Var}(b_{2k}) = (1/N)8k(k+2) . \tag{12.31}$$

so that a standardized z-statistic to test the hypothesis $\beta_{2k} = k(k+2)$ is

$$Z_k = \frac{b_{2k} - E(b_{2k})}{\sqrt{(1/N)8k(k+2)}} \tag{12.32}$$

Based on further transformation of these quantities, they developed a more accurate formula that can be used for testing this hypothesis. Let

$$d = 6 + \sqrt{\frac{8k(k+2)N}{(k+8)^2}} \left\{ \sqrt{\frac{k(k+2)N}{2(k+8)^2}} + \sqrt{1 + \frac{k(k+2)N}{2(k+8)^2}} \right\} \tag{12.33}$$

Then

$$Z_k = 3\sqrt{d/2} \left\{ 1 - (2/9d) - \left[\frac{1 - (2/d)}{1 + e\sqrt{2/(d-4)}} \right]^{1/3} \right\} , \tag{12.34}$$

where e is the standardized z-statistic in (12.32). This has an approximate standard normal distribution under the hypothesis.

As in the univariate case, there is an omnibus test for testing the joint hypothesis $\beta_{1k} = 0$ and $\beta_{2k} = k(k+2)$ This is

$$C_{sk} = Z_s^2 + Z_k^2 , \tag{12.35}$$

which has an approximate χ^2-distribution under multivariate normality.

Chapter 13

Appendix C: Computational Notes on Censored Regression

13.1 Computational Notes on Univariate Censored Regression

The estimation of a censored regression equation is described in Chapter 6 of Maddala (1983) for the case of a variable that is censored below at 0. The development outlined here covers the cases when the observed variable y is censored below, censored above, censored both below and above, and not censored at all. It also covers the case when there are no regressors in which case one can estimate the mean and standard deviation of y.

Changing notation slightly from Section 2.5.2, consider the estimation of the regression equation

$$y^\star = \alpha^\star + \boldsymbol{\gamma}^{\star\prime}\mathbf{x} + z \,, \tag{13.1}$$

where $\alpha^\star$ is the intercept term, $\boldsymbol{\gamma}^\star$ is the vector of regression coefficients, and $\mathbf{x}$ the regressors. The error term z is assumed to be normally distributed with mean 0 and variance $\psi^{\star 2}$. If there are no regressors, the second term in (13.1) is not included.

The observed variable

$$
\begin{aligned}
y &= c_1 \text{ if } y^\star \leq c_1 \\
&= y^\star \text{ if } c_1 < y^\star < c_2 \\
&= c_2 \text{ if } y^\star \geq c_2 \,,
\end{aligned}
$$

where c_1 and c_2 are constants. If y is censored below set $c_2 = +\infty$. If y is censored above set $c_1 = -\infty$. If y is not censored set both $c_1 = -\infty$ and $c_2 = \infty$.

Let $(y_i, \mathbf{x}_i)$ be the observed values of y and $\mathbf{x}$ of case i in a random sample of N independent observations. The likelihood of $(y_i, \mathbf{x}_i)$ is

$$L_i = \left[\Phi\left(\frac{c_1 - \alpha^\star - \boldsymbol{\gamma}^{\star\prime}\mathbf{x}_i}{\psi^\star}\right)\right]^{j_{1i}} \left[\frac{1}{\sqrt{2\pi}\psi^\star}e^{-\frac{1}{2}(\frac{y_i - \alpha^\star - \boldsymbol{\gamma}^{\star\prime}\mathbf{x}_i}{\psi^\star})^2}\right]^{1-j_{1i}-j_{2i}} \left[1 - \Phi\left(\frac{c_2 - \alpha^\star - \boldsymbol{\gamma}^{\star\prime}\mathbf{x}_i}{\psi^\star}\right)\right]^{j_{2i}} \,,$$

where $j_{i1} = 1$ if $y = c_1$ and $j_{i1} = 0$ otherwise and $j_{i2} = 1$ if $y = c_2$ and $j_{i2} = 0$ otherwise. Note that j_{i1} and j_{i2} cannot be 1 simultaneously.

© Springer International Publishing Switzerland 2016
K.G. Jöreskog et al., *Multivariate Analysis with LISREL*, Springer Series in Statistics,
DOI 10.1007/978-3-319-33153-9_13

The log likelihood is

$$\ln L = \sum_{i=1}^{N} \ln L_i \; .$$

This is to be maximized with respect to the parameter vector $\boldsymbol{\theta}^{\star\prime} = (\alpha^{\star}, \boldsymbol{\gamma}^{\star\prime}, \psi^{\star})$.

First and second derivatives of $\ln L$ with respect to $\boldsymbol{\theta}^{\star}$ are very complicated. They will be considerably simplified and the maximization of $\ln L$ will be considerably more efficient if another parameterization due to Tobin (1958) is used.

This parameterization uses the parameter vector $\boldsymbol{\theta}^{\prime} = (\alpha, \boldsymbol{\gamma}^{\prime}, \psi)$ instead of $\boldsymbol{\theta}^{\star}$, where $\alpha = \alpha^{\star}/\psi^{\star}$, $\boldsymbol{\gamma} = \boldsymbol{\gamma}^{\star}/\psi^{\star}$, and $\psi = 1/\psi^{\star}$.

Multiplication of (13.1) by $\psi = 1/\psi^{\star}$ gives

$$\psi y^{\star} = \alpha + \boldsymbol{\gamma}^{\prime} \mathbf{x} + v \; , \tag{13.2}$$

where $v = \psi z = z/\psi^{\star}$ which is $N(0,1)$. Then

$$y = c_1 \leftrightarrow y^{\star} \leq c_1 \leftrightarrow \psi y^{\star} \leq \psi c_1 \leftrightarrow v \leq \psi c_1 - \alpha - \boldsymbol{\gamma}^{\prime} \mathbf{x} \; ,$$

$$y = c_2 \leftrightarrow y^{\star} \geq c_2 \leftrightarrow \psi y^{\star} \geq \psi c_2 \leftrightarrow v \geq \psi c_2 - \alpha - \boldsymbol{\gamma}^{\prime} \mathbf{x} \; .$$

Hence the likelihood L_i becomes

$$L_i = \left[\Phi(\psi c_1 - \alpha - \boldsymbol{\gamma}^{\prime} \mathbf{x}_i) \right]^{j_{1i}} \left[\frac{1}{\sqrt{2\pi}} \psi e^{-\frac{1}{2}(\psi y_i - \alpha - \boldsymbol{\gamma}^{\prime} \mathbf{x}_i)^2} \right]^{1 - j_{1i} - j_{2i}} \left[1 - \Phi(\psi c_2 - \alpha - \boldsymbol{\gamma}^{\prime} \mathbf{x}_i) \right]^{j_{2i}} \; .$$

Let

$$\delta_i = \psi y_i - \alpha - \boldsymbol{\gamma}^{\prime} \mathbf{x}_i \; . \tag{13.3}$$

Then $\ln L_i$ becomes

$$\ln L_i = -\ln \sqrt{2\pi} + (1 - j_{1i} - j_{2i})(\ln \psi - \tfrac{1}{2}\delta_i^2) + j_{1i} \ln \Phi(\delta_i) + j_{2i} \ln[1 - \Phi(\delta_i)] \; . \tag{13.4}$$

First and second derivatives of $\ln L_i$ are straightforward by noting that $\partial \delta_i/\partial \alpha = -1$, $\partial \delta_i/\partial \boldsymbol{\gamma} = -\mathbf{x}_i$, and $\partial \delta_i/\partial \psi = y_i$. Furthermore, $\Phi^{\prime}(t) = \phi(t)$, $\phi^{\prime}(t) = -t\phi(t)$ and if $A(t) = \phi(t)/\Phi(t)$, then $A^{\prime}(t) = -A(t)[t + A(t)] = B(t)$, say.

Omitting index i, the required derivatives are

$$\partial \ln L/\partial \alpha = (1 - j_1 - j_2)\delta - j_1 A(\delta) + j_2 A(-\delta)$$

$$\partial \ln L/\partial \boldsymbol{\gamma} = (1 - j_1 - j_2)\delta \, \mathbf{x} - j_1 A(\delta)\mathbf{x} + j_2 A(-\delta)\mathbf{x}$$

$$\partial \ln L/\partial \psi = (1 - j_1 - j_2)(1/\psi - \delta y) + j_1 A(\delta)y - j_2 A(-\delta)y$$

$$\partial^2 \ln L/\partial \alpha \partial \alpha = -(1 - j_1 - j_2) + j_1 B(\delta) + j_2 B(-\delta)$$

$$\partial^2 \ln L/\partial \boldsymbol{\gamma} \partial \alpha = -(1 - j_1 - j_2)\mathbf{x} + j_1 B(\delta)\mathbf{x} + j_2 B(-\delta)\mathbf{x}$$

$$\partial^2 \ln L/\partial \boldsymbol{\gamma} \partial \boldsymbol{\gamma}^{\prime} = -(1 - j_1 - j_2)\mathbf{x}\mathbf{x}^{\prime} + j_1 B(\delta)\mathbf{x}\mathbf{x}^{\prime} + j_2 B(-\delta)\mathbf{x}\mathbf{x}^{\prime}$$

$$\partial^2 \ln L/\partial \psi \partial \alpha = (1 - j_1 - j_2)y - j_1 B(\delta)y - j_2 B(-\delta)y$$

$$\partial^2 \ln L/\partial \psi \partial \boldsymbol{\gamma}^{\prime} = (1 - j_1 - j_2)y\mathbf{x}^{\prime} - j_1 B(\delta)y\mathbf{x}^{\prime} - j_2 B(-\delta)y\mathbf{x}^{\prime}$$

$$\partial^2 \ln L/\partial \psi \partial \psi = -(1 - j_1 - j_2)(1/\psi^2 + y^2) + j_1 B(\delta)y^2 + j_2 B(-\delta)y^2$$

Maximizing $\ln L$ is equivalent to minimizing the fit function $F(\boldsymbol{\theta}) = -\ln L$. Let $\mathbf{g}(\boldsymbol{\theta}) = \partial F/\partial\boldsymbol{\theta}$ be the gradient vector and $\mathbf{H}(\boldsymbol{\theta}) = \partial^2 F/\partial\boldsymbol{\theta}\partial\boldsymbol{\theta}'$ be the Hessian matrix. Amemiya (1973) proved that $\mathbf{H}$ is positive definite everywhere.

The fit function $F(\boldsymbol{\theta})$ is minimized using a Newton-Raphson procedure which converges very fast. The starting values $\boldsymbol{\theta}_0$ are the parameters estimated by OLS. Successive estimates are obtained by the formula

$$\boldsymbol{\theta}_{s+1} = \boldsymbol{\theta}_s - \mathbf{H}_s^{-1}\mathbf{g}_s \;, \tag{13.5}$$

where $\mathbf{g}_s = \mathbf{g}(\boldsymbol{\theta}_s)$ and $\mathbf{H}_s = \mathbf{H}(\boldsymbol{\theta}_s)$.

Let $\hat{\boldsymbol{\theta}} = (\hat{\alpha}, \hat{\boldsymbol{\gamma}}, \hat{\psi})$ be the maximum likelihood estimates of $\boldsymbol{\theta}$. The asymptotic covariance matrix of $\hat{\boldsymbol{\theta}}$ is $\mathbf{E} = \mathbf{H}^{-1}(\boldsymbol{\theta})$ evaluated at the true parameter $\boldsymbol{\theta}$. Since the transformation from $\boldsymbol{\theta}$ to $\boldsymbol{\theta}^\star$ is one-to-one, the maximum likelihood estimates of $\boldsymbol{\theta}^\star$ is $\hat{\boldsymbol{\theta}}^\star = (\hat{\alpha}^\star, \hat{\boldsymbol{\gamma}}^\star, \hat{\psi}^\star)$, where $\hat{\alpha}^\star = \hat{\alpha}/\hat{\psi}$, $\hat{\boldsymbol{\gamma}}^\star = \boldsymbol{\gamma}/\hat{\psi}$, and $\hat{\psi}^\star = 1/\hat{\psi}$.

To obtain the asymptotic covariance matrix of $\hat{\boldsymbol{\theta}}^\star$, we evaluate the matrix $\partial\boldsymbol{\theta}^\star/\partial\boldsymbol{\theta}'$. This is

$$\partial\boldsymbol{\theta}^\star/\partial\boldsymbol{\theta}' = (1/\psi^2)\begin{pmatrix} \psi & \mathbf{0}' & -\alpha \\ 0 & \psi\mathbf{1}' & \boldsymbol{\gamma} \\ 0 & \mathbf{0}' & -1 \end{pmatrix} = \mathbf{A}(\boldsymbol{\theta}), \text{say} \;, \tag{13.6}$$

where $\mathbf{0}$ and $\mathbf{1}$ are column vectors of zeros and ones, respectively. The asymptotic covariance matrix of $\hat{\boldsymbol{\theta}}^\star$ is $\mathbf{AEA}'$, where $\mathbf{A}$ and $\mathbf{E}$ are evaluated at the true parameter values. An estimate of the asymptotic covariance matrix of $\hat{\boldsymbol{\theta}}^\star$ is $\mathbf{AEA}'$ obtained by using the estimated parameter values in $\mathbf{A}$ and $\mathbf{E}$. Asymptotic standard error estimates of the parameter estimates are obtained as the square roots of the diagonal elements of this matrix.

13.2   Computational Notes on Multivariate Censored Regression

Let $U^\star$ and $V^\star$ have a bivariate standard normal distribution with correlation ρ. The density and distribution functions are

$$\phi_2(u, v; \rho) = \frac{1}{2\pi\sqrt{(1-\rho^2)}} e^{-\frac{1}{2(1-\rho^2)}(u^2 - 2\rho uv + v^2)} \;,$$

and

$$\Phi_2(u, v; \rho) = \int_{-\infty}^{u}\int_{-\infty}^{v}\phi_2(s, t; \rho)ds dt \;,$$

respectively. Note that

$$\phi_2(u, v; \rho) = \phi(u)\phi\left(\frac{v - \rho u}{\sqrt{1 - \rho^2}}\right) = \phi(v)\phi\left(\frac{u - \rho v}{\sqrt{1 - \rho^2}}\right) \;.$$

Furthermore, let

$$\Delta\Phi_2(a, b, c, d; \rho) = \int_{a}^{b}\int_{c}^{d}\phi_2(s, t; \rho)ds dt \;.$$

Let U and V be censored both above and below, such that

$$\begin{aligned} U &= a \text{ if } U^\star \leq a \\ &= U^\star \text{ if } a < U^\star < b \\ &= b \text{ if } U^\star \geq b \;, \end{aligned}$$

$$
\begin{aligned}
V &= c \text{ if } V^\star \le c \\
  &= V^\star \text{ if } c < V^\star < d \\
  &= d \text{ if } V^\star \ge d ,
\end{aligned}
$$

where a, b, c, and d are constants. The density of U and V is divided into nine regions as shown in following figure.

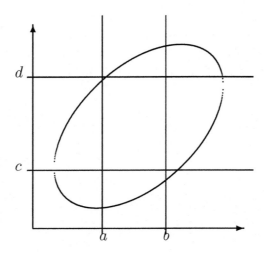

The density of U and V is

$$
\begin{aligned}
f(u,v) &= \Phi_2(a,c,\rho) && \text{if } U = a , V = c \\
       &= \alpha\phi(u)\Phi(\tfrac{c-\rho u}{\alpha}) && \text{if } a < U < b , V = c \\
       &= \Delta\Phi_2(b,+\infty,-\infty,c;\rho) && \text{if } U = b , V = c \\
       &= \alpha\phi(v)\Phi(\tfrac{a-\rho v}{\alpha}) && \text{if } U = a , c < V < d \\
       &= \phi_2(u,v;\rho) && \text{if } a < U < b , c < V < d \\
       &= \alpha\phi(v)[1 - \Phi(\tfrac{b-\rho v}{\alpha})] && \text{if } U = b , c < V < d \\
       &= \Delta\Phi_2(-\infty,a,d,+\infty;\rho) && \text{if } U = a , V = d \\
       &= \alpha\phi(u)[1 - \Phi(\tfrac{d-\rho u}{\alpha})] && \text{if } a < U < b , V = d \\
       &= \Delta\Phi_2(b,+\infty,d,+\infty;\rho) && \text{if } U = b , V = d
\end{aligned}
$$

where $\alpha = \sqrt{1 - \rho^2}$ and $\Phi(x)$ is the standard normal distribution function.

Chapter 14

Appendix D: Normal Scores

Consider a variable z, *i.e.*, any column of the data matrix, and suppose the sample values are $z_1, z_2, \ldots, z_N$. The normal scores are computed as follows. Suppose there are n distinct values among the N sample values and let these be ordered such that

$$z_{(1)} < z_{(2)}, < \ldots, z_{(n)} \ .$$

Each $n_i \geq 1$ and $\sum_{i=1}^{n} n_i = N$. The *normal score* s_i corresponding to $z_{(i)}$ is computed by the formula

$$s_i = (N/n_i)[\phi(\hat{\alpha}_{i-1}) - \phi(\hat{\alpha}_i)] \qquad i = 1, \ 2, \ \ldots, \ n \ , \tag{14.1}$$

where $\hat{\alpha}_0 = -\infty$, $\hat{\alpha}_k = +\infty$, and

$$\hat{\alpha}_i = \Phi^{-1}\Big(\sum_{j=1}^{i} n_j/N \Big), \qquad i = 1, \ 2, \ \ldots, \ n-1 \ .$$

Here ϕ is the standard normal density function and Φ^{-1} is the inverse of the standard normal distribution function. These normal scores are then scaled so that they have the same sample mean and standard deviation as the original variable. Thus, the normal score is a monotonic transformation of the original score with the same mean and standard deviation but with much reduced skewness and kurtosis. The rank ordering of cases are the same, *i.e.*, if $z_i \leq z_j$ then $s_i \leq s_j$.

© Springer International Publishing Switzerland 2016
K.G. Jöreskog et al., *Multivariate Analysis with LISREL*, Springer Series in Statistics,
DOI 10.1007/978-3-319-33153-9_14

Chapter 15

Appendix E: Asessment of Fit

The testing of the model and the assessment of fit is one of the most debated and discussed issues in structural equation models. In this Appendix we consider what this is all about.

Structural equation models are often used to test a theory about relationships between theoretical constructs. In this chapter we give a short and non-technical account of the main issues involved in the translation of a theory to a structural equation model and the fitting and testing of this model by empirical data. For a fuller account of these issues, see Bollen (1989a), Jöreskog (1993) and other chapters in Bollen and Long (1993). In particular, we discuss issues of model evaluation and assessment of fit that we have only touched upon in the examples of the previous chapters.

15.1 From Theory to Statistical Model

Most theories in the social and behavioral sciences are formulated in terms of hypothetical constructs that cannot be observed or measured directly. Examples of such constructs are prejudice, radicalism, alienation, conservatism, trust, self-esteem, discrimination, motivation, ability, and anomia. The measurement of a hypothetical construct is accomplished indirectly through one or more observable indicators, such as responses to questionnaire items, that are *assumed* to represent the construct adequately .

In theory, the researcher defines the hypothetical constructs by specifying the dimensions of each (see, e.g., Bollen, 1989a, pp. 179–184). The theory further specifies how the various constructs are postulated to be interrelated. This includes first the classification of the constructs as dependent (caused, criterion, endogenous) or independent (causal, explanatory, exogenous). Secondly, for each dependent construct, the theory specifies which of the other constructs it is postulated to depend on. The theory may also include a statement about the sign and/or relative size of the direct effect of one construct on another.

Since the theoretical constructs are not observable, the theory cannot be tested directly. All one can do is examine the *theoretical validity* of the postulated relationships in a given context. Before the theory can be tested empirically, a set of observable indicators must be defined for each dimension of each construct. There must be clear rules of correspondence between the indicators and the constructs, such that each construct and dimension is distinct (Costner, 1969).

© Springer International Publishing Switzerland 2016
K.G. Jöreskog et al., *Multivariate Analysis with LISREL*, Springer Series in Statistics,
DOI 10.1007/978-3-319-33153-9_15

The theoretical relationships between the constructs constitute the structural equation part of the model, and the relationships between the observable indicators and the theoretical constructs constitute the measurement part of the model. In order to test the conceptual model, each of these parts must first be formulated as a *statistical* model.

The statistical model requires the specification of the form of the relationship, linear or nonlinear. The structural relationships are usually assumed to be linear but nonlinear models can also be used. If the observed indicators are continuous, i.e., measured on interval or ratio scales, the measurement equations are also assumed to be linear, possibly after a logarithmic, exponential, or other nonlinear type of transformation of the observable indicators. If the observed indicators are ordinal, one usually assumes that there is a continuous variable underlying each ordinal variable and formulates the measurement model in terms of these underlying variables.

It is not expected that the relationships in the model are exact deterministic relationships. Most often, the independent constructs in the model account for only a fraction of the variation and covariation in the dependent constructs, because there may be many other variables that are associated with the dependent constructs, but are not included in the model. The aggregation of all such omitted variables is represented in the model by a set of stochastic error terms, one for each dependent construct. By definition, these error terms represent the variation and covariation in the dependent constructs left unaccounted for by the independent constructs. *The fundamental assumption in structural equation models is that the error term in each relationship is uncorrelated with all the independent constructs.* Studies should be planned and designed, and variables should be chosen so that this is the case. Failure to do so will lead to biased and inconsistent estimates (omitted variable bias) of the structural coefficients in the linear equations and thus invalidate the testing of the theory. Omitted variables bias is one of the most difficult specification errors to test.

The relationships in the measurement model contain also stochastic error terms that are usually interpreted to be the *sum* of specific factors and random measurement errors in the observable indicators. It is only in specially designed studies such as panel models or multitrait-multimethod designs that one can separate these two components of the error terms. In cross-sectional studies, the error terms should be uncorrelated from one indicator to another. This is part of the definition of indicators of a construct. Because, if the error terms for two or more indicators correlate, it means that the indicators measure *something else* or *something in addition* to the construct they are supposed to measure. If this is the case, the meaning of the construct and its dimensions may be different from what is intended. It is a widespread misuse of structural equation modeling to include correlated error terms[1] in the model for the sole purpose of obtaining a better fit to the data. Every correlation between error terms must be justified and interpreted substantively.

The testing of the structural model, i.e., the testing of the initially specified theory, may be meaningless unless it is first established that the measurement model holds. If the chosen indicators for a construct do not measure that construct, the specified theory must be modified before it can be tested. Therefore, the measurement model should be tested before the structural relationships are tested. It may be useful to do this for each construct separately, then for the constructs taken two at a time, and then for all constructs simultaneously. In doing so, one should let the constructs themselves be freely correlated, i.e., the covariance matrix of the constructs should be unconstrained.

---

[1]We do not include here the situation when error terms correlate for the same variable over time, or other designed conditions due to specific factors or method factors

15.2 Nature of Inference

Once a model has been carefully formulated, it can be confronted with empirical data and, if all assumptions hold, various techniques for covariance structure analysis, briefly reviewed in the next section, can be used to test if the model is consistent with the data.

The inference problem is of the following general form. Given assumptions A, B, C, ..., test model M against a more general model M_A. Most researchers will take assumptions A, B, C, ..., for granted and proceed to test M formally. But the assumptions should be checked before the test is made, whenever this is possible.

If the model is rejected by the data, the problem is to determine what is wrong with the model and how the model should be modified to fit the data better.

If the model *fits* the data, it does not mean that it is *the "correct" model or even the "best" model*. In fact, there can be many *equivalent* models, all of which will fit the data equally well as judged by any goodness-of-fit measure. This holds not just for a particular data set but for any data set. The direction of causation and the causal ordering of the constructs cannot be determined by the data[2]. To conclude that the fitted model is the "best," one must be able to exclude all models equivalent to it on logical or substantive grounds. For a good discussion of causal models, see Bollen (1989a, Chapter 3). Further discussion of equivalent models is found in Stelzl (1986) and Lee and Hershberger (1990). For a fuller and deeper account of these issues, see Bekker, Merckens, and Wansbeek (1994).

Even outside the class of models equivalent to the given model, there may be many models that fit the data almost as well as the given model. Covariance structure techniques can be used to distinguish sharply between models that fit the data *very badly* and those that fit the data *reasonably well*. To discriminate further between models that fit the data almost equally well requires a careful power study (see Satorra & Saris, 1985, and Matsueda & Bielby, 1986).

To be strictly testable, the theory should be overidentified in the sense that the structural equation part of the model is overidentified. If the covariance matrix of the construct variables is unconstrained by the model, any test of the model is essentially just a test of the measurement models for the indicators of the constructs. Given that the measurement models hold, the only way to test a saturated structural equation model is to examine whether estimated parameters agree with a priori specified signs and sizes and whether the strengths of the relationships are sufficiently large.

15.3 Three Situations

We distinguish between three situations:

SC In a *strictly confirmatory* situation the researcher has formulated one single model and has obtained empirical data to test it. The model should be accepted or rejected.

---

[2]Different equivalent models will give different parameter estimates, and some may give estimates that are not meaningful. This fact may be used to distinguish some equivalent models from others.

AM The researcher has specified several *alternative models* or *competing models* and on the basis of an analysis of a single set of empirical data, one of the models should be selected.

MG The researcher has specified a tentative initial model. If the initial model does not fit the given data, the model should be modified and tested again using the same data. Several models may be tested in this process. The goal may be to find a model which not only fits the data well from a statistical point of view, but also has the property that every parameter of the model can be given a substantively meaningful interpretation. The re-specification of each model may be theory-driven or data-driven. Although a model may be tested in each round, the whole approach is *model generating* rather than model testing.

In practice, the **MG** situation is by far the most common. The **SC** situation is very rare because few researchers are content with just rejecting a given model without suggesting an alternative model. The **AM** situation is also rare because researchers seldom specify the alternative models a priori.

We consider the **SC** situation in this section, the **AM** situation in Section 4.4, and the **MG** situation in Section 4.5.

Once the relationships in the theoretical model have been translated into a statistical model for a set of linear stochastic equations among random observable variables (the indicators) and latent variables (the theoretical constructs), the model can be estimated and tested on the basis of empirical data using statistical methods.

The statistical model and its assumptions imply a covariance structure $\Sigma(\boldsymbol{\theta})$ for the observable random variables, where $\boldsymbol{\theta} = (\theta_1, \theta_2, \ldots, \theta_t)$ is a vector of parameters in the statistical model. The testing problem is now conceived of as testing the model $\Sigma(\boldsymbol{\theta})$. It is assumed that the empirical data is a random sample of size N of cases (individuals) on which the observable variables have been observed or measured. From this data a sample covariance matrix $\mathbf{S}$ is computed, and it is this matrix that is used to fit the model to the data and to test the model.

The model is fitted by minimizing a fit function $F[\mathbf{S}, \Sigma(\boldsymbol{\theta})]$ of $\mathbf{S}$ and $\Sigma(\boldsymbol{\theta})$ which is non-negative and zero only if there is a perfect fit, in which case $\mathbf{S}$ equals $\Sigma(\boldsymbol{\theta})$. Although the family of fit functions F includes all the fit functions that are used in practice, e.g., ULS, GLS, ML, DWLS, and WLS, we will be concerned only with ML here.

Suppose that $\mathbf{S}$ converges in probability to Σ_0 as the sample size increases, and let $\boldsymbol{\theta}_0$ be the value of $\boldsymbol{\theta}$ that minimizes $F[\Sigma_0, \Sigma(\boldsymbol{\theta})]$. We say that the model holds if $\Sigma_0 = \Sigma(\boldsymbol{\theta}_0)$. Furthermore, let $\hat{\theta}$ be the value of $\boldsymbol{\theta}$ that minimizes $F[\mathbf{S}, \Sigma(\boldsymbol{\theta})]$ for the given sample covariance matrix $\mathbf{S}$.

To test the model, one may use

$$c = NF[\mathbf{S}, \Sigma(\hat{\theta})] \, . \tag{15.1}$$

If the model holds and is identified, c is approximately distributed in large samples as χ^2 with $d = s - t$ degrees of freedom, where $s = k(k+1)/2$ and t is the number of independent parameters estimated. To use this test formally, one chooses a significance level α, draws a random sample of observations on $z_1, z_2, \ldots, z_k$, computes $\mathbf{S}$, estimates the model, and rejects the model if c exceeds the $(1 - \alpha)$ percentile of the χ^2 distribution with d degrees of freedom. As will be argued in the next section, this is a highly limited view of the testing problem.

If the model does not hold, $\Sigma_0 \neq \Sigma(\boldsymbol{\theta}_0)$ and c in (4.1) is distributed as non-central χ^2 with $s - t$ degrees of freedom and non-centrality parameter (Browne, 1984)

$$\lambda = nF[\Sigma_0, \Sigma(\boldsymbol{\theta}_0)] \, , \tag{15.2}$$

an unknown population quantity that may be estimated as

$$\hat{\lambda} = \mathsf{Max}\{(c-d), 0\} , \tag{15.3}$$

cf., Browne and Cudeck (1993). One can also set up a confidence interval for λ. Let $\hat{\lambda}_L$ and $\hat{\lambda}_U$ be the solutions of

$$G(c|\lambda_L, d) = 0.95 \tag{15.4}$$

$$G(c|\lambda_U, d) = 0.05 , \tag{15.5}$$

where $G(x|\lambda, d)$ is the distribution function of the non-central chi-square distribution with non-centrality parameter λ and d degrees of freedom. Then $(\hat{\lambda}_L; \hat{\lambda}_U)$ is a 90 percent confidence interval for λ (see Browne & Cudeck, 1993).

Once the validity of the model has been established, one can test structural hypotheses about the parameters $\boldsymbol{\theta}$ in the model such that

- certain θ's have particular values (fixed parameters)

- certain θ's are equal (equality constraints)

- certain θ's are specified linear or nonlinear functions of of other parameters.[3]

Each of these types of hypotheses leads to a model with fewer parameters, u, say, for $u < t$ and with a parameter vector $\boldsymbol{\nu}$ of order $(u \times 1)$ containing a subset of the parameters in $\boldsymbol{\theta}$. In conventional statistical terminology, the model with parameters $\boldsymbol{\nu}$ is called the *null hypothesis* H_0 and the model with parameters $\boldsymbol{\theta}$ is called the *alternative hypothesis* H_1. Let c_0 and c_1 be the value of c for models H_0 or H_1, respectively. The *likelihood ratio test statistic* for testing H_0 against H_1 is then

$$D^2 = c_0 - c_1 \tag{15.6}$$

which is used as χ^2 with $t - u$ degrees of freedom. The degrees of freedom can also be computed as the difference between the degrees of freedom associated with c_0 and c_1.

To use the tests formally, one chooses a significance level α (probability of a type 1 error) and rejects H_0 if D^2 exceeds the $(1 - \alpha)$ percentile of the χ^2 distribution with $t - u$ degrees of freedom.

15.4 Selection of One of Several Specified Models

The previous sections have focused on the testing of a single specified model including the checking of the assumptions on which the test is based. In this section, we consider the **AM** situation, in which the researcher has specified a priori a number of alternative or competing models $M_1, M_2, \ldots, M_k$, say, and wants to select one of these. Here it is assumed that all models are identified and can be tested with a valid chi-square test as in Section 4.3. It is also assumed that each model gives reasonable results apart from fit considerations. Any model that gives unreasonable results can be eliminated from further consideration.

---

[3]Linear and nonlinear constraints cannot be tested using the SIMPLIS language; the LISREL language is required for these tests.

We begin with the case in which the models are nested so they can be ordered in decreasing number of parameters (increasing degrees of freedom). M_1 is the most flexible model, i.e., the one with most parameters (fewest degrees of freedom). M_k is the most restrictive model, i.e., the one with fewest parameters (largest degrees of freedom). If the models are nested, each model is a special case of the preceding model, i.e., M_i is obtained from M_{i-1} by placing restrictions on the parameters of M_{i-1}.

We assume that model M_1 has been tested by a valid chi-square test (as discussed in previous sections), and found to fit the data and to be interpretable in a meaningful way. We can then test sequentially M_2 against M_1, M_3 against M_2, M_4 against M_3, and so on. Each test can be a likelihood ratio test, a Lagrangian multiplier test, or a Wald test. If M_i is the first model rejected, one takes M_{i-1} to be the "best" model. Under this procedure, the probability of rejecting M_i when M_j is true is not known. There is no guarantee that the tests are independent, not even asymptotically. In general, what is required is a multiple decision procedure that takes explicitly into account the probability of rejecting each hypothesis when a particular one is true. The mathematical theory for such a procedure is so complicated that its practical value is very limited.

Another approach is to compare the models on the basis of some criteria that takes parsimony (in the sense of number of parameters) into account as well as fit. This approach can be used regardless of whether or not the models can be ordered in a nested sequence. Three strongly related criteria are available in **LISREL**: AIC, BIC, and ECVI. To apply these measures to the decision problem, one estimates each model, ranks them according to one of these criteria and chooses the model with the smallest value. Browne and Cudeck (1993) point out that one can also take the precision of the estimated value of ECVI into account. For example, a 90 percent confidence interval for ECVI is

$$\left(\frac{\hat{\lambda}_L + s + t}{N} ; \frac{\hat{\lambda}_U + s + t}{N} \right) . \tag{15.7}$$

For a given data set, N is the same for all models. Therefore, AIC and ECVI will give the same rank order of the models, whereas the rank ordering of AIC and BIC can differ. To see how much they can differ, consider the case of two nested models M_1 and M_2 differing by one degree of freedom. Let $D^2 = c_2 - c_1$. The conventional testing procedure will reject M_2 in favor of M_1 if D^2 exceeds the $(1 - \alpha)$ percentage point of the chi-square distribution with one degree of freedom. If $\alpha = 0.05$ this will lead to accepting M_1 if D^2 exceeds 3.84. AIC will select M_1 if D^2 exceeds 2, whereas BIC will select M_1 if D^2 exceeds $1 + \ln N$.

15.5   Model Assessment and Modification

We now consider the **MG** case. The problem is not just to accept or reject a specified model or to select one out of a set of specified models. Rather, the researcher has specified an initial model that is not assumed to hold exactly in the population and may only be tentative. Its fit to the data is to be evaluated and assessed in relation to what is known about the substantive area, the quality of the data, and the extent to which various assumptions are satisfied. The evaluation of the model and the assessment of fit is not entirely a statistical matter. If the model is judged not to be good on substantive or statistical grounds, it should be modified within a class of models suitable for the substantive problem. The goal is to find a model within this class of models that not only fits the data well statistically, taking all aspects of error into account, but that also has the property of every parameter having a substantively meaningful interpretation.

The output from a structural equations program provides much information useful for model evaluation and assessment of fit. It is helpful to classify this information into the three groups

- Examination of the solution

- Measures of overall fit

- Detailed assessment of fit

and to proceed as follows.

1. Examine the parameter estimates to see if there are any unreasonable values or other anomalies. Parameter estimates should have the right sign and size according to theory or a priori specifications. Examine the squared multiple correlation R^2 for each relationship in the model. The R^2 is a measure of the strength of linear relationship. A small R^2 indicates a weak relationship and suggests that the model is not effective.

2. Examine the measures of overall fit of the model, particularly the chi-squares. A number of other measures of overall fit (see the sections that follow), all of which are functions of chi-square, may also be used. If any of these quantities indicate a poor fit of the data, proceed with the detailed assessment of fit in the next step.

3. The tools for examining the fit in detail are the residuals and standardized residuals, the modification indices, as well as the expected change. Each of these quantities may be used to locate the source of misspecification and to suggest how the model should be modified to fit the data better.

15.6   Chi-squares

In previous sections, we regarded c as a chi-square for testing the model against the alternative that the covariance matrix of the observed variables is unconstrained. This is valid if all assumptions are satisfied, if the model holds and the sample size is sufficiently large. In practice it is more useful to regard chi-square as a *measure* of fit rather than as a *test statistic*. In this view, chi-square is a measure of overall fit of the model to the data. It measures the distance (difference, discrepancy, deviance) between the sample covariance (correlation) matrix and the fitted covariance (correlation) matrix. Chi-square is a badness-of-fit measure in the sense that a small chi-square corresponds to good fit and a large chi-square to bad fit. Zero chi-square corresponds to perfect fit. Chi-square is calculated as N times the minimum value of the fit function, where N is the sample size.

Even if all the assumptions of the chi-square test hold, it may not be realistic to assume that the model holds exactly in the population. In this case, chi-square should be compared with a non-central rather than a central chi-square distribution (see Browne, 1984).

A number of other goodness-of-fit measures have been proposed and studied in the literature. For a summary of these and the rationale behind them, see Bollen (1989a). All of these measures are functions of chi-square.

15.7   Goodness-of-Fit Indices

Chi-square tends to be large in large samples if the model does not hold. A number of goodness-of-fit measures have been proposed to eliminate or reduce its dependence on sample size. This is a hopeless task, however, because even though a measure does not depend on sample size explicitly in its calculation, its sampling distribution will depend on N. The goodness-of-fit measures GFI and AGFI of Tanaka and Huba (1985) do not depend on sample size explicitly and measure how much *better* the model fits as compared to no model at all. For a specific class of models, Maiti and Mukherjee (1990) demonstrate that there is an exact monotonic relationship between GFI and chi-square.

The goodness-of-fit index (GFI) is defined as

$$\mathsf{GFI} = 1 - \frac{F[\mathbf{S}, \boldsymbol{\Sigma}(\hat{\theta})]}{F[\mathbf{S}, \boldsymbol{\Sigma}(\mathbf{0})]} \; . \tag{15.8}$$

The numerator in (15.8) is the minimum of the fit function after the model has been fitted; the denominator is the fit function before any model has been fitted, or when all parameters are zero.

The goodness-of-fit index adjusted for degrees of freedom, or the adjusted GFI, AGFI, is defined as

$$\mathsf{AGFI} = 1 - \frac{k(k+1)}{2d}(1 - \mathsf{GFI}) \; , \tag{15.9}$$

where k is the number of observed variables and d is the degrees of freedom of the model. This corresponds to using mean squares instead of total sums of squares in the numerator and denominator of $1 - \mathsf{GFI}$. Both of these measures should be between zero and one, although it is theoretically possible for them to become negative. This should not happen, of course, since it means that the model fits worse than no model at all.

15.8   Population Error of Approximation

The use of chi-square as a central χ^2-statistic is based on the assumption that the model holds exactly in the population. As already pointed out, this may be an unreasonable assumption in most empirical research. A consequence of this assumption is that models which hold approximately in the population will be rejected in large samples. Browne and Cudeck (1993) proposed a number of fit measures which take particular account of the error of approximation in the population and the precision of the fit measure itself. They define an estimate of the *population discrepancy function* (PDF) as

$$\hat{F}_0 = \mathsf{Max}\{\hat{F} - (d/n), 0\} \; , \tag{15.10}$$

where $\hat{F}$ is the minimum value of the fit function, $n = N - 1$[4] and d is the degrees of freedom, and suggest the use of a 90 percent confidence interval

$$(\hat{\lambda}_L/n; \hat{\lambda}_U/n) \tag{15.11}$$

to assess the error of approximation in the population.

Since $\hat{F}_0$ generally decreases when parameters are added in the model, Browne and Cudeck (1993) suggest using Steiger's (1990) Root Mean Square Error of Approximation (RMSEA)

$$\epsilon = \sqrt{\hat{F}_0/d} \tag{15.12}$$

as a measure of *discrepancy per degree of freedom*. Browne and Cudeck (1993) suggest that a value of 0.05 of ϵ indicates a close fit and that values up to 0.08 represent reasonable errors of approximation in the population. A 90 percent confidence interval of ϵ and a test of $\epsilon < 0.05$ give quite useful information for assessing the degree of approximation in the population. A 90 percent confidence interval for ϵ is

$$\left(\sqrt{\frac{\hat{\lambda}_L}{nd}} ; \sqrt{\frac{\hat{\lambda}_U}{nd}} \right) \tag{15.13}$$

and the P-value for test of $\epsilon < 0.05$ is calculated as

$$P = 1 - G(c|0.0025nd, d) \tag{15.14}$$

(see Browne & Cudeck, 1993).

15.9   Other Fit Indices

Another class of fit indices measures how much *better* the model fits as compared to a baseline model, usually the independence model. The first indices of this kind were developed by Tucker and Lewis (1973) and Bentler and Bonett (1980) (NNFI, NFI). Other variations of these have been proposed and discussed by Bollen (1986, 1989a, b) (RFI, IFI) and Bentler (1990) (CFI). These indices are supposed to lie between 0 and 1, but values outside this interval can occur, and, since the independence model almost always has a huge chi-square, one often obtains values very close to 1. James, Mulaik, and Brett (1982, p. 155) suggest taking parsimony (degrees of freedom) into account and define a parsimony normed fit index (PNFI), and Mulaik et al. (1989) suggest a parsimony goodness-of-fit index (PGFI)

For completeness, we give the definition of each of these measures here. Let F be the minimum value of the fit function for the estimated model, let F_i be the minimum value of the fit function for the independence model, and let d and d_i be the corresponding degrees of freedom. Furthermore, let $f = nF/d$, $f_i = nF_i/d_i$, $\tau = \mathsf{max}(nF - d, 0)$, and $\tau_i = \mathsf{max}(nF_i - d_i, nF - d, 0)$. Then

$$\mathsf{NFI} = 1 - F/F_i \tag{15.15}$$

$$\mathsf{PNFI} = (d/d_i)(1 - F/F_i) \tag{15.16}$$

---

[4]LISREL 9 uses $n = N$, not $n = N - 1$ in all definitions that follow.

$$\mathsf{NNFI} = \frac{f_i - f}{f_i - 1} \tag{15.17}$$

$$\mathsf{CFI} = 1 - \tau/\tau_i \tag{15.18}$$

$$\mathsf{IFI} = \frac{nF_i - nF}{nF_i - d} \tag{15.19}$$

$$\mathsf{RFI} = \frac{f_i - f}{f_i} \tag{15.20}$$

$$\mathsf{PGFI} = (2d/k(k+1))\mathsf{GFI} \tag{15.21}$$

Hoelter (1983) proposed a critical N (CN) statistic:

$$\mathsf{CN} = \frac{\chi^2_{1-\alpha}}{F} + 1 \,, \tag{15.22}$$

where $\chi^2_{1-\alpha}$ is the $1 - \alpha$ percentile of the chi-square distribution. This is the sample size that would make the obtained chi-square just significant at the significance level α. For a discussion of this statistic, see Bollen and Liang (1988) and Bollen (1989a).

Chapter 16

Appendix F: General Statistical Theory

16.1 Continuous Variables

16.1.1 Data and Sample Statistics

Multivariate data can be viewed as a data matrix of order $N \times k$ where the rows correspond to independent observations and the columns as variables:

$$\mathbf{Z} = \begin{bmatrix} z_{11} & z_{12} & \cdots & z_{1k} \\ z_{21} & z_{22} & \cdots & z_{2k} \\ . & . & \cdots & . \\ . & . & \cdots & . \\ z_{N1} & z_{N2} & \cdots & z_{Nk} \end{bmatrix} = \begin{bmatrix} \mathbf{z}'_1 \\ \mathbf{z}'_2 \\ . \\ . \\ \mathbf{z}'_N \end{bmatrix}$$

The sample mean vector is

$$\bar{\mathbf{z}} = (1/N) \sum_{i=1}^{N} \mathbf{z}_i = (1/N)\mathbf{Z}'\mathbf{1} \tag{16.1}$$

and the sample covariance matrix is

$$\mathbf{S} = (1/N) \sum_{i=1}^{N} (\mathbf{z}_i - \bar{\mathbf{z}})(\mathbf{z}_i - \bar{\mathbf{z}})' = (1/N)\mathbf{Z}'\mathbf{Z} - \bar{\mathbf{z}}\bar{\mathbf{z}}' \tag{16.2}$$

These play a fundamental role in the estimation of structural equation models. The standard deviations of the variables are the square roots of the diagonal elements of $\mathbf{S}$, i.e., $s_i = \sqrt{s_{ii}}, i = 1, 2, \ldots, k$. Sometimes a correlation matrix is used:

$$\mathbf{R} = (\text{diag}\mathbf{S})^{-\frac{1}{2}} \mathbf{S} (\text{diag}\mathbf{S})^{-\frac{1}{2}} , \tag{16.3}$$

where $\text{diag}\mathbf{S}$ is a diagonal matrix with $s_{11}, s_{22}, \ldots, s_{kk}$ in the diagonal. A typical element of $\mathbf{R}$ is $r_{ij} = s_{ij}/s_i s_j$, where s_{ij} is a typical element of $\mathbf{S}$.

16.1.2 The Multivariate Normal Distribution

The univariate normal density function is

$$p(z; \mu, \sigma) = \frac{1}{\sqrt{2\pi}} e^{-\frac{1}{2}(\frac{z-\mu}{\sigma})^2} . \tag{16.4}$$

© Springer International Publishing Switzerland 2016
K.G. Jöreskog et al., *Multivariate Analysis with LISREL*, Springer Series in Statistics,
DOI 10.1007/978-3-319-33153-9_16

It is characterized by the mean $\mu = E(z)$ and the variance $\sigma^2 = E(z - \mu)^2$.

The multivariate normal density function is a direct generalization of the univariate normal distribution. Let $\mathbf{z}' = (z_1, z_2, \ldots, z_k)$ be a vector of k random variables. The multivariate normal density function is

$$p(\mathbf{z}; \boldsymbol{\mu}, \boldsymbol{\Sigma}) = (2\pi)^{-\frac{1}{2}} \mid \boldsymbol{\Sigma} \mid^{-\frac{1}{2}} exp\{-\frac{1}{2}(\mathbf{z} - \boldsymbol{\mu})'\boldsymbol{\Sigma}^{-1}(\mathbf{z} - \boldsymbol{\mu})\} \ , \tag{16.5}$$

where $\mid \boldsymbol{\Sigma} \mid$ is the determinant of $\boldsymbol{\Sigma}$. It is characterized by the mean vector $\boldsymbol{\mu} = E(\mathbf{z})$ and the covariance matrix $\boldsymbol{\Sigma} = E(\mathbf{z} - \boldsymbol{\mu})(\mathbf{z} - \boldsymbol{\mu})'$, a positive definite matrix. $\mid \boldsymbol{\Sigma} \mid$ is the determinant of $\boldsymbol{\Sigma}$ and $\boldsymbol{\Sigma}^{-1}$ is the inverse of $\boldsymbol{\Sigma}$. To specify that $\mathbf{z}$ is multivariate normal, the notation is $\mathbf{z} \sim N(\boldsymbol{\mu}, \boldsymbol{\Sigma})$. The following properties of the multivariate normal distribution are useful in what follows:

- Any subset of $z_1, z_2, \ldots, z_k$ is also multivariate normal.

- The conditional distribution of any subset given the other variables is also multivariate normal

In formulas this can be written as follows. Let $\mathbf{z}$ be partitioned as $\mathbf{z} = (\mathbf{z}_1, \mathbf{z}_2)$ with

$$\boldsymbol{\mu} = \begin{pmatrix} \boldsymbol{\mu}_1 \\ \boldsymbol{\mu}_2 \end{pmatrix} \quad \boldsymbol{\Sigma} = \begin{pmatrix} \boldsymbol{\Sigma}_{11} & \boldsymbol{\Sigma}_{12} \\ \boldsymbol{\Sigma}_{21} & \boldsymbol{\Sigma}_{22} \end{pmatrix} . \tag{16.6}$$

Then

$$\mathbf{z}_1 \sim N(\boldsymbol{\mu}_1, \boldsymbol{\Sigma}_{11}) \quad \mathbf{z}_2 \sim N(\boldsymbol{\mu}_2, \boldsymbol{\Sigma}_{22}) \tag{16.7}$$

$$\mathbf{z}_2 \mid \mathbf{z}_1 \sim N(\boldsymbol{\mu}_2 + \boldsymbol{\Sigma}_{21}\boldsymbol{\Sigma}_{11}^{-1}(\mathbf{z}_1 - \boldsymbol{\mu}_1), \boldsymbol{\Sigma}_{22} - \boldsymbol{\Sigma}_{21}\boldsymbol{\Sigma}_{11}^{-1}\boldsymbol{\Sigma}_{12}) \tag{16.8}$$

$$\mathbf{z}_1 \mid \mathbf{z}_2 \sim N(\boldsymbol{\mu}_1 + \boldsymbol{\Sigma}_{12}\boldsymbol{\Sigma}_{22}^{-1}(\mathbf{z}_2 - \boldsymbol{\mu}_2), \boldsymbol{\Sigma}_{11} - \boldsymbol{\Sigma}_{12}\boldsymbol{\Sigma}_{22}^{-1}\boldsymbol{\Sigma}_{21}) \tag{16.9}$$

16.1.3   The Multivariate Normal Likelihood

Let $\mathbf{z}_1, \mathbf{z}_2, \ldots, \mathbf{z}_N$, be independently and identically (iid) distributed each with $\mathbf{z}_i \sim N(\boldsymbol{\mu}, \boldsymbol{\Sigma})$ with $\boldsymbol{\Sigma}$ positive definite. The assumption of identical distribution may be relaxed but the assumption of independence is fundamental.

The likelihood of $\mathbf{z}_i$ is

$$L_i = (2\pi)^{-\frac{1}{2}} \mid \boldsymbol{\Sigma} \mid^{-\frac{1}{2}} exp\{-\frac{1}{2}(\mathbf{z}_i - \boldsymbol{\mu})'\boldsymbol{\Sigma}^{-1}(\mathbf{z}_i - \boldsymbol{\mu})\} \tag{16.10}$$

This is considered as a function of $\boldsymbol{\mu}$ and $\boldsymbol{\Sigma}$ for given data. In most cases it is convenient to work with the natural logarithm of L_i instead of L_i. This is

$$\ln L_i = -\frac{1}{2}\ln(2\pi) - \frac{1}{2}\ln \mid \boldsymbol{\Sigma} \mid -\frac{1}{2}(\mathbf{z}_i - \boldsymbol{\mu})'\boldsymbol{\Sigma}^{-1}(\mathbf{z}_i - \boldsymbol{\mu}) \tag{16.11}$$

Since the observations are independent, the log-likelihood of all the observations is

$$\ln L = \sum_{i=1}^{N} \ln L_i \tag{16.12}$$

So $\ln L$ becomes

$$\ln L = c - \frac{N}{2}\ln|\boldsymbol{\Sigma}| - \frac{1}{2}\left\{\sum_{i=1}^{N}(\mathbf{z}_i - \boldsymbol{\mu})'\boldsymbol{\Sigma}^{-1}(\mathbf{z}_i - \boldsymbol{\mu})\right\} \tag{16.13}$$

$$= c - \frac{N}{2}\{\ln|\boldsymbol{\Sigma}| + \text{tr}(\mathbf{S}\boldsymbol{\Sigma}^{-1}) + (\bar{\mathbf{z}} - \boldsymbol{\mu})'\boldsymbol{\Sigma}^{-1}(\bar{\mathbf{z}} - \boldsymbol{\mu})\}, \tag{16.14}$$

where $c = -(N/2)\ln(2\pi)$ and $\text{tr}(\mathbf{A})$ is the sum of the diagonal elements of a square matrix $\mathbf{A}$. For fixed N, c is a constant independent of both data and parameters. So it will be ignored in what follows. Equation (16.14) shows that $\bar{\mathbf{z}}$ and $\mathbf{S}$ are sufficient statistics, which means that all information about the multi-normal likelihood is contained in $\bar{\mathbf{z}}$ and $\mathbf{S}$. To get from (16.13) to (16.14), consider that

$$\sum_{i=1}^{N}(\mathbf{z}_i - \boldsymbol{\mu})'\boldsymbol{\Sigma}^{-1}(\mathbf{z}_i - \boldsymbol{\mu})$$

$$= \sum_{i=1}^{N}\text{tr}\{\boldsymbol{\Sigma}^{-1}(\mathbf{z}_i - \boldsymbol{\mu})(\mathbf{z}_i - \boldsymbol{\mu})'\}$$

$$= \text{tr}\{\boldsymbol{\Sigma}^{-1}\sum_{i=1}^{N}(\mathbf{z}_i - \boldsymbol{\mu})(\mathbf{z}_i - \boldsymbol{\mu})'\}$$

$$= \text{tr}\{\boldsymbol{\Sigma}^{-1}\sum_{i=1}^{N}[(\mathbf{z}_i - \bar{\mathbf{z}}) + (\bar{\mathbf{z}} - \boldsymbol{\mu})][(\mathbf{z}_i - \bar{\mathbf{z}}) + (\bar{\mathbf{z}} - \boldsymbol{\mu})]'\}$$

$$= \text{tr}\{\boldsymbol{\Sigma}^{-1}N\mathbf{S} + N(\bar{\mathbf{z}} - \boldsymbol{\mu})(\bar{\mathbf{z}} - \boldsymbol{\mu})'\}$$

$$= N\{\text{tr}(\mathbf{S}\boldsymbol{\Sigma}^{-1}) + (\bar{\mathbf{z}} - \boldsymbol{\mu})'(\bar{\mathbf{z}} - \boldsymbol{\mu})\}$$

A saturated model is one where both $\boldsymbol{\mu}$ and $\boldsymbol{\Sigma}$ are unconstrained. If $\boldsymbol{\mu}$ is unconstrained, then $\ln L$ in (16.14) is maximized with respect to $\boldsymbol{\mu}$ if $\boldsymbol{\mu} = \bar{\mathbf{z}}$ so that the maximum likelihood estimate of $\boldsymbol{\mu}$ is $\hat{\boldsymbol{\mu}} = \bar{\mathbf{z}}$. Then $\ln L$ becomes

$$\ln L = -\frac{N}{2}\{\ln|\boldsymbol{\Sigma}| + \text{tr}(\mathbf{S}\boldsymbol{\Sigma}^{-1})\} \tag{16.15}$$

Maximizing this with respect to $\boldsymbol{\Sigma}$ gives the maximum likelihood $\hat{\boldsymbol{\Sigma}}$ estimate of $\boldsymbol{\Sigma}$. It can be shown that $\hat{\boldsymbol{\Sigma}} = \mathbf{S}$. Then $\ln L$ becomes

$$\ln L = -\frac{N}{2}\{\ln|\mathbf{S}| + k\}. \tag{16.16}$$

A LISREL model for a single group is in general a mean and covariance structure in the sense that the mean vector $\boldsymbol{\mu}(\boldsymbol{\theta})$ and the covariance matrix $\boldsymbol{\Sigma}(\boldsymbol{\theta})$ are functions of a parameter vector $\boldsymbol{\theta}' = (\theta_1, \theta_2, \ldots, \theta_t)$ consisting of t independent parameters. The maximum likelihood estimate of $\boldsymbol{\theta}$ is obtained by maximizing the log likelihood function

$$\ln L(\boldsymbol{\theta}) = -\frac{N}{2}\{\ln|\boldsymbol{\Sigma}(\boldsymbol{\theta})| + \text{tr}(\mathbf{S}\boldsymbol{\Sigma}^{-1}(\boldsymbol{\theta})) + (\bar{\mathbf{z}} - \boldsymbol{\mu}(\boldsymbol{\theta}))'\boldsymbol{\Sigma}^{-1}(\bar{\mathbf{z}} - \boldsymbol{\mu}(\boldsymbol{\theta}))\} \tag{16.17}$$

with respect to $\boldsymbol{\theta}$. The maximum likelihood estimate $\hat{\boldsymbol{\theta}}$ of $\boldsymbol{\theta}$ is determined by an iterative procedure described in the appendix in Chapter 17.

16.1.4   Likelihood, Deviance, and Chi-square

Consider a model M_0 with t_0 independent parameters and an alternative model M_1 with $t_1 > t_0$ independent parameters. The likelihood $L = L(\mathbf{Z}, \boldsymbol{\theta})$ is a function of both data $\mathbf{Z}$ and parameters $\boldsymbol{\theta}$ but in likelihood theory it is considered as a function of parameters for given data. It is convenient to work with $D = -2 \ln L$, sometimes called deviance, instead of L. D is to be minimized with respect to $\boldsymbol{\theta}$. Maximizing L is equivalent to minimizing D. Let

$$L_0 = max L \text{ under } M_0 \tag{16.18}$$

$$L_1 = max L \text{ under } M_1 \tag{16.19}$$

Then

$$L_1 \geq L_0 \Longrightarrow \log L_1 \geq \log L_0 \Longrightarrow D_1 \leq D_0 \tag{16.20}$$

and define

$$\chi^2 = D_0 - D_1 = 2 \log L_1 - 2 \log L_0 = 2 \log \frac{L_1}{L_0} = 2 \log(LR) \tag{16.21}$$

According to a general theorem of Box (1949), if model M_0 holds, then χ^2 has a chi-square distribution with

$$\text{Degrees of freedom } d = t_1 - t_0 \tag{16.22}$$

This χ^2 is called a likelihood ratio statistic.

Suppose now that M_1 is a saturated model and M_0 is an estimated model with t independent parameters. From (16.16) we have

$$D_1 = N\{\ln \mid \mathbf{S} \mid + k\} , \tag{16.23}$$

and from (16.17) we have

$$D_0 = N\{\ln \mid \hat{\boldsymbol{\Sigma}} \mid + \mathrm{tr}(\mathbf{S}\hat{\boldsymbol{\Sigma}}^{-1}) + (\bar{\mathbf{z}} - \hat{\boldsymbol{\mu}})'\hat{\boldsymbol{\Sigma}}^{-1}(\bar{\mathbf{z}} - \hat{\boldsymbol{\mu}})\} , \tag{16.24}$$

where $\hat{\boldsymbol{\mu}} = \boldsymbol{\mu}(\hat{\boldsymbol{\theta}})$ and $\hat{\boldsymbol{\Sigma}} = \boldsymbol{\Sigma}(\hat{\boldsymbol{\theta}})$. Then

$$\chi^2 = N\{\ln \mid \hat{\boldsymbol{\Sigma}} \mid + \mathrm{tr}(\mathbf{S}\hat{\boldsymbol{\Sigma}}^{-1}) + (\bar{\mathbf{z}} - \hat{\boldsymbol{\mu}})'\hat{\boldsymbol{\Sigma}}^{-1}(\bar{\mathbf{z}} - \hat{\boldsymbol{\mu}}) - \ln \mid \mathbf{S} \mid + k\} . \tag{16.25}$$

This is the likelihood ratio chi-square statistic C_1 reported in the output from **LISREL**. The degrees of freedom is

$$d = k + \frac{1}{2}k(k + 1) - t . \tag{16.26}$$

A special case of this is if the mean vector is unconstrained for then

$$\chi^2 = N\{\ln \mid \hat{\boldsymbol{\Sigma}} \mid + \mathrm{tr}(\mathbf{S}\hat{\boldsymbol{\Sigma}}^{-1}) - \ln \mid \mathbf{S} \mid + k\} , \tag{16.27}$$

with degrees of freedom

$$d = \frac{1}{2}k(k + 1) - t . \tag{16.28}$$

16.1.5 General Covariance Structures

The majority of LISREL models are estimated using maximum likelihood (ML) method. This estimation method is appropriate for variables that are continuous, and at least approximately normally distributed. In many research problems the variables under study are neither normal nor even approximately continuous and the use of ML is not valid. An important technical development has been in extending the class of estimation methods to procedures that are correct when used with many different kinds of variables. This more general approach to estimation includes ML, GLS, and ULS and other methods as special cases. But it also applies to very different statistical distributions.

The statistical inference problem associated with all kinds of structural equation models, including factor analysis models, can be formulated very generally and compactly as follows. For the orginal formulation see, e.g., Browne (1984) and Satorra (1989).

Let $\mathsf{Vec}(\mathbf{S})$ be the column vector formed by the columns of $\mathbf{S}$ stringed under each other. Thus, $\mathsf{Vec}(\mathbf{S})$ is of order $k^2 \times 1$. Since $\mathbf{S}$ is a symmetric matrix, the covariance matrix of $\mathsf{Vec}(\mathbf{S})$ is singular. Therefore, it is convenient to work with $\mathbf{s}$ instead, where $\mathbf{s}$ is a vector of the non-duplicated elements of $\mathbf{S}$:

$$\mathbf{s}' = (s_{11}, s_{21}, s_{22}, s_{31}, s_{32}, \ldots,) \ . \tag{16.29}$$

The relationship between $\mathbf{s}$ and $\mathsf{Vec}(\mathbf{S})$ is

$$\mathbf{s} = \mathbf{K}'\mathsf{Vec}(\mathbf{S}) \ , \tag{16.30}$$

where $\mathbf{K}$ is a matrix of order $k^2 \times \frac{1}{2}k(k+1)$. Each column of $\mathbf{K}$ has one nonzero value which is 1 for a diagonal element and $\frac{1}{2}$ for a non-diagonal element. The reverse relationship of (16.30) is

$$\mathsf{Vec}(\mathbf{S}) = \mathbf{K}(\mathbf{K}'\mathbf{K})^{-1}\mathbf{s} = \mathbf{D}\mathbf{s} \ , \tag{16.31}$$

where $\mathbf{D} = \mathbf{K}(\mathbf{K}'\mathbf{K})^{-1}$ is the duplication matrix, see Magnus and Neudecker (1988).

Similarly, let $\boldsymbol{\sigma} = \mathbf{K}'\mathsf{Vec}(\boldsymbol{\Sigma})$ be the corresponding vector of the non-duplicated elements of $\boldsymbol{\Sigma}$ and suppose that

$$\boldsymbol{\sigma} = \boldsymbol{\sigma}(\boldsymbol{\theta}) \ , \tag{16.32}$$

is a differentiable function of a parameter vector $\boldsymbol{\theta}$. For example, in a LISREL model, $\boldsymbol{\theta}$ is a vector of all independent parameters in all parameter matrices $\boldsymbol{\Lambda}_y$, $\boldsymbol{\Lambda}_x$, $\mathbf{B}$, $\boldsymbol{\Gamma}$, $\boldsymbol{\Phi}$, $\boldsymbol{\Psi}$, $\boldsymbol{\Theta}_\epsilon$, $\boldsymbol{\Theta}_\delta$, and $\boldsymbol{\Theta}_{\delta\epsilon}$ but models which are not LISREL models can also be used.

The sample vector $\mathbf{s}$ has a limiting normal distribution when $N \to \infty$. Browne (1984) showed that

$$N^{\frac{1}{2}}(\mathbf{s} - \boldsymbol{\sigma}) \xrightarrow{d} N(\mathbf{0}, \boldsymbol{\Omega}) \ , \tag{16.33}$$

where $\xrightarrow{d}$ denotes convergence in distribution. Under general assumptions about the distribution of the observed variables, the elements of the covariance matrix $\boldsymbol{\Omega}$ are given by (Browne, 1984, eq. 2.2)

$$\omega_{ghij} = \sigma_{ghij} - \sigma_{gh}\sigma_{ij} \ , \tag{16.34}$$

where

$$\sigma_{ghij} = E[(z_g - \mu_g)(z_h - \mu_h)(z_i - \mu_i)(z_j - \mu_j)] \ , \tag{16.35}$$

is a fourth order central moment, and

$$\sigma_{gh} = E[(z_g - \mu_g)(z_h - \mu_h)] \ . \tag{16.36}$$

Under normality

$$\omega_{ghij} = \sigma_{gi}\sigma_{hj} + \sigma_{gj}\sigma_{hi} \, , \tag{16.37}$$

which can be written in matrix form as

$$\boldsymbol{\Omega} = 2\mathbf{K}'(\boldsymbol{\Sigma} \otimes \boldsymbol{\Sigma})\mathbf{K} \, , \tag{16.38}$$

where $\otimes$ denotes a Kronecker product. Note that under normality $\boldsymbol{\Omega}$ is a function of $\boldsymbol{\Sigma}$ only so that no fourth order moments are involved.

Let $\mathbf{W}$ be a consistent estimate of $\boldsymbol{\Omega}$. In analogy with (16.34) – (16.36), the elements of $\mathbf{W}$ are obtained as

$$w_{gh,ij} = m_{ghij} - s_{gh}s_{ij} \, , \tag{16.39}$$

where

$$m_{ghij} = (1/N) \sum_{a=1}^{N} (z_{ag} - \bar{z}_g)(z_{ah} - \bar{z}_h)(z_{ai} - \bar{z}_i)(z_{aj} - \bar{z}_j) \tag{16.40}$$

is a fourth-order central sample moment.

Under normality we use the estimate

$$\mathbf{W} = 2\mathbf{K}'(\hat{\boldsymbol{\Sigma}} \otimes \hat{\boldsymbol{\Sigma}})\mathbf{K} \, , \tag{16.41}$$

in analogy with (16.38). Whenever it is necessary to distinguish the general $\mathbf{W}$ defined by (16.39) from the specific $\mathbf{W}$ in (16.41), the notation $\mathbf{W}_{NT}$ is used for $\mathbf{W}$ in (16.41) and $\mathbf{W}_{NNT}$ for the general $\mathbf{W}$ in (16.39).

To estimate the model, consider the minimization of the fit function

$$F(\mathbf{s}, \boldsymbol{\theta}) = [\mathbf{s} - \boldsymbol{\sigma}(\boldsymbol{\theta})]'\mathbf{V}[\mathbf{s} - \boldsymbol{\sigma}(\boldsymbol{\theta})] \tag{16.42}$$

where $\mathbf{V}$ is either a fixed positive definite matrix or a random matrix converging in probability to a positive definite matrix $\overline{\mathbf{V}}$. The fit functions available in LISREL are Unweighted Least Squares (ULS)[1], Generalized Least Squares (GLS), Maximum Likelihood (ML), Diagonally Weighted Least Squares (DWLS), and Weighted Least Squares (WLS). They correspond to taking $\mathbf{V}$ in (16.42) as

$$\text{ULS}: \quad \mathbf{V} = \quad \mathbf{K}'(\mathbf{I} \otimes \mathbf{I})\mathbf{K} \tag{16.43}$$

$$\text{GLS}: \quad \mathbf{V} = \quad \mathbf{D}'(\mathbf{S}^{-1} \otimes \mathbf{S}^{-1})\mathbf{D} \tag{16.44}$$

$$\text{ML}: \quad \mathbf{V} = \quad \mathbf{D}'(\hat{\boldsymbol{\Sigma}}^{-1} \otimes \hat{\boldsymbol{\Sigma}}^{-1})\mathbf{D} \tag{16.45}$$

$$\text{DWLS}: \quad \mathbf{V} = \quad \mathbf{D}_{\mathrm{W}} = [diag\mathbf{W}]^{-1} \tag{16.46}$$

$$\text{WLS}: \quad \mathbf{V} = \quad \mathbf{W}^{-1} \tag{16.47}$$

The matrix $\bar{\mathbf{V}}$ is unknown for all methods except ULS but can be estimated by $\hat{\mathbf{V}}$, where

$$\text{ULS}: \quad \hat{\mathbf{V}} = \quad \mathbf{V} \tag{16.48}$$

$$\text{GLS}: \quad \hat{\mathbf{V}} = \quad \mathbf{W}_{NT}^{-1} \tag{16.49}$$

$$\text{ML}: \quad \hat{\mathbf{V}} = \quad \mathbf{W}_{NT}^{-1} \tag{16.50}$$

$$\text{DWLS}: \quad \hat{\mathbf{V}} = \quad [diag\mathbf{W}_{NNT}]^{-1} \tag{16.51}$$

$$\text{WLS}: \quad \hat{\mathbf{V}} = \quad \mathbf{W}_{NNT}^{-1} \tag{16.52}$$

---

[1]The ULS fit function was originally defined as $\sum_{i=1}^{k} \sum_{j=1}^{k} (s_{ij} - \sigma_{ij})^2$, which translate the matrix $\mathbf{V}$ in (16.43).

Note that

$$\mathbf{W}_{\mathrm{NT}}^{-1} = [2\mathbf{K}'(\hat{\boldsymbol{\Sigma}} \otimes \hat{\boldsymbol{\Sigma}})\mathbf{K}]^{-1} = \frac{1}{2}\mathbf{D}'(\hat{\boldsymbol{\Sigma}}^{-1} \otimes \hat{\boldsymbol{\Sigma}}^{-1})\mathbf{D} = \frac{1}{2}\mathbf{V}_{\mathrm{ML}} , \tag{16.53}$$

as shown by Browne (1977, equations 20 and 21).

The fit function for ML is usually written

$$\mathsf{F}[\mathbf{S}, \boldsymbol{\Sigma}(\boldsymbol{\theta})] = \ln|\boldsymbol{\Sigma}| + \mathsf{tr}(\mathbf{S}\boldsymbol{\Sigma}^{-1}) - \ln|\mathbf{S}| - k \tag{16.54}$$

but Browne (1977) showed that minimizing F with $\mathbf{V}$ in (16.45) and minimizing (16.54) are equivalent. Minimizing F with $\mathbf{V}$ in (16.45) can be interpreted as ML estimated by means of iteratively re-weighted least squares in which $\hat{\boldsymbol{\Sigma}}$ is updated in each iteration. Both of these fit functions have a minimum at the same point in the parameter space, namely at the ML estimates. However, the minimum value of the functions are not the same.

All fit functions are non-negative and equal to zero only for a saturated model, where $\hat{\boldsymbol{\Sigma}} = \mathbf{S}$.

Under multivariate normality of the observed variables, ML and GLS estimates are asymptotically equivalent in the sense that they have the same asymptotic covariance matrix. There is no advantage in using GLS except when $\boldsymbol{\sigma}(\boldsymbol{\theta})$ is a linear function. In practice ML is most often used.

Under non-normality, WLS, also called ADF, see Browne (1984), is in principle the best method since it is valid for any non-normal distribution for continuous variables. But in practice this method does not work well because it is difficult to determine $\mathbf{W}$ (and hence $\mathbf{W}^{-1}$) accurately unless N is huge.

Let $\hat{\boldsymbol{\theta}}$ be the minimizer of $F(\mathbf{s}, \boldsymbol{\theta})$ and let $\boldsymbol{\theta}_0$ be a unique minimizer of $F(\boldsymbol{\sigma}, \boldsymbol{\theta})$. We assume here that the model holds so that $F(\boldsymbol{\sigma}, \boldsymbol{\theta}_0) = 0$. See Browne (1984), Satorra (1989), and Foss, Jöreskog, and Olsson (2011) for the case where the model does not hold.

Let

$$\boldsymbol{\Delta} = \left[\frac{\partial \boldsymbol{\Sigma}}{\partial \boldsymbol{\theta}'}\right]_{\boldsymbol{\theta}_0} . \tag{16.55}$$

Then

$$NACov(\hat{\boldsymbol{\theta}}) = (\boldsymbol{\Delta}'\overline{\mathbf{V}}\boldsymbol{\Delta})^{-1}\boldsymbol{\Delta}'\overline{\mathbf{V}}\boldsymbol{\Omega}\overline{\mathbf{V}}\boldsymbol{\Delta}(\boldsymbol{\Delta}'\overline{\mathbf{V}}\boldsymbol{\Delta})^{-1} , \tag{16.56}$$

which can be estimated as

$$NEst[ACov(\hat{\boldsymbol{\theta}})] = (\hat{\boldsymbol{\Delta}}'\hat{\mathbf{V}}\hat{\boldsymbol{\Delta}})^{-1}\hat{\boldsymbol{\Delta}}'\hat{\mathbf{V}}\mathbf{W}\hat{\mathbf{V}}\hat{\boldsymbol{\Delta}}(\hat{\boldsymbol{\Delta}}'\hat{\mathbf{V}}\hat{\boldsymbol{\Delta}})^{-1} , \tag{16.57}$$

where $\hat{\boldsymbol{\Delta}}$ is $\boldsymbol{\Delta}$ evaluated at $\hat{\boldsymbol{\theta}}$. The standard errors reported by LISREL for each parameter are obtained from the diagonal elements of (16.57).

Two special cases are of particular interest:

- Under normality and with methods GLS or ML, $\hat{\mathbf{V}} = \mathbf{W}_{\mathrm{NT}}^{-1}$ and $\mathbf{W} = \mathbf{W}_{\mathrm{NT}}$, so that (16.57) reduces to

$$NEst[ACov(\hat{\boldsymbol{\theta}}) = (\hat{\boldsymbol{\Delta}}'\hat{\mathbf{V}}\hat{\boldsymbol{\Delta}})^{-1} , \tag{16.58}$$

 which is the estimated Fisher Information Matrix.

- Under non-normality and with method WLS, $\hat{\mathbf{V}} = \mathbf{W}_{\mathrm{NNT}}^{-1}$ and $\mathbf{W} = \mathbf{W}_{\mathrm{NNT}}$, so that (16.57) reduces also to (16.58).

To test the model under multivariate normality, one can use

$$C_1 = N[\log | \hat{\boldsymbol{\Sigma}} | + \text{tr}(\mathbf{S}\hat{\boldsymbol{\Sigma}}^{-1}) - \log | \mathbf{S} | - k] . \tag{16.59}$$

Although C_1 can be computed for any $\mathbf{V}$, *i.e.*, any fit function in (16.43)–(16.47), its natural use is with ML. Then C_1 is the likelihood ratio χ^2 statistic. Under multivariate normality C_1 has an asymptotic χ^2 distribution with

$$d = s - t \tag{16.60}$$

degrees of freedom if the model holds. Recall that $s = (1/2)k(k+1)$ and t is the number of independent parameters in the model.

To test the model for any fit function, Browne (1984) developed a general formula:

$$C_2(\mathbf{W}) = N(\mathbf{s} - \hat{\boldsymbol{\sigma}})'[\mathbf{W}^{-1} - \mathbf{W}^{-1}\hat{\boldsymbol{\Delta}}(\hat{\boldsymbol{\Delta}}'\mathbf{W}^{-1}\hat{\boldsymbol{\Delta}})^{-1}\hat{\boldsymbol{\Delta}}'\mathbf{W}^{-1}](\mathbf{s} - \hat{\boldsymbol{\sigma}}) \tag{16.61}$$

This is Browne's ADF chi-square statistic (Browne, 1984, equation 2.20b). If $\mathbf{W}_{\text{NNT}}$ is available, LISREL computes both $C_2(\mathbf{W}_{\text{NT}})$ and $C_2(\mathbf{W}_{\text{NNT}})$. Otherwise, LISREL gives only $C_2(\mathbf{W}_{\text{NT}})$. $C_2(\mathbf{W}_{\text{NT}})$ is valid for all methods with $\hat{\mathbf{V}}$ defined in (16.48)–(16.51) under normality. $C_2(\mathbf{W}_{\text{NNT}})$ is valid for the same methods under non-normality. Under general assumptions $C_2(\mathbf{W}_{\text{NNT}})$ has an asymptotic χ^2 distribution with d degrees of freedom if the model holds. With the ML method, C_1 and $C_2(\mathbf{W}_{\text{NT}})$ are asymptotically equivalent under multivariate normality. It has been found in simulation studies that $C_2(\mathbf{W}_{\text{NNT}})$ does not work well under non-normality, see *e.g*, Curran, West, and Finch (1996). This is probably because the matrix $\mathbf{W}_{\text{NNT}}$ is often unstable unless the sample size N is huge

As a remedy for the poor behavior of C_1 and $C_2(\mathbf{W}_{\text{NT}})$ under non-normality, Satorra and Bentler (1988) developed an alternative test procedure. Let

$$\mathbf{U} = \hat{\mathbf{V}} - \hat{\mathbf{V}}\hat{\boldsymbol{\Delta}}(\hat{\boldsymbol{\Delta}}'\mathbf{V}\hat{\boldsymbol{\Delta}})^{-1}\hat{\boldsymbol{\Delta}}'\hat{\mathbf{V}} . \tag{16.62}$$

Under non-normality the asymptotic distribution of C_1 and $C_2(\mathbf{W}_{\text{NT}})$ is not known but Satorra and Bentler (1988) conclude that C_1 and $C_2(\mathbf{W}_{\text{NT}})$ are asymptotically distributed as a linear combination of χ^2's with one degree of freedom where the coefficients of the linear combination are the non-zero eigenvalues of $\mathbf{UW}$. Based on this result Satorra and Bentler (1988) suggest a scale factor for C_1 and $C_2(\mathbf{W}_{\text{NT}})$ such that these statistics have the correct asymptotic mean d. Let

$$h_1 = \text{tr}(\mathbf{UW}_{\text{NNT}}) . \tag{16.63}$$

The scale factor is d/h_1. In LISREL 9 we apply this scaling to C_1 if the ML method is used and define[2]

$$C_3 = (d/h_1)C_1 . \tag{16.64}$$

For all other methods we apply the scale factor to $C_2(\mathbf{W}_{\text{NT}})$ and define

$$C_3 = (d/h_1)C_2(\mathbf{W}_{\text{NT}}) . \tag{16.65}$$

Although C_3 does not have an asymptotic χ^2 distribution under non-normality, it is often used as an approximate chi-square. C_3 is called the Satorra-Bentler scaled chi-square or the Satorra-Bentler mean adjusted chi-square. It has been found to work well in practice as it outperforms $C_2(\mathbf{W}_{\text{NNT}})$ under non-normality, see *e.g*, Curran, West, and Finch (1996).

---

[2]In previous versions of LISREL we applied this scaling to $C_2(\mathbf{W}_{\text{NT}})$ for all methods.

To get an even better approximation to a χ^2 distribution Satorra and Bentler (1988) suggest a mean and variance adjusted statistic. This is a Satterthwaite (1941) type of adjustment. Let

$$h_2 = \text{tr}(\mathbf{U}\mathbf{W}_{\text{NNT}}\mathbf{U}\mathbf{W}_{\text{NNT}}) \, . \tag{16.66}$$

Then

$$C_4 = (h_1/h_2)C \, , \tag{16.67}$$

where $C = C_1$ for ML and $C = C_2(\mathbf{W}_{\text{NT}})$ for all other methods. Regard C_4 as a an approximate chi-square with

$$d' = [\text{tr}(\mathbf{U}\mathbf{W}_{\text{NNT}})]^2/h_2 \tag{16.68}$$

degrees of freedom. In LISREL 9 we give d' with three decimals and use this fractional degrees of freedom to compute the P-value for C_4.

16.1.6 The Independence Model

Let $\mathbf{z} \sim N(\boldsymbol{\mu}, \boldsymbol{\Sigma})$. The independence model states that $\boldsymbol{\Sigma}$ is diagonal. If the variables in $\mathbf{z}$ are multi-normal they would be independent, hence the name *independence model*. In the independence model all covariances in $\boldsymbol{\Sigma}$ are 0. Hence all variables in $\mathbf{z}$ are uncorrelated. Let $\boldsymbol{\Delta} = diag\boldsymbol{\Sigma}$. Intuitively one would estimate $\boldsymbol{\Delta}$ as $\hat{\boldsymbol{\Delta}} = diag\mathbf{S}$. As can easily be shown $\hat{\boldsymbol{\Delta}} = diag\mathbf{S}$ is an ML estimate in the sense that it minimizes the ML fit function (16.45). It is also an ULS estimate since it minimizes the ULS fit function (16.43). Similarly, it is a DWLS estimate since it minimizes the DWLS fit function (16.51). However, interestingly enough, $\hat{\boldsymbol{\Delta}} = diag\mathbf{S}$ **is not** a GLS estimator nor a WLS estimator. Fortunately, both the GLS and the WLS fit functions are quadratic in $\boldsymbol{\Delta}$, so the GLS and WLS estimate can be easily obtained.

No matter which method of estimation is used LISREL provides a chi-square test of the independence model by evaluating N times the value of

$$\log \|\boldsymbol{\Sigma}\| + tr(\mathbf{S}\boldsymbol{\Sigma}^{-1}) - \log \|\mathbf{S}\| - k \tag{16.69}$$

at $\boldsymbol{\Sigma} = \hat{\boldsymbol{\Delta}}$.

The independence model must be rejected, for if all observed variables are uncorrelated it is not meaningful to model the covariance structure.

Table 7.6 lists the chi-square for the independence model for all fit functions for the CFA model for the NPV data.

16.1.7 Mean and Covariance Structures

There are two possible approaches to analyze the covariance matrix and the mean vector jointly. One is described in this section and the other in the next section.

In principle, the approach of the previous section can be applied to any vector of moments. So we just extend the vector $\mathbf{s}$ with the mean vector $\bar{\mathbf{z}}$ and then use the same definitions as before:

$$s = \frac{1}{2}k(k+1) + k \tag{16.70}$$

$$t = \text{as before} \tag{16.71}$$

$$\beta(\boldsymbol{\theta}) = \begin{bmatrix} \boldsymbol{\sigma}(\boldsymbol{\theta}) \\ \boldsymbol{\mu}(\boldsymbol{\theta}) \end{bmatrix} \tag{16.72}$$

$$\Delta = \frac{\partial \boldsymbol{\beta}}{\partial \boldsymbol{\theta}'} \tag{16.73}$$

$$\mathbf{b} = \begin{bmatrix} \mathbf{s} \\ \bar{\mathbf{z}} \end{bmatrix} \tag{16.74}$$

$\boldsymbol{\Omega}$ is now N times the asymptotic covariance matrix of $\mathbf{b}$ and $\mathbf{W}$ a consistent estimate of it. We assume that $\mathbf{s}$ and $\bar{\mathbf{z}}$ are asymptotically independent so that $\boldsymbol{\Omega}$ and $\mathbf{W}$ are block diagonal.

Under normal theory NT, $\mathbf{W}$ is estimated as

$$\mathbf{W}_{\mathsf{NTE}} = \begin{bmatrix} \mathbf{W}_{\mathsf{NT}} & \\ \mathbf{0} & \hat{\boldsymbol{\Sigma}} \end{bmatrix} \tag{16.75}$$

and under non-normal theory NNT, $\mathbf{W}$ is estimated as

$$\mathbf{W}_{\mathsf{NNTE}} = \begin{bmatrix} \mathbf{W}_{\mathsf{NNT}} & \\ \mathbf{0} & \hat{\boldsymbol{\Sigma}} \end{bmatrix} \tag{16.76}$$

where $\mathbf{W}_{\mathsf{NT}}$ and $\mathbf{W}_{\mathsf{NNT}}$ are defined as in the previous section.

The fit functions are defined by

$$\mathsf{F} = (\mathbf{b} - \boldsymbol{\beta})'\mathbf{V}(\mathbf{b} - \boldsymbol{\beta}) \tag{16.77}$$

$$\text{ULS:} \quad \mathbf{V} = \begin{bmatrix} \mathbf{I}^* & \\ \mathbf{0} & \hat{\boldsymbol{\Sigma}}^{-1} \end{bmatrix} \tag{16.78}$$

$$\text{GLS:} \quad \mathbf{V} = \begin{bmatrix} \mathbf{D}'(\mathbf{S}^{-1} \otimes \mathbf{S}^{-1})\mathbf{D} & \\ \mathbf{0} & \mathbf{S}^{-1} \end{bmatrix} \tag{16.79}$$

$$\text{ML:} \quad \mathbf{V} = \begin{bmatrix} \mathbf{D}'(\hat{\boldsymbol{\Sigma}}^{-1} \otimes \hat{\boldsymbol{\Sigma}}^{-1})\mathbf{D} & \\ \mathbf{0} & \hat{\boldsymbol{\Sigma}}^{-1} \end{bmatrix} \tag{16.80}$$

$$\text{WLS:} \quad \mathbf{V} = \begin{bmatrix} \mathbf{W}_{\mathsf{NNT}}^{-1} & \\ \mathbf{0} & \hat{\boldsymbol{\Sigma}}^{-1} \end{bmatrix} \tag{16.81}$$

$$\text{DWLS:} \quad \mathbf{V} = \begin{bmatrix} \mathbf{D}_{\mathsf{W}}^{-1} & \\ \mathbf{0} & \hat{\boldsymbol{\Sigma}}^{-1} \end{bmatrix} \tag{16.82}$$

The same results as in Section 16.1.5 apply but with $\mathbf{W}_{\mathsf{NT}}$ replaced with $\mathbf{W}_{\mathsf{NTE}}$ and $\mathbf{W}_{\mathsf{NNT}}$ replaced with $\mathbf{W}_{\mathsf{NNTE}}$.

16.1.8   Augmented Moment Matrix

In the previous section it was assumed that $\mathbf{s}$ and $\bar{\mathbf{z}}$ are asymptotically independent. Under non-normality this may not hold. A way to avoid this assumption is to use the augmented moment matrix. This is the matrix of sample moments about zero for the vector $\mathbf{z}$ augmented with a variable which is constant equal to 1 for every case. The population augmented moment matrix and the sample augmented moment matrix are defined in equations (16.85) and (16.86) that follow, where $\mathbf{a}$ is a vector of the non-duplicated elements $\mathbf{A}$. Because the last element of $\mathbf{a}$ is constant equal to 1, its covariance matrix $\mathbf{W}_a$ is singular. However the inverse of $\mathbf{W}_a$ is only used with WLS in which case it is replaced by its generalized inverse. Under non-normal theory $\mathbf{W}_a$ is estimated as $\mathbf{W}_{a\mathsf{NNT}}$ whose elements are

$$w_{gh,ij} = \mathsf{Est}\left[\mathsf{ACov}(a_{gh}, a_{ij})\right] = n_{ghij} - a_{gh}a_{ij} \; , \tag{16.83}$$

where

$$n_{ghij} = (1/N) \sum_{a=1}^{N} z_{ag} z_{ah} z_{ai} z_{aj} \tag{16.84}$$

is a fourth-order sample moment about zero.

$$\boldsymbol{\Upsilon} \;=\; \mathsf{E}\left[\begin{pmatrix} \mathbf{z} \\ 1 \end{pmatrix}\right] [\mathbf{z}' \quad 1] = \begin{bmatrix} \boldsymbol{\Sigma} + \boldsymbol{\mu}\boldsymbol{\mu}' & \\ \boldsymbol{\mu}' & 1 \end{bmatrix} \tag{16.85}$$

$$\mathbf{A} \;=\; \frac{1}{N}\sum_{c=1}^{N}\begin{bmatrix} \mathbf{z}_c \\ 1 \end{bmatrix}[\mathbf{z}'_c \quad 1] = \begin{bmatrix} \mathbf{S} + \bar{\mathbf{z}}\bar{\mathbf{z}}' & \\ \bar{\mathbf{z}}' & 1 \end{bmatrix} \tag{16.86}$$

$$\mathbf{a} \;=\; \mathbf{K}'\mathsf{vec}(\mathbf{A}) \qquad \mathsf{vec}(\mathbf{A}) = \mathbf{Da} \tag{16.87}$$

$$\boldsymbol{\alpha} \;=\; \mathbf{K}'\mathsf{vec}(\boldsymbol{\Upsilon}) \tag{16.88}$$

$$\mathbf{W}_a \;=\; \mathsf{Est}[\mathsf{ACov}(\mathbf{a})] \quad \text{singular} \tag{16.89}$$

$$\mathbf{W}_{a\mathsf{NT}} \;=\; 2\mathbf{K}'(\hat{\boldsymbol{\Upsilon}} \otimes \hat{\boldsymbol{\Upsilon}})\mathbf{K} \tag{16.90}$$

$$\mathbf{W}_{a\mathsf{NNT}} \;=\; \text{defined by (16.83)} \tag{16.91}$$

The fit functions are

$$\mathsf{F} = (\mathbf{a} - \boldsymbol{\alpha})'\mathbf{V}(\mathbf{a} - \boldsymbol{\alpha}) \tag{16.92}$$

with $\mathbf{V}$ defined as in (16.43)–(16.47) but with $\mathbf{A}$ instead of $\mathbf{S}$ and $\hat{\boldsymbol{\Upsilon}}$ instead of $\hat{\boldsymbol{\Sigma}}$.

The same results as in Section 16.1.5 apply but with $\mathbf{W}_{\mathsf{NT}}$ replaced with $\mathbf{W}_{a\mathsf{NT}}$ and $\mathbf{W}_{\mathsf{NNT}}$ replaced with $\mathbf{W}_{a\mathsf{NNT}}$.

16.1.9   Multiple Groups

To generalize the results in Section 16.1.5 to multiple groups we need only consider that the samples from different groups are supposed to be independent. The presentation here follows Satorra (1993).

Suppose there are G groups. The sample size in group g is N_g and the total sample for all groups is $N = N_1 + N_2 + \cdots + N_G$. The fit function in (16.95) is a weighted sum of the fit functions for each group. The asymptotic covariance matrix in (16.96) can be estimated for each group separately under NT and NNT as in Section 16.1.5. The total asymptotic covariance matrix is defined in (16.97) and the $\mathbf{V}$-matrix is defined in (16.99) where each $\mathbf{V}_g$ is N_g/N times the $\mathbf{V}$ in Section 16.1.5.

We use the following definitions

$$\mathbf{s}' = (\mathbf{s}_1', \mathbf{s}_2', \ldots, \mathbf{s}_G') \tag{16.93}$$

$$\boldsymbol{\sigma}' = (\boldsymbol{\sigma}_1', \boldsymbol{\sigma}_2', \ldots, \boldsymbol{\sigma}_G') \tag{16.94}$$

$$\mathsf{F}(\mathbf{s}, \boldsymbol{\sigma}) = \sum \frac{N_g}{N} \mathsf{F}_g(\mathbf{s}_g, \boldsymbol{\sigma}_g) \tag{16.95}$$

$$\mathbf{W}_g = \mathsf{Est}[N_g \mathsf{ACov}(\mathbf{s}_g)] \tag{16.96}$$

$$\mathbf{W} = \begin{bmatrix} \mathbf{W}_1 & & & \mathbf{0} \\ & \mathbf{W}_2 & & \\ & & \ddots & \\ \mathbf{0} & & & \mathbf{W}_G \end{bmatrix} \quad \text{block diagonal} \tag{16.97}$$

$$\mathbf{V}_g = \frac{N_g}{N} \times \mathbf{V} \text{defined in} (16.43) - (16.47) \tag{16.98}$$

different for different methods

$$\mathbf{V} = \begin{bmatrix} \mathbf{V}_1 & & & \mathbf{0} \\ & \mathbf{V}_2 & & \\ & & \ddots & \\ \mathbf{0} & & & \mathbf{V}_G \end{bmatrix} \quad \text{block diagonal} \tag{16.99}$$

$$\tag{16.100}$$

Let $\boldsymbol{\theta}$ be the vector of all independent parameters in all groups and let

$$\boldsymbol{\Delta} = \frac{\partial \boldsymbol{\sigma}}{\partial \boldsymbol{\theta}'} \tag{16.101}$$

$\boldsymbol{\Delta}$ can be partitioned as

$$\boldsymbol{\Delta} = \begin{bmatrix} \dfrac{\partial \boldsymbol{\sigma}_1}{\partial \boldsymbol{\theta}'} \\ \dfrac{\partial \boldsymbol{\sigma}_2}{\partial \boldsymbol{\theta}'} \\ \vdots \\ \dfrac{\partial \boldsymbol{\sigma}_G}{\partial \boldsymbol{\theta}'} \end{bmatrix} = \begin{bmatrix} \boldsymbol{\Delta}_1 \\ \boldsymbol{\Delta}_2 \\ \vdots \\ \boldsymbol{\Delta}_G \end{bmatrix} \tag{16.102}$$

$$\tag{16.103}$$

With these definitions, the same formulas as in Section 16.1.5 apply, but note that

$$\mathbf{E} = \boldsymbol{\Delta}'\mathbf{V}\boldsymbol{\Delta} = \sum_g \boldsymbol{\Delta}'_g\mathbf{V}_g\boldsymbol{\Delta}_g \tag{16.104}$$

$$\boldsymbol{\Delta}'\mathbf{V}\mathbf{W}\mathbf{V}\boldsymbol{\Delta} = \sum_g \boldsymbol{\Delta}'_g\mathbf{V}_g\mathbf{W}_g\mathbf{V}_g\boldsymbol{\Delta}_g \tag{16.105}$$

$$\tag{16.106}$$

The generalization of the results in Section 16.1.7 to multiple groups follows by using a **b** vector instead of **s** in (16.96) and definitions in analogy to Sections 16.1.7.

The generalization of the results in Section 16.1.8 to multiple groups follows by using an **a** vector instead of **s** in (16.95) and definitions in analogy to Sections 16.1.8. For further details, see Satorra (1993).

16.1.10 Maximum Likelihood with Missing Values (FIML)

Let **z** be a random vector of order k with a multivariate normal distribution with mean vector $\boldsymbol{\mu}$ and covariance matrix $\boldsymbol{\Sigma}$ both of which are functions of a parameter vector $\boldsymbol{\theta}$. Suppose we have a sample of N independent observations of **z**. If there are missing data we assume they are missing at random. Let $\mathbf{z}_i$ of order k_i be the subset of **z** observed in unit i. Then

$$\mathbf{z}_i \sim N(\boldsymbol{\mu}_i, \boldsymbol{\Sigma}_i) ,$$

where $\boldsymbol{\mu}_i$ and $\boldsymbol{\Sigma}_i$ consist of the rows and columns of $\boldsymbol{\mu}$ and $\boldsymbol{\Sigma}$ for which there are observations in $\mathbf{z}_i$. The logarithm of the likelihood for the total sample is

$$\ln L = -\frac{1}{2}\sum_{i=1}^{N}\{\ln(2\pi) + \ln|\boldsymbol{\Sigma}_i| + \text{tr}[\boldsymbol{\Sigma}_i^{-1}(\mathbf{z}_i - \boldsymbol{\mu}_i)(\mathbf{z}_i - \boldsymbol{\mu}_i)']\} \tag{16.107}$$

Instead of maximizing $\ln L$, maximum likelihood estimates of $\boldsymbol{\theta}$ can be obtained by minimizing the function

$$F(\boldsymbol{\theta}) = \sum_{i=1}^{N}\{\ln|\boldsymbol{\Sigma}_i| + \text{tr}\boldsymbol{\Sigma}_i^{-1}\mathbf{G}_i\} , \tag{16.108}$$

where

$$\mathbf{G}_i = (\mathbf{z}_i - \boldsymbol{\mu}_i)(\mathbf{z}_i - \boldsymbol{\mu}_i)' . \tag{16.109}$$

The first and second derivatives of $F(\boldsymbol{\theta})$ required for the minimization can be obtained in the usual way. The deviance for the saturated model can be obtained by minimizing

$$\sum_{i=1}^{N}\{\ln|\boldsymbol{\Sigma}_i| + \text{tr}\boldsymbol{\Sigma}_i^{-1}\mathbf{G}_i\} , \tag{16.110}$$

with respect to $\boldsymbol{\mu}$ and $\boldsymbol{\Sigma}$.

16.1.11   Multiple Imputation

EM Algorithm

Suppose $\mathbf{z} = (z_1, z_2, ..., z_p)'$ is a vector of random variables normally distributed with mean $\boldsymbol{\mu}$ and covariance matrix $\boldsymbol{\Sigma}$ and that $\mathbf{z}_1, \mathbf{z}_2, ..., \mathbf{z}_N$ is a random sample from $\mathbf{z}$. For each $i = 1, 2, ..., N$, partition $\mathbf{z}_i'$ as $(\mathbf{z}_{imiss}' \mid \mathbf{z}_{iobs}')$, where $\mathbf{z}_{imiss}'$ consists of those values of $\mathbf{z}_i$ that are missing and $\mathbf{z}_{iobs}'$ consists of those values of $\mathbf{z}_i$ that are observed. If all values of $\mathbf{z}_i$ are missing, one can choose to delete this case or replace it with mean values.

Step 0: Initial Estimates

Obtain initial estimates $\hat{\boldsymbol{\mu}}$ and $\hat{\boldsymbol{\Sigma}}$ of $\boldsymbol{\mu}$ and $\boldsymbol{\Sigma}$ using listwise or pairwise deletion.

Step 1: (E-Step)

For $i = 1, 2, \ldots, N$, replace $\mathbf{z}_{imiss}$ by $E(\mathbf{z}_{imiss} \mid \mathbf{z}_{iobs}; \hat{\boldsymbol{\mu}}, \hat{\boldsymbol{\Sigma}})$, see (16.8). This gives a complete data set.

Step 2: (M-Step)

Use the complete data set from Step 1 to obtain an update of $\boldsymbol{\mu}$ and $\boldsymbol{\Sigma}$. Repeat Steps 1 and 2 until convergence

MCMC Algorithm

The estimates of $\boldsymbol{\mu}$ and $\boldsymbol{\Sigma}$ obtained from the EM-algorithm are used as initial parameters of the distributions used in Step 1 of the MCMC procedure

Step 1: (P-Step)

Simulate an estimate $\boldsymbol{\mu}_k$ of $\boldsymbol{\mu}$ and an estimate $\boldsymbol{\Sigma}_k$ of $\boldsymbol{\Sigma}$ from a multivariate normal and an inverted Wishart distribution respectively.

Step 2: (I-Step)

Simulate $\mathbf{z}_{imiss} \mid \mathbf{z}_{iobs}$, $i = 1, 2, ..., N$ from conditional normal distributions with parameters based on $\boldsymbol{\mu}_k$ and $\boldsymbol{\Sigma}_k$.

Replace the missing values with simulated values and calculate $\boldsymbol{\mu}_{k+1} = \bar{\mathbf{z}}$ and $\boldsymbol{\Sigma}_{k+1} = \mathbf{S}$ where $\bar{\mathbf{z}}$ and $\mathbf{S}$ are the sample mean vector and covariance matrix of the completed data set respectively. Repeat Steps 1 and 2 m times. In LISREL, missing values in row i are replaced by the average of the simulated values over the m draws, after an initial burn-in period.

For further details, see Schafer (1997).

16.2   Ordinal Variables

A model for p ordinal manifest variables $\mathbf{x}' = (x_1, x_2, \ldots, x_p)$ and k continuous latent variables $\boldsymbol{\xi}' = (\xi_1, \xi_2, \ldots, \xi_k)$ may be formulated as follows, see Jöreskog and Moustaki (2001) and references given there. Let x_i be an ordinal variable with $m_i \geq 2$ ordered categories $c_i = 1, 2, \ldots, m_i$. The models are defined by the cumulative response function

$$Pr\{x_i \leq c_i \mid \boldsymbol{\xi}\} = F(\alpha_{c_i}^{(i)} + \sum_{j=1}^{k} \beta_{ij}\xi_j) , \tag{16.111}$$

where F is a distribution function and

$$-\infty = \alpha_0^{(i)} < \alpha_1^{(i)} < \alpha_2^{(i)} \cdots < \alpha_{m_i-1}^{(i)} < \alpha_{m_i}^{(i)} = \infty \ .$$

In factor analysis terminology the $\alpha_s^{(i)}$ are threshold parameters and the β_{ij} are factor loadings.

In principle, $F(t)$ can be any distribution function defined for $-\infty < t < +\infty$. The choices of F in LISREL are

$$\text{PROBIT}: \ F(t) = \Phi(t) = \int_{-\infty}^{t} \frac{1}{\sqrt{2\pi}} e^{-\frac{1}{2}u^2} du \ , \tag{16.112}$$

$$\text{LOGIT}: \ F(t) = \Psi(t) = \frac{e^t}{1 + e^t} \tag{16.113}$$

$$\text{LOGLOG} \ F(t) = e^{-e^{-t}} \tag{16.114}$$

$$\text{CLL}: \ F(t) = 1 - e^{-e^t} \tag{16.115}$$

There are $M = \prod_i^p m_i$ possible response patterns. The model defines the probability of all these response patterns. Let $\mathbf{x}_r = (x_1 = c_1, x_2 = c_2, \ldots, x_p = c_p)$ be a general response pattern, where $c_i = 1, 2, \ldots, m_i$ and let

$$\pi_{c_i}^{(i)}(\boldsymbol{\xi}) = F(\alpha_{c_i}^{(i)} - \sum_{j=1}^{k} \beta_{ij}\xi_j) - F(\alpha_{c_i-1}^{(i)} - \sum_{j=1}^{k} \beta_{ij}\xi_j) \tag{16.116}$$

be the conditional probability that variable x_i falls in the ordered category c_i. Then the conditional probability of $\mathbf{x}_r$ is

$$\pi_r(\boldsymbol{\xi}) = \prod_{i=1}^{p} \pi_{c_i}^{(i)}(\boldsymbol{\xi}) = \prod_{i=1}^{p} [F(\alpha_{c_i}^{(i)} - \sum_{j=1}^{k} \beta_{ij}\xi_j) - F(\alpha_{c_i-1}^{(i)} - \sum_{j=1}^{k} \beta_{ij}\xi_j)] \ . \tag{16.117}$$

Let $h(\boldsymbol{\xi})$ be the density of $\boldsymbol{\xi}$. This is usually taken to be multivariate normal with means zero and variances one so that $\boldsymbol{\xi} \sim N(\mathbf{0}, \mathbf{P})$, where $\mathbf{P}$ is a correlation matrix. For exploratory factor analysis we take $\mathbf{P} = \mathbf{I}$, so that $h(\boldsymbol{\xi}) = \prod_{j=1}^{k} \phi(\xi_j)$. The unconditional probability π_r is

$$\pi_r(\boldsymbol{\theta}) = \int_{-\infty}^{+\infty} \pi_r(\boldsymbol{\xi}) h(\boldsymbol{\xi}) d\boldsymbol{\xi} \ , \tag{16.118}$$

where the integral is a k-dimensional multiple integral which can evaluated numerically by adaptive quadrature.

The probability π_r is a function of the parameter vector $\boldsymbol{\theta}$ consisting of all the intercepts $\alpha_{c_i}^{(i)}$, all the off-diagonal elements of $\mathbf{P}$, and all the factor loadings β_{ij}. Some of the β_{ij} may be specified to be zero.

16.2.1 Estimation by FIML

Let n_r be the frequency of occurrence of the response pattern $\mathbf{x}_r$ and let $p_r = n_r/N$, where N is the sample size. The corresponding probability π_r is defined by (16.118). The logarithm of the likelihood function is

$$\ln L = \sum_r n_r \ln \pi_r(\boldsymbol{\theta}) = N \sum_r p_r \ln \pi_r(\boldsymbol{\theta}) \ , \tag{16.119}$$

where the sum runs over all response patterns occurring in the sample, *i.e.*, over all r with $n_r > 0$.

Standardized and Unstandardized Parameters

The parameters $\alpha_{c_i}^{(i)}$ and β_{ij} are *unstandardized parameters*. These are the parameters that are iterated in the algorithm that maximizes the log-likelihood. After convergence the FIML estimates are denoted $\hat{\alpha}_{c_i}^{(i)}$, $\hat{\mathbf{P}}$, and $\hat{\beta}_{ij}$ For purposes of interpretation it may be more convenient to consider the *standardized parameters* $\hat{\tau}_{c_i}^{(i)}$ and $\hat{\lambda}_{ij}$ defined by

$$\hat{\tau}_a^{(i)} = \hat{\alpha}_a^{(i)}/(1 + \sum_{j=1}^{k} \hat{\beta}_{ij}^2)^{-\frac{1}{2}} \,, \tag{16.120}$$

$$\hat{\lambda}_{ij} = \hat{\beta}_{ij}/(1 + \sum_{j=1}^{k} \hat{\beta}_{ij}^2)^{-\frac{1}{2}} \,. \tag{16.121}$$

The standardized and unstandardized parameters are in a one-to-one correspondence so one can obtain the unstandardized parameters from the standardized parameters by the formulas:

$$\hat{\alpha}_a^{(i)} = \hat{\tau}_a^{(i)}/(1 - \sum_{j=1}^{k} \hat{\lambda}_{ij}^2)^{-\frac{1}{2}} \,, \tag{16.122}$$

$$\hat{\beta}_{ij} = \hat{\lambda}_{ij}/(1 - \sum_{j=1}^{k} \hat{\lambda}_{ij}^2)^{-\frac{1}{2}} \,. \tag{16.123}$$

Using the Jacobian of the transformation one can obtain an estimate of the asymptotic covariance matrix of the standardized parameters from the estimate of the asymptotic covariance matrix of the unstandardized parameters. The standard errors of parameter estimates that appear in the **LISREL** output are the square roots of the diagonal elements of these asymptotic covariance matrices.

Model Test

Let

$$F(\boldsymbol{\theta}) = \sum_r p_r \ln(p_r/\hat{\pi}_r) \,. \tag{16.124}$$

This is a function of $\boldsymbol{\theta}$ consisting of all the thresholds $\alpha_a^{(i)}$ and all the factor loadings β_{ij}.

If $F(\boldsymbol{\theta})$ has been minimized with respect to $\boldsymbol{\theta}$, one can use the likelihood ratio (LR) test statistic

$$\chi_{\text{LR}}^2 = 2 \sum_r n_r \ln(p_r/\hat{\pi}_r) = 2N \sum_r p_r \ln(p_r/\hat{\pi}_r) = 2NF(\hat{\boldsymbol{\theta}}) \,, \tag{16.125}$$

to test the model, where $\hat{\boldsymbol{\theta}}$ is the estimated parameter vector and $\hat{\pi}_r = \pi_r(\hat{\boldsymbol{\theta}})$. Hence, this χ^2 is $2N$ times the minimum value of the fit function (16.124). If the model holds, this is distributed approximately as χ^2 with degrees of freedom equal to the number of different response patterns minus one minus the number of independent elements of $\boldsymbol{\theta}$.

Alternatively, one can use the goodness-of-fit (GF) test statistic

$$\chi_{\text{GF}}^2 = \sum_r [(n_r - N\hat{\pi}_r)^2/(N\hat{\pi}_r)] = N \sum_r (p_r - \hat{\pi}_r)^2/\hat{\pi}_r \,. \tag{16.126}$$

If the model holds, both statistics (16.125) and (16.126) have the same asymptotic distribution under H_0.

16.2.2 Estimation via Polychorics

Another approach, more commonly used, is based on underlying latent variables, see *i.e.*, Muthen (1984), Lee, Poon, and Bentler (1990), and Jöreskog (1990). It is assumed that there is a continuous variable $x_i^\star$ *underlying* the ordinal variable x_i. This continuous variable $x_i^\star$ represents the attitude underlying the ordered responses to x_i and is assumed to have a range from $-\infty$ to $+\infty$. It is the underlying variable $x_i^\star$ that is assumed to have a **LISREL** model.

The underlying variable $x_i^\star$ is unobservable. Only the ordinal variable x_i is observed. For an ordinal variable x_i with m_i categories, the connection between the ordinal variable x_i and the underlying variable $x_i^\star$ is

$$x_i = c_i \quad \Longleftrightarrow \quad \tau_{c_i-1}^{(i)} < x_i^\star < \tau_{c_i}^{(i)}, \ c_i = 1, 2, \ldots, m_i , \tag{16.127}$$

where

$$\tau_0^{(i)} = -\infty , \ \tau_1^{(i)} < \tau_2^{(i)} < \ldots < \tau_{m_i-1}^{(i)} , \ \tau_{m_i}^{(i)} = +\infty , \tag{16.128}$$

are threshold parameters. For variable x_i with m_i categories, there are $m_i - 1$ strictly increasing threshold parameters $\tau_1^{(i)}, \tau_2^{(i)}, \ldots, \tau_{m_i-1}^{(i)}$.

Because only ordinal information is available about x_i, the distribution of $x_i^\star$ is determined only up to a monotonic transformation. It is convenient to let $x_i^\star$ have the standard normal distribution with density function $\phi(.)$ and distribution function $\Phi(.)$. Then the probability $\pi_{c_i}^{(i)}$ of a response in category c_i on variable x_i, is

$$\pi_{c_i}^{(i)} = Pr[x_i = c_i] = Pr[\tau_{c_i-1}^{(i)} < x_i^\star < \tau_{c_i}^{(i)}] = \int_{\tau_{c_i-1}^{(i)}}^{\tau_c^{(i)}} \phi(u)du = \Phi(\tau_{c_i}^{(i)}) - \Phi(\tau_{c_i-1}^{(i)}) , \tag{16.129}$$

for $c_i = 1, 2, \ldots, m_{i-1}$, so that

$$\tau_{c_i}^{(i)} = \Phi^{-1}(\pi_1^{(i)} + \pi_2^{(i)} + \cdots + \pi_{c_i}^{(i)}) , \tag{16.130}$$

where Φ^{-1} is the inverse of the standard normal distribution function. The quantity $(\pi_1^{(i)} + \pi_2^{(i)} + \cdots + \pi_{c_i}^{(i)})$ is the probability of a response in category c_i or lower.

The probabilities $\pi_c^{(i)}$ are unknown population quantities. In practice, $\pi_c^{(i)}$ can be estimated consistently by the corresponding percentage $p_c^{(i)}$ of responses in category c on variable x_i. Then, estimates of the thresholds can be obtained as

$$\hat{\tau}_{c_i}^{(i)} = \Phi^{-1}(p_1^{(i)} + p_2^{(i)} + \cdots + p_{c_i}^{(i)}), \ c_i = 1, \ldots, m_i - 1 . \tag{16.131}$$

The quantity $(p_1^{(i)} + p_2^{(i)} + \cdots + p_{c_i}^{(i)})$ is the proportion of cases in the sample responding in category c_i or lower on variable x_i.

Let x_i and x_j be two ordinal variables with m_i and m_j categories, respectively. Their marginal distribution in the sample is represented by a contingency table

$$\begin{pmatrix} n_{11}^{(ij)} & n_{12}^{(ij)} & \cdots & n_{1m_j}^{(ij)} \\ n_{21}^{(ij)} & n_{22}^{(ij)} & \cdots & n_{2m_j}^{(ij)} \\ \vdots & \vdots & \vdots \vdots \vdots & \vdots \\ n_{m_i1}^{(ij)} & n_{m_i2}^{(ij)} & \cdots & n_{m_im_j}^{(ij)} \end{pmatrix} , \tag{16.132}$$

where $n_{ab}^{(ij)}$ is the number of cases in the sample in category a on variable x_i and in category b on variable x_j. The underlying variables $x_i^\star$ and $x_j^\star$ are assumed to be bivariate normal with zero means, unit variances, and with correlation ρ_{ij}, the polychoric correlation.

Let $\tau_1^{(i)}, \tau_2^{(i)}, \ldots, \tau_{m_i-1}^{(i)}$ be the thresholds for variable $x_i^\star$ and let $\tau_1^{(j)}, \tau_2^{(j)}, \ldots, \tau_{m_j-1}^{(j)}$ be the thresholds for variable $x_j^\star$.

The polychoric correlation can be estimated by maximizing the log-likelihood of the multinomial distribution, see Olsson (1979)

$$\ln L = \sum_{a=1}^{m_i} \sum_{b=1}^{m_j} n_{ab}^{(ij)} log\pi_{ab}^{(ij)} , \tag{16.133}$$

where

$$\pi_{ab}^{(ij)} = Pr[x_i = a, x_j = b] = \int_{\tau_{a-1}^{(i)}}^{\tau_a^{(i)}} \int_{\tau_{b-1}^{(j)}}^{\tau_b^{(j)}} \phi_2(u, v) du dv , \tag{16.134}$$

and

$$\phi_2(u, v) = \frac{1}{2\pi\sqrt{1-\rho^2}} e^{-\frac{1}{2(1-\rho^2)}(u^2 - 2\rho uv + v^2)} , \tag{16.135}$$

is the standard bivariate normal density with correlation ρ_{ij}. Maximizing $\ln L$ gives the sample polychoric correlation denoted r_{ij}.

The polychoric correlation can be estimated by a two-step procedure, see Olsson (1979). In the first step, the thresholds are estimated from the univariate marginal distributions by (16.131). In the second step, the polychoric correlations are estimated from the bivariate marginal distributions by maximizing $\ln L$ for given thresholds. The parameters can also be estimated by a one-step procedure which maximizes $\ln L$ with respect to the thresholds and the polychoric correlation simultaneously but this is not necessary because the estimates are almost the same as with the two-step procedure and it is not practical because it would yield different thresholds for the same variable when paired with different variables.

Jöreskog (1994) showed that the polychoric correlation r_{ij} is asymptotically linear in the bivariate marginal proportions $\mathbf{P}_{ij}$, where $\mathbf{P}_{ij}$ is a matrix of order $m_i \times m_j$ whose elements are $p_{ab}^{(ij)} = n_{ab}^{(ij)}/N$, where N is the sample size. Thus, $r_{ij} \simeq tr(\mathbf{\Gamma}_{ij}' \mathbf{P}_{ij})$. The elements of the matrix $\mathbf{\Gamma}_{ij}$ are given in Jöreskog (1994), equation (16). Using this result one can estimate the asymptotic covariance $NACov(r_{gh}, r_{ij})$ for all $g \neq h$ and $i \neq j$, see Jöreskog (1994) for details.

Since the models in LISREL are mean and covariance structures $\boldsymbol{\mu}(\boldsymbol{\theta})$ and $\boldsymbol{\Sigma}(\boldsymbol{\theta})$, see Section 16.1.7, it is better to work with an alternative parameterization.

Recall that the variable $x^\star$ underlying the ordinal variable x (x may be an x or y variable in the LISREL model) is determined only up to a monotonic transformation. If we want to retain normality of the underlying variable, the transformation must be linear. In principle, one can make an arbitrary linear transformation of the underlying variable. If the number of categories $m \geq 3$, one such transformation is obtained by specifying that $\tau_1 = 0$ and $\tau_2 = 1$. Then the mean μ and standard deviation σ of $x^\star$ can be defined instead.

The parameters of the two parameterizations are given in the following table.

Parameterization	Mean	St.Dev.	Thresholds				
Standard	0	1	τ_1	τ_2	τ_3	$\ldots$	τ_{m-1}
Alternative	$\mu^\star$	$\sigma^\star$	0	1	$\tau_3^\star$	$\ldots$	$\tau_{m-1}^\star$

where

$$\mu^\star = -\tau_1/(\tau_2 - \tau_1), \qquad \sigma^\star = 1/(\tau_2 - \tau_1),$$
$$\tau_i^\star = (\tau_i - \tau_1)/(\tau_2 - \tau_1), \qquad i = 3, 4, \ldots, m - 1.$$

It should be emphasized that the two parameterizations are equivalent in the sense that there is a one-to-one correspondence between the two sets of parameters.

For the Alternative Parameterization to be meaningful there must be at least three categories. If there are only two categories there is only one threshold and it is impossible to estimate both the mean and the standard deviation of the underlying variable. In this case, LISREL will fix the threshold at 0, the standard deviation at 1 and estimate the mean.

Let $\bar{\mathbf{x}}$ and $\mathbf{S}$ be the mean vector and covariance matrix obtained from the alternative parameterization. The matrix $\mathbf{S}$ may be called the polychoric covariance matrix.

To estimate the model, one can use any of the fit functions defined in Section 16.1.7. Based on the results of Yang-Wallentin, Jöreskog, and Luo (2010) or Forero, Maydeu-Olivares, and Gallardo-Pujol (2009), the recommended methods are *Unweighted least squares* (ULS), and *Diagonally weighted least squares* (DWLS).

One can also evaluate the fit of the model to the univariate and bivariate marginals in the sample. The files **BIVFITS.POM** or **BIVFITS.NOR** give the contributions associated with each univariate bivariate margins.

Consider bivariate probabilities, let g and h be two pairs of variables and let a index the categories of variable g and b index the categories of variable h. Since the variables are conditionally independent for given $\boldsymbol{\xi}$, they are also pairwise conditionally independent. Thus, the conditional probability of $(x_g = a,\ x_h = b)$ is

$$\pi_{ab}^{(gh)}(\boldsymbol{\xi}) = [F(\alpha_a^{(g)} - \sum_{j=1}^{k} \beta_{gj}\xi_j) - F(\alpha_{a-1}^{(g)} - \sum_{j=1}^{k} \beta_{gj}\xi_j)][F(\alpha_b^{(h)} - \sum_{j=1}^{k} \beta_{hj}\xi_j) - F(\alpha_{b-1}^{(h)} - \sum_{j=1}^{k} \beta_{hj}\xi_j)] \tag{16.136}$$

The unconditional probability is

$$\pi_{ab}^{(gh)}(\boldsymbol{\theta}) = \int_{-\infty}^{+\infty} \pi_{ab}^{(gh)}(\boldsymbol{\xi})h(\boldsymbol{\xi})d\boldsymbol{\xi} \tag{16.137}$$

Let $n_{ab}^{(gh)}$ be the number of cases with $x_g = a$ and $x_h = b$ and let $p_{ab}^{(gh)} = n_{ab}^{(gh)}/N$ be the corresponding proportion. To obtain a measure of bivariate fit for the pair of variables g and h one can compute the LR statistic

$$F_{LR}^{(gh)}(\hat{\boldsymbol{\theta}}) = 2N \sum_{a=1}^{m_g} \sum_{b=1}^{m_h} p_{ab}^{(gh)} \ln[p_{ab}^{(gh)}/\hat{\pi}_{ab}^{(gh)}] \tag{16.138}$$

The corresponding GF statistic is

$$F_{GF}^{(gh)}(\hat{\boldsymbol{\theta}}) = N \sum_{a=1}^{m_g} \sum_{b=1}^{m_h} (p_{ab}^{(gh)} - \hat{\pi}_{ab}^{(gh)})^2/\hat{\pi}_{ab}^{(gh)} \tag{16.139}$$

To get an overall measure of bivariate fit one can sum these statistics over all pairs of variables.

It should be pointed out that the bivariate statistics in (16.138) and (16.139) do not have asymptotic chi-square distributions because the bivariate likelihoods have not been maximized. Neither do their sums have asymptotic chi-square distributions. Nevertheless, the statistics are useful as measures of fit.

Chapter 17

Appendix G: Iteration Algorithms

17.1   General Definitions

Since many of the methods in **LISREL** are based on the maximum likelihood method, one can build on general theory for maximizing likelihood functions.

Let $\mathbf{y}_1, \mathbf{y}_2, \ldots, \mathbf{y}_N$ be independently distributed with density function $f(\mathbf{y}_i, \boldsymbol{\theta})$ depending on a parameter vector $\boldsymbol{\theta} = (\theta_1, \theta_2, \ldots, \theta_t)$. The likelihood function is

$$L(\mathbf{Y}, \boldsymbol{\theta}) = \prod_{i=1}^{n} f(\mathbf{y}_i, \boldsymbol{\theta}) . \tag{17.1}$$

In most cases it is convenient to work with $\ln L$ instead of L:

$$F(\mathbf{Y}, \boldsymbol{\theta}) = \ln L(\mathbf{Y}, \boldsymbol{\theta}) = \sum_{i=1}^{N} \ln f(\mathbf{y}_i, \boldsymbol{\theta}) \tag{17.2}$$

The maximum likelihood estimator $\hat{\boldsymbol{\theta}}$ is defined such that

$$F(\mathbf{Y}, \hat{\boldsymbol{\theta}}) \geq F(\mathbf{Y}, \boldsymbol{\theta}) \tag{17.3}$$

for all $\boldsymbol{\theta}$.

To determine the maximum likelihood estimate for a given data set $\mathbf{Y}$, some iterative procedure is usually applied. Let

$$\mathbf{g}(\mathbf{Y}, \boldsymbol{\theta}) = \partial F(\mathbf{Y}, \boldsymbol{\theta})/\partial \boldsymbol{\theta} , \tag{17.4}$$

be the gradient vector. The Hessian $\mathbf{H}$ and the Information Matrix $\mathbf{E}$ are defined as

$$\mathbf{H}(\mathbf{Y}, \boldsymbol{\theta}) = \partial^2 F(\mathbf{Y}, \boldsymbol{\theta})/\partial \boldsymbol{\theta} \partial \boldsymbol{\theta}' , \tag{17.5}$$

and

$$\mathbf{E}(\boldsymbol{\theta}) = E(\mathbf{g}\mathbf{g}') = -E(\partial^2 F(\mathbf{Y}, \boldsymbol{\theta})/\partial \boldsymbol{\theta} \partial \boldsymbol{\theta}') . \tag{17.6}$$

In **LISREL** the fit function is defined such that it should be minimized instead of maximized. For example, the ML fit function is the negative of the logarithm of a likelihood function.

The iterative algorithm defines successive points in the parameter space $\boldsymbol{\theta}_0, \boldsymbol{\theta}_1, \boldsymbol{\theta}_2, \ldots$ such that

$$F(\boldsymbol{\theta}_{s+1}) < F(\boldsymbol{\theta}_s) , \tag{17.7}$$

© Springer International Publishing Switzerland 2016
K.G. Jöreskog et al., *Multivariate Analysis with LISREL*, Springer Series in Statistics,
DOI 10.1007/978-3-319-33153-9_17

using the iteration formula

$$\boldsymbol{\theta}_{s+1} = \boldsymbol{\theta}_s - \mathbf{M}_s \mathbf{g}_s , \tag{17.8}$$

where

$$\mathbf{g}_s = (\partial F / \partial \boldsymbol{\theta})_{\boldsymbol{\theta} = \boldsymbol{\theta}_s} , \tag{17.9}$$

and where

$$\mathbf{M}_s = \mathbf{E}_s^{-1} , \tag{17.10}$$

for the Fisher scoring algorithm and

$$\mathbf{M}_s = -\mathbf{H}_s^{-1} , \tag{17.11}$$

for the Newton-Raphson algorithm, where $\mathbf{H}_s$ and $\mathbf{E}_s$ are the Hessian and Information matrix evaluated at $\boldsymbol{\theta}_s$, respectively.

In practice the Newton-Raphson algorithm can be exceedingly time consuming because the Hessian and its inverse is often complicated to compute. For small problems the Fisher Scoring algorithm may work well. But for large models with many parameters and for complex models where it is difficult to determine good starting values the Fisher Scoring algorithm can be exceedingly time consuming because the information matrix $\mathbf{E}$ must be evaluated and inverted in each iteration. Although the Fisher Scoring algorithm may be specified in **LISREL** the Davidon-Fletcher-Powell (DFP) procedure (see Fletcher & Powell, 1963) is used by default. This updates $\mathbf{E}$ in each iteration using a simple formula. Although this procedure may require more iterations than the Fisher Scoring algorithm, the total time required to reach the solution is often much less. The DFP algorithm described in details in the following sections has been found to work well with models having over hundred parameters.

The *information matrix* is defined as

$$\mathbf{E} = plim \, \partial^2 F / \partial \boldsymbol{\theta} \partial \boldsymbol{\theta}' . \tag{17.12}$$

This corresponds to Fisher's information matrix (see Silvey, 1970, p.41). We will use the term information matrix also for the other methods. If the model is identified, i.e., if all its parameters are identified, the information matrix is positive definite. This is a mathematical statement. In practice, the information matrix must be evaluated at a point in the parameter space estimated from the data, and the positive definiteness of $\mathbf{E}$ can only be assessed within the numerical accuracy by which computations are performed. In the program, $\mathbf{E}$ is inverted using an ingenious variant of the square root method due to Dickman and Kaiser (1961). In this method successive pivotal quantities $|\mathbf{E}_{11}|, |\mathbf{E}_{22}|, |\mathbf{E}_{33}|, \ldots, |\mathbf{E}_{tt}|$ are computed, where $\mathbf{E}_{ii}$ is the submatrix of $\mathbf{E}$ formed by the first i rows and columns and $|\mathbf{E}_{ii}|$ is the determinant of $\mathbf{E}_{ii}$. If $\mathbf{E}$ is positive definite all the pivotal quantities are positive. In the program, the parameter IC, with default value 5×10^{-11}, has the following function. If $|\mathbf{E}_{i-1,i-1}| \geq$ IC and $|\mathbf{E}_{ii}| <$ IC, then the matrix $\mathbf{E}_{i-1,i-1}$ is considered positive definite, $\mathbf{E}_{ii}$ is considered singular, and the program prints a message suggesting that the parameter θ_i may not be identified. This is usually an indication that θ_i is involved in an indeterminacy with one or more of the parameters $\theta_1, \theta_2, \ldots, \theta_{i-1}$ and it may well be that some of the parameters involved in this indeterminacy are not identified.

17.2   Technical Parameters

Several technical parameters control the *minimization algorithm*. The default values for these parameters have been chosen after considerable experimentation so as to optimize the algorithm. The technical parameters and their default values are

Parameter	Default Value
IT	$3 \times$ number of parameters
EPS	0.000001
IM	2
IS	2
IC	5×10^{-11}
MT	20

To describe the function of each of these technical parameters we will briefly describe the minimization algorithm. Let $\boldsymbol{\theta}(t \times 1)$ be the vector of independent model parameters to be estimated and let $F(\boldsymbol{\theta})$ be a general function to be minimized. It is assumed that $F(\boldsymbol{\theta})$ is continuous and twice differentiable. Let $\boldsymbol{\theta}^{(0)}$ represent initial estimates. The minimization algorithm generates successive points $\boldsymbol{\theta}^{(1)}$, $\boldsymbol{\theta}^{(2)}$, ..., in the parameter space such that

$$F[\boldsymbol{\theta}_{s+1}] < F[\boldsymbol{\theta}_s] .$$

The process ends when the convergence criterion (defined in Section 17.4) is satisfied or when s = IT, whichever occurs first. For s = 0, 1, 2,..., let $\mathbf{g}_s$ be the gradient vector at $\boldsymbol{\theta}=\boldsymbol{\theta}_s$, let α_s be a sequence of positive scalars, and let $\mathbf{E}_s$ be a sequence of positive definite matrices. Then the minimization algorithm is

$$\boldsymbol{\theta}_{s+1} = \boldsymbol{\theta}_s - \alpha_s \mathbf{E}_s \mathbf{g}_s . \tag{17.13}$$

The parameter IS controls the choice of α_s and the parameter IM controls the choice of $\mathbf{E}_s$ as follows:

IS = 1 α_s = 1 if $F[\boldsymbol{\theta}_{s+1}] < F[\boldsymbol{\theta}_s]$, otherwise α_s is determined by *line search*; see Section 17.5.

IS = 2 α_s is determined by line search; see Section 17.5.

IM = 1 $\mathbf{E}_{s+1} = \mathbf{E}_s$, i.e., $\mathbf{E}_s$ is held constant.

IM = 2 $\mathbf{E}_{s+1}$ is determined from $\mathbf{E}_s$ by the method of Davidon-Fletcher-Powell described in the next section.

IM = 3 $\mathbf{E}_{s+1}$ = The inverse of the information matrix (defined below) evaluated at $\boldsymbol{\theta} = \boldsymbol{\theta}_{s+1}$.

In all three cases of IM, $\mathbf{E}^{(0)}$ equals the inverse of the *information matrix* evaluated at $\boldsymbol{\theta}= \boldsymbol{\theta}^{(0)}$.

Combinations of IS and IM give rise to six alternative minimization algorithms that can be used. The combination IS=2 and IM=2 is the Davidon-Fletcher-Powell algorithm. This is used by default. The combination IS=1 and IM=3 is the Fisher's Scoring algorithm. This gives the fewest iterations but takes more time in each iteration due to the computation and inversion of the information matrix. The combination IM=1 with IS=1 takes the least time per iteration but usually requires more iterations. Our experience is that no combination is optimal for all problems. The behavior of the iterative procedure depends strongly on how good or bad the initial estimates are and on how well-behaved the function is.

17.3   The Davidon-Fletcher-Powell Method

Let

$$\mathbf{d}_s = -\mathbf{E}_s\mathbf{g}_s \ . \tag{17.14}$$

Then $\mathbf{d}_s$ is the direction of search from $\boldsymbol{\theta}_s$. Determine λ_s such that $F(\boldsymbol{\theta}_s + \lambda_s\mathbf{d}_s)$ is minimum, see Section 17.5. Let

$$\boldsymbol{\theta}_{s+1} = \boldsymbol{\theta}_s + \lambda_s\mathbf{d}_s \ , \tag{17.15}$$

and note that $\mathbf{d}_s'\mathbf{g}_{s+1} = 0$. Let

$$\mathbf{c}_s = \mathbf{g}_{s+1} - \mathbf{g}_s \ , \tag{17.16}$$

and

$$\mathbf{e}_s = \mathbf{E}_s\mathbf{c}_s \ . \tag{17.17}$$

Then

$$\mathbf{E}_{s+1} = \mathbf{E}_s + \lambda_s^2\mathbf{d}_s\mathbf{d}_s' - \mathbf{e}_s\mathbf{e}_s'/\mathbf{e}_s'\mathbf{E}_s\mathbf{e}_s \ . \tag{17.18}$$

Thus $\mathbf{E}_{s+1}$ can be computed rapidly. It can be shown that $\mathbf{E}_{s+1}$ is positive definite if $\mathbf{E}_s$ is positive definite and that $\mathbf{E}_s$ converges to an approximation of the information matrix evaluated at the minimum.

17.4   Convergence Criterion

The convergence criterion is satisfied if for all $i = 1, 2, \ldots, t$

$$|\partial F/\partial \theta_i| < \ \text{EPS if } |\theta_i| \leq 1 \tag{17.19}$$

and

$$|(\partial F/\partial \theta_i)/\theta_i| < \ \text{EPS if } |\theta_i| > 1 \ , \tag{17.20}$$

where the bars indicate absolute values. The default value, 0.000005, of EPS has been chosen so that the solution is usually accurate to three significant digits. However, this cannot be guaranteed to hold for all problems. If a less accurate solution is sufficient, EPS may be increased and if a more accurate solution is required, EPS should be decreased. This does not necessarily mean, however, that EPS = 0.00005 gives a solution with two correct digits, nor that EPS = 0.0000005 gives a solution correct to four digits.

17.5   Line Search

Let $\boldsymbol{\theta}$ be a given point in the parameter space and let $\mathbf{g}$ and $\mathbf{E}$ be the corresponding gradient vector and positive definite weight matrix, respectively. In this section we consider the problem of minimizing $F(\boldsymbol{\theta})$ along the line

$$\boldsymbol{\theta} - \alpha\mathbf{E}\mathbf{g}, \ \ \alpha \geq 0 \ .$$

Along this line the function $F(\boldsymbol{\theta})$ may be regarded as a function $f(\alpha)$, of the distance α from the point $\boldsymbol{\theta}$, i.e.,

$$f(\alpha) = F(\boldsymbol{\theta} - \alpha\mathbf{E}\mathbf{g}), \ \ \alpha \geq 0 \ . \tag{17.21}$$

The slope of $f(\alpha)$ at any point α is given by

$$s(\alpha) = -\mathbf{g}'\mathbf{E}\mathbf{g}_\alpha \, , \tag{17.22}$$

where $\mathbf{g}_\alpha$ is the gradient vector of $F(\boldsymbol{\theta})$ at $\boldsymbol{\theta} - \alpha\mathbf{E}\mathbf{g}$. In particular, the slope at $\alpha = 0$ is $s(0) = -\mathbf{g}'\mathbf{E}\mathbf{g}$, which is negative unless $\mathbf{g} = \mathbf{0}$. If $\mathbf{g} = \mathbf{0}$, the minimum of $F(\boldsymbol{\theta})$ is located at $\boldsymbol{\theta}$. If $s(0) < 0$, $f(\alpha)$ has a minimum for some $\alpha > 0$, since $F(\boldsymbol{\theta})$ is continuous and non-negative. Figure 17.1 shows a typical example of $f(\alpha)$ and $s(\alpha)$. In most cases the fit function $F(\boldsymbol{\theta})$ is convex at least in a region around the minimum, so that $f(\alpha)$ will be as in Figure 17.1 . However, convexity may not hold for all data and models at all points in the parameter space. In fact, situations like Figure 17.2 may occur. The procedure to be described is capable of handling such cases as well.

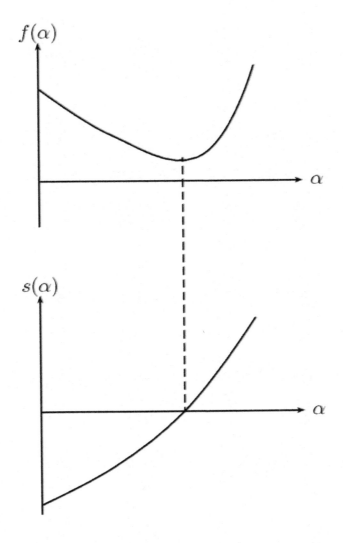

Figure 17.1: A Typical Example of $f(\alpha)$ and $s(\alpha)$

The minimizing α may be approximated by various interpolation and extrapolation procedures. For example, one takes a trial value α^* of α and determines $f(\alpha^*)$ and $s(\alpha^*)$. If $s(\alpha^*)$ is positive,

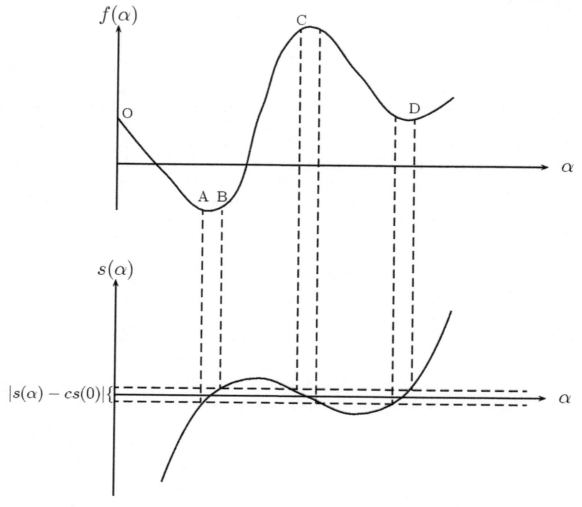

Figure 17.2: Various Regions of $f(\alpha)$ and $s(\alpha)$ Curves

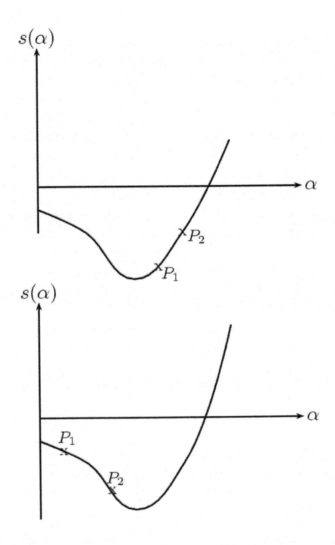

Figure 17.3: Two Slope Curves Showing Different Locations of P_1 and P_2

one interpolates cubically for the minimum, using function values and slope values at $\alpha = 0$ and $\alpha = \alpha^*$. If $s(\alpha^*)$ is negative, one extrapolates linearly for the zero of $s(\alpha)$ using only slope values at $\alpha = 0$ and $\alpha = \alpha^*$. Although this procedure is satisfactory in most cases, a more complicated procedure is necessary if a very accurate determination of the minimum is required or if $f(\alpha)$ and $s(\alpha)$ have more irregular forms than those of Figure 17.1. The following procedure is capable of locating the minimum to any desired degree of accuracy, within machine capacity, and it can also deal with various irregular shapes of the curves $f(\alpha)$ and $s(\alpha)$ as shown in Figures 17.2 and 17.3.

The behavior of $f(\alpha)$ is investigated at a sequence of test points $P^* = [\alpha^*, s(\alpha^*), f(\alpha^*)]$. Three triples $P_1 = [\alpha_1, s(\alpha_1), f(\alpha_1)]$, $P_2 = [\alpha_2, s(\alpha_2), f(\alpha_2)]$ and $P_3 = [\alpha_3, s(\alpha_3), f(\alpha_3)]$ are used to save information about the function. The value α_3 is the smallest value for which $s(\alpha_3) > 0$, α_2 is the largest value with $s(\alpha_2) < 0$, and α_1 is the second largest value with $s(\alpha_1) < 0$. If only one point with negative slope is known, P_1 and P_2 are assumed to be the same point. By these definitions, $\alpha_1 \leq \alpha_2 < \alpha_3$ and α_1 and α_2 cannot decrease and α_3 cannot increase. At the beginning only one point is known namely $P_0 = [0, s(0), f(0)]$, where $s(0) < 0$, so that P_1 and P_2 are both equal to P_0 and no point P_3 is known.

Each test point $P^* = [\alpha^*, s(\alpha^*), f(\alpha^*)]$ is examined as follows. First the truth values of each of the following five logical statements are determined:

- B_1: $s(\alpha^*) < cs(0)$, where c is a small positive constant

- B_2: $f(\alpha^*) > f(0)$

- B_3: $s(\alpha^*) > 0$

- B_4: $(B_1.OR..NOT.B_3).AND.B_2$

- B_5: $[s(\alpha^*) > s(\alpha_2)].AND.[s(\alpha_2) \geq s(\alpha_1)]$

Statements B_1, B_2, and B_3 involve relations between P^* and P_0 only, B_4 is a function of B_1, B_2, and B_3, and B_5 involves P^*, P_1, and P_2 only. The eight possible outcomes of B_1, B_2 and B_3 and the consequent outcome of B_4 are shown in Table 17.1.

Table 17.1: All Possible Truth Values of B1, B2, B3, and B4

T = True			F = False
line	B1	B2	B3	B4
1	T	T	T	T
2	T	T	F	T
3	T	F	T	F
4	T	F	F	F
5	F	T	T	F
6	F	T	F	T
7	F	F	T	F
8	F	F	F	F

Statement B_4 is examined first. If B_4 is true (lines 1, 2 and 6) it means that P^* is in a region "too far out" as shown by region CD in Figure 17.2. The trial step α^* is therefore decreased by multiplying by a constant scale factor $b < 1$ and starting anew with the decreased α^*, disregarding information from previous test points, if any. If B_4 is false, B_1 is examined. If B_1 is true (lines 3 and 4), the minimum is at P^* as illustrated by region AB in Figure 17.2. If B_1 is false, B_3 is examined. If B_3 is true (lines 5 and 7), the test point P^* has positive slope, yielding a P_3, and a new test point is determined by cubic interpolation using the information provided by P_2 and P_3. This is illustrated by region BC in Figure 17.2. The interpolation formulas are given in Section 17.6. If B_3 is false (line 8), the test point P^* has negative slope (region OA), yielding a P_2. Then, if a previous P_3 is available, a new test point is interpolated between P_2 and P_3, as above; otherwise B_5 is examined. If B_5 is true, a new test point is obtained by extrapolation using P_1 and P_2. Otherwise, the step size α^* is increased by multiplying by a constant scale factor $a > 1$. Figure 17.3 illustrate the two cases where B_5 is true and B_5 is false.

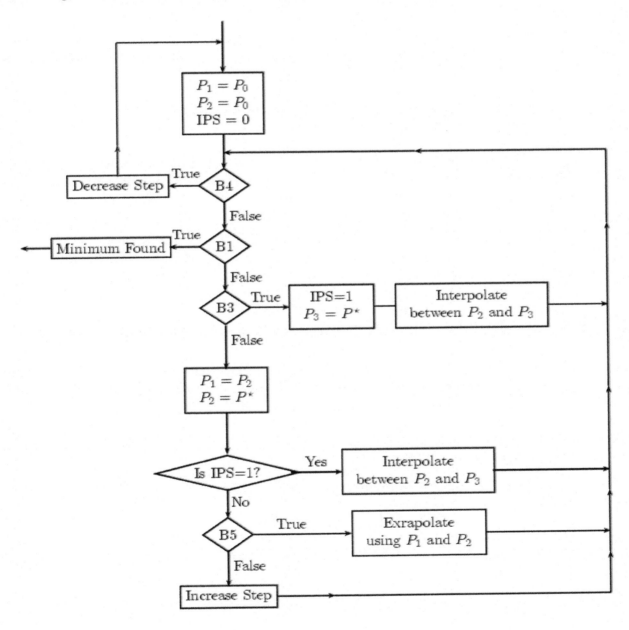

Figure 17.4: Flow Diagram Showing How Test Points are Examined

Figure 17.4 shows a flow diagram of how test points are examined. Successive test points are taken in this way until B_4 is false and B_1 is true. This usually occurs after two test points but, if the search is done in a region far away from the minimum of $F(\boldsymbol{\theta})$, more test points may be required. If a point where B_1 is true is found, this is the new point $\boldsymbol{\theta}_{s+1}$. The information matrix is then updated to $\mathbf{E}_{s+1}$ as explained in Section 17.3.

During the line search, the point with the smallest function value is saved. At most MT (default value $= 20$) test points are permitted in each direction. After MT test points, the program examines the smallest function value obtained. If this is smaller than $f(0)$, the point with the smallest function value is taken as the new point $\boldsymbol{\theta}_{s+1}$ and $\mathbf{E}_{s+1}$ is computed by inverting the information matrix at that point.

If no test point with function value smaller than $f(0)$ is found, the program changes the direction of search to the steepest descent direction represented by the line $\boldsymbol{\theta} - \alpha\mathbf{g}$. If the line search along this direction is successful this yields a new point $\boldsymbol{\theta}_{s+1}$. If no point with function value smaller than $f(0)$ is found along the steepest descent direction, the program gives up and prints the message:

```
W_A_R_N_I N_G: Serious problems encountered during minimization.
              Unable to continue iterations. Check your data and model.
```

In most cases, the problem is in the data or the model. However, it can also occur in ill-conditioned problems (model being nearly non-identified, information matrix being nearly singular), where it is impossible to obtain the accuracy required by the convergence criterion due to insufficient arithmetic precision. The latter case is often characterized by a solution which is very close to the minimum—sufficiently close for most practical purposes.

The constants a, b and c referred to in this section are fixed in the program at 2.0, 0.75, and 0.1, respectively.

17.6   Interpolation and Extrapolation Formulas

Interpolation is used whenever two points $P_2 = [\alpha_2, s(\alpha_2), f(\alpha_2)]$, with negative slope $s(\alpha_2)$, and $P_3 = [\alpha_3, s(\alpha_3), f(\alpha_3)]$, with positive slope $s(\alpha_3)$, $\alpha_2 < \alpha_3$, are known. Extrapolation is used whenever two points $P_1 = [\alpha_1, s(\alpha_1), f(\alpha_1)]$ and $P_2[\alpha_2, s(\alpha_2), f(\alpha_2)]$, both with negative slopes and with $s(\alpha_2) > s(\alpha_1)$, $\alpha_1 < \alpha_2$, are known. In both cases we determine the "smoothest" curve $f^*(\alpha)$ satisfying the four boundary conditions and approximate the minimum of $f(\alpha)$ by an appropriate minimum of $f^*(\alpha)$.

Let $A = [a, s_a, f_a]$ and $B = [b, s_b, f_b]$ be two points with $a < b$. The curve $f^*(\alpha)$ is defined as the one which minimizes

$$\int (d^2 f^* / d\alpha^2)^2 d\alpha$$

and satisfies $f^*(a) = f_a$, $s^*(a) = s_a$, $f^*(b) = f_b$ and $s^*(b) = s_b$. By means of calculus of variations it can be shown that the resulting curve is a cubic, whose slope at any point α, $a \leq \alpha \leq b$, is given by

$$s^*(\alpha) = s_a - 2[(\alpha - a)/\lambda](s_a + z) + [(\alpha - a)/\lambda]^2(s_a + s_b + 2z) , \qquad (17.23)$$

where

$$\lambda = b - a , \qquad (17.24)$$

and

$$z = (3/\lambda)(f_a - f_b) + s_a + s_b . \tag{17.25}$$

If $s_a + s_b + 2z = 0$, $s^*(\alpha)$ degenerates into a linear function, whose zero is located at

$$\alpha = a - \frac{s_a}{s_b - s_a}(b - a) . \tag{17.26}$$

Otherwise, $s^*(\alpha)$ has the two roots

$$\alpha_1 = a + \frac{s_a + z + \sqrt{Q}}{s_a + s_b + 2z}(b - a) , \tag{17.27}$$

$$\alpha_2 = a + \frac{s_a + z - \sqrt{Q}}{s_a + s_b + 2z}(b - a) , \tag{17.28}$$

where

$$Q = z^2 - s_a s_b . \tag{17.29}$$

If s_a and s_b have different signs, Q is positive. If s_a and s_b are both negative, Q may be positive or negative. If Q is negative, the two roots (17.27) and (17.28) are complex and cannot be used. In this case the linear slope formula is used. This yields a value $\alpha > b$.

If Q is positive, both roots are real. Then if $s_a + s_b + 2z > 0$, the largest of the two roots, which is α_1, corresponds to a minimum of $f^*(\alpha)$. On the other hand, if $s_a + s_b + 2z < 0$, the smallest of the two roots, which is α_1 corresponds to the minimum. Thus, in both cases, α_1 is the root of interest. It is readily verified that α_1 is between a and b if $s_a < 0$ and $s_b > 0$ and larger than b if $s_b < s_a < 0$.

If $s_b \approx -s_a$, the form (17.27) of α_1 is not suitable for computation since considerable accuracy may be lost in taking the difference of nearly equal quantities. For this reason, the following mathematically equivalent form is used in the computations:

$$\alpha = a + \left[1 - \frac{s_b + \sqrt{Q} - z}{s_b - s_a + 2\sqrt{Q}} \right] . \tag{17.30}$$

Bibliography

Agresti, A. (1990). *Categorical data analysis.* New York: Wiley.

Aish, A. M., & Jöreskog, K. G. (1990). A panel model for political efficacy and responsiveness: An application of LISREL 7 with weighted least squares. *Quality and Quantity, 24,* 405–426.

Aitkin, M. A., & Longford, N. T. (1986). Statistical modeling issues in school effectiveness studies. *Journal of the Royal Statistical Society A, 149,* 1–43.

Ajzen, I. (1991). The theory of planned behavior. *Organizational Behavior and Human Decision Processes, 50,* 179–211.

Akaike, H. (1974). A new look at statistical model identification. *IEEE Transactions on Automatic Control, 19,* 716–723.

Amemiya, T. (1973). Regression analysis when the dependent variable is truncated normal. *Econometrica, 41,* 997–1016.

Amemiya, Y., & Anderson, T. W. (1990). Asymptotic chi-square tests for a large class of factor analysis models. *The Annals of Statistics, 3,* 1453–1463.

Anderson, T. W. (1960). Some stochastic process models for intelligence test scores. In K. J. Arrow, S. Karlin, & P. Suppes (Eds.), *Mathematical methods in the social sciences.* Stanford: Stanford University Press.

Anderson, T. W. (1984). *An introduction to multivariate statistical analysis.* New York: Wiley.

Anderson, T.W., and Rubin, H. (1956) Statistical inference in factor analysis. In *Proceedings of the Third Berkeley Symposium* (Vol. V). Berkeley: University of California Press.

Archer, C. O., & Jennrich, R. I. (1973). Standard errors for rotated factor loadings. *Psychometrika, 38,* 581–592.

Barnes, S. H., & Kaase, M. (Eds.). (1979). *Political action: Mass participation in five western democracies.* Beverly Hills and London: Sage Publications.

© Springer International Publishing Switzerland 2016
K.G. Jöreskog et al., *Multivariate Analysis with LISREL,* Springer Series in Statistics,
DOI 10.1007/978-3-319-33153-9

Bartholomew, D. (1995). Spearman, and the origin and development of factor analysis. *British Journal of Mathematical and Statistical Psychology, 48*, 211–220.

Bartholomew, D., & Knott, M. (1999). *Latent variable models and factor analysis.* London: Arnold.

Bartholomew, D. J., Steele, F., Moustaki, I., & Galbraith, J. I. (2002). *The analysis and interpretation og multivariate data for social scientists.* Boca Raton: Chapman & Hall.

Bartlett, M. S. (1937) The statistical conception of mental factors. Bekker, P. A., Merckens, A., & Wansbeek, T. J. (1994). *Identification, equivalent models and computer algebra.* Boston: Academic Press.

Bentler, P. M. (1990). Comparative fit indexes in structural models. *Psychological Bulletin, 107*, 238–246.

Bentler, P. M., & Bonett, D. G. (1980). Significance tests and goodness of fit in the analysis of covariance structures. *Psychological Bulletin, 88*, 588–606.

Biblarz, T. J., & Raftery, A. E. (1993). The effects of family disruption on social mobility. *American Sociological Review, 58*, 97–109.

Bishop, Y. M. M., Fienberg, S. E., & Holland, P. W. (1975). *Discrete multivariate analysis: Theory and practice.* Cambridge MA: MIT Press.

Bock, R. D., & Lieberman, M. (1970). Fitting a response model for n dichotomously scored items. *Psychometrika, 35*, 179–197.

Bollen, K. A. (1989a). *Structural equations with latent variables.* New York: Wiley.

Bollen, K. A. (1989b). A new incremental fit index for general structural equation models. *Sociological Methods and Research, 17*, 303–316.

Bollen, K. A., & Liang, J. (1988). Some properties of Holter's CN. In *Sociological Methods and Research, 16*, 492–503.

Bollen, K. A., & Arminger, G. (1991). Observational residuals in factor analysis and structural equation models. *Sociological Methodology* (pp. 235–262).

Bollen, K. A., & Curran, P. J. (2006). *Latent curve models: A structural equation approach.* Hoboken: Wiley.

Bollen, K. A., & Liang, J. (1988). Some properties of Hoelter's CN. *Sociological Methods and Research, 16*, 492–503

Bollen, K. A., & Long, J. S. (Eds.). (1993). *Testing structural equation models.* Newbury Park: Sage Publications.

Boomsma, A., & Hoogland, J. J. (2001). The robustness of LISREL modeling revisited. In R. Cudeck, S. Du Toit, & D. Sörbom (Eds.), *Structural equation modeling: Present and future* (pp. 139–168). Lincolnwood, IL: Scientific Software International.

Bortkiewicz, L. (1898). *Das Gesetz der kleinen Zahlen (The law of small numbers) Leipzig.* Germany: B.G. Teubner.

Box, G. E. P. (1949). A general distributions theory for a class of likelihood criteria. *Biometrika, 36,* 317–346.

Breslow, N. E., & Day, N. E. (1987). *Statistical methods in cancer research, volume 2: The sesign and analysis of cohort studies.* Lyon: Intenational Agency for Research on Cancer.

Browne, M. W. (1968). A comparison of factor analytic techniques. *Psychometrika, 33,* 267–334.

Browne, M. W. (1977). Generalized least-squares estimators in the analysis of covariance structures. In D. J. Aigner & A. S. Goldberger (Eds.), *Latent variables in socio-economic models* (pp. 205–226). Amsterdam: North-Holland.

Browne, M. W. (1984). Asymptotically distribution-free methods for the analysis of covariance structures. *British Journal of Mathematical and Statistical Psychology, 37,* 62–83.

Browne, M. W. (2001). An overview of analytic rotation in exploratory factor analysis. *Multivariate Behavioral Research, 36,* 111–150.

Browne, M. A., & Shapiro, A. (1988). Robustness of normal theory methods in the analysis of linear latent variable models. *British Journal of Mathematical and Statistical Psychology., 41,* 193–208.

Browne, M. W., & Cudeck, R. (1993). Alternative ways of assessing model fit. In K. A. Bollen & J. S. Long (Eds.), *Testing structural equation models.* Newbury Park: Sage Publications.

Bryant, F. B., & Jöreskog, K. G. (2016). Confirmatory factor analysis of ordinal data using full-information adaptive quadrature. *Australian & New Zealand Journal of Statistics, 58*(2), 173-196.

Bryant, F. B., & Satorra, A. (2012). Principles and practice of chi-square difference testing. *Structural Equation modeling, 19,* 372–398.

Bryk, A. S., & Raudenbush, S. W. (1992). *Hierarchical linear models.* Thousand Oaks: Sage Publications.

Campbell, A., Gurin, G., & Miller, W. A. (1954). *The voter decides.* New York: Wiley.

Carroll, J. B. (1953). An analytical solution for approximating simple structure in factor analysis. *Psychometrika, 18*, 23–38.

Cattell, R. B. (1966). The scree test for the number of factors. *Multivariate Behavioral Research, 1*, 245–276.

Christoffersson, A. (1975). Factor analysis of dichotomized variables. *Psychometrika, 40*, 5–32.

Costner, H. L. (1969). Theory, deduction, and rules of correspondence. *American Journal of Sociology, 75*, 245–263.

Cragg, J. G., & Uhler, R. (1970). The demand for automobiles. *Canadian Journal of Economics, 3*, 386–406.

Craig, S. C., & Maggiotto, M. A. (1981). Measuring political efficacy. Political. *Methodology, 8*, 85–109.

Cramer, H. (1957). *Mathematical methods of statistics.* Seventh Printing: Princeton University Press.

Crowder, M. J., & Hand, D. J. (1990). *Analysis of repeated measures.* Monographs on statistics and applied probability New York: Chapman and Hall/CRC.

Cudeck, R. (1989). Analysis of correlation matrices using covariance structure models. *Psychological Bulletin, 105*, 317–327.

Cudeck, R., & Browne, M. W. (1983). Cross-validation of covariance structures. *Multivariate Behavioral Research, 18*, 147–157.

Curran, P. J., West, S. G., & Finch, J. F. (1996). The robustness of test statistics to nonnormality and specification error in confirmatory factor analysis. *Psychological Methods, 1*, 16–29.

Dagnelie, P. (1975). *Analyse Statistique à Plusieurs Variables.* Brussels: Vander.

D'Agostino, R. B. (1970). Transformation to normality of the null distribution of g_1. *Biometrika, 57*, 679–681.

D'Agostino, R. B. (1971). An omnibus test of normality for moderate and large sample size. *Biometrika, 58*, 341–348.

D'Agostino, R. B. (1986). Tests for the normal distribution. In R. B. D'Agostino & M. A. Stephens (Eds.), *Goodness-of-fit techniques* (pp. 367–419). New York: Marcel Dekker.

D'Agostino, R. B., Belanger, A., & D'Agostino, R. B, Jr. (1990). A suggestion for using powerful and informative tests of normality. *The American Statistician, 44*, 316–321.

DeCarlo, L. (1997). On the meaning and use of kurtosis. *Psychological Methods, 2*, 292–397.

Dickman, K., & Kaiser, H. F. (1961). Program for inverting a Gramian matrix. *Educational and Psychological Measurement, 21*, 721–727.

Dobson, A. J., & Barnett, A. G. (2008). *An introduction to generalized linear models* (3rd ed.). ARC Press.

Duncan, O. D. (1975). *Introduction to structural equation models*. New York: Academic Press.

Duncan, O. D., Haller, A. O., & Portes, A. (1968). Peer influence on aspiration: A reinterpretation. *American Journal of Sociology, 74*, 119–137.

Du Toit S. H. C. & Cudeck, R. (2001). In R. Cudeck, S. Du Toit, & D. Sörbom (Eds.), *Structural equation modeling: Present and future* (pp. 11–38). Lincolnwood, IL: Scientific Software International.

Fair, R. (1978). A theory of extramarital affairs. *Journal of Political Economy, 86*, 45–61.

Finn, J. D. (1974). *A general model for multivariate analysis*. New York: Holt, Reinhart and Winston.

Fisher, R. A. (1930). The moments of the distribution for normal samples of measures of departures from normality. *Proceedings of the Royal Society, London, Series A, 130*, 16.

Fisher, R. A. (1973). *Statistical methods for research workers*. New York: Hafner Publishing. Fourteenth Printing.

Fletcher, R., & Powell, M. J. D. (1963). A rapidly convergent descent method for minimization. *Computer Journal, 6*, 163–168.

Forero, C. G., Maydeu-Olivares, A., & Gallardo-Pujol, D. (2009). Factor analysis with ordinal indicators: A Monte Carlo study comparing DWLS and ULS estimation. *Structural Equation Modeling, 17*, 392–423. Please check and confirm the inserted year in the Ref. Forero et al., and correct if necessary.

Foldnes, N. & Olsson, U. H. (2015). Correcting too much or too little? The performance of three chi-squares corrections. Multivariate Behavioral Research. ISSN 0027-3171 print. 1532-7906 online.

Forero, C. G., Maydeu-Olivares, A., & Gallardo-Pujol, D. (2009). Factor analysis with ordinal indicators: A Monte Carlo study comparing DWLS and ULS estimation. *Structural Equation Modeling, 17*, 392–423.

Fouladi, R. (2000). Performance of modified test statistics in cavariance and correlation structure analysis under conditions of multivariate nonnormality. *Structural Equation Modeling: A Multidisciplinary Journal, 7*, 356–410.

Foss, T., Jöreskog, K. G., & Olsson, U. H. (2011). Testing structural equation models: The effect of kurtosis. *Computational Statistics and Data Analysis, 55*, 2263–2275.

French, J. V. (1951). The description of aptitude and achievement tests in terms of rotated factors. *Psychometric Monographs, 5.*

Gallo, J. J., Rabins, P. V., & Anthony, J. C. (1999). Sadness in older persons: 13 year follow-up of a community sample in Baltimore, Maryland. *Psychological Medicine, 29*, 341–350.

Goldberger, A. S. (1964). *Econometric theory.* New York: Wiley.

Goldberger, A. S. (1972). Structural equation methods in the social sciences. *Econometrica, 40*, 979–1001.

Goodman, L. A. (1978). *Analysing qualitative/categorical data.* London: Addison Wesley.

Goldstein, H. (2003). *Multilevel statistical models* (3rd ed.). London: Arnold.

Greene, W. H. (2000). *Econometric analysis* (4th ed.). Upper Saddle River: Prentice Hall International.

Guilford, J. P. (1956). The structure of intellect. *Psychological Bulletin, 53*, 267–293.

Gujarati, D. N. (1995). *Basic econometrics* (3rd ed.). New York: McGraw-Hill International.

Guttman, L. (1940). Multiple rectilinear prediction and the resolution into components. *Psychometrika, 5*, 75–99.

Guttman, L. (1944). General theory and methods for matrix factoring. *Psychometrika, 9*, 1–16.

Guttman, L. (1954a). Some necessary conditions for common-factor analysis. *Psychometrika, 19*, 149–161.

Guttman, L. (1954b). A new approach to factor analysis: The radix. In P. F. Lazarsfeld (Ed.), *Mathematical thinking in the social sciences.* New York: Columbia University Press.

Guttman, L. (1956). 'Best possible' systematic estimates of communalities. *Psychometrika, 21,* 273–285.

Guttman, L. (1957). A necessary and sufficient formula for matrix factoring. *Psychometrika, 22,* 79–81.

Haberman, S. J. (1974). *The analysis of frequency data.* Chicago: University of Chicago Press.

Harman, H. H. (1967). *Modern factor analysis* (3rd ed.). Chicago: University of Chicago Press.

Hauser, R. M., & Goldberger, A. S. (1971). The treatment of unobservable variables in path analysis. In H. L. Costner (Ed.), *Sociological methodology 1971* (pp. 81–117). Jossey-Bass.

Hoelter, J. W. (1983). The analysis of covariance structures: Goodness-of-fit indices. *Sociological Methods and Research, 11,* 325–344.

Hayduk, L. A. (2006). Blocked-error-R2: A conceptually improved definition of the proportion of explained variance in models containing loops or correlated residuals. *Quality and Quantity, 40,* 629–649.

Hendrickson, A. E., & White, P. O. (1964). Promax: A quick method for rotation to oblique simple structure. *British Journal of Mathematical and Statistical Psychology, 17,* 65–70.

Hershberger, S. L. (2003). The growth of structural equation modeling. *Structural Equation Modeling, 10,* 35–46.

Holt, D., Scott, A. J., & Ewings, P. D. (1980). Chi-squared tests with survey data. *Journal of the Royal Statistical Society A, 143,* 303–320.

Holzinger, K., & Swineford, F. (1939). *A study in factor analysis: The stability of a bifactor solution.* Supplementary educational monograph no. 48 Chicago: University of Chicago Press.

Howe, H. G. (1955). Some contributions to factor analysis. Report ORNL-1919. Oak Ridge, Tenn.: Oak Ridge National Laboratory.

Hox, J. (2002). *Multilevel analysis: techniques and applications.* Mahwah, N.J.: Lawrence Erlbaum Associates.

Hu, L., Bentler, P. M., & Kano, Y. (1992). Can test statistics in covariance structure analysis be trusted? *Psychological Bulletin, 112,* 351–362.

Humphreys, L. G. (1968). The fleeting nature of college academic success. *Journal of Educational Psychology, 59*, 375–380.

Hägglund, G. (2001). Milestones in the history of factor analysis. In R. Cudeck, S. Du Toit, & D. Sörbom (Eds.), *Structural equation modeling: Present and future* (pp. 11–38). Lincolnwood, IL: Scientific Software International.

Härnqvist, K. (1962). *Manaul till DBA (Manual for DBA)*. Stockholm: Skandinaviska Testförlaget. (In Swedish).

James, L. R., Mulaik, S. A., & Brett, J. M. (1982). *Causal analysis: Assumptions, models, and data*. Beverly Hills: Sage.

Jennrich, R. I. (1973). Standard errors for obliquely rotated factor loadings. *Psychometrika, 38*, 593–604.

Jennrich, R. I. (1986). A Gauss-Newton algorithm for exploratory factor analysis. *Psychometrika, 51*, 277–284.

Jennrich, R. I. (2007). Rotation methods, algorithms, and standard errors. In R. Cudeck & R. C. MacCallum (Eds.), *Factor analysis at 100* (pp. 315–335). Mahwah, New Jersey: Lawrence Erlbaum Associates Publishers.

Jennrich, R. I., & Robinson, S. M. (1969). A Newton-Raphson algorithm for maximum likelihood factor analysis. *Psychometrika, 34*, 111–123.

Johnson, R. A., & Wichern, D. W. (2002). *Applied multivariate statistical analysis*. Upper Saddle River, New Jersey: Prentice-Hall.

Joliffe, J. (1986). *Principal components analysis*. Berlin: Springer.

Jöreskog, K. G. (1967) Some contributions to maximum likelihood factor analysis. *Psychometrika, 32*, 443–482.

Jöreskog, K. G. (1969). A general approach to confirmatory maximum likelihood factor analysis. *Psychometrika, 34*, 183–202.

Jöreskog, K. G. (1970). Estimation and testing of simplex models. *British Journal of Mathematical and Statistical Psychology, 23*, 121–145.

Jöreskog, K. G. (1971). Simultaneous factor analysis in several populations. *Psychometrika, 57*, 409–426.

Jöreskog, K. G. (1973). A general method for estimating a linear structural equation system. In A. S. Goldberger & O. D. Duncan (Eds.), *Structural equation models in the social sciences* (pp. 85–112). New York: Seminar Press.

Jöreskog, K. G. (1974). Analyzing psychological data by structural analysis of covariance matrices. In R. C. Atkinson, et al. (Eds.), *Contemporary developments in mathematical psychology* (Vol. II, pp. 1–56). San Francisco: W.H. Freeman.

Jöreskog, K. G. (1977). Factor analysis by least-squares and maximum-likelihood methods. In K. Enslein, A. Ralston, & H. S. Wilf (Eds.), *Statistical methods for digital computers* (pp. 125–153). New York: Wiley.

Jöreskog, K. G. (1981). Analysis of covariance structures. *Scandinavian Journal of Statistics, 8,* 65–92.

Jöreskog, K. G. (1990). New developments in LISREL: Analysis of ordinal variables using polychoric correlations and weighted least squares. *Quality and Quantity, 24*(4), 387–404.

Jöreskog, K. G. (1993). Testing structural equation models. In K. A. Bollen & J. S. Long (Eds.), *Testing structural equation models.* Newbury Park: Sage Publications.

Jöreskog, K. G. (1994). On the estimation of polychoric correlations and their asymptotic covariance matrix. *Psychometrika, 59,* 381–389.

Jöreskog, K. G. (1998). Interaction and nonlinear modelling: Issues and approaches. In R. E. Schumacker & G. A. Marcoulides (Eds.), *Interaction and nonlinear effects in structural equation modeling* (pp. 239–250). Publishers: Lawrence Erlbaum Associates.

Jöreskog, K. G., & Goldberger, A. S. (1975). Estimation of a model with multiple indicators and multiple causes of a single latent variable. *Journal of the American Statistical Association, 10,* 631–639.

Jöreskog, K. G., & Goldberger, A. S. (1972). Factor analysis by generalized least squares. *Psychometrika, 37,* 243–250.

Jöreskog, K. G., & Lawley, D. N. (1968). New methods in maximum likelihood factor analysis. *British Journal of Mathematical and Statistical Psychology, 21,* 85–96.

Jöreskog, K. G., & Moustaki, I. (2001). Factor analysis of ordinal variables: A comparison of three approaches. *Multivariate Behavioral Research, 36,* 347–387.

Jöreskog, K. G., & Yang, F. (1996). Nonlinear structural equation models: The Kenny-Judd model with interaction effects. In G. A. Marcoulides & R. E. Schumacker (Eds.), *Advanced structural equation modeling: Issues and techniques* (pp. 57–88). Mahwah: Lawrence Erlbaum Associates Publishers.

Kaiser, H. F. (1958). The varimax criterion for analytical rotation in factor analysis. *Psychometrika, 23,* 187–200.

Kaiser, H. F. (1970). A second generation little jiffy. *Psychometrika, 35,* 401–415.

Kanfer, R., & Ackerman, P. L. (1989). Motivation and cognitive abilities: An integrative/aptitude-treatment interaction approach to skill acquisition. *Journal of Applied Psychology, Monograph, 74*, 657–690.

Kendall, M. G., & Stuart, A. (1952). *The advanced theory of statistics, volume 1: Distribution theory* (5th ed.). London: Griffin & Co.

Kenny, D. A., Kashy, D. A., & Cook, W. L. (2006). *Dyadic data analysis.* New York: Guilford Press.

Klein, L. R. (1950). *Economic fluctuations in the United States 1921–1941* (Vol. 11). Cowles commission monograph. New York: Wiley.

Klein, L. R., & Goldberger, A. S. (1955). *An econometric model of the United States 1929–1952.* Amsterdam: North-Holland Publishing Company.

Kluegel, J. R., Singleton, R., & Starnes, C. E. (1977). Subjective class identification: A multiple indicators approach. *American Sociological Review, 42*, 599–611.

Kroonenberg, P. M., & Lewis, C. (1982). Methodological issues in the search for a factor model: Exploration through confirmation. *Journal of Educational Statistics, 7*, 69–89.

Kreft, I. G. G., de Leeuw, J., & Aiken, L. (1995). The effect of different forms of centering in hierarchical linear models. *Multivariate Behavioral Research, 30*, 1–22.

Lawley, D. N. (1940). The estimation of factor loadings by the method of maximum likelihood. *Proceedings of the Royal Society Edinburgh, 60*, 64–82.

Lawley, D. N., & Maxwell, A. E. (1971). *Factor analysis as a statistical method* (2nd ed.). London: Butterworths.

Lee, S., & Hershberger, S. (1990). A simple rule for generating equivalent models in covariance structure modeling. *Multivariate Behavioral Research, 25*, 313–334.

Lee, S.-Y., Poon, W.-Y., & Bentler, P. (1990). Full maximum likelihood analysis of structural equation models with polytomous variables. *Statistics and Probability Letters, 9*, 91–97.

Lord, F. M., & Novick, M. E. (1968). *Statistical theories of mental test scores.* Reading: Addison-Wesley Publishing Co.

Maddala, G. S. (1983). *Limited dependent and qualitative variables in econometrics.* Cambridge: Cambridge University Press.

Maiti, S. S., & Mukherjee, B. N. (1990). A note on distributional properties of the Jöreskog-Sörbom fit indices. *Psychometrika, 55*, 721–726.

Mardia, K. V. (1970). Measures of multivariate skewness and kurtosis with applications. *Biometrika, 57,* 519–530.

Mardia, K. V. (1980). Tests of univariate and multivariate normality. In P. R. Krishnaiah (Ed.), *Handbook of statistics* (Vol. 1, pp. 279–320). Amsterdam: North Holland.

Mardia, K. V. (1985). Mardia's test of multinormality. In S. Kotz & N. L. Johnson (Eds.), *Encyklopedia of statistical sciences* (Vol. 5, pp. 217–221). New York: Wiley.

Mardia, K. V. & Foster, K. (1983). Omnibus tests of multinormality based on skewness kurtosis. *Communications in Statistics, 12,* 207–221.

Mardia, K. V., Kent, J. T., & Bibby, J. M. (1980). *Multivariate analysis.* New York: Academic Press.

Matsueda, R. L., & Bielby, W. T. (1986). Statistical power in covariance structure models. In N. B. Tuma (Ed.), *Sociological methodology 1986* (pp. 120–158). San Francisco: Jossey Bass.

McFadden, D. (1973). Conditional logit analysis of qualitative choice behavior. In P. Zarembka (Ed.), *Frontiers in econometrics* (pp. 105–142).

McFadden, D., Powers, J., Brown, W., & Walker, M. (2000). Vehicle and drivers attributes affecting distance from the steering wheel in motor vehicles. *Human Factors, 42,* 676–682.

McCullagh, P., & Nelder, J. A. (1983). *Generalized linear models.* London: Chapman & Hall.

McDonald, J. A., & Clelland, D. A. (1984). Textile workers and union sentiment. *Social Forces, 63,* 502–521.

Melenberg, B., & van Soest, A. (1996). Parametric and semi-parametric modeling of vacation expenditures. *Journal of Applied Econometrics, 11,* 59–76.

Meredith, W. (1964). Rotation to achieve factorial invariance. *Psychometrika, 29,* 187–206.

Miller, W. E., Miller, A. H., & Schneider E. J. (1980). *American national election studies data source book: 1952–1978.*

Millsap, R. E., & Meredith, W. (2007). Factorial invariance: Historical perspectives and new problems. In R. Cudeck & R. C. MacCallum (Eds.), *Factor analysis at 100* (pp. 131–152). Mahwah: Lawrence Erlbaum Publishers.

Mortimore, P., Sammons, P., Stoll, L., Lewis, D., & Ecob, R. (1988). *School matters*. Wells: Open Books.

Mulaik, S. A. (1972). *The foundation of factor analysis*. New York: McGraw-Hill.

Mulaik, S. A. (1986). Factor analysis and Psychometrika: Major developments. *Psychometrika, 51*, 23–33.

Mulaik, S., James, L., Van Alstine, J., Bennett, N., Lind, S., & Stilwell, C. (1989). Evaluation of goodness-of-fit indices for structural equation models. *Psychological Bulletin, 105*, 430–445.

Muthén, B. (1984). A general structural equation model with dichotomous, ordered categorical, and continuous latent variable indicators. *Psychometrika, 49*, 115–132.

Nagelkerke, N. J. D. (1991). A note on a general definition of the coefficient of determination. *Biometrika, 78*, 691–693.

Neale, C. M., & Cardon, L. R. (1992). *Methodology for genetic studies of twins and families*. Dordrecht: Kluwer Academic Publishers.

Nelder, J. A., & Wedderburn, R. W. M. (1972). General linear models. *Journal of the Royal Statistical Society, Series A, 135*, 370–384.

Olsson, U. (1979). Maximum likelihood estimation of the polychoric correlation coefficient. *Psychometrika, 44*, 443–460.

Quester, A., & Greene, W. (1982). Divorce risk and wifes' labor supply behavioral. *Social Science Quarterly*, 16–27.

Radelet, M. L., & Pierce, G. L. (1991). Choosing those who will die. Race and the death penalty in Florida. *Florida law review, 43*, 1–34.

Reinecke, J. (2002). Nonlinear structural equation models with the theory of planned behavior: Comparison of multiple group and latent product term analysis. *Quality and quantity, 36*, 93–112.

Roberts, G., Martin, A. L., Dobson, A. J., & McCarthy, W. H. (1981). Tumor thickness and histological type in malignant melanoma in New South Wales, Australia, 1970–76. *Pathology, 13*, 763–770.

Rock, D. A., Werts, C. E., Linn, R. L., & Jöreskog, K. G. (1977). A maximum likelihood solution to the errors in variables and errors in equation models. *Journal of Multivariate Behavioral Research, 12*, 187–197.

Romney, D. M., & Bynner, J. M. (1992). *The structure of personal characteristics*. London: Praeger.

Satorra, A. (1989). Alternative test criteria in covariance structure analysis: A unified approach. *Psychometrika, 54*, 131–151.

Satorra, A. (1993). Multi-sample analysis of moment structures: Asymptotic validity of inferences based on second-order moments. In K. Haagen, D. J. Bartholomew, & M. Deistler (Eds.), *Statistical modelling and latent variables* (pp. 283–298). Elsevier Science Publishers.

Satorra, A., & Bentler, P. M. (1988). Scaling corrections for chi-square statistics in covariance structure analysis. *Proceedings of the Business and Economic Statistics Section of the American Statistical Association*, 308–313.

Satorra, A., & Bentler, P. M. (1990). Model conditions for asymptotic robustness in the analysis of linear relations. *Computational Statistics and Data Analysis, 10*, 235–249.

Satorra, A., & Bentler, P. M. (2001). A scaled difference chi-square test statistic for moment structure ananlysis. *Psychometrika, 66*, 507–514.

Satorra, A., & Saris, W. E. (1985). Power of the likelihood ratio test in covariance structure analysis. *Psychometrika, 50*, 83–90.

Satterthwaite, F. E. (1941). Synthesis of variance. *Psychometrika, 6*, 309–316.

Schafer, J. L. (1997). *Analysis of incomplete data.* London: Chapman & Hall/CRC.

Schmidt, J., & Leiman, J. M. (1957). The development of hierarchical factor solutions. *Psychometrika, 22*, 53–61.

Schilling, S., & Bock, R. D. (2005). High-dimensional maximum marginal likelihood item factor analysis by adaptive quadrature. *Psychometrika, 70*, 533–555.

Schwarz, G. E. (1978). Estimating the dimension of a model. *Annals if Statistics, 6*, 461–464.

Shapiro, A. (1987). Robustness properties of the MDF analysis of moment structures. *South African Journal, 21*, 39–62.

Sharma, S. (1996). *Applied multivariate techniques.* New York: Wiley.

Silvey, S. D. (1970). *Statistical inference.* Middlesex, U.K.: Penguin Books.

Smith, D. A., & Patterson, E. B. (1984). Applications and generalization of MIMIC models to criminological research. *Journal of Research in Crime and Delinquency, 21*, 333–352.

Snijders, T., & Bosker, R. (1999). *Multilevel analysis.* Thousand Oakes, CA: Sage.

Spearman, C. (1904). General intelligence objectively determined and measured. *American Journal of Psychology, 15*, 201–293.

Steiger, J. H. (1990). Structural model evaluation and modification: An interval estimation approach. *Multivariate Behavioral Research, 25*, 173–180.

Stelzl, I. (1986). Changing causal relationships without changing the fit: Some rules for generating equivalent LISREL models. *Multivariate Behavioral Research, 21*, 309–331.

Sörbom, D. (1974). A general method for studying differences in factor means and factor structures between groups. *British Journal of Mathematical and Statistical Psychology, 27*, 229–239.

Sörbom, D. (1989). Model modification. *Psychometrika, 54*, 371–384.

Tanaka, J. S., & Huba, G. J. (1985). A fit index for covariance structure models under arbitrary GLS estimation. *British Journal of Mathematical and Statistical Psychology, 42*, 233–239.

Theil, H. (1971). *Principles of econometrics.* New York: Wiley.

Thomson, G. H. (1939). *The factorial nature of human ability.* New York: Houghton-Mifflin.

Thurstone, L. L. (1938). Primary mental abilities. *Psychometric Monographs, 1*.

Timm, N. H. (1975). *Multivariate analysis with applications in education and psychology.* Belmont, California: Wadsworth Publishing Company.

Tintner, G. (1952). *Econometrics.* New York: Wiley.

Tobin, J. (1958). Estimation of relationships for limited dependent variables. *Econometrica, 26*, 24–36.

Tucker, L. R., & Lewis, C. (1973). A reliability coefficient for maximum likelihood factor analysis. *Psychometrika, 38*, 1–10.

Tukey, J. W. (1977). *Exploratory data analysis.* Reading, MA: Addison-Wesley Publishing Company.

Van de Ven, W. P. M. M., & Van der Gaag, J. (1982). Health as an unobservable: A MIMIC model of demand for health care. *Journal of Health Economics, 1*, 157–183.

Vasdekis, V. G. S., Cagnone, S., & Moustaki, I. (2012). A composite likelihood inference in latent variable models for ordinal longitudinal responses. *Psychometrika, 77*, 423–441.

Vinci, V. E., et al. (Eds.). (2010). *Handbook of partial least squares*. Springer handbooks of computational statistics. Berlin Heidelberg: Springer.

Votaw, D. F, Jr. (1948). Testing compound symmetry in a normal multivariate distribution. *Annals of Mathematical Statistics, 19*, 447–473.

Warren, R. D., White, J. K., & Fuller, W. A. (1974). An errors in variables analysis of managerial role performance. *Journal of the American Statistical Association, 69*, 886–893.

Wheaton, B., Muthén, B., Alwin, D., & Summers, G. (1977). Assessing reliability and stability in panel models. In D. R. Heise (Ed.), *Sociological methodology 1977*. San Francisco: Jossey-Bass.

White, H. (1980). A heteroskedasticity-consistent covariance matrix estimator and a direct test for heteroskedasticity. *Econometrica, 48*, 817–838.

Witte, A. (1980). Estimating an economic model of crime with individual data. *Quarterly Journal of Economics, 94*, 57–84.

Wold, H. (1982). Soft modeling: The basic design and some extensions. Chapter 1. In K. G. Jöreskog & H. Wold (Eds.), *Systems under indirect observation, part II*. Amsterdam: North Holland Publishing Company.

Wolfle, L. M. (2003). The introduction of path analysis to the social sciences, and some emergent themes: An annotated bibliography.

Yang-Wallentin, F., Jöreskog, K. G., & Luo, H. (2010). Confirmatory factor analysis of ordinal variables with misspecified models. *Structural Equation Modeling, 17*, 392–423.

Subject Index

Adaptive quadrature, 221, 318-323, 333, 370, 459, 462, 517

ADF, 302, 509, 510

Analysis of correlation matrix, 305, 359

Analysis of variance (ANOVA), 58, 174, 347

Analysis of covariance (ANCOVA), 58, 347

Asymptotic chi-square, 302, 521

Asymptotic covariance matrix (ACM), 11, 131, 300, 302, 306, 323, 489, 509-520

Asymptotic robustness, 300

Asymptotic variance, 243

Asymptotically distribution free, 302

Augmented moment matrix, 37, 43, 513

Binary variable, 3, 41, 112, 113, 120, 125, 274

Binomial distribution, 115, 136, 237, 149, 152

Bivariate distribution, 159, 160

Bivariate normal distribution

Browne's ADF chi-square

Censored regression

Censored variable

Chi-square

      C1, 506, 510

      C2, 510

      C3, 510

      C4, 504

Command language, 80, 86

© Springer International Publishing Switzerland 2016
K.G. Jöreskog et al., *Multivariate Analysis with LISREL*, Springer Series in Statistics,
DOI 10.1007/978-3-319-33153-9

Printed by Printforce, the Netherlands